Turfgrass Biology, Genetics, and Breeding

Turfgrass Biology, Genetics, and Breeding

Edited by

Michael D. Casler
&
Ronny R. Duncan

John Wiley & Sons, Inc.

Copyright © 2003 by John Wiley & Sons, Inc. All rights reserved

Published by John Wiley & Sons, Inc., Hoboken, New Jersey
Published simultaneously in Canada

No part of this publication may be reproduced, stored in a retrieval system, or transmitted in any form or by any means, electronic, mechanical, photocopying, recording, scanning, or otherwise, except as permitted under Section 107 or 108 of the 1976 United States Copyright Act, without either the prior written permission of the Publisher, or authorization through payment of the appropriate per-copy fee to the Copyright Clearance Center, Inc., 222 Rosewood Drive, Danvers, MA 01923, (978) 750-8400, fax (978) 750-4470, or on the web at www.copyright.com. Requests to the Publisher for permission should be addressed to the Permissions Department, John Wiley & Sons, Inc., 111 River Street, Hoboken, NJ 07030, (201) 748-6011, fax (201) 748-6008, e-mail: permcoordinator@wiley.com.

Limit of Liability/Disclaimer of Warranty: While the publisher and author have used their best efforts in preparing this book, they make no representations or warranties with respect to the accuracy or completeness of the contents of this book and specifically disclaim any implied warranties of merchantability or fitness for a particular purpose. No warranty may be created or extended by sales representatives or written sales materials. The advice and strategies contained herein may not be suitable for your situation. You should consult with a professional where appropriate. Neither the publisher nor author shall be liable for any loss of profit or any other commercial damages, including but not limited to special, incidental, consequential, or other damages.

For general information on our other products and services or for technical support, please contact our Customer Care Department within the United States at (800) 762-2974, outside the United States at (317) 572-3993 or fax (317) 572-4002.

Wiley also publishes its books in a variety of electronic formats. Some content that appears in print may not be available in electronic books. For more information about Wiley products, visit our web site at www.wiley.com.

Library of Congress Cataloging-in-Publication Data:

Turfgrass biology, genetics, and breeding / [edited by] M.D. Casler, R.R. Duncan.
    p. cm.
Includes bibliographical references.
ISBN 1-57504-159-6
1. Turfgrasses. 2. Turf management. I. Casler, Michael Darwin. II. Duncan, Ronny R.
SB433 .T8225 2002
635.9'642--dc21
                                        2002008678

ABOUT THE EDITORS

Michael D. Casler

Michael D. Casler is a Professor in the Department of Agronomy, University of Wisconsin-Madison. He is also a member of the Plant Breeding and Plant Genetics program at the University of Wisconsin, where he served as chair from 1987 to 1990. He studied at the University of Illinois, where he received a B.S. degree in Agronomy, and the University of Minnesota, where he received an M.S. degree in 1979 and a Ph.D. degree in 1980 in Plant Breeding.

Dr. Casler's research focuses on the breeding and genetics of four types of perennial grasses: turfgrasses, ornamental grasses, biofuel/bioenergy grasses, and forage/pasture grasses. His interests include statistics, quantitative and population genetics, selection theory and methodology, evolution and natural selection, and domestication. He has trained 19 graduate students, several of whom hold important and prestigious positions in the public and private sector. He has produced over 100 peer-reviewed publications and has been invited to speak at several international conferences. He has developed and registered four improved cultivars of perennial grasses.

R.R. Duncan

Ronny R. Duncan is Professor of Turfgrass Breeding (Paspalum, Tall Fescue) and Stress Physiology, Crop and Soil Sciences Department, University of Georgia, Griffin. Research emphasis has been on developing turfgrasses with enhanced tolerance to abiotic, edaphic, and biotic stresses, improved compatibility with the environment, and minimal requirements for maintenance.

Dr. Duncan has coauthored two books including *Seashore Paspalum* and *Salt-Affected Turfgrass Sites,* both published by Ann Arbor Press.

CONTRIBUTING AUTHORS

Sharon J. Anderson
Texas Agricultural Experiment Station
17360 Coit Rd.
Dallas, TX 75252-6599
email: une_perle@yahoo.com

A. Douglas Brede
Simplot/Jacklin Seed
5300 W. Riverbend Ave.
Post Falls, ID 83854
email: dbrede@simplot.com

Leah Brilman
Seed Research of Oregon
27630 Llewellyn Rd.
Corvallis, OR 97333
email: srofarm@attglobal.net

Sarah J. Browning
University of Nebraska, Coop. Extension
1206 W. 23 St.
Fremont, NE 68025
email: sbrowning2@unl.edu

Suleiman Bughrara
Department of Crop and Soil Sciences
Plant and Soil Sciences Bldg.
Wilson Rd. & Bogue St.
Michigan State University
East Lansing, MI 48825-1325
email: bughrara@msu.edu

Philip Busey
Environmental Horticulture
  Department
University of Florida
3205 College Ave.
Ft. Lauderdale, FL 33314
email: turf@ufl.edu

Michael D. Casler
Department of Agronomy
University of Wisconsin-Madison
1575 Linden Dr.
Madison, WI 53706-1597
email: mdcasler@facstaff.wisc.edu

Ronny R. Duncan
Department of Crop and Soil Sciences
University of Georgia
1109 Experiment St.
Griffin, GA 30223-1797
email: rduncan@gaes.griffin.peachnet.edu

Milt Engelke
Texas Agricultural Experiment Station
17360 Coit Rd.
Dallas, TX 75252-6599
email: m-engelke@tamu.edu

Wayne W. Hanna
USDA-ARS
Coastal Plain Experiment Station
P.O. Box 748
Tifton, GA 31793
email: hanna@tifton.cpes.peachnet.edu

Kenneth Hignight
Advanta Seeds Pacific, Inc.
33725 Columbus St. S.E.
Albany, OR 97321
email: aspfestuca@msn.com

David R. Huff
Department of Agronomy
The Pennsylvania State University
116 ASI Bldg.
State College, PA 16802-3504
email: drh15@psu.edu

Richard H. Hurley
Center for Turfgrass Science
Cook College, Rutgers University
Foran Hall, Room 364
59 Dudley Road
New Brunswick, NJ 08901-8520
e-mail: hurlrich@aol.com

J. Liu
Institute of Botany, Jiangsu Province
and Chinese Academy of Science
Nanjing 210014 China
email: tang20@jlonline.com

William A. Meyer
Department of Plant Science
Cook College, Rutgers University
Foran Hall, 59 Dudley Rd.
New Brunswick, NJ 08901
email: wmeyer@aesop.rutgers.edu

Terry Riordan
Horticulture Department
University of Nebraska
377F Plant Science
East Campus
Lincoln, NE 68583-0724
email: triordan@unlnotes.unl.edu

Bridget A. Ruemmele
Department of Plant Sciences
University of Rhode Island
213 Woodward Hall
9 East Alumni Ave., Suite 7
Kingston, RI 02881
email: bridgetr@uri.edu

Mark J. Sellmann
Simplot/Jacklin Seed
5300 W. Riverbend Ave.
Post Falls, ID 83854
email: msellman@simplot.com

Charles M. Taliaferro
Department of Plant and Soil Sciences
Oklahoma State University
481 Agriculture Hall
Stillwater, OK 74078-6028
email: cmt@soilwater.agr.okstate.edu

Daniel Thorogood
IGER
Welsh Plant Breeding Station
Plas Gogerddan, Aberystwyth
SY23 3EB United Kingdom
email: danny.thorogood@bbsrc.ac.uk

Scott E. Warnke
Oregon State University
3450 S.W. Campus Way
Corvallis, OR 97331
email: warnkes@ucs.orst.edu

Eric Watkins
Department of Plant Science
Cook College, Rutgers University
Foran Hall, 59 Dudley Rd.
New Brunswick, NJ 08901
email: ericw@rci.rutgers.edu

Joseph Wipff
Pure Seed Testing, Inc.
P.O. Box 449
Hubbard, OR 97032
email: joseph@turf-seed.com

CONTENTS

Part 1
Introduction

1. Origins of the Turfgrasses 5
   M.D. Casler and R.R. Duncan

Part 2
Cool-Season Grasses

2. Kentucky Bluegrass 27
   D.R. Huff
3. Annual Bluegrass 39
   D.R. Huff
4. Supina Bluegrass 53
   S. Bughrara
5. Texas Bluegrass 61
   J.C. Read and S.J. Anderson
6. Rough Bluegrass 67
   R. Hurley
7. Perennial Ryegrass 75
   D. Thorogood
8. Tall Fescue 107
   W.A. Meyer and E. Watkins
9. Fine-Leaved *Festuca* Species 129
   B.A. Ruemmele, J. Wipff, L. Brilman, and K. Hignight
10. Creeping Bentgrass 175
    S.E. Warnke
11. Colonial Bentgrass 187
    B.A. Ruemmele
12. Velvet Bentgrass 201
    L. Brilman
13. Three Minor *Agrostis* Species: Redtop, Highland Bentgrass, and
    Idaho Bentgrass 207
    A.D. Brede and M.J. Sellmann
14. Hairgrasses 225
    L. Brilman and E. Watkins

Part 3
Warm-Season Grasses

15. Bermudagrass 235
    C.M. Taliaferro
16. Buffalograss 257
    T.P. Riordan and S.J. Browning

17. Zoysiagrasses 271
    M.C. Engelke and S. Anderson
18. Centipedegrass 287
    W.W. Hanna and J. Liu
19. Seashore Paspalum 295
    R.R. Duncan
20. St. Augustinegrass 309
    P. Busey
21. Bahiagrass 331
    P. Busey

Index 349

though
Turfgrass Biology, Genetics, and Breeding

INTRODUCTION

The evolution of turfgrass science is one of the remarkable stories of the twentieth century. What started as a small number of professors with limited resources and even less visibility has evolved into a dynamic, multibillion dollar, worldwide industry. Turfgrasses evolved from forage and pasture grasses in many ways. The grasses themselves were the foundation of livestock agriculture from its beginning. Many sports evolved from play on rough pastures to finer-textured meadows, as rules, skills, and goals also evolved. Turfgrass cultivars were developed from germ plasm used largely for forage and fodder uses. Many turfgrass scientists, ourselves included, began work on turfgrasses after establishing ourselves on forage or fodder crops. Similarly, the grass seed industry adapted to a turf-based economy as consumers demanded better products. Turfgrass breeders responded with improved commercial products, including finer textured grasses and closer mowing heights for the golfer, better and more persistent (wear tolerant) sports turfs for sports facility owners and players, and more cosmetically appealing home lawns/landscapes for homeowners, public officials, and business people.

In these days when agricultural science is experiencing a severe identity crisis, turfgrass science serves to enhance the visibility and identity of agriculture. Turfgrasses have improved the quality of life for most people in developed parts of the world. They enhance our recreation, leisure, and competitive experiences, as well as our homes. They provide greenspace to break up urban sprawl and industrial complexes. Turfgrasses have evolved from a nonessential luxury item to an existential component of our daily lives. The turfgrass industry has transitioned from initial light green, coarse-leaved grasses to high-quality dark green and fine-leaved grasses. The evolution has also transcended development of grasses in an initial era when fertilizer was cheap and plentiful and water issues were not a concern to a currently emerging era of environmental stewardship, water quality and quantity issues, and adaptation/persistence under challenging environmental extremes. Genetic manipulation has become the cornerstone for enhancement of any 'new and improved' cultivars.

Much has been written about turfgrasses in recent years. There are some excellent books describing the management and physiology of turfgrasses. For this reason, we have purposely avoided the general subject of turfgrass management, focusing on basic biology and its relationship to adaptation, genetics, and breeding of turfgrasses. Each author presents detailed information on the natural and economic distribution of a turfgrass species, including interactions with the environment and with humans. They address the when, how, where, and why each species was adapted for use as a turfgrass, bringing us up-to-date on the progress that has been made in domesticating these grasses for turf.

We have assembled a world-class and highly respected group of authors to write this book. These colleagues have dedicated their time, effort, and expertise to help produce an excellent resource for turfgrass students and scientists. We thank them for agreeing to assist us with this project and for sharing their knowledge and experiences on turfgrass science.

Chapter 1

Origins of the Turfgrasses

M.D. Casler and R.R. Duncan

Three types of selective forces act to modify populations of organisms (Darwin, 1875), including all turfgrass species. Natural selection is the process whereby those individuals best fitted to a particular environment, a range of fluctuating environments, or a human-imposed management system have the greatest survival rate or contribute the greatest numbers of viable progeny to succeeding generations. Unconscious selection is the process by which humans save the phenotypically most valuable or desirable individuals, or their seed, and destroy or ignore the less valuable or desirable individuals. This process allows humans to facilitate genetic changes insofar as genetic variation allows, without the need to define or predetermine selection criteria or potential correlated traits. Methodical selection comprises the forces that are applied by humans in their systematic attempts to create predetermined changes to populations. All three of these selective forces have acted and continue to act to create the worldwide pool of turfgrass germ plasm.

Turfgrasses are used in a range of highly artificial and contrived environments, many of which bear no similarity to natural environments in which turfgrasses evolved over many millennia. Turf environments are human-defined and are often highly stressful, typically combining a large number of biotic and abiotic stress factors (Duncan and Carrow, 1999) that are seldom combined, or may be altogether absent, in natural habitats. Ironically, all turfgrasses arose from natural habitats, including mountain meadows, prairie and steppe grasslands, coastal plains, riparian and estuarine habitats, and many other habitats. Prior to human intervention and our discovery that perennial grasses can enhance our quality of life, these grasses served other purposes in the grand evolutionary scheme of Planet Earth.

NATURAL SELECTION

Perennial grasses coevolved with herbivores during the Miocene epoch. Fossil records suggest that evolution of the extensive North American grasslands was largely responsible for the great evolutionary advances in the horse, including increase in body size, strengthening of teeth, and loss of toes (Thomasson, 1979). Likewise, perennial grasses have evolved numerous defense mechanisms to survive and/or thrive under grazing pressure, including axillary meristems, rhizomes, stolons, trichomes, siliceous dentations, alkaloids, phenolic compounds, and associations with endophytic fungi (Casler et al., 1996). Many grasses have become dependent on herbivores—grazing pressure is often required for their survival (McNaughton, 1979). Perennial grasses respond to grazing by increasing their photosynthetic rate, leaf growth rate, and protein concentration, making them more desirable and nutritionally valuable to graziers (McNaughton et al., 1982).

The evolution of grasslands may have partially directed the evolution of humans, forcing the upright gait, the tool-using hand, and the heightened intellect required for survival in such a demanding habitat (Pohl, 1986). Whether grasses contributed to the prehistorical evolution of humans is highly debatable. However, it is scarcely debatable that perennial grasses are an integral component of the evolution of modern humans and our societies. Perennial grasses have evolved a large number of traits that collectively serve to make them useful for human enterprise and recreation. As human needs change, so too will turfgrasses, resulting in future coevolution with humans.

The evolution of perennial grasses generated tremendous inter- and intraspecific variability for adaptation. It requires little scientific expertise to observe and identify adaptive variation among some species, e.g., poor winter survival of many warm-season grasses in cold climates and poor summer survival of many cool-season grasses in warm climates. The tremendous differences among turfgrass species in adaptation to diverse climates and biotomes are illustrated throughout this text. What is less obvious to untrained observers is the tremendous variability that natural selection pressures have generated within species. Natural selection has resulted in ecotypic variability within all turfgrasses, providing the genetic foundation for all turfgrass cultivars that have been developed to date.

Most perennial turfgrass plants are highly heterozygous and originate from highly heterogeneous populations. These species are generally highly self-incompatible (with facultative apomicts as the obvious exceptions) and often exist as a polyploid series. Furthermore, many turfgrass species, as reviewed later in this text, freely exchange genes with their closest relatives due to similar genomic structures and frequent high levels of cross-compatibility. These characteristics, combined with the ever-present possibility of new mutations, act to create highly heterogeneous populations with huge numbers of potential genotypes and the possibility of creating new genotypes with each pollination event.

The natural environments in which these heterogeneous populations evolved are typically characterized by fluctuating environmental conditions, often including annual cycles of fluctuating temperature, humidity, precipitation, and sunlight. Furthermore, these geophysical and geoclimatic cycles are not constant, varying across decades and centuries (Ruzmaikin, 1999; Cai and Whetton, 2000; Cockell, 2000; Robertson et al., 2000). Climatic fluctuation creates natural selection pressures that vary in intensity and direction, resulting in populations with mixtures of interbreeding phenotypes favored by different environmental conditions. This process, a form of disruptive selection, maintains a large amount of genetic variation within populations.

Biotic Stresses

The high degree of uniformity, the presence of numerous predisposing abiotic stresses, the constant presence of litter and/or decomposing tissue, and the use of mechanical mowers in turfgrass environments often leads to disease epidemics and epiphytotics unlike those that occur in nature (Endo, 1972). Nevertheless, turfgrasses have coevolved with many disease-causing pathogens, resulting in alleles for resistance in many natural populations. Some turfgrasses, such as perennial ryegrass (*Lolium perenne* L.), have evolved loci that act in a gene-for-gene manner with highly virulent and variable pathogens, such as the rust fungi (*Puccinia* spp.) (Wilkins, 1978, 1991). In many other pathosystems, resistance is less obvious, often measured on a quantitative scale and likely controlled by loci with small individual effects (Casler and Pederson, 1996). Natural selection has also created morphological and chemical barriers that act as resistance mechanisms to various insects (Chamberlain and Evans, 1979; Lancashire et al., 1977).

Abiotic Stresses

The variability present within most perennial grasses has allowed them to colonize a large portion of the Earth's land mass. The Park Grass Experiments at Rothamstead, reviewed by Snaydon (1973, 1978), have shown that natural selection can create rapid genetic shifts in natural populations of perennial grasses. Natural populations of perennial grasses contain loci that confer resistance or tolerance to a wide array of abiotic stresses, often including opposite extremes of the same stress factor, such as soil pH or fertility. Less than 50 years were required to bring about dramatic changes in populations for limed vs. unlimed or high- vs. low-P soils. In some cases evolutionary changes to an environmental extreme resulted in a population unable to survive at the opposite environmental extreme. The infrequent cutting management that allowed surviving plants to undergo sexual reproduction, combined with human intervention to create a more-or-less constant environment, may have been partly responsible for these rapid changes.

Cold or freezing tolerance shows considerable variation among accessions of several turfgrass species. In perennial ryegrass, freezing tolerance of natural populations is closely associated with mean temperature of the coldest month at their site of origin (Humphreys and Eagles, 1988; Tcacenco et al., 1989). Crown height, as indicated by subcrown internode length, is inversely related to cold tolerance in perennial ryegrass (Wood and Cohen, 1984). In tall fescue (*Festuca arundinacea* Schreb.), genome size (amount of heterochromatin and frequency of repeated DNA sequences) was positively correlated with latitude of origin, suggesting that structural changes in DNA may play a role in environmental adaptation (Ceccarelli et al., 1992).

Natural variation exists for drought tolerance in species such as perennial ryegrass, which possess a wide geographic and climatic range. Accessions collected from consistently dry habitats have greater drought tolerance than commercial cultivars (Reed et al., 1987). Numerous drought-tolerance or drought-avoidance mechanisms have evolved in perennial ryegrass, including genetic variation for stomatal resistance (via differences in stomatal size and/or frequency), depth of epidermal ridging, leaf water conductance, and leaf osmotic potential (reviewed by Casler et al., 1996).

Tolerance to salinity has evolved naturally in numerous turfgrasses, including members of *Agrostis, Festuca, Lolium,* and *Poa* (Venables and Wilkins, 1978; Wu, 1981; Humphreys et al., 1986; Acharya et al., 1992). The evolution of salt tolerance in turfgrass has occurred in coastal marshes and rocky, alpine habitats. Natural selection pressure for salt tolerance drives selection coefficients, the rate at which alleles change in natural populations, and the eventual extent to which salt tolerance evolves (Ahmad et al., 1981; Wu, 1981). Natural populations can vary dramatically in salt tolerance across distances as little as 10 m. Salt tolerant genotypes of turfgrasses can evolve a form of salt dependence, as illustrated by greater increases in root and shoot growth under high- vs. low-saline growth conditions (Ashraf et al., 1986). A comprehensive listing of salinity tolerance levels across turf species is available (Carrow and Duncan, 1998).

Seashore paspalum (*Paspalum vaginatum* Swatrz) has genetic variation for salinity tolerance varying from bermudagrass [*Cynodon dactylon* (L.) Pers.] tolerance levels to ocean-salt levels (Lee, 2000).

UNCONSCIOUS SELECTION

Domestication of Livestock and Grasses

Domestication is evolution under human influence (Harlan, 1975). In a strict sense, domesticated species should be phenotypically distinct from their wild forms, such that domesticated forms are clearly more useful to humans (Isaac, 1970). With one exception, perennial forage grasses are not domesticated; "wild" collections are generally phenotypically indistinct from cultivated forms. The single exception is Italian ryegrass (*Lolium multiflorum* Lam.), which was developed by unconscious selection prior to the twelfth century in the Lombardy and Piedmont plains of Italy (Beddows, 1953). Ryegrass (*Lolium* spp.) once existed as a "huge hybrid swarm" with perennial and Italian phenotypes representing opposite extremes of a continuum (Tyler et al., 1987). Hay harvesting, followed by reseeding with shattered seed, resulted in the tall, sparse-tillered, semiannual phenotype that has since been elevated to species status (Breese and Tyler, 1986).

The initial events in the domestication of turfgrasses likely were associated with the domestication of livestock for agriculture. Animals were herded, corralled, or tethered in tight groups, creating closely-grazed sods that may have enticed humans to create games of sport, skill, and strength (Roberts et al., 1992). Animals were kept close to homes to protect them from wild carnivores, resulting in grazed sods in close proximity to homesteads, perhaps leading to an evolution of human needs and desires for lawns and landscaping. Many graziers, even in the highly mechanized rural societies of contemporary industrialized countries, continue to use sheep and goats to maintain their home lawns.

The practice of herding animals, particularly if accompanied by a human desire to maintain a closely-grazed sod for long periods of time, set in motion selective forces unlike those that exist in nature. Human needs for uniformity and consistency of sward management put pressure on plants to survive under long periods of stress, pressured by stresses that may not have occurred in nature. Grazing pressure can create rapid shifts in the phenotype of grass populations, with phenotypic changes proportional to grazing intensity and duration (Brougham et al., 1960; Charles, 1964; Brougham and Harris, 1967). Indeed, perennial ryegrass appears specifically adapted to survive in association with grazers (Beddows, 1953; Breese, 1983) and is rarely found in natural habitats devoid of grazing pressure (Davies et al., 1973). Perennial ryegrass appears to have spread throughout the Mediterranean Basin in direct association with the development and spread of livestock agriculture via human migration (Balfourier et al., 2000).

Frequent defoliation of perennial grass populations results in genetic shifts toward shorter basal internodes, a more prostrate growth habit, higher tiller densities, later flowering dates, and greater longevity of individual plants (Casler et al., 1996). In some cases, long-term defoliation by either herbivores or mechanical means (constant and low mowing heights) can bring about such changes (Van Dijk, 1955). Frequent defoliation can lead to proliferation of genotypes that are extremely long-lived, with estimates of a single *Holcus mollis* L. clone covering a distance of 800 m (Harberd, 1967) and a single *Festuca rubra* L. clone over 1,000 years old (Harberd, 1961).

Human Ball Games and Grass

The genesis of human ball games was at least 5,000 years ago, dating to the Egyptian game of ninepins and the Greek form of hockey (Altham and Swanton, 1926). Soccer dates to China in the third century B.C. as a form of military training called 'Tsu Chu' (Green, 1953). According to Green, the expression Tsu Chu derives from Tsu, to kick the ball with feet, and Chu, a stuffed ball made of leather. Other forms of 'football' derive from the Romans about the same time (Marples, 1954). Early versions of football games were violent, undisciplined, and relatively rule-free, suggesting that grass playing fields were highly desirable for the participants' well-being. (Imagine playing modern rugby or American football on bare soil or gravel without modern protective equipment, medications, or antibiotics.) The Roman Circus and ancient Mayan ball fields suggest that dedicated and permanent ball fields, likely including a perennial grass cover, are at least 2,500 years old. Natural selection likely played a role in the evolution of grass cover on these ancient ball fields.

Golf dates to early fifteenth century Scotland (Browning, 1955; Campbell, 1952). Campbell credits nature, beasts, and humans as the early architects of golf courses. By spreading seed and manure, birds and rodents assisted in the vegetation of dunes, coastlines, and riparian areas. Rabbit furthered the process by creating burrows and runs connecting their burrows. Foxes and other game enlarged these runs to tracks. Humans eventually found these tracked, grassy warrens to be ideal for their newly popular game of golf (Campbell, 1952). Holes on the earliest links were organized according to rabbit enclaves, with areas of highest (presumably grazing) activity serving as greens and game trails serving to connect tees and greens. Natural grasslands on volcanic peaks in the Azores, dominated by *Agrostis canina* L., are similarly characterized by heavily-grazed rabbit enclaves (Casler, 2000, personal observations). It requires little imagination to visualize these areas as prototypical putting or bowling greens.

As human ball games evolved in concert with the human race, so too did perennial grasses. As sport and recreation became more important to humans, rules evolved and increased in complexity in order to develop uniformity and preserve heritage of the games. Maintenance of a uniform turf with characteristics that support the game of choice gradually became an integral part of the caretakers' and communities' responsibilities. Sheep and goats would eventually replace rabbits for golf turf maintenance, largely due to the proximity of golf links to grazing lands and the ability of humans to control grazing pressure and duration, i.e., "mowing" height. Greater control over turf maintenance resulted in turf that remained more uniform for longer periods of time, supporting human needs for uniform application of newly evolving rules and perhaps the development of new rules. In an elegant irony, which is parallel to Miocene evolution, domesticated racehorses were likely made responsible for maintenance of the turf on which they would race for human recreation and benefit.

The development of rules for ball games, the subsequent desire to maintain more uniform and high quality turf, and the increasing importance of natural variation in agriculture eventually increased the levels of unconscious selection of turfgrasses. Natural variation for traits that control adaptation and usefulness of grain crops in agriculture has been recognized by humans for thousands of years (Harlan, 1975). However, the earliest evidence of human recognition for ecotypic variation and the concept of superior strains in forage grasses is barely 200 years old (Beddows, 1953). Collection, propagation, and selection of superior turf strains in Europe and North America likely began about the same time, although the level of activity varied considerably among species. Commercial sod farming began in Japan in 1701 (Busey, 1989).

Although the changes these practices effected in natural populations of turfgrasses cannot be documented, they undoubtedly increased the rate at which allele frequencies were modified by selection. Selection pressures brought about by the advent of turf management practices were parallel to natural selection pressures by herbivores, but represented an increase in selection intensity, a more stable selection pressure over time, and a possible reduction or elimination of sexual recombination in turf populations that were maintained at a vegetative growth stage. A microcosm of these effects can be observed on creeping bentgrass (*Agrostis palustris* L.) putting greens, which are typically composed of a patchwork of clones varying in color and texture (Madison, 1982). European colonialism and the efforts of numerous European botanists in the sixteenth to nineteenth centuries resulted in collection and dissemination of many interesting turf ecotypes (see later chapters in this book). Successful introductions resulted in long-lived turfs and non-native species became naturalized to a new locale. Variation thus exploited was created by a combination of natural selection and unconscious selection by humans and provided the foundation for methodical selection of turfgrasses in the twentieth century.

Human Effects on the Environment

While the measurement and magnitude of human influences on climate and atmosphere are hotly debated in the literature and popular press, it is scarcely debatable that human industrialization has affected Earth's climate, atmosphere, and soil. Air pollutants can be directly absorbed by stomata, causing direct toxicities, and can combine with water to create acid rain. Both point and nonpoint source pollutants can rapidly contaminate soils and can spread to uncontaminated sites by movement of surface and groundwater. While many industrial pollutants are also naturally occurring compounds (O_3, CO, CO_2, NO_x, SO_x, and heavy metals), their role as forces creating selection pressures on turfgrass populations is entirely human-derived.

Tolerance to toxic atmospheric gases has evolved in several turfgrasses used for urban parks and recreation, including perennial ryegrass, Kentucky bluegrass (*Poa pratensis* L.), and red fescue (*Festuca rubra* L.). Rapid evolution of populations with tolerance to acute SO_2 injury, in less than 5 years, suggests that alleles for SO_2 tolerance exist in natural populations of perennial ryegrass (Wilson and Bell, 1985). Evolution of SO_2 tolerance is driven by selection pressures, which are proportional to local atmospheric SO_2 concentrations (Horsman et al., 1979; Wilson and Bell, 1986). Tolerance to SO_2 may carry a physiological cost, as relaxation of selection pressure (caused by reduction in atmospheric pollutions levels) may cause turfgrass populations to lose tolerance (Wilson and Bell, 1986). Avoidance mechanisms appear to regulate tolerance to acute injury, caused by short-term exposure to high levels of toxic gases, while internal mechanisms appear to regulate tolerance to chronic injury, caused by long-term exposure to low levels of toxic gases (Ayazloo et al., 1982).

Numerous turfgrasses have evolved tolerance to a wide array of heavy metals, including arsenate, cadmium, copper, lead, nickel, and zinc. As with other abiotic stress tolerances, alleles conferring tolerance exist in extremely low frequencies within natural populations, rapidly increasing in frequency as exposure and selection pressure are applied to populations. Most documented tolerance has evolved in close proximity to mine spoils, smelters, areas contaminated with windblown mine waste, or galvanized (Zn-coated) electricity pylons, the latter caused by Zn solubilization from rainwater. As with tolerance to atmospheric gases, heavy-metal tolerances often carry a physiological cost. Heavy-metal tolerant populations may be extremely poor in vigor in uncontaminated soils and may revert to intolerant after relaxation of selection pressure (Cook et al., 1972; Hickey and McNeilly, 1975; Baker et al., 1986). Cultivars of red fescue and colonial bentgrass (*Agrostis capillaris* L.) have been developed from plants collected at heavy-metal contaminated sites (Humphreys and Bradshaw, 1977). Seashore paspalum also has an innate capability to tolerate heavy metals, due to an extremely efficient uptake system (Duncan and Carrow, 2000).

Finally, global climate changes may continue to create selection pressures of value to humans. Atmospheric loading of CO_2 and sulfate aerosols are capable of creating localized and global changes in surface air temperatures (Santer et al., 1995). The interrelationship of atmospheric aerosols and terrestrial plants is illustrated by the effect of temperature changes on agricultural pest populations (Patterson et al., 1999) and by the potential mitigating effect of healthy cropping systems (Robertson et al., 2000). In the short term, temperature changes will begin to cause genetic shifts within populations, favoring those plants that respond best to the change. These changes will probably occur much more slowly than those brought about by exposure to air and soil pollutants. In the long term, temperature changes will likely change geographic adaptation of turfgrass species, unless populations contain sufficient genetic variation for temperature response that they can respond, in situ, with changes in allele frequencies.

METHODICAL SELECTION

Natural Variation and Domestication

The practice of methodical selection of turfgrasses has undergone a dramatic evolution. Early efforts to apply plant breeding and selection concepts to turfgrasses directly utilized genetic variation created by natural selection. Early cultivars of highly stoloniferous species, such as bermudagrass, were identified and propagated from turf areas as early as 1907 (Roux, 1969). This practice was first applied to the cool-season apomictic species, Kentucky bluegrass, in the 1930s with the release of 'Merion' (Huff, Chapter 2; Meyer and Funk, 1989). Natural selection from old turf areas was used to select parents for synthetic cultivars of cool-season turfgrasses beginning in the 1960s. 'Manhattan' perennial ryegrass, derived from plants growing in Central Park, Manhattan, NY, was the most prominent of these early cultivars (Funk et al., 1969).

It could be argued that the value of natural selection in old turf sods will gradually fade as most turf areas have been planted to new cultivars. Logic suggests that new cultivars are more homogeneous and more highly adapted to the stresses imposed by a turf environment. However, transgressive segregation and high selection pressures within turf environments, particularly those subject to severe biotic or abiotic stresses, will ensure that valuable and useful genotypes will continue to appear in old turf sods. The creeping bentgrass breeding programs for dollar-spot resistance (caused by *Sclerotinia homeocarpa* F.T. Bennet) at Rutgers University (Bonos and Meyer, 2000) and snow-mold resistance (caused by *Typhula* spp.) at the University of Wisconsin (Wang et al., 2000) are based exclusively on germ plasm collected from old golf-course turfs, often of undetermined origin. Furthermore, as turf management changes with human preferences or the edaphic environment changes due to cyclic weather patterns, the stresses imposed on turf environments may also change. Thus, the genotypes favored by natural selection may vary over time, warranting long-term commitments to collecting in old turf areas.

Contrary to the vast majority of forage grasses, most cultivars of turfgrasses can (and should) be considered as domesticated by humans. Cultivars that cannot reproduce themselves without human management are domesticated from their wild forms (Harlan, 1992). For synthetic turfgrass cultivars, which are highly heterogeneous mixtures of numerous genotypes, seed production must be carefully controlled to preserve the genotypic composition (and phenotype) of the cultivar. Synthetic cultivars of perennial grasses can undergo rapid shifts in genotypic composition if seed production occurs under unique, stressful, or poorly defined environmental and/or management conditions (Evans et al., 1961; Crossley and Bradshaw, 1968). Uncontrolled genotypic changes to cultivars during seed production will result in unpredictable changes to phenotype, compromising the reliability of products marketed to consumers.

Grasses are characterized by allometric development—organ size is directly proportional to primary meristem size (Sinnott, 1921; Grafius, 1978). Thus, an inherent correlation exists between the size of all structures arising from a single meristem. As turf breeders strive to find plants that can be mown closer to the ground and maintain a dense, high-quality turf, the traits that contribute to such a phenotype probably respond to selection in concert with each other. Finer (narrower) leaf blades, shorter leaf blades, finer (thinner) stems, and shorter internodes are traits that contribute to a dense and high-quality turf and, without coincidence, leaf blades, sheaths, and stems all arise from a single apical meristem within a tiller. This basic physiological characteristic of grasses has probably contributed to the speed and efficiency with which turfgrass breeders have made advancements by methodical selection (see later chapters in this book).

Grasses in general, and perennial turfgrasses in particular, are highly plastic, capable of responding to fluctuating environments and stresses by morphological or physiological changes to their phenotype (Bradshaw, 1965). This *phenotypic plasticity* allows turfgrass plants to survive under a wide range of environmental conditions, including spatial and temporal environmental heterogeneity (Bradley, 1982). This phenomenon is the basis of the successful turfgrass seed industry. Seed production of even the finest-leaved, densest-tillering genotypes (used for high-quality putting greens, bowling greens, or cricket infields) is successful because these plants respond to relatively noncompetitive conditions by increasing their lateral spread, producing larger tillers with wider and longer leaves, and producing elongated culms with large floral structures, i.e., a phenotype completely unlike their turf phenotype. Phenotypic plasticity is under genetic control and can be modified by selection (Jain, 1978; Khan and Bradshaw, 1976; Morishima and Oka, 1975). Because breeders of all seeded turfgrasses must be concerned with both turf quality traits and seed production, breeding for high levels of phenotypic plasticity is an essential (although probably unconscious) objective of nearly all turfgrass breeding programs. In this respect, turfgrasses are unique among cultivated crops—cultivated crops generally have reduced phenotypic plasticity compared to their wild progenitors (Morishima and Oka, 1975).

As turfgrass breeding programs become more mature and more species advance beyond the first two or three cycles of selection from "wild" germ plasm, phenotypic plasticity will probably continue to increase in turf cultivars. Plasticity profiles of plant populations represent the accumulated historical integration of numerous evolutionary processes (Counts, 1993; Marshall and Jain, 1968; Sultan, 1987). Habitats dominated by short-term environmental instability tend to be dominated by plants that have high levels of phenotypic plasticity, i.e., numerous stress tolerances (Duncan, 2000). Contrasting and fluctuating environments tend to select plants that possess multiple stress tolerances, increasing longevity of individual genotypes as well as the turf sward. Selection strategies designed to broaden the adaptation of turfgrass cultivars must incorporate screening against multiple stresses, a process which should fortify the plasticity profile of turfgrass cultivars (Duncan, 2000).

Recurrent Selection

Most new turfgrass cultivars are created by phenotypic and/or genotypic recurrent selection. Phenotypic recurrent selection involves an evaluation of a large number of individual plants or replicated clones, followed by selection and intercrossing of the best individuals. Selection is based on plant or clone performance per se, i.e., phenotype. Genotypic recurrent selection requires an additional step—plants or clones to be evaluated are put into a crossing block and seed is produced on each individual, resulting in a half-sib (polycross if the parents are replicated and randomized) family of seed produced from each plant. Families can be tested in turf plots followed by selection of the best families. Intercrossing, to complete the cycle, is conducted either with the original parents of the best families (among-family selection) or the phenotypically best plants dug from each of the best families (among- and within-family selection). Numerous recurrent selection schemes have been described by Hallauer and Miranda (1981).

In many species, five or six cycles of recurrent selection have been completed since the original collections were made from old turf areas. This level of activity and number of cycles exceeds that of all perennial forage grasses, with the exception of perennial and Italian ryegrasses (Casler et al., 1996). Synthetic cultivars are created either as a composite of polycross families from a selection cycle or as a polycross of parents deriving from various recurrent selection programs that may be in different germ plasms and cycles of selection. Apomictic and vegetatively propagated cultivars are also derived by recurrent selection, although the breeder need only identify a single superior genotype to represent a cycle of selection with a new cultivar. New cycles of apomictic or vegetatively propaged cultivars are typically initiated from a small number of hand crosses.

The progress made in improving perennial forage crops by recurrent selection has lagged severely behind that for annual grain crops (Humphreys, 1999). For turfgrasses, it is difficult to measure progress using the traditional quantitative scale that expresses genetic gain as a percentage of the previous cycle or the original population. Turfgrass cultivars are generally ranked on the basis of the semiquantitative rating scale termed *turf quality* (1 to 9, where 9 is best and 6 is considered adequate). The performance of a given cultivar is always evaluated in relation to others in a given trial and the rating scale is not linear, i.e., the difference between 2 and 3 may bear no relation to the difference between 7 and 8. Thus, the incremental numerical (and surely the percentage) increase in turf quality of one cultivar relative to another is subjective and arbitrary. Perhaps the best measure of breeding progress is to track the ranking of an individual cultivar in successive trials, often planted 3 or 4 years apart, a numerical value that often shows dramatic changes (e.g., Meyer and Watkins, Chapter 8). If crop cultivars were evaluated in this manner, the progress made by turfgrass breeders between 1960 and 2000, relative to grain, forage, fruit, and fiber crops, would be comparatively astounding. Furthermore, much of this progress has been achieved without the serious and severe reductions in genetic variation and consequential increases in genetic vulnerability that have occurred in many agronomic and horticultural crops.

Form vs. Function of Turfgrasses

Form and function of turfgrasses are not mutually exclusive. Historically, governmental testing agencies have largely emphasized visual appearance of turfgrasses (Shildrick, 1980; Voigt, 1996). Numerous visual traits are typically combined into the single variable *turf quality*. Turf quality generally combines all important traits, including color, texture, density, and pest resistance. Turf quality may incorporate functional stress tolerances insofar as stresses are manifested at trial sites. However, documentation of stress levels is not specifically required and is often lacking (Voigt, 1996). Winners and losers of national trials are largely based on turf quality, fueling breeder's efforts to develop turfgrass cultivars with better quality.

Huge gains have been made to improve the visual appearance of turfgrasses, but more effort is required to make them more functional. Turfgrass breeders have undoubtedly made progress toward improved functionality (e.g., environmental stress tolerances, pest resistances, improved longevity, and wear tolerance), but these are poorly documented at best. Conversely, hundreds of turfgrass cultivars have been developed and released with little thought or effort toward environmental stress tolerances, except what may naturally occur in breeder's test plots. Nevertheless, it is likely that considerable genetic variation exists for many important stress tolerances within improved turfgrass germ plasm pools. Natural variability for tolerance to numerous biotic and abiotic stresses, described earlier in this chapter, likely exists within many turfgrass germ plasm pools. Discovery, characterization, and use of this variability should be a relatively simple matter to redesign breeding programs around functional stress tolerance tests, combined with evaluation of elite germ plasm. In some cases, natural stress tolerances may have been lost due to drift (random loss of alleles) during selection. These can be recovered in crosses of elite, high-quality genotypes with nonelite, stress-tolerant genotypes.

The emphasis on national trials with broad regional adaptations, particularly in the United States where a trial may be planted in 20 to 30 highly contrasting environments, limits the type of data that can be collected. Quantification of various stress tolerances and pest resistances is expensive and prohibitive, particularly on trials that are beginning to approach 200 entries. Environmental stresses such as drought, heat, and cold, must be evaluated under highly controlled conditions in which the number of entries may be highly restricted and little or no environmental replication is possible. Furthermore, some of the leading scientific experts in testing turfgrasses for stress tolerances are turfgrass breeders themselves, creating the appearance and potential for conflict-of-interest, which should be avoided (Voigt, 1996). The gains made in turf quality, particularly in the mainstream species, will soon begin to plateau as new cultivars approach levels of color, tiller density, leaf size, and uniformity that will be difficult to surpass. To survive in this highly competitive industry, turfgrass breeders must adapt to improve functionality of these high quality turfgrasses. Development of standards, testing procedures, and evaluation criteria related to important stress tolerances, to be used by governmental testing agencies, will provide a mechanism for turfgrass breeders to develop sound marketing strategies for more functional turfgrass cultivars. Standard tests to characterize pest resistance of alfalfa (*Medicago sativa* L.) cultivars were highly successful, resulting in a high level of participation and uniform descriptions of cultivar resistance levels to multiple pests (Barnes et al., 1974).

New Technologies

The technology exists to make pinpoint genetic changes to turfgrasses, using marker-assisted selection or transgenic technology (Sticklen and Kenna, 1998). Marker-assisted selection (MAS) for quantitative traits consists of (1) identifying putative loci for the trait (quantitative trait loci, QTL) by correlating phenotypic data to markers within a linkage map, (2) determining which putative QTL explain the largest amount of phenotypic variance and have the desired effect (positive or negative) on the phenotypic trait, and (3) selecting plants on the basis of the molecular marker(s) linked to the putative QTL (Dudley, 1993). While the concept of MAS dates back to 1967 (Smith, 1967), it has received very little attention in turfgrasses.

Gresshoff et al. (1998) describe turfgrass researchers as being in a "catch-up" mode. Because many turfgrasses are complex polyploids with genomes that derive from multiple progenitors or with polysomic inheritance, development of reliable linkage maps in many turfgrasses is more complicated and expensive than in diploids. Part of the lag observed in use of molecular markers in turfgrass selection strategies may also relate to the "yield lag" phenomenon in crop plants. As limited resources are devoted to development of new selection strategies, the loss of resources devoted to ongoing cultivar development reduces the rate of gain and the frequency with which new cultivars are produced. Organizations dependent on cultivar royalties for their economic survival cannot devote the resources necessary to build efficient, repeatable, and broadly applicable MAS systems. Marker-assisted selection will most likely become useful in some turfgrass breeding programs, but it may not be widely available to all breeders.

Molecular marker technology has resulted in several important scientific advancements toward understanding the genetics and genomics of turfgrasses (Sticklen and Kenna, 1998). Molecular markers have served to identify genetic similarities and dissimilarities among cultivars, illustrating patterns of genetic variation that relate to geographic and climatic origin, breeding history, germ plasm collection and exploration, and phenotype. Markers have also been used to help identify modes of genetic inheritance, information crucial to the plant breeder, but often difficult to identify by conventional means. Molecular markers have the potential as tools to identify turfgrass cultivars from unknown seed lots or herbage, but this technology is much closer to reality for species which are characterized by homogeneous cultivars (Barker and Warnke, 2001; Huff, 1998, 2001). A small number of molecular markers can be used to determine if collections of Kentucky bluegrass represent "new" genotypes or existing cultivars (D.R. Huff, 1998, personal communication). This technology saves the breeder from evaluating and possibly selecting a genotype that has already been released as a cultivar.

Transgenic technology allows specific blocks of chromosomes, the functional genes themselves, to be transferred among species or among genotypes within a species. While there are many potential ecological and sociological arguments against transgenic technology, one argument for this technology is too seldom stated. Transgenic technology will allow turfgrass breeders greater ability to focus on functionally meaningful traits, such as edaphic and climatic stress tolerances, that will improve the adaptation and persistence of a turf cultivar. Transgenic technology is a reality in turfgrasses and it will soon become commercialized as it has in numerous grain and food crops. Duncan and Carrow (1999, 2001) have provided thorough reviews of potential physiological mechanisms and candidate genes that may be useful in transforming turfgrasses with improved stress tolerances. These authors emphasize the importance of improving root plasticity as a prerequisite to improving stress tolerance of perennial turfgrasses. Care must be taken by breeders to utilize different genes and mechanisms across species and cultivars within species, avoiding the potential for genetic vulnerability associated with a small number of loci present in a large proportion of cultivated germ plasm.

NEW GERM PLASM SOURCES

Interspecific Hybrids

Numerous turfgrass cultivars are a direct or indirect result of interspecific hybridizations. Interspecific and intergeneric hybrids occur naturally in many tribes of the

Poaceae. Most polyploid turfgrasses evolved from natural interspecific hybrids into stable amphidiploids (allopolyploids). These plants carry the diploid chromosome number from both parent genomes, resulting from spontaneous doubling of chromosomes in the progeny of an (often) unstable and infertile first-generation hybrid. Amphidiploids often possess vigor, competitive ability, and longevity surpassing their parents (Rieger et al., 1991). The multiple-genome characteristic of amphidiploids results in genetic redundancy—multiple copies of functional genes—which may provide these species with superior phenotypic plasticity. Taxonomists and agronomists agree that amphidiploidy is a mechanism of speciation and recognize amphidiploids as separate species from their parents. Examples of amphidiploids include several bluegrasses (*Poa* spp.), fescues (*Festuca* spp.), and bentgrasses (*Agrostis* spp.). Many naturalized collections of *Zoysia* appear to represent natural interspecific hybrids within *Zoysia*. These plants are classified as hybrids because they possess half the chromosome complement of their parents and represent intermediate types relative to their parents (Engelke and Anderson, Chapter 17).

Interspecific and intergeneric hybrids have also been created in numerous breeding programs and have contributed to numerous turfgrass cultivars. Perhaps the most prominent of these is in bermudagrass, *Cynodon dactylon* (L.) Pers. var. *dactylon* x *C. transvaalensis* Burtt-Davy. Crosses between tetraploid *C. dactylon* and diploid *C. transvaalensis* produce sterile triploid hybrids which can be vegetatively propagated by stolons (Taliaferro, Chapter 15; Busey, 1989). Numerous sterile, triploid, hybrid bermudagrass cultivars have been developed for use in the golf industry. An interspecific hybrid between Kentucky bluegrass and Texas bluegrass (*Poa arachnifera* Torr.) shows potential for extending the adaptation range of both cool-season grasses, which have dramatically different optimal temperature requirements (Read and Anderson, Chapter 5). Numerous interspecific hybrids are possible within *Agrostis* (Brede and Sellmann, Chapter 13).

Surprisingly, one of the most important interspecific hybrid groups among cool-season forage grasses, fescue-ryegrass hybrids, has not been utilized for turf. Tall fescue and meadow fescue (*Festuca pratensis* Huds.) will hybridize with perennial or Italian ryegrass to produce stable amphiploids after chromosome doubling. Fescue-ryegrass amphiploids can combine the rapid establishment of the ryegrass parent with the drought and cold tolerance of the fescue parent (Thomas and Humphreys, 1991). Fescue-ryegrass hybrids have also been useful in backcrossing programs to introgress a limited number of genes from one genus into the other, such as drought tolerance from tall fescue into Italian ryegrass (Humphreys and Pasakinskiene, 1996). Fescue-ryegrass hybrids represent a vast untapped resource that could be used to combine the favorable traits of tall fescue and perennial ryegrass. To date, introgression of the stay-green gene from meadow fescue into turf-type perennial ryegrass represents the only example of fescue-ryegrass hybridizations used for turf (Thorogood, Chapter 7).

Novel Species

The identification and development of novel turfgrass species is an ongoing and evolutionary process in itself. As discussed throughout this book, all turfgrasses were derived from "wild" species that dominated natural meadows and pastures, largely forage and pasture grasses that humans recognized as having multiple cultural purposes. Natural variation is often sufficient for humans to recognize the turf potential of a species. A single cycle of selection is often sufficient to create a reasonably homogeneous population that has commercial potential for particular turf uses. However, numerous cycles of selection or more sophisticated techniques are often required to convert the "wild" species into something that has multiple turf uses, an expanded range of turf usage, or adaptation to multiple turfgrass managements. Screening genotypes for tolerance to specific and severe edaphic stresses can identify previously unknown sources of tolerance (Ashraf et al., 1986; Lehman et al., 1998; Torello and Symington, 1984). Transformation may impart stress tolerances not previously observed within a particular species (Duncan and Carrow, 1999, 2000).

While most novel species may have limited individual adaptation during the early years of the domestication process, they serve to broaden the adaptation of turfgrass in general. Many novel species are niche-adapted, such as the alkali grasses, represented by *Puccinellia distans* (L.) Parl. This species is used as a component of turf mixtures for roadsides, boulevards, and rights-of-way where winter snows bring about frequent salt applications and plowed snow is typically deposited on turf, creating localized salinity problems. 'Fairway' crested wheatgrass [*Agropyron cristatum* (L.) Gaertn.] is the first seeded turfgrass cultivar, released in 1932 and used extensively for multiple dryland purposes, including range (grazing), turf, and revegetation (Asay and Jensen, 1996). Redtop (*Agrostis gigantea* Roth) has a combination of superior seeding vigor and establishment capacity, salt tolerance, and heavy metal tolerance, making it superior to most turfgrasses as a component for bioremediation of disturbed lands (Brede and Sellmann, Chapter 13).

Of the great grasslands of the world, relatively few have contributed the mainstream turfgrass species, which largely originate in Europe (cool-season grasses) or Asia and Africa (warm-season grasses). However, the growing list of turfgrass species that are used for various purposes, as evidenced by several minor species discussed in this book, indicates that the list will continue to grow. As more people with diverse needs and opinions become involved in turfgrass research, demonstration, and applications, there is considerable potential that grasslands from additional parts of the world may become represented among domesticated turfgrasses.

Native species hold special value for many people, particularly those who are involved in bioremediation of disturbed lands, eradication of invasive species, and reintroduction of native plants to recreate natural habitats. Many regions of the world are not naturally inhabited by the mainstream turfgrass species. In these regions, industrial and urban development leads directly to introduction of non-native turfgrass species. Non-native species and/or cultivars are essential components of high-quality turf in all industrialized parts of the world. Unfortunately, this has led many people to view these non-natives as invasive, although humans are entirely responsible for this invasion. While non-native turfgrasses are not necessarily biologically or botanically invasive, there is a growing need for development of native germ plasm to be used by landscapers. Native species provide land developers and landscapers additional options to beautify landscapes with plants that are pleasing to the eye, to the touch, and to the mind.

Buffalograss (*Buchloe dactyloides* (Nutt.) Engelm) is the most successful native species domesticated for turf usage in North America (Riordan and Browning, Chapter 16). Buffalograss is native from central Mexico to southern Canada and cultivars can be used for low-maintenance turf throughout that range. Carpetgrass (*Axonopus affinis* Chase), possibly native to the southeastern United States, is adapted to moderate shade (Busey, 1989). Texas bluegrass has an extremely limited natural range in Texas and Oklahoma, but appears to have considerable potential as a cool-season turfgrass adapted to warm climates (Read and Anderson, Chapter 5). Idaho bentgrass (*Agrostis idahoensis* Nash) is native to the western United States and has potential to replace non-native species for turf applications under moderately dryland conditions (Brede and Sellmann, Chapter 13). Grasses native to prairie and alpine regions of western

Canada have potential as salt-tolerant turfgrasses (Acharya et al., 1992). Many other grasses native to North America have potential as turfgrass species, either as low-maintenance ecovars, regionally deployed seed increases of local ecotypes, or as germ plasm for breeding programs aimed at domestication (Fraser and Anderson, 1980; Smith and Mintenko, 2000).

CONCLUSIONS

The turfgrasses are a remarkably diverse and adaptable group of species. Together they inhabit every continent on the planet and nearly every terrestrial ecosystem. Turfgrasses have been an integral component of human recreation for thousands of years. They represent a classless commodity, equally enjoyed by and accessible to any class of any society (albeit in different forms). Human consciousness, behavior, and habit are linked to Planet Earth as tightly or more so by turfgrasses than by food crops. While food is essential for survival, most humans in industrialized countries have lost their appreciation for the social and economic value, as well as the true source, of food. Conversely, the pervasiveness of turfgrasses in both urban and rural settings creates a level of consciousness and caring that many humans reserve only for their most treasured possessions. In a true expression of capitalistic legerdemain, these feelings translate to consumerism in industrialized countries, powering growth of the turf industry and, completing the cycle, delivery of ever-improving products and technologies to the consumer.

Turfgrasses exist as a result of a unique combination of forces that have driven genetic evolution, including climatic and soil factors; diseases, insects, ungulates, and domestic livestock; and humans. Thousands of years of natural evolution, grazing pressure, and human domestication events have formed this germ plasm that is so valuable to the human race. As human recreation and reclamation needs increase and evolve, so too will turfgrasses continue to evolve. Turfgrass phenotypes are highly plastic and pliable, lending themselves to modification by breeding and selection in relatively short periods of time. The current germ plasm base of many turfgrasses contains considerable genotypic breadth, and genetic vulnerability appears to be low. However, intensive breeding of some species has focused breeding programs toward a common phenotype that has undoubtedly restricted the germ plasm base, creating the potential for future genetic vulnerability. This effect may be magnified as plant transformation becomes more common and efforts focus on a restricted germ plasm base and relatively few transgenes. Turfgrass breeders and germ plasm explorers must remain diligent and continue to explore, collect, evaluate, preserve, and utilize germ plasm from natural habitats and domestic turfs throughout the world.

REFERENCES

Acharya, S.N., B.A. Darroch, R. Hermesh, and J. Woosaree. 1992. Salt stress tolerance in native Alberta populations of slender wheatgrass and alpine bluegrass. *Can. J. Plant Sci.* 72:785–792.

Ahmad, I., S.J. Wainwright, and G.R. Stewart. 1981. The solute and water relations of *Agrostis stolonifera* ecotypes differing in their salt tolerance. *New Phytol.* 87:615–629.

Altham, H.S. and E.W. Swanton. 1926. *A History of Cricket.* George Allen & Unwin Ltd., London.

Asay, K.H. and K.B. Jensen. 1996. Wheatgrasses. pp. 691–724. In L.E. Moser et al., Eds. *Cool-Season Forage Grasses.* American Society of Agronomy, Madison, WI.

Ashraf, M., T. McNeilly, and A.D. Bradshaw. 1986. The response of selected salt-tolerant and normal lines of four grass species to NaCl in sand culture. *New Phytol.* 104:453–461.

Ayazloo, M., S.G. Garsed, and J.N.B. Bell. 1982. Studies on the tolerance to sulphur dioxide of grass populations in polluted areas. II. Morphological and physiological investigations. *New Phytol.* 90:109–126.

Baker, A.J.M., C.J. Grant, M.H. Martin, S.C. Shaw, and J. Whitebrook. 1986. Induction and loss of cadmium tolerance in *Holcus lanatus* L. and other grasses. *New Phytol.* 102:575–587.

Balfourier, F., C. Imbert, and G. Charmet. 2000. Evidence for phylogeographic structure in *Lolium* species related to the spread of agriculture in Europe. A cpDNA study. *Theor. Appl. Genet.* 101:131–138.

Barker, R.E. and S.E. Warnke. 2001. Application of molecular markers to genetic diversity and identity in forage crops. pp. 135–148. In G. Spangenberg, Ed. *Molecular Breeding of Forage Crops.* Kluwer Academic Publ. Dordrecht, The Netherlands.

Barnes, D.K., F.I. Frosheiser, E.L. Sorensen, J.H. Elgin Jr., M.W. Nielson, W.F. Lehman, K.T. Leath, R.H. Ratcliffe, and R.J. Buker. 1974. Standard Tests to Characterize Pest Resistance in Alfalfa Varieties. USDA Agric. Res. Service, ARS-NC-19. U.S. Govt. Print. Office, Washington, DC.

Beddows, A.R. 1953. The Ryegrasses in British Agriculture: A Survey. Welsh Pl. Breed. Stn., Bulletin Series H, No. 17.

Bonos, S.A. and W.A. Meyer. 2000. Evaluation and heritability of dollar spot resistance in creeping bentgrass genotypes. p. 161. *In Agronomy Abstracts.* ASA-CSSA-SSSA, Madison, WI.

Bradley, B.P. 1982. Models for physiological and genetic adaptation to variable environments. pp. 33–50. In H. Dingle and J.P. Hegmann, Eds. *Evolution and Genetics of Life Histories.* Springer-Verlag, New York.

Bradshaw, A.D. 1965. Evolutionary significance of phenotypic plasticity in plants. *Adv. Genet.* 13:115–155.

Breese, E.L. 1983. Exploitation of the genetic resources through breeding: *Lolium* species. pp. 275–288. In J.G. McIvor and R.A. Bray, Eds. *Genetic Resources of Forage Plants.* CSIRO, East Melbourne, Australia.

Breese, E.L. and B.F. Tyler. 1986. Patterns of variation and the underlying genetic and cytological architecture in grasses with particular reference to *Lolium.* pp. 53–69. In B.T. Styles, Ed. *Infraspecific Classification of Wild and Cultivated Plants.* Clarendon Press, Oxford, England.

Brougham, R.W., A.C. Glenday, and S.O. Fejer. 1960. The effects of frequency and intensify of grazing on the genotypic structure of a ryegrass population. *N. Z. J. Agric. Res.* 3:442–453.

Brougham, R.W. and W. Harris. 1967. Rapidity and extent of changes in genotypic structure induced by grazing in a ryegrass population. *N. Z. J. Agric. Res.* 10:56–65.

Browning, R. 1955. *A History of Golf, the Royal and Ancient Game.* E.P. Dutton & Co., Inc. New York.

Busey, P. 1989. Progress and benefits to humanity from breeding warm-season grasses for turf. pp. 49–70. In D.A. Sleper et al., Eds. Contributions from Breeding Forage and Turf Grasses. CSSA Spec. Publ. 15, American Society of Agronomy, Madison, WI.

Cai, W. and P.H. Whetton. 2000. Evidence for a time-varying pattern of greenhouse warming in the Pacific Ocean. *Geophys. Res. Lett.* 27:2577–2580.

Campbell, G. 1952. *A History of Golf in Britain.* Cassell & Co. Ltd. London.

Carrow, R.N. and R.R. Duncan. 1998. *Salt-affected Turfgrass Sites: Assessment and Management.* Ann Arbor Press, Chelsea, MI.

Casler, M.D. J.F. Pedersen, G.C. Eizenga, and S.D. Stratton. 1996. Germ plasm and cultivar development. pp. 413–469. In L.E. Moser et al., Eds. *Cool-season Forage Grasses.* American Society of Agronomy, Madison, WI.

Casler, M.D. and G.A. Pederson. 1996. Host resistance and tolerance and its deployment. pp. 475–507. In S. Chakraborty et al., Eds. Pasture and Forage Crop Pathology. American Society of Agronomy, Madison, WI.

Ceccarelli, S., E. Falistocco, and P.G. Cionini. 1992. Variation of genome size and organization within hexaploid *Festuca arundinacea. Theor. Appl. Genet.* 83:273–278.

Chamberlain, J.A. and P.E. Evans. 1979. Relationship between depth of leaf ridging and numbers of a possible mite vector. pp. 197–198. In *Welsh Plant Breed. Stn. Ann. Rep. for 1978.* Aberystwyth, Wales.

Charles, A.H. 1964. Differential survival of plant types in swards. *J. Br. Grassl. Soc.* 19:198–204.
Cockell, C.S. 2000. The ultraviolet history of terrestrial planets: Implications for biological evolution. *Planet. Space Sci.* 48:203–214.
Cook, S.C.A., C. Lefébvre, and T. McNeilly. 1972. Competition between metal tolerant and normal plant populations on normal soil. *Evolution* 26:366–372.
Counts, R.L. 1993. Phenotypic plasticity and genetic variability in annual *Zizania* spp. along a latitudinal gradient. *Can. J. Bot.* 71:145–154.
Crossley, G.K. and A.D. Bradshaw. 1968. Differences in response to mineral nutrients of populations of ryegrass, *Lolium perenne* L., and orchardgrass, *Dactylis glomerata* L. *Crop Sci.* 8:383–387.
Darwin, C. 1875. *The Variation of Animals and Plants under Domestication*. 2nd ed. Murray, London.
Davies, W.E., B.F. Tyler, M. Borrill, J.P. Cooper, H. Thomas, and E.L. Breese. 1973. Plant introduction at the Welsh Plant Breeding Station. pp. 143–162 In *Welsh Pl. Breed. Stn. Ann. Rep. for 1972*. Aberystwyth, Wales.
Dudley, J.W. 1993. Molecular markers in plant improvement: Manipulation of genes affecting quantitative traits. *Crop Sci* 33:660–668.
Duncan, R.R. 2000. Plant tolerance to acid soil constraints: Genetic resources, breeding methodology, and plant improvement. pp. 1–38. In R.E. Wilkinson Ed. *Plant-Environment Interactions*. 2nd ed. Marcel Dekker, Inc. New York.
Duncan, R.R. and R.N. Carrow. 1999. Turfgrass molecular genetic improvement for abiotic/edaphic stress resistance. *Advan. Agron.* 67:233–305.
Duncan, R.R., and R.N. Carrow. 2000. *Seashore Paspalum, the Environmental Turfgrass*. Ann Arbor Press, Chelsea, MI.
Duncan, R.R., and R.N. Carrow. 2001. Molecular breeding for tolerance to abiotic/edaphic stresses in forage and turfgrass. pp. 251–260. In G. Spangenberg, Ed. *Molecular Breeding of Forage Crops*. Kluwer Academic Publ. Dordrecht, The Netherlands.
Endo, R.M. 1972. The turfgrass community as an environment for the development of facultative fungal parasites. pp. 171–202 In V.B. Youngner and C.M. McKell, Eds. *The Biology and Utilization of Grasses*. Academic Press, New York.
Evans, G., W.E. Davies, and A.H. Charles. 1961. Shift and the production of authenticated seed of herbage cultivars. pp. 99–105. In *Welsh Pl. Breed. Stn. Ann. Rep. 1960*, Aberystywth.
Fraser, J.G. and J.E. Anderson. 1980. Wear tolerance and regrowth between cuttings of some native grasses under two moisture levels. *New Mexico Agric. Exp. Stn. Res. Rep. 418*.
Funk, C.R., R.E. Engel, and P.M. Halisky. 1969. Registration of Manhattan perennial ryegrass. *Crop Sci.* 9:679–680.
Grafius, J.E. 1978. Multiple characters and correlated response. *Crop Sci.* 18:931–934.
Green, G. 1953. *Soccer: the World Game. A Popular History*. Phoenix House Ltd., London.
Gresshoff, P.M., L.M Callahan, F. Ghassemi, and G. Caetano-Anollés. 1998. Molecular genetic analysis of turfgrass. pp. 3–18. In M.B. Sticklen and M.P. Kenna, Eds. *Turfgrass Biotechnology. Cell and Molecular Approaches to Turfgrass Improvement*. Ann Arbor Press, Chelsea, MI.
Hallauer, A.R., and J.B. Miranda, Fo. 1981. *Quantitative Genetics in Maize Breeding*. Iowa State University Press, Ames, IA.
Harberd, D.J. 1961. Observations on population structure and longevity of *Festuca rubra* L. *New Phytol.* 60:184–206
Harberd, D.J. 1967. Observations on natural clones in *Holcus mollis*. *New Phytol.* 66:401–408.
Harlan, J.R. 1975. Crops and Man. Amer. Soc. Agron., Madison.
Harlan, J.R. 1992. Origins and processes of domestication. pp. 159–175. In G.P. Chapman, Ed. *Grass Evolution and Domestication*. Cambridge University Press, Cambridge, England.
Hickey, D.A. and T. McNeilly. 1975. Competition between metal tolerant and normal plant populations; a field experiment on normal soil. *Evolution* 29:458–464.
Horsman, D.C., T.M. Roberts, and A.D. Bradshaw. 1979. Studies on the effect of sulphur dioxide on perennial ryegrass (*Lolium perenne* L.). II. Evolution of sulphur dioxide tolerance. *J. Exp. Bot.* 30:495–501.

Huff, D.R. 1998. Genetic characterization of open-pollinated turfgrass cultivars. pp. 19–30. In M.B. Sticklen and M.P. Kenna, Eds. *Turfgrass Biotechnology. Cell and Molecular Approaches to Turfgrass Improvement*. Ann Arbor Press, Chelsea, MI.

Huff, D.R. 2001. Genetic characterization of heterogeneous plant populations in forage, turf, and native grasses. pp. 149–160. In G. Spangenberg, Ed. *Molecular Breeding of Forage Crops*. Kluwer Academic Publ. Dordrecht, The Netherlands.

Humphreys, M.O. 1999. The contribution of conventional plant breeding to forage crop improvement. pp. 54–63. In J.G. Buchanon-Smith, Ed. *Proc. XVIII Intl. Grassl. Congr.* 8–19 June 1997. Winnipeg, Manitoba and Saskatoon, Saskatchewan.

Humphreys, M.O. and A.D. Bradshaw. 1977. Genetic potentials for solving problems of soil mineral stress: heavy metal toxicities. pp. 95–123. In M.J. Wright, Ed. *Plant Adaptation to Mineral Stress in Problem Soils*. Cornell University, Ithaca, NY.

Humphreys, M.O. and C.F. Eagles. 1988. Assessment of perennial ryegrass (*Lolium perenne* L.) for breeding. I. Freezing tolerance. *Euphytica* 38:75–84.

Humphreys, M.O., M.P. Kraus, R.G. Wyn-Jones. 1986. Leaf-surface properties in relation to tolerance of salt spray in *Festuca rubra* ssp. *litoralis* (G.F.W. Meyer) Auquier. *New Phytol.* 103:717–723.

Humphreys, M.W. and I. Pasakinskiene. 1996. Chromosome painting to locate genes for drought resistance transferred from *Festuca arundinacea* into *Lolium multiflorum*. *Heredity* 77:530–534.

Isaac, E. 1970. *Geography of Domestication*. Prentice Hall, Englewood Cliffs, NJ.

Jain, S.K. 1978. Inheritance of phenotypic plasticity in soft chess, *Bromus mollis* L. (Gramineae). *Experientia* 34:835–836.

Khan, M.A. and A.D. Bradshaw. 1976. Adaptation to heterogeneous environments. II. Phenotypic plasticity in response to spacing in *Linum*. *Aust. J. Agric. Res.* 27:519–531.

Lancashire, J.A., D. Wilson, R.W. Bailey, M.J. Ulyatt, and P. Singh. 1977. Improved summer performance of a 'Low Cellulose' selection from 'Grasslands Ruanui' hybrid perennial ryegrass. *N. Z. J. Agric. Res.* 20:63–67.

Lee, G. 2000. Comparative Salinity Tolerance and Salt Tolerance Mechanisms of Seashore Paspalum Ecotypes. Ph.D. Dissertation. University of Georgia.

Lehman, V.G., M.C. Engelke, K.B. Marcum, P.F. Colbaugh, J.A. Reinert, B.A. Ruemmele, and R.H. White. 1998. Registration of 'Mariner' creeping bentgrass. *Crop Sci.* 38:537.

Madison, J.H. 1982. *Principles of Turfgrass Culture*. Robert E. Krieger Publ. Co., Malabar, FL.

Marples, M. 1954. *A History of Football*. Secker & Warburg, London.

Marshall, D.R. and S.K. Jain. 1968. Phenotypic plasticity of *Avena fatua* and *Avena barbata*. *Amer. Nat.* 102:457–467.

McNaughton, S.J. 1979. Grassland-herbivore dynamics. pp. 46–81 In A.R.E. Sinclair and M. Norton-Griffiths, Ed. *Serengeti. Dynamics of an Ecosystem*. University of Chicago Press, Chicago.

McNaughton, S.J., M.B Coughenour, and L.L. Wallace. 1982. Interactive processes in grassland ecosystems. pp. 167–193. In J.R. Estes et al., Ed. *Grasses and Grasslands. Systematics and Ecology*. University of Oklahoma Press, Norman.

Meyer, W.A. and C.R. Funk. 1989. Progress and benefits to humanity from breeding cool-season grasses for turf. pp. 31–48. In D.A. Sleper et al., Eds. Contributions from Breeding Forage and Turf Grasses. CSSA Spec. Publ. 15, American Society of Agronomy, Madison, WI.

Morishima, H. and H.I. Oka. 1975. Physiological and genetic basis of adaptability. pp. 133–171. In T. Matsuo, Ed. *Adaptability in Plants*. Comm. Intl. Biol. Program, University of Tokyo Press, Tokyo.

Patterson, D.T., J.K. Westbrook, R.J.V. Joyce, P.D. Lindgren, and J. Rogasik. 1999. Weeds, insects, and diseases. *Climate Change* 43:711–727.

Pohl, R.W. 1986. Man and grasses: A history. pp. 355–358. In T.R. Soderstrom et al., Eds. *Grass Systematics and Evolution*. Smithsonian Inst. Press, Washington, DC.

Reed, K.F.M., P.J. Cunningham, J.T. Barrie, and J.F. Chin. 1987. Productivity and persistence of cultivars and Algerian introductions of perennial ryegrass (*Lolium perenne* L.) in Victoria. *Aust. J. Exp. Agric.* 27:267–274.

Rieger, R., A. Michaelis, and M.M. Green. 1991. *Glossary of Genetics, Classical and Molecular.* 5th ed. Springer-Verlag, New York.

Roberts, E.C., W.W. Huffine, F.V. Grau, and J.J. Murray. 1992. Turfgrass science—historical overview. pp. 1–27. In D.V. Waddington et al., Eds. *Turfgrass.* American Society of Agronomy, Madison, WI.

Robertson, G.P., E.A. Paul, and R.R. Harwood. 2000. Greenhouse gases in intensive agriculture: Contributions of individual gases to the radiative forcing of the atmosphere. *Science* 289: 1922–1925.

Roux, W. M. 1969. *Grass, a Story of Frankenwald.* Oxford University Press, Cape Town, South Africa.

Ruzmaikin, A. 1999. Can El Niño amplify the solar forcing of climate? *Geophys. Res. Lett.* 26:2255–2258.

Santer, B.D., K.E. Taylor, T.M.L. Wigley, J.E. Penner, P.D. Jones, and U. Cubasch. 1995. Towards the detection and attribution of an anthropogenic effect on climate. *Climate Dynamics* 12:77–100.

Shildrick, J.P. 1980. Species and cultivar selection. pp. 69–97. In I.H. Rorison and R. Hunt, Eds. *Amenity Grassland. An Ecological Perspective.* John Wiley & Sons, New York.

Sinnott, E.W. 1921. The relation between body size and organ size in plants. *Am. Nat.* 55:385–403.

Smith, C. 1967. Improvement of metric traits through specific genetic loci. *Anim Prod.* 9:349–358.

Smith, S.R. and A. Mintenko. 2000. Developing and evaluating North American native grasses for turf use. *Diversity* 16:43–45.

Snaydon, R.W. 1973. Ecological factors, genetic variation and speciation in plants. pp. 1–29 In V.H. Heywood, Ed. *Taxonomy and Ecology.* Academic Press, New York.

Snaydon, R.W. 1978. Genetic changes in pasture populations. pp. 253–269. In J.R. Wilson, Ed. *Plant Relations in Pastures.* CSIRO, East Melbourne.

Sticklen, M.B., and M.P. Kenna. 1998. *Turfgrass Biotechnology. Cell and Molecular Approaches to Turfgrass Improvement.* Ann Arbor Press, Chelsea, MI.

Sultan, S.E. 1987. Evolutionary implications of phenotypic plasticity in plants. *Evol. Biol.* 21:127–173.

Tcacenco, F.A., C.F. Eagles, and B.F. Tyler. 1989. Evaluation of winter hardiness in Romanian introductions of *Lolium perenne. J. Agric. Sci., Camb.* 112:249–255.

Thomas, H., and M.O. Humphreys. 1991. Progress and potential of interspecific hybrids of *Lolium* and *Festuca. J. Agric. Sci., Camb.* 117:1–8.

Thomasson, J.R. 1979. Late Cenozoic grasses and other angiosperms from Kansas, Nebraska, and Colorado: Biostratigraphy and relationships to living taxa. *Kansas Geological Surv. Bull.* 218.

Torello, W.A., and A.G. Symington. 1984. Screening of turfgrass species and cultivars for NaCl tolerance. *Plant Soil* 82:155–161.

Tyler, B.F., K.H. Chorlton, and I.D. Thomas. 1987. Collection and field-sampling techniques for forages. pp. 3–10. In B.F. Tyler, Ed. Collection, Characterization and Utilization of Genetic Resources of Temperate Forage Grass and Clover. IBPGR Training Courses: Lecture series 1. Intl. Board Pl. Gen. Resources, Rome.

Van Dijk, G.E. 1955. The influence of sward-age and management on the type of timothy and cocksfoot. *Euphytica* 4:83–93.

Venables, A.V. and D.A. Wilkins. 1978. Salt tolerance in pasture grasses. *New Phytol.* 80:613–622.

Voigt, T. 1996. NTEP: Scientific selections. *Grounds Maint.* 31(August):14–16.

Wang, Z., M.D. Casler, J.C. Stier, J. Gregos, S.M. Millett, and D.P. Maxwell. 2000. Speckled snow mold resistance in creeping bentgrass. p. 151. In *Agronomy Abstracts.* ASA-CSSA-SSSA, Madison, WI.

Wilkins, P.W. 1978. Specialization of crown rust on highly and moderately resistant plants of perennial ryegrass. *Ann. Appl. Biol.* 88:179–184.

Wilkins, P.W. 1991. Breeding perennial ryegrass for agriculture. *Euphytica* 52:201–214.

Wilson, G.B. and J.N.B. Bell. 1985. Studies on the tolerance to SO_2 of grass populations in polluted areas. III. Investigations on the rate of development of tolerance. *New Phytol.* 100:63–77.

Wilson, G.B. and J.N.B. Bell. 1986. Studies on the tolerance to SO_2 of grass populations in polluted areas. IV. The spatial relationship between tolerance and a point source of pollution. *New Phytol.* 102:563–574.

Wood, G.M. and R.P. Cohen. 1984. Predicting cold tolerance in perennial ryegrass from subcrown internode length. *Agron. J.* 76:516–517.

Wu, L. 1981. The potential for evolution of salinity tolerance in *Agrostis stolonifera* L. and *Agrostis tenuis* Sibth. *New Phytol.* 89:471–486.

Part 2
Cool-Season Grasses

Chapter 2

Kentucky Bluegrass

David R. Huff

Kentucky bluegrass (*Poa pratensis* L.) is one of the most popular, and hence, widely propagated turfgrass species for amenity uses in the northern United States and Canada. Its overall aesthetic appeal as a lawn grass sets the standard against which most other turfgrasses are compared due to its combination of softness, medium- to fine-leaf texture, high shoot density, dark green color, and persistence. In addition to its use as turf, Kentucky bluegrass is also extensively cultivated worldwide as a forage grass and for conservation purposes protecting against soil erosion. The use and management of Kentucky bluegrass for turf purposes has been extensively reviewed by Beard (1982) and Turgeon (1998) and for forage purposes by Smith (1981), Duell (1985), and Wedin and Huff (1996).

Kentucky bluegrass is a widely adapted and highly variable perennial which spreads by slender, extensive, determinate rhizomes. As a species, it is considered to be extremely vigorous in its sod forming ability and its persistence. Commercial sod production in the northern United States has increased markedly because the extensive rhizome production of Kentucky bluegrass produces solid dense sod, which when harvested has strong tensile strength. The holding characteristics of this sod are of great value to sod growers, who mix it with other turfgrass species to utilize the strength that Kentucky bluegrass rhizomes impart to sod blocks and rolls. Persistency is another important attribute of Kentucky bluegrass. It is often found to occur along highway roadsides in the central and northeastern United States, even though it is not normally sown for roadsides or, if so, in very small amounts when compared to other grass species.

Kentucky bluegrass was most likely introduced to North America by the earliest settlers as an agronomic crop, as animal feed and bedding, or as a contaminant of these (Bashaw and Funk, 1987). From its first introduction there was a rapid spreading westward across the continent with large stands established that preceded pioneers moving westward. Most Kentucky bluegrass occurring in pastures and meadows of the central and northeastern United States likely arrived at these sites through natural dispersal mechanisms such as birds, animals, wind, and rain. Occasionally, reference is made that Kentucky bluegrass is indigenous to the United States, particularly to the state of Kentucky. Early settlers of the Appalachian mountains discovered grassy areas that they described as Kentucky bluegrass on limestone-derived soils (Duell, 1985). Yet

Kentucky bluegrass is such a variable species with circumpolar distribution that it is difficult to determine whether or not some strains found in northern and western North America are indigenous (Gray, 1908). Certain ecotypes identifiable as Kentucky bluegrass may be native to the Rocky Mountains, having possibly originated through interspecific hybridization (Soreng, 1985). Nevertheless, most naturalized germ plasm growing in the United States is likely the result of introductions from Europe (Bashaw and Funk, 1987).

A botanical description of Kentucky bluegrass is as follows: Plants are strongly rhizomatous with leaf blades mostly 2 to 4 mm wide and 5 to 15 cm long, though the basal leaf blades of unmowed plants are often much longer. Sheaths are closed ca. 1/3 to 1/2 its length and have membranous ligules approximately 1.5 mm in length. Tillers of a critical size vernalize under periods of short days and low temperatures producing flowering culms 30 to 120 cm long. The pyramidal panicle often has 4 or more branches at it lowest node. Spikelets are 3 to 6 mm in length, and have 3 to 5 florets and strongly nerved lemmas with a hairy or cottony base. The pyramidal panicle often has 4 or more branches at it lowest node (Hitchcock, 1950; Soreng, 1985).

Kentucky bluegrass produces seed both sexually and asexually by a process called apomixis. Seed formed by apomictic reproduction is genetically identical to the seed-bearing "mother" plant. Apomixis is therefore an excellent means of maintaining genetic purity of a cultivar from one generation to the next. However, it also makes the crossing and subsequent selection of Kentucky bluegrass a difficult and often frustrating process. Clones of Kentucky bluegrass vary as to the level of apomixis. Meyer (1982) reported that the cultivar 'Merion' had a level of apomixis of 96% or higher, while 'A-20' had an apomixis level of around 25%. Thus, Merion's high level of apomixis made stable seed production possible, while 'A-20' was increased mainly by vegetative propagation as sod because of its high level of sexuality.

Prior to 1950, most of the Kentucky bluegrass seed was produced in Kentucky, westward into eastern Kansas (southern limit), in southern Iowa, Wisconsin, Minnesota, and upward to eastern North Dakota (northern limit). Seed was collected by "stripping" existing common Kentucky bluegrass stands, which in many cases were also grazed in the same growing season. These practices have largely stopped, but for nearly 75 years the Kentucky bluegrass seed industry was dependent on this type of cropping system.

Present-day seed production of Kentucky bluegrass in the United States is localized in Northern Minnesota and the Pacific Northwest. Kentucky bluegrass seed is also produced in Canada and Europe (Denmark and Netherlands). Today's seed production is oriented to specific cultivars, with less Common Kentucky bluegrass in production. Seed is harvested from designated seed fields of broadcast or row plantings intensively managed with irrigation, fertilizers, herbicides, fungicides, and insecticides as needed to maximize economic yields of high quality seed. At harvest, fields are swathed prior to combining. The remaining stubble and straw residue has in the past been burned to effect control of diseases, insects, and weeds; however, the practice of burning is now highly restricted in most areas or has been prohibited.

The majority of all cultivars of Kentucky bluegrass were developed specifically for turf use. Higher maintenance is often required for improved cultivars of Kentucky bluegrass. Common types collected from old pastures and high-cut turfs often outperform improved cultivars under low-maintenance turfgrass conditions. However, some improved cultivars have shown adaptation to drought stress and low soil fertility by exhibiting reduced leaf surface area and enhanced rooting characteristics, like lower shoot/root ratios and deeper distribution of root production (Burt and Christians, 1990).

Initially, Kentucky bluegrass is slower to establish than many other cool-season turfgrasses, due to an approximate 14-d germination time and a generally longer juvenile stage. Once established, however, it rapidly colonizes meadows, pastures, and turfs through extensive rhizome production. Its highly persistent nature likely results from the inherent adaptability of apomictic lines which have been preselected for survival under a particular set of environmental conditions. Thus, locally adapted apomictic strains often persist better than introduced apomictic cultivars that have been developed under a set of contrasting environmental conditions.

Some of the more commonly known physiological limitations of Kentucky bluegrass to turfgrass management are its susceptibility to drought and disease. However, according to Funk (2000), nearly every characteristic needed for an ideal lawn is present in Kentucky bluegrass. These characteristics include enhanced tolerance to drought, heat, shade, close mowing, excessive wear, acid soils, salinity, as well as resistance to many major turfgrass diseases. The problem is that we currently lack the necessary breeding techniques required to combine these characteristics from various genotypes into a single cultivar (Funk, 2000).

DISTRIBUTION AND CYTOTAXONOMY

The center of origin of *Poa*, based on morphological, cytological, and species diversity is considered to be EuroAsia, though many species have worldwide distribution. Kentucky bluegrass has a circumpolar distribution, ranging in occurrence from about 30°N lat to above 83°N lat and from sea level to an altitude of 4000 m in alpine habits in the Sierra Nevada mountains of California (Clausen, 1961). A natural limitation to its distribution is an extended mean monthly temperature of 24°C and higher (Hartley, 1961). Given this limitation, Kentucky bluegrass occurs in most, if not all, regions of Europe, Siberia, the Far East, Central Asia, Scandinavia, Mongolia, Japan, China, North Africa, North America, and mountainous ranges in the Mediterranean Basin, Asia Minor, Iran, South America, and the Himalayans. The buffering capacity of polyploidy and the asexual propagation of the internal heterozygosity is one explanation for the wide distribution and adaptation of Kentucky bluegrass in many temperate regions of the world (Grazi et al., 1961; Clausen, 1961). In addition, the ability to perpetuate a single genotype over extended periods of time through apomixis may add another dimension enabling Kentucky bluegrass to have a circumpolar distribution (Kellogg, 1990). Because of its forage and turf uses, Kentucky bluegrass has also been introduced and is propagated in many other countries including Australia, New Zealand, and South Africa.

Difficulties arise in identifying and distinguishing species of bluegrass (*Poa*) because morphological variation often overlaps, the species have a wide range of adaptability, and the retention of pollen recognition systems between species enables a variety of interspecific hybridizations to occur in nature (Clausen, 1961). Interspecific hybridization results in many *Poa* species being introgressed with each other to such an extent that species classification is often difficult (Stebbins, 1950). Kentucky bluegrass (*Poa pratensis* L.) is a member of the tribe Poeae in the Pooideae subfamily of grasses. As many as 10 subspecies have been recognized by different botanists (Hitchcock, 1950; Hubbard, 1984; Tsvelev, 1976). The most common subspecies are *P. pratensis* subsp. *angustifolia* L. Lej. (narrow-leaf meadowgrass), which ranges from 15 to 80 cm in height, has narrow leaves, 1 to 2 mm wide, generally forms dense tufts, and is found on limestone soils, and subsp. *irrigata* (Lindm) H. Lindb., which is more rhizomatous and grows in moist to marshy soils. To date, no information is available concerning the range of chromosome numbers or the hybridization potential among these subspecies.

A complex series of polyploidy and aneuploidy exists among strains of Kentucky bluegrass, suggesting an allopolyploid origin for the species. Within *Poa* the base number of chromosomes (x) is 7. Within Kentucky bluegrass, however, chromosome numbers range continuously from 24 to 124 (Love and Love, 1975). Within this distributional range, modal peaks in the 49 to 56, 63 to 70, and 84 to 91 chromosome-number classes have been observed among strains (Nielsen, 1946). Classical cytogenetic analysis is made even more difficult because chromosome number may differ by 3 to more than 30 among somatic cells within a single genotype (Wu and Jampates, 1986; McDonnell and Conger, 1984) and from one generation to the next (Clausen, 1961). However, average chromosome number in most strains seems to be consistent, suggesting constancy of chromosome number among germ cells within a cultivar (McDonnell and Conger, 1984; Hanson, 1972). Results from 2C nuclear DNA content determination using flow cytometry suggest that an average Kentucky bluegrass chromosome contains approximately 0.13 to 0.14 pg of DNA (Huff and Bara, 1993; Arumuganathan et al., 1999)

Differences in the number of chromosomes and in the frequency of nonbivalent chromosome pairing are easily propagated through the asexual apomictic breeding system (Nielsen, 1946). Apomixis is a complex form of asexual reproduction that occurs extensively in *Poa* (for reviews of apomixis see: Asker, 1979; van Dijk, 1991). The breeding system of Kentucky bluegrass is a form of apomixis termed pseudogamous apospory (Müntzing, 1933; Tinney, 1940; Grazi et al., 1961). To varying degrees, Kentucky bluegrass also is capable of reproducing sexually and thus is referred to as a facultative apomict. For commercial seed production purposes, Kentucky bluegrass cultivars are expected to maintain a level of apomixis of 95% or higher.

COLLECTION, SELECTION, AND BREEDING HISTORY

Ecotype Selection

All cultivars of Kentucky bluegrass used before 1970 in the United States were derived from composites collected from naturalized stands located in the Midwest or from individual apomictic clones found in old turf areas (Meyer, 1982). 'Merion' was the first cultivar of Kentucky bluegrass possessing a low, turf-type of growth habit and exhibiting an improved resistance to leaf spot disease (a serious disease of Kentucky bluegrass, particularly common types, caused by *Drechslera poae* [Baudys.] Shoem. or *Bipolaris sorokiniana* [Sacc.] Shoem.). Merion Kentucky bluegrass was originally collected by golf course superintendent Joe Valentine on Merion CC located in Ardmore, PA. All other cultivars available until the late 1960s were highly susceptible to leaf spot disease when maintained under close mowing heights and high fertility (Meyer, 1982).

An extensive amount of morphological and physiological variation exists in old pastures and turf areas, natural meadows, and naturalized grasslands. This is the most widely available source of variation from which a breeder can commercialize cultivars through the apomictic process. Ecotype breeding involves extensive germ plasm collecting and subsequent screening in comparative trials with existing cultivars. In selecting ecotypes, breeders utilize the adaptation of individual genotypes to various natural or man-made environments which have occurred through natural selection. For example, selections from subartic regions are more likely to exhibit a compact, low shoot density growth habit. They also show a long winter dormancy and are more slow to green up in the early springtime. Selections collected from areas with long dry summers tend to produce a more open, less dense, turf and have deep extensive rhizomes. Common types from old pastures and high-cut turfs in temperate climates often exhibit erect, upright growth patterns and have narrow-bladed leaves.

As a result of the complications that apomixis imposes on the breeding process, ecotype breeding was the only method, prior to 1970, that had produced commercially successful cultivars (Bashaw and Funk, 1987). However, the success of ecotype breeding in Kentucky bluegrass may be its own demise. The extensive collection efforts by past breeders in Europe and the United States likely have begun to exhaust the pool of unique ecotypes available for cultivar development (Bashaw and Funk, 1987; van Dijk, 1991). Many of the most successful ecotypes seem to have wide distribution and are repeatedly collected from different geographical locations (Duyvendak and Luesink, 1979). This difficulty in collecting unique types from the wild has led to an increased emphasis on using hybridization for developing new distinct cultivars.

The Role and Rule of Apomixis

Successful cultivar development and breeding in Kentucky bluegrass is heavily dependent on an understanding of its apomictic reproductive system. Apomixis is an ideal system of seed production once a superior genotype has been identified because the cultivar can be increased directly as seed without recombination or segregation, in essence, maintaining stable hybrid vigor. Thus, apomixis also provides high levels of uniformity and stability required for certified seed production and commercialization. However, a disadvantage of apomixis is that it competes with a breeder's ability to effectively perform hybridizations because it masks or inhibits the sexual process. Moreover, when sexual recombination does occur, the resulting F1 hybrid progeny typically segregate in extreme fashion, vastly exceeding parental character values, a process known as transgressive segregation, making the task of interpreting quantitative genetic data more complex. Therefore, developing improved cultivars of Kentucky bluegrass using traditional breeding techniques and approaches is often a frustrating experience.

An important contribution toward breeding Kentucky bluegrass would be a method of manipulating the apomictic breeding system. Manipulation of apomixis in Kentucky bluegrass has had some limited success. Hovin et al. (1976) observed a slightly greater frequency of aberrants among field-grown plants in field locations promoting a longer flowering period, but found that cultivar uniformity was not seriously impaired. Grazi et al. (1961) and Han (1969) observed a greater tendency toward sexuality under greenhouse conditions than in the field. A major development in the breeding of Kentucky bluegrass occurred at Rutgers University in the late 1960s. Dr. C. Reed Funk and his associates developed a greenhouse crossing technique that resulted in an increased frequency of F1 hybrids compared to field crosses. They found that most Kentucky bluegrass flowers opened and shed pollen between 1:00 and 4:00 a.m. in the greenhouse and that foreign pollen had to be applied during this time to obtain a maximum number of hybrids (Funk and Han, 1967). Plants are allowed to flower early in the spring under greenhouse conditions and to pollinate either mechanically or by hand as early as possible after the stigmas emerge and are receptive. Pollinating at this time has been suggested to increase chances of fertilizing eggs (either sexual or aposporous) (Bashaw and Funk, 1987) because the apomictic proembryo often begins development at or slightly before anthesis (Akerberg and Bingefors, 1953). Improvements to this technique were later made by Hintzen and van Wijk (1985), who used artificial lighting to promote long daylengths in order to further increase the frequency of Kentucky bluegrass hybridizations. Several methods of mechanical pollination have also been described to facilitate hybridization in Kentucky bluegrass (Hintzen and van Wijk, 1985; Riordan et al., 1988).

Hybridization in an Apomictic System

Hybridization in a facultative apomictic system is possible only to the extent that sexuality is present, or that apomixis can be broken down. In order to retain high levels of apomixis in hybrid progeny for seed production and commercialization purposes, the parent plants of Kentucky bluegrass, particularly the female "seed-bearing" parents, should be highly apomictic (Bashaw and Funk, 1987). Because the frequency of hybrid (aberrant) plants is inherently low among such parents, large numbers of progeny per cross must be produced and evaluated.

Progeny that differ from their seed-bearing "maternal" parent are referred to as off-types or aberrants. However, not all maternally-deviating aberrant progeny are the product of a sexual process. Typically, during the time between egg mother cell reductional division and anthesis, the reduced meiotic egg dies or is outcompeted by an unreduced apospory embryo sac or by a rapidly developing apomictic proembryo. The unreduced (2n) apospory embryo sac initiates from somatic tissue, usually of the nucellar region, and is therefore independent of the germ cell line. However, in some instances, the meiotically derived embryo sac, containing a reduced (n) egg, develops and may even coexist alongside one or more unreduced apospory sacs within the same ovule (polyembryony). Given the complexities of embryo sac formation and origin within ovules of Kentucky bluegrass, many different genetic origins exist for aberrant progeny of a facultative apomict (Huff and Bara, 1993). Four main types of aberrants are: (1) the psuedogamous development of a reduced (n) egg results in a progeny with only 1/2 the chromosomes of the seed-bearing parent and none of the pollen parent, referred to as B_I (i.e., polyhaploid) individuals; (2) the union of a reduced (n) egg with a reduced (n) pollen nucleus (i.e., "normal" sexual reproduction) results in a progeny with 1/2 the chromosomes of each parent and is referred to as a B_{II} hybrid; (3) the union of an unreduced (2n) egg with a reduced (n) pollen nucleus results in a progeny with all the chromosomes of the seed-bearing parent and 1/2 those of the pollen parent, referred to as a B_{III} hybrid (polytriploid; note that either of the parents may contribute the unreduced gamete and therefore there are maternal B_{III} and paternal B_{III} aberrants, depending); and (4) in rare instances, the union of an unreduced egg and an unreduced pollen nucleus may occur resulting in a B_{IV} hybrid (polytetraploid). Thus, aberrant progeny plants that do not resemble the maternal seed-bearing parent have several completely different genetic origins. The two main factors determining genetic origin of aberrants are ploidy level and self- vs. cross-fertilization. In addition, aberrants may also arise from the possibility of irregular meiosis leading to the production of fertilized or unfertilized aneuploids (Grazi et al., 1961). Therefore, the facultative apomictic breeding system of Kentucky bluegrass is capable of generating and perpetuating extreme chromosomal variation. As a result of these different genetic origins, aberrant progeny should be genetically examined (using a combination of flow cytometry and molecular markers, see below) before inclusion into research or intensive breeding programs.

The presence of polyembryonic seed and the production of identical and nonidentical twin, triplet, and quadruplet seedlings, has been a useful tool for apomixis research and breeding in Kentucky bluegrass (Andersen, 1927; Akerberg, 1939; Nielsen, 1946; Duich and Musser, 1959; Huff and Bara, 1993). On two occasions, subhaploid plants (2n=16 or 18) of Kentucky bluegrass were found to resemble *Poa trivialis* L. (Kiellander, 1942; Akerberg and Bingefors, 1953), indicating that *P. trivialis* might be one of the component progenitor species of Kentucky bluegrass. However, not all subhaploids (2n=14) resemble *P. trivialis* (Neilsen, 1945).

Intraspecific Hybridization

Intraspecific hybridization has been demonstrated to be a successful breeding method for Kentucky bluegrass, resulting in the production of numerous successful cultivars, including 'Adelphi,' 'America,' 'Eclipse,' 'Midnight,' and 'P-104' (Bashaw and Funk, 1987).

Large parental differences exist in the frequency of production and the performance of aberrant progeny. The frequency of aberrant progeny has been found to be mostly dependent on the female seed-bearing parent (Hintzen, 1979), though some influence of the male pollen parent has been detected (Gates, 1997). Grazi et al. (1961) presented evidence suggesting that male parents also influence seed set. The agronomic performance of hybrid aberrants often depends on both the male and female parents (Hintzen, 1979; Gates, 1997). Because the frequency of aberrants is inherently low among highly apomictic parents, large numbers of progeny must be examined in order to retrieve progeny resulting from sexual hybridization. The most common method of screening such large numbers of progeny plants is the use of spaced-plant progeny nurseries (Tinney and Aamodt, 1940). Individual seedlings resulting from potential hybridization crosses are typically planted into the field as isolated spaced-plants and the seed of each spaced-plant is collected and sown for turf evaluation and uniformity test as spaced-plants.

Many, if not most, selections from intraspecific hybridizations are found to be B_{III} hybrids, resulting from the fertilization of unreduced eggs (Pepin and Funk, 1971; Funk et al., 1973; Hintzen, 1979). The high occurrence of B_{III} hybrids among selected hybrids may reflect the relatively higher frequency of this hybridization event, particularly in highly apomictic lines. Or, it may reflect the preferential selection of this type of hybrid over other hybrid types due to associated gigas characteristics of a more robust plant with darker green color, due to an increased number of chromosomes. The increased number of chromosomes of B_{III} selections raises a concern among breeders as to the optimal number of chromosomes for best cultivar performance. Pepin and Funk (1974) suggest that 2n=100 may be the upper limit of that optimum; while currently, some breeders consider 2n=100 to be too high (Funk, 1992, personal communication).

Interspecific Hybridization

In nature, interspecific hybridization among *Poa* species results in introgression to such an extent that species classification is often difficult (Stebbins, 1950). The retention of pollen recognition systems within *Poa* species, combined with the asexual nature of apomixis, present many possibilities for apomixis research and breeding applications through interspecific hybridization (Clausen et al., 1962; Funk and Han, 1967). For example, analysis of chloroplast-DNA by Soreng (1990) suggests that Canada bluegrass (*P. compressa* L.) may be a result of hybridization between Kentucky bluegrass and an unspecified diploid species with compressed culms. The long-term commitment required by interspecific breeding objectives, however, has generally resulted in a lack of substantial progress (Bashaw and Funk, 1987). Even so, some progress has been accomplished. Dr. James Read, Texas A&M University, Dallas, has recently released the cultivar 'Reveille' for turf use in the semiarid regions of the southern United States. Reveille is an F1 hybrid between Texas bluegrass (*Poa arachnifera*) and Kentucky bluegrass (*Poa pratensis*).

Müntzing (1940) found that interspecific hybridization often leads to a breakdown of apomixis in progeny plants, resulting in partially or completely sexual F1 hybrids. This was the basis of his hypothesis that the genetics of apomixis in Kentucky bluegrass was controlled by a delicate balance of gene interactions which is capable of being disrupted by the slightest genomic shock. For breeding purposes, apomixis may be regained from interspecific crosses by screening for aposporous recombinates in succeeding F2 and F3 generations (Akerberg and Bingefors, 1953). Examples of interspecific hybridizations involving Kentucky bluegrass include: *P. pratensis* X *P. compressa* (Dale et al., 1975); *P. pratensis* X *P. alpina* (Akerberg, 1942; Akerberg and Bingefors, 1953); *P. ampla* X *P. pratensis* and *P. scabrella* X *P. pratensis* (Hiesey and Nobs, 1982); *P. longifolia* X *P. pratensis* (Almgard, 1966; Williamson and Watson, 1981; van Dijk and Winklehorst, 1982), and *P. arachnifera* X *P. pratensis* (referred to in Oliver, 1910; but see above).

Future Breeding

The difficulties inherent in breeding improved cultivars of Kentucky bluegrass might best be demonstrated by examining those cultivars that perform best in any recent National Turfgrass Evaluation Program (NTEP) evaluation trial. Currently, the cultivars Midnight and America are among the top performers even though these cultivars were developed in the 1970s (Funk, 2000). As analogized by Funk (2000), breeding apomictic Kentucky bluegrass is unlike that of cross-pollinated turfgrasses where each cycle of phenotypic improvement builds upon the previous cycles. However, breeding Kentucky bluegrass is more akin to playing the lottery, where if you have the winning ticket (genotype) you will win and win big, but most tickets (genotypes) are losers. Funk (2000) goes further to suggest, and rightly so, that each new collection trip and each cycle of selection from progenies of a hybridization program is essentially a brand-new lottery game.

A deeper understanding of the underlying genetic control of apomixis in Kentucky bluegrass would be a valuable asset toward breeding Kentucky bluegrass. However, many researchers have suggested that the underlying genetic system of apomixis in Kentucky bluegrass may be too complicated to solve or understand (Müntzing, 1940; Grazi et al., 1961; Bashaw and Funk, 1987). Müntzing (1940), studying F2 and F3 progenies of sexual, haploid, and triploid plants, concluded that apomictic seed formation is recessive to sexual seed formation. Pepin and Funk (1974) suggested that apomixis is mostly under dominant control from their study of an F1 family between two highly apomictic varieties segregating for sexual and apomictic seed formation. Similar contradictory results from a study by Akerberg and Bingefors (1953) were interpreted by Almgard (1966, p. 54) as "Evidently apomixis as a functioning system can here be described as completely recessive, while its basic prerequisite, the presence in this case of an aposporous embryosac, rather showed a dominant pattern of inheritance in F1."

In the future, the genetic mechanism of apomixis might best be investigated by examining its component parts, including the formation of the unreduced megagametophyte and the pseudogamous development of the endosperm, rather than the overall frequency of aberrant progeny which is the whole of the apomictic process in Kentucky bluegrass. To this end, Huff and Bara (1993) and Barcaccia et al. (1997) have shown that a combination of flow cytometry and molecular markers such as random amplified polymorphic DNA (RAPD) are powerful tools for determining the origin of aberrant progeny. The combined use of these techniques has shown that flow cytometry accurately distinguishes progeny ploidy levels, while the molecular markers distinguish progeny resulting from cross-fertilization. Breeding tools like molecular markers have only recently been applied, and to a limited extent, to Kentucky bluegrass. A recent effort by Barcaccia et al. (1998) to reveal the genetic basis of apomixis in Kentucky bluegrass suggests a monogenic form of inheritance. However, contradictory to this and all other simple-inheritance mutant-gene models is the duplicate-gene asynchrony model proposed by Carman (1997) which contends that apomixis is caused by the "interference" of developmental pathways that results from hybridization of divergent genetic backgrounds. Hopefully, in the future, the dramatic new developments in molecular biology and genetic engineering techniques will open up new frontiers in our understanding and exploitation of apomixis in Kentucky bluegrass.

Germ Plasm Resources

According to Funk (2000), an estimated 90% of the Kentucky bluegrass growing on pastures, roadways, and rangelands established without ever being intentionally planted. Thus, these germ plasm resources represent an enormous and valuable source of genetic variability. It is for this reason that Johnston and Johnson (2000) suggest that the USDA Kentucky bluegrass collection, located at the Western Regional Plant Introduction Station at Pullman, Washington, is relatively small, containing only 300 accessions from 27 countries. Therefore, every effort should be made to expand and update the Kentucky bluegrass collection to ensure that diverse germ plasm reserves are maintained and publicly available.

There is a tremendous amount of genotypic and phenotypic variability that exists within Kentucky bluegrass. For instance, there is more variability among entries in an NTEP Kentucky bluegrass test than among entries within any other cool-season turfgrass species. This is partly because the apomictic breeding system ensures that nearly all the genetic variability resides between entries of Kentucky bluegrass, whereas in cross-pollinated turfgrass species, most genetic variability resides within entries (see Huff, 1997). However, it is the genomic complexity of the entries, being a series of polyploid and aneuploid genotypes, that most certainly is the cause for the large differences among entries of Kentucky bluegrass. Work at Rutgers University has developed an extensive classification system of Kentucky bluegrass cultivars, based primarily on agronomic and turf characteristics. The classifications are:

- BVMG types (ex. 'Baron' and 'Victa')
- Bellevue types (ex. 'Banff' and 'Dawn')
- mid-Atlantic types (ex. 'Julia' and' Ikone')
- Compact types (ex. 'Midnight' and 'Nublue')
- Aggressive types (ex. 'Mystic' and 'Washington')
- Midwest types (ex. 'Park' and 'Kenblu')
- Other types (This is a catchall that contain a wide range of variability unable to fit into any other above categories, ex. 'Coventry' and 'Eclipse')

Induced Variation

Additional methods used to generate genetic variation other than hybridization include mutation and tissue culture. Although irradiation has been shown to produce leaf spot resistance in a susceptible variety, the process also disabled resistance to *Helminthosporium* in the same material (Hanson and Juska, 1959). To date, no commercial cultivars have been produced from irradiation in Kentucky bluegrass. Regeneration of whole plants from callus culture has been reported (McDonnell and Conger,

1984; Wu and Jampates, 1986) but has not as yet resulted in any useful somaclonal variation.

REFERENCES

Akerberg, E. 1939. Apomictic and sexual seed formation in *Poa pratensis*. *Hereditas* 25:359–370.
Akerberg, E. 1942. Cytogenetic studies in *Poa pratensis* and its hybrid with *Poa alpina*. *Hereditas* 28:1–26.
Akerberg, E. and S. Bingefors. 1953. Progeny studies in *Poa pratensis* and its hybrid with *Poa alpina*. *Hereditas* 28:1–126.
Almgard, G. 1966. Experiments with *Poa*. III. Further studies of *Poa longifolia* Trin. with special reference to its cross with *Poa pratensis* L. Lantbrukshogsk. *Ann.* 32:3–64.
Andersen, A.M. 1927. Development of the female gametophyte and caryopsis of *Poa pratensis* and *Poa compressa*. *Journ. Agr. Research* 34:1001–1008.
Arumuganathan, K., S.P. Tallury, M.L. Fraser, A.H. Bruneau, and R. Qu. 1999. Nuclear DNA content of thirteen turfgrass species by flow cytometry. *Crop Sci.* 39:1518–1521.
Asker, S. 1979. Progress in apomixis research. *Hereditas* 91:231–240.
Barcaccia, G., A. Mazzucato, A. Belardinelli, M. Pezzotti, S. Lucretti, and M. Falcinelli. 1997. Inheritance of parental genomes in progenies of *Poa pratensis* L. from sexual and apomictic genotypes as assessed by RAPD markers and flow cytometry. *Theor. Appl. Genet.* 95:516–524.
Barcaccia, G., A. Mazzucato, E. Albertini, J. Zethof, A. Gerats, M. Pezzotti, and M. Falcinelli. 1998. Inheritance of parthenogenesis in *Poa pratensis* L.: auxin and AFLP linkage analysis support monogenic control. *Theor. Appl. Genet.* 97:74–82.
Bashaw, E.C. and C.R. Funk. 1987. Apomictic grasses. In *Principles of Cultivar Development*. Vol. 2 Crop Sciences. Walter R. Fehr, Ed. Macmillan Publishing Company, New York.
Beard, J.B. 1982. *Turfgrass Management for Golf Courses*. Burgess Publication Co., Minneapolis, MN.
Burt, M.G. and N.E. Christians. 1990. Morphological and growth characteristics of low- and high-maintenance Kentucky bluegrass cultivars. *Crop Sci.* 30:1239–1243.
Carman, J.G. 1997. Asynchronous expression of duplicate genes in angiosperms may cause apomixis, bispory, tetraspory, and polyembryony. *Biol. J. Linnean Soc.* 61:51–94.
Clausen, J. 1961. Introgression facilitated by apomixis in polyploid *Poas*. *Euphytica* 10:87–94.
Clausen, J., N. Hiesey, and M.A. Nobs. 1962. Studies in *Poa* hybridization. *Carnegie Institute Washington Yearbook* 61:325–333.
Dale, M.R., M.K. Ahmed, G. Jelenkovic, and C.R. Funk. 1975. Characteristics and performance of interspecific hybrids between Kentucky bluegrass and Canada bluegrass. *Crop Science* 15:797–799.
Dijk, G.E. van. 1991. *Advances in Plant Breeding*. Vol. 2. A.K. Mandal, P.K. Ganguli, and S.P. Banerjee, Eds. CBS Publishers and Distributors, Delhi, India.
Dijk, G.W. van and G.D. Winkelhorst. 1982. Interspecific crosses as a tool in breeding *Poa pratensis* L. I. *P. longifolia* Trin. x *P. pratensis* L. *Euphytica* 31:215–223.
Duell, R.W. 1985. The bluegrasses. In M. E. Heath et al., Eds. *Forages*. 4th ed. pp. 188–197. Iowa State University Press, Ames, IA.
Duich, J.M. and H.B. Musser. 1959. The extent of aberrants produced by Merion Kentucky bluegrass, *Poa pratensis* L. as determined by first and second generation progeny test. *Agron. J.* 51:421–424.
Duyvendak, R. and B. Luesink. 1979. Preservation of genetic resources in grasses. *Proc. Cont. Broadening Genetic Base of Crops*. pp. 67–73. Pudoc, Wageningen.
Funk, C.R. 2000. Long live Kentucky bluegrass, the king of grasses! *Diversity* 16:26–28.
Funk, C.R. and S.J. Han. 1967. Recurrent Interspecific Hybridization: A proposed method of breeding Kentucky bluegrass, *Poa pratensis*. *New Jersey Agricultural Experiment Station Bulletin* 818:3–14.
Funk, C.R., R.E. Engel, G.W. Pepin, A.M. Radko, and R.J. Peterson. 1973. Registration of Bonnieblue Kentucky bluegrass. *Crop Science* 14:906.

Gates, M.J. 1997. Seed Set Variation in the Presence or Absence of Foreign Pollen in Apomictic Kentucky Bluegrass, *Poa pratensis* L. Master of Science thesis. The Pennsylvania State University.

Gray, A. 1908. *Gray's New Manual of Botany*. 7th ed. In B.L. Robinson and M.L. Fernald, Eds. pp. 154–157. Am. Book Co., New York.

Grazi, F., M. Umaerus, and E. Akerberg. 1961. Observations on the mode of reproduction and the embryology of *Poa pratensis* L. *Hereditas* 47:489–541.

Han, S.J. 1969. Effects of Genetic and Environmental Factors on Apomixis and the Characteristics of Nonmaternal Plants in Kentucky Bluegrass (*Poa pratensis* L.). Ph.D. dissertation. Rutgers University, New Brunswick, New Jersey.

Hanson, A.A. 1972. Grass Varieties in the United States. USDA Agric. Handbook. 170. U.S. Gov. Print. Office, Washington, DC.

Hanson, A.A. and F.V. Juska. 1959. A "progressive" mutation induced in *Poa pratensis* L. by ionizing radiation. *Nature* (London) 184:1000–1001.

Hartley, W. 1961. Studies on the origin, evolution, and distribution of the *Gramineae*. IV. The genus *Poa* L. *Aust. J. Bot.* 9:152–161.

Hiesey, W.M. and M.A. Nobs. 1982. Experimental Studies on the Nature of Species. Carnegie Instit. of Washington. Pub. 636.

Hintzen, J.J. 1979. Methods of apomitic species. In *Plant Breeding Perspectives* J. Sneep and A.J.T. Henderiksan. Eds. Pudoc, Wageningen.

Hintzen, J.J. and A.J.P. van Wijk. 1985. Ecotype breeding and hybridization in Kentucky bluegrass (*Poa pratensis* L.). In F. Lemarie, Ed. Proc. 5th International Turfgrass Research Conference, Avignon, France.

Hitchcock, A.S. 1950. Manual of the Grasses of the United States. U.S.D.A. Misc. Pub. No. 200. Revised by A. Chase. Dover Edition, 1971.

Hovin, A.W., C.C. Berg, E.C. Bashaw, R.C. Buckner, D.R. Dewey, G.M. Dunn, C.S. Hoveland, C.M. Rineker, and G.M. Wood. 1976. Effects of geographic origin and seed production environments on apomixis in Kentucky bluegrass. *Crop Science* 16:635–638.

Hubbard, C.E. 1984. *Grasses: A Guide to Their Structure, Identification, Uses, and Distribution in the British Isles*. 3rd Ed. Reviewed by J.C.E. Hubbard. Penguin Books, Middlesex, England.

Huff, D.R. 1997. RAPD characterization of heterogeneous perennial ryegrass cultivars. *Crop Science* 37:557–564.

Huff, D.R. and J.R. Bara. 1993. Determining genetic origins of aberrant progeny from facultative apomictic Kentucky bluegrass using a combination of flow cytometry and silver-stained RAPD markers. *Theor. Appl. Genet.* 87:201–208.

Johnston, W.J. and R.C. Johnson. 2000. Washington state and USDA work to preserve and clarify the rich diversity of Kentucky bluegrass. *Diversity* 16:30–32.

Kellogg, E.A. 1990. Variation and species limits in agamosphermous grasses. *Syst. Bot.* 15:112–123.

Kiellander, C.L. 1942. A subhaploid *Poa pratensis* L. with 18 chromosomes and its progeny. *Svensk. Bot. Tidskr.* 36:200–220.

Love, A. and D. Love. 1975. *Cytotaxonomical Atlas of the Artic Flora*. Strauss and Cramer, Leutershausen, Germany.

McDonnell, R.E. and B.V. Conger. 1984. Callus induction and plantlet formation from mature embryo explants of Kentucky bluegrass. *Crop Science* 24:573–578.

Meyer, W.A. 1982. Breeding disease-resistant cool-season turfgrass cultivars for the United States. *Plant Disease* 66:341–344.

Müntzing, A. 1933. Apomictic and sexual seed production in *Poa*. *Hereditas* 17:131–154.

Müntzing, A. 1940. Further studies on apoximis and sexuality in *Poa*. *Hereditas* 27:115–190.

Nielsen, E.L. 1945. Cytology and breeding behavior of selected plants of *Poa pratensis*. *Botanical Gazette* 106:357–382.

Nielsen, E.L. 1946. Breeding behavior and chromosome numbers in progenies from twin and triplet plants of *Poa pratensis*. *Botanical Gazette* 108:26–40.

Oliver, G.W. 1910. New Methods of Plant Breeding. USDA Bulletin No. 167.

Pepin, G.W. and C.R. Funk. 1971. Intraspecific hybridization as a method of breeding Kentucky bluegrass (*Poa pratensis* L.) for turf. *Crop Science* 11:445–448.

Pepin, G.W. and C.R. Funk. 1974. Evaluation of turf, reproductive, and disease-response characteristics in crossed and selfed progenies of Kentucky bluegrass. *Crop Science* 14:356–359.

Riordan, T.P., R.C. Shearman, J.E. Watkins, and J.P. Behling. 1988. Kentucky bluegrass automatic hybridization apparatus. *Crop Science* 28:183–185.

Smith, D. 1981. *Forage Management in the North*. 4th ed. Kendall/Hunt Publ. Co., Toronto, Ontario, Canada.

Soreng, R.J. 1985. *Poa* in New Mexico, with a key to middle and southern Rocky Mountain species (*Poaceae*). *Great Basin Natur.* 45:395–422.

Soreng, R.J. 1990. Chloroplast-DNA phylogenetics and biogeography in a reticulating group: Study in *Poa* (*Poaceae*). *American Journal of Botany* 77:1383–1400.

Stebbins, G.L. 1950. *Variation and Evolution in Plants*. Columbia University Press, New York.

Tinney, F.W. 1940. Cytology of parthenogenesis in *Poa pratensis*. *Journal of Agricultural Research* 60:351–360.

Tinney, F.W. and G.S. Aamodt. 1940. The progeny test as a measure of the type of seed development in *Poa pratensis* L. *Hereditas* 31:457–464.

Tsvelev, N.N. 1976. *Grasses of the Soviet Union*, Zlaki SSSR. Nauka Publishers, Lenengrad. English translation, Smithsonian Inst.: Oxonian Press, Pvt. Ltd, New Delhi, 1983.

Turgeon, A.J. 1998. *Turfgrass Management*. 5th ed. Prentice Hall, Englewood Cliffs, NJ.

Van Dijk, G.E. 1991. *Advances in Plant Breeding*. Vol. 2. A.K. Mandal et al., Eds. CBS Publ. Distrib., Dehli, India

Wedin, W.J. and D.R. Huff. 1996. Bluegrass. In *Cool-Season Forage Grasses*. L.E. Moser, Ed. American Society of Agronomy Monograph Series 34, pp. 665–691. ASA-CSSA-SSSA, Madison, WI.

Williamson, C.J. and P.J. Watson. 1981. Production and description of interspecific hybrids between *Poa pratensis* and *Poa longifolia*. *Euphytica* 29:715–725.

Wu, L. and R. Jampates. 1986. Chromosome number and isoenzyme variation in Kentucky bluegrass cultivars and plants regenerated from tissue culture. *Cytologia* 51:125–132.

Chapter 3

Annual Bluegrass (*Poa annua* L.)

David R. Huff

Annual bluegrass (*Poa annua* L.; also known as annual meadow-grass in Europe) is a common grass worldwide that is often observed as an unsown component of many different types of turf. For over a century, the utility of *Poa annua* as turf has been a controversial topic. As early as 1877 it was described as a valuable turfgrass for lawns and golf greens (see Watson, 1903). In 1927, noted turfgrass experts Piper and Oakley (1927) also described the value of *Poa annua* for golf course putting greens. Since then, some turfgrass agronomists have encouraged the use and cultivation of *Poa annua* as turf (Zontek, 1973; Vermeulen, 1989; Huff, 1996, 1998) while others have focused on its eradication as a weed pest (Kane, 1988; Breuninger, 1993; McCarty, 1999; Johnson, 1999).

Over the years, several turfgrass researchers have recognized that whether the aim is to eradicate *Poa annua* or to find use for it in a turf program, it is necessary to learn about the growth habit and other characteristics of the species (Hovin, 1957; Lush, 1989; Johnson et al., 1993). Beard (1970) went further and suggested that "unfortunately, too much emphasis has been placed on the control of annual bluegrass and not enough thought given to the characteristics, adaptation and cultural requirements of this species" and that "to achieve either alternative, it is important to have a basic understanding of the environmental and cultural practices which either enhance or impair the growth and development of annual bluegrass."

Thus it is important to understand that valuable information for the utilization and subsequent improvement of *Poa annua* for turf purposes is to be found in two types of literature: (1) literature that teaches toward its use [i.e., *Poa annua* as turf] and (2) literature that teaches away from its use [i.e., *Poa annua* as a weed]. For those who are interested in eradicating *Poa annua*, the following references are provided: Watschke et al., 1995; Fermanian et al., 1997). The present chapter will focus on the utility and genetic improvement of *Poa annua* for turf purposes.

Poa annua is widely recognized to provide high quality turf of fine texture and high shoot density that is uniform and tolerant of close mowing (Beard, 1970; Beard et al., 1978; Warwick, 1979; Danneberger and Vargas, 1984). Beard (1970) observed that *Poa annua* forms a very fine-textured turf of high shoot density, uniformity, and overall turfgrass quality when maintained under optimum cultural, environmental, and soil conditions. If grown in monoculture, *Poa annua* provides an excellent putting surface

(Warwick, 1979). Watson (1903) described the lawns at Kew Gardens, UK, as "owing much of their brightness" to *Poa annua* and that "this grass is also a feature of the golflinks in the Old Deer Park adjoining, where the 'greens' are exceptionally good and pile-like." In addition, Watson (1903) recommended at least 20% of grass seed mixtures for lawns be composed of *Poa annua*.

BOTANICAL CHARACTERIZATION

The taxonomic classification of annual bluegrass is:

Family	Gramineae (Poaceae)
Subfamily	Pooideae
Supertribe	Poodae
Tribe	Poeae
Genus	*Poa*
Species	*annua*
Authority	Linnaeus

In general, the following botanical description applies to *Poa annua*: Plants may be bunch-type or weakly-stoloniferous, folded vernation, prominent "train-track" midrib on adaxial leaf surface, and a "boat-shaped" leaf tip. Ligule 0.8 to 3 mm long acute, auricles absent, broad-divided collar. Panicle-shaped inflorescences generally apparent throughout most of the growing season but particularly abundant in springtime. Lemmas not webbed at base, distinctly 5-nerved.

As a species, *Poa annua* displays highly variable plant types. The two main morphological types are the annual form (*Poa annua* f. *annua* L.) and the perennial form (*Poa annua* f. *reptans* [Hausskn.] T. Koyama). The annual form behaves as a winter annual in hot climates like the central valley of California and the southern United States. In general, it is the perennial form that is desirable as a turfgrass. The existence of such divergent forms occurring within a single species is a challenge for the purposes of botanical nomenclature. The annual and perennial forms of *Poa annua* were initially described as varieties (i.e., *P. annua* var. *reptans* [Hausskn.]) then were reclassified as subspecies by Timm [1965] as suggested by Tutin [1957] (i.e., *P. annua* spp. *reptans* [Hausskn.] Timm), and most recently as forms by Koyama (1987) (i.e., *P. annua* f. *reptans* [Hausskn.] T. Koyama). The use of the forma subspecific classification name was verified on 8 May, 1996 by the Systematic Botany Laboratory, New York Botanical Garden. The confusion and waffling on subspecific naming suggests that taxonomic differences between annual and perennial forms will likely continue to be an area of dispute.

The two forms of *Poa annua* represent extremes in life history characteristics. Basically, the annual form has a bunch-type, upright growth habit and is found in open fields, orchards, and meadows (Figure 3.1). Plants of the annual form tend to behave more as annuals in that they are noncreepers and are prolific seed producers. The perennial form has a more prostrate, spreading growth habit and is capable of rooting and producing new shoots from the upper nodes of the decumbent shoots (Younger, 1959). Gibeault and Goetze (1973) and Till-Bottraud et al. (1990) detailed the production of secondary flowering tillers on perennial *Poa annua* and also demonstrated an increased number of nodes (up to five) per vegetative tiller on perennial plants compared to annual plants. In addition, the perennial form produces more tillers which contributes to the appearance of a rather tight turf (Hovin, 1957). As a result, the perennial form is often found growing in closely mowed turf such as bowling greens

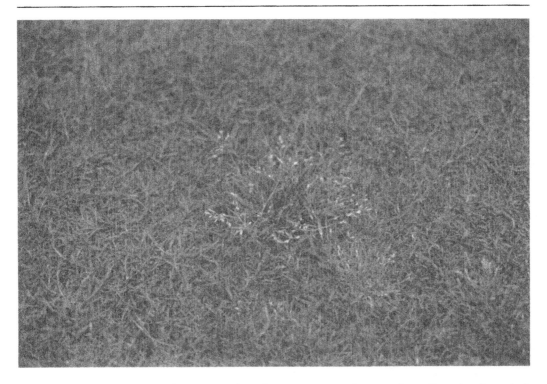

Figure 3.1. An annual type of *Poa annua* invading open or weak areas of an otherwise dense turf.

and long-established golf greens. These prostrate, creeping plants are more restricted in the degree of seed formation and they behave more like perennials (Beard, 1970), characteristically putting more of their resources (photosynthates) into vegetative growth rather than seed production (Law, 1977).

DISTRIBUTION AND CYTOTAXONOMY

Poa annua is one of the five most widely distributed plants in the world (Fenner, 1985). While *Poa annua* is normally limited to areas of human inhabitation, the grass apparently has been newly introduced to Heard Island (first seen 1986/87), although its introduction to the island by human activity seems unlikely (www.wcmc.org.uk/protected_areas/data/wh/himi.html). Heard Island is located in the south Indian Ocean and is considered one of the most inaccessible places on earth. *Poa annua* is an opportunistic grass that frequently becomes established in turfs weakened by biotic or abiotic stresses or from the lack of competition from planted species. It frequently invades heavily trafficked, closely mowed, intensively managed sports turfs. In particular, it will frequently invade golf tees and greens in bare areas resulting from divots and ball marks.

Poa annua is a polyploid grass species meaning that each of its cells carries multiple genomes of its ancestral species. Polyploidy is rare among animals and insects but is common among plants (see Lewis, 1979). *Poa annua* is considered to be of allotetraploid origin between two parental species: *Poa infirma* H.B.K. (*P. exilis* [Thomm]. Murb.) and *Poa supina* Schrad. (Tutin, 1952). Each of these parental species is a diploid organism carrying 14 chromosomes (denoted as $2n=2x=14$ chromosomes). The haploid genomes (basic number = 7 chromosomes) of these ancestral species are evolutionarily divergent and, through interspecific hybridization, result in sterile offspring due to the nonhomology of their chromosomes. In plants, however, and probably due to their modularity, cytological events may occur that restore fertility by the doubling of cellular DNA, giving each chromosome a homologous pairing partner for meiosis to proceed normally. This is potentially the situation with *Poa annua*. The two parental species, *Poa supina* and *Poa infirma* hybridized to produce a sterile plant ($1n=2x=14$ chromosomes) whose cellular DNA content spontaneously doubled to yield *Poa annua* ($2n=4x=28$ chromosomes). However, there may be some discrepancies in this understanding of the evolutionary origin of *Poa annua*. Koshy (1968) observed that *Poa supina* and *Poa infirma* each carries three distinctly "large" chromosomes. Thus, an allotetraploid derivative between these species would, theoretically, carry six such "large" chromosomes. However, Koshy (1968) observed only three such "large" chromosomes in *Poa annua*. He concluded that either some chromosomal modification has occurred over the course of evolution or only one of these diploids is a parental species with the other parent being an unknown species (Koshy, 1968). In addition, chromosomal modifications within the genome of *Poa annua* might be an ongoing process. The picogram (pg) content of nuclear DNA at a constant chromosome number ($2n=28$) has been observed to vary as much as 80% between families of seed progeny derived from established plants of *Poa annua* (Mowforth and Grime, 1989).

Occasionally, diploid ($2n=14$) and/or male-sterile plants of *Poa annua* have been reported in the literature (see Johnson et al., 1993). These observations are most likely errors in reporting the occurrence of amphihaploid (dihaploid) plants of *Poa annua* ($1n=2x=14$ chromosomes). These unique specimens have only one-half the amount of nuclear DNA compared to a normal *Poa annua* and are completely sterile due to an absence of bivalent chromosome pairing at meiosis. Dihaploid plants of *Poa annua* are most often found on old golf course putting greens (Hovin, 1958; Huff, 1999).

ENVIRONMENTAL LIMITATION OF GROWTH

Poa annua is a C3 grass species that patterns a cool-season growth cycle. The primary events in the life cycle of *Poa annua* are as follows (Bigelow and Chalmers, 1995): resumption of growth in the spring months, with a period of profuse tillering followed by seedhead production. Seedhead production occurs mostly in the late spring months. During the hot summer months a reduced stand density often is observed and large patches sometimes fail. In the early fall, a large pulse of germination occurs from the soil seed bank. This seed bank is in a constant state of renewal with the yearly spring seed production. Although this is a traditional cycle of *Poa annua*, these events are not completely limited to these specific times of year.

Both high and low temperatures represent the major environmental limitations to distribution and growth of *Poa annua* (Beard et al., 1978). In general, this lack of tolerance to extreme temperatures makes *Poa annua* a weak turf for some part of the year in various climates. Despite this general observation, strains of *Poa annua* have been observed to perform well in irrigated turf areas subjected to the heat of Arizona (D. Kopec, 1998, personal communication). Duff (1978) also reported significant differences among strains for heat tolerance. At the other extreme, Dionne et al. (2001) reported finding significant differences among strains for tolerance to freezing temperatures. *Poa annua* is also widely known for its susceptibility to many turfgrass diseases, including dollar spot (*Sclerotinia homoeocarpa* F.T. Bennett), anthracnose (*Colletotrichum graminicola* [Ces] G.W. Wils.), and pink snow mold (*Monographella nivalis* [Schaffnit] E. Muller). However, the *Poa annua* breeding program at Pennsylvania State University has identified strains exhibiting excellent field resistance to anthracnose and dollar spot (Huff, 1996, 1999). Thus, while most scientific efforts regarding *Poa annua* have been directed toward its eradication or control, those research efforts aimed at determining and identifying strains possessing unique tolerances have been successful.

REPRODUCTIVE BIOLOGY

Poa annua has a self-pollinating type of breeding system that results in true breeding strains through inbreeding. Eggs within the flowers (florets) are often fertilized before the florets ever open (cleistogamy). Such an inbreeding system of sexual reproduction is unique among those grasses typically used for turf (most are cross-pollinated) and is more similar to that of wheat (*Triticum aestivum* L.) or soybean (*Glycine max* [L.] Merrill).

For plant species that inbreed, the female and male gametes (eggs and sperm) that fuse to create new individuals are both derived from the same parent plant (genotype). As a result, any heterozygosity occurring within a genotype rapidly becomes partitioned among families (sometimes referred to as lines, strains, or races). The result of this partitioning is that a typical population consists of numerous family lines that exhibit little to no variation within a family but large amounts of variability between families (Hayes et al., 1955; Briggs and Knowles, 1967).

Thus a population of *Poa annua* is considered to be a mixture of homozygous family lines, each of which will breed true. Should an outcrossing event occur, the level of heterogeneity generated will be reduced by one-half in each subsequent generation of self-pollination. Tutin (1957) described self-pollination as the rule for *Poa annua* and therefore most plants are practically homozygous (true-breeding, uniform, and stable). Furthermore, Tutin (1957) suggested that "even if self-pollination occasionally fails [in *Poa annua*], the chances of true outbreeding, in the sense of crossing between 'races.' seems remote." Hovin (1957), Younger (1959), and Law (1977) also observed that due to a high degree of self-pollination and good seed fertility in *Poa annua*, various morphologically deviating lines, that show promise as turf, have become established. Johnson et al. (1993) demonstrated that several strains of the perennial form, as observed through several generations of sexually produced seed, were almost exclusively selfing in nature and bred true-to-type.

Warwick and Briggs (1978) grew out seed samples of individual *Poa annua* plants from bowling greens and observed that 94 to 98% of the (seeded) individuals were prostrate, with only a small portion (2–6%) growing into erect plants. Their data suggested that pure lines have been established as a consequence of self-fertilization. Darmency et al. (1992), took this one step further and, using molecular markers, demonstrated the homogeneity within and the genetic differences between the prostrate (perennial) and erect (annual) forms of *Poa annua* at the esterase enzyme gene locus.

The ability to reproduce true-breeding, uniform, and stable seed progeny within family lines has long been exploited by biologists and breeders studying the nature and utility of *Poa annua*. These scientists have cultivated, demonstrated, and utilized the true-breeding, uniform, and stable nature of individual plant selections of *Poa annua* for nearly half a century.

EVOLUTION OF TURF-TYPES

When *Poa annua* first invades a turf area, it generally does so as seed of the annual form. The seedlings become established in damaged or weakened open areas of turf and through phenotypic plasticity, adjust to the given management conditions of that particular turf, i.e., mowing height, irrigation schedule, and fertility program. *Poa annua* has a unique ability to adjust the height of its flowering culms such that it is capable of flowering and seed-set under nearly any mowing height, i.e., as low as 2.5 mm. Even though *Poa annua* is primarily a self-pollinated species, occasionally seed is set as a result of cross-pollination. Levels of cross-pollination among various strains have been observed to range between 0 and 15% (Ellis, 1973).

Outcrossing events among inbred parents produce a range of genetically-based morphological variation. Turf management programs act as a strong selection force on this variation. Over time, the subsequent generations of *Poa annua* begin to take on the characteristics of a more perennial form and ultimately adapt to the particular turfgrass management program; be it a home lawn, athletic field, or golf course putting green. For example, a hybridization event between two inbred lines would typically generate a large amount of variability, i.e., 100%. The level of this variability is reduced by 50% within a particular strain after every generation of self-pollination. After the first generation of inbreeding, the amount of variability is reduced to 50%; the second reduces it to 25%; the third to 12.5%, and so on. A similar reduction in variability is necessary after making artificial hybrids between distinct, true breeding strains of *Poa annua*. Thus after mating two different strains of *Poa annua*, it takes a minimum of six to eight generations to regain lines that are uniform, stable, and true breeding. However, these advanced progeny generations may be morphologically and genetically quite different from the original parent plants. McNeilly (1981) has shown that populations of *Poa annua* differentially evolve as a result of different mowing and competition stresses. Similar observations were reported by Adams and Bryan (1980) and Peel (1982). Thus, with every generation, *Poa annua* evolves and adapts in response to the specific cultivation and management practices of a given turf.

On old golf course putting greens this evolutionary process results in strains known as greens-type *Poa annua* (Huff, 1996; Dionne et al., 2001; Turgeon, 2002). These greens-type *Poa annua* are perennials that possess a short stature, extremely high shoot densities, and are vegetatively aggressive (Figure 3.2). Greens-type *Poa annua* may begin to appear on golf greens as young as 10 years of age. Such a "rapid" evolutionary event is an indication of the extreme selection forces existing on golf greens (primarily mowing height and wear). In fact, the selection pressures of the green environment are so intense, that on greens as young as 60 years of age, it is common to observe a "reverse" evolutionary process that results in the appearance of amphihaploid plant types, possibly resulting from the development of unfertilized eggs. These amphihaploids are cytologically similar to the original interspecific sterile hybrid ($1n=2x=14$ chromosomes) between *P. supina* and *P. infirma* and have only one-half the amount of DNA of *Poa annua*. Amphihaploids represent some of the densest, finest, and highest turf quality strains (Huff, 1999).

The evolution of *Poa annua* from wild, weedy, annual forms to the perennial forms adapted to golf turf, lawn turf, and athletic field turf has been documented by plant ecologists and plant evolutionists as a classic example of rapid microevolution (Law et al., 1977; Law, 1977, 1979, and 1981; McNeilly, 1981). A key feature of this evolution is the type of selection that occurs under density-independent environments versus density-dependent environments. Density-independent environments are those environments where few grass plants live, i.e., disturbed sites and waste areas. In such environments a constant supply of open, bare ground allows seed to germinate and subsequently grow and develop into adult plants. Density-independent environments favor those individual *Poa annua* plants that allocate more of their limited resources (nutrients, water, products of photosynesis, etc.) to seed production and less to vegetative growth.

Conversely, density-dependent environments are those environments that are fully crowded with individuals (i.e., golf greens and other maintained turfs). In density-dependent environments, seed germination and seedling growth and development is

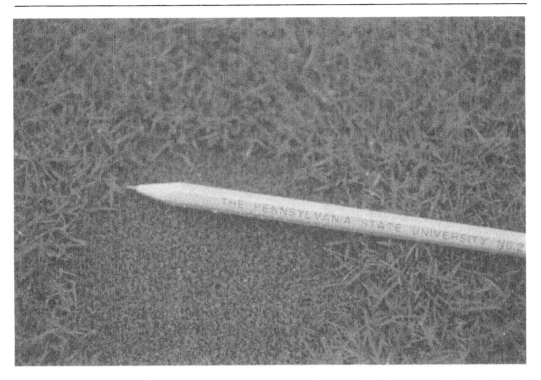

Figure 3.2. A cup-cut of a greens-type *Poa annua* line in a creeping bentgrass (*Agrostis palustris* Huds.) sod.

difficult because very little open space within the turf canopy occurs. Density-dependent environments favor individual *Poa annua* plants that allocate more resources to vegetative growth and less to seed production. What the ecologists, evolutionists, and turfgrass scientists have taught us is that the *Poa annua* adapted to turf areas (be it golf turf, lawn turf, athletic turf, or so on) are perennial and restrict their flowering and seed production when compared to *Poa annua* adapted to nonturf areas. This knowledge has become important for focusing our attempts to eradicate *Poa annua* as a weed of turf, but also for focusing our efforts at cultivating, managing, and breeding *Poa annua* for improved turf cultivars.

BREEDING CULTIVARS OF PERENNIAL *POA ANNUA* FOR TURF

Many turfgrass breeders have successfully demonstrated and practiced the art of selecting and breeding uniform and stable lines of perennial *Poa annua* f. *reptans* with improved turf qualities and determinant flowering (by both definition or design). Duff (1978) detailed his methods of breeding of *Poa annua* f. *reptans* for turf by demonstrating his selection and propagation techniques for 11 seeded selections. The statistically significant differences observed among his selections for turf quality and heat tolerance is evidence for the uniformity of these selections. Younger (1959) also suggested that his preliminary breeding studies indicated that it may be possible to obtain relatively true breeding selections of perennial *Poa annua* of some turf value. Law (1977) has shown that perennial *Poa annua* "characteristically put their resources into vegetative growth and appear to be potentially very useful to the amenity [turf] breeder. Thus perennial *Poa annua* should provide a sound basis for further breeding work. In addition, *P. annua* is self-compatible and appears generally to self-pollinate in nature. This

Figure 3.3. A field experiment designed to evaluate numerous collections of *Poa annua* for turf quality at putting green mowing height. Note the variation in color, density, texture, turf cover, and overall quality.

simplifies the breeding techniques that can be applied." Hovin (1957) suggested that since so much morphological variation is present in *Poa annua*, improved strains may be obtained that may show promise as a future turfgrass. Moreover, Adams and Bryan (1980) and Wu and Harivandi (1993) have demonstrated the presence of large genetic differences for shoot growth and developmental characteristics in perennial forms that would be readily exploitable by plant breeders. Variation in morphology, growth habit, texture, and genetic color have been observed within the Pennsylvania State University collection of *Poa annua* (Figure 3.3).

Limitations of Developing Cultivated Varieties of *Poa annua* for Turf

Three main obstacles need to be overcome for successful cultivation of *Poa annua* seed for the turfgrass commercial market; namely, (1) low seed yield, (2) the indeterminacy of seed maturity, and (3) the control of undesirable forms of *Poa annua* within seed production fields. Overcoming the first two obstacles will be important goals for breeders interested in developing commercial cultivars of *Poa annua* for the turfgrass seed market (lawns, athletic fields, or golf courses). Unfortunately, at present, no chemical control measures are available for eradicating undesirable strains of *Poa annua* in production fields of desirable strains. Moreover, some herbicide resistance has been reported among plants of the annual form (Kelly et al., 1999; Hanson and Smith, 2000).

Low seed yield is a major limiting factor that has likely prevented many seed growers and breeders from producing cultivars of *Poa annua* for the commercial turf seed market. Johnson et al. (1993) suggested that the quantity of seed produced by cultivars of *Poa annua* plays a major role in the development of an improved cultivar. Fortunately, Johnson et al. (1993) have indicated that UM-184, a typical perennial type with very seasonal flowering, had significantly greater yield than other perennial forms. Their work demonstrated that the variation exists and that potential for breeding strains with higher seed yield existed among strains of the perennial form of *Poa annua*.

Overcoming the second obstacle requires an obvious focus on enhancing the timing of flowering and seed set in order to overcome the indeterminacy of seed maturity. In addition, the disdain of turf professionals for the unsightly seedheads in the turf, as taught by Danneberger and Vargas (1984) also suggested that a narrowing of flowering timing would be beneficial. Fortunately, a restricted flowering habit is inherent in the perennial forms of *Poa annua* selected by turf cultivation and maintenance practices. For example, Johnson et al. (1993) have identified superior types of the perennial form of *Poa annua* with improved turf qualities and determinant (i.e., restricted) flowering for breeding improved cultivars.

Restricted Flowering Habit of Poa annua

Many different limitations of the flowering period have been observed for the perennial form of *Poa annua*. In general, *Poa annua* flowers in the spring (McBurney and Kaufman, 1982; Danneberger and Vargas, 1984). The exact period of flowering, however, has been shown to be strain specific. McBurney and Kaufman (1982) observed *Poa annua* seedheads to develop at East Lansing, Michigan, from 13 to 21 May (a period of approximately 1–2 weeks in the spring). Danneberger and Vargas (1984), also at East Lansing, Michigan, showed that while maximum seedhead production occurred for a period of 14 to 18 days at all of their locations, 95% of the seedheads for the HRC83 perennial biotype population were produced over a period of May 1 through June 15 (46 days or 6.5 weeks) and for the HRC82 perennial biotype population were produced over a period of May 3 through June 5 (34 days or 4.8 weeks). However, Law (1981), at Liverpool, England, showed that flowering periods of some strains were comprised of two flushes separated by periods of less than four weeks (in four family lines) or between 4 to 8 weeks (in one family line). Some variation in flowering potential in *Poa annua* has been demonstrated to result from variations in soil pH (Juska and Hanson, 1969), and thus, some environmental variation exists in the flowering potential of *Poa annua*.

Specific limitations of the flowering response of *Poa annua* to various photoperiods (daylengths) have also been described. Beard et al. (1978) described that the flower induction of *Poa annua* can occur over a wide range of photoperiods, including 6, 9, 12, or 24 hours of light per day, however it usually becomes induced to flower under short-daylength conditions. Beard et al. (1978) also provided evidence that some strains of *Poa annua* were unaffected by daylength but that flowering was generally greater under daylengths greater than 5 hours. Wu and Harivandi (1993) observed that when daylength increased in May, the golf green biotype of *Poa annua* remained vegetative and did not flower. Johnson and White (1992) suggested that those *Poa annua* genotypes exhibiting spring-only flowering characteristics offered great potential for improvement of the species.

Furthermore, Johnson and White (1993) indicated that photoperiod and vernalization requirements for flowering varied among University of Minnesota *Poa annua* genotypes. They demonstrated day-neutrality for PA-2283, long-day induction for PA-234, vernalization (cold) requirements for PA-42, 117, and 184 and that short days substituted for vernalization in PA-117. They also showed that day-neutral types produce flowers any time environmental conditions were favorable for growth and development; that long-day induced genotypes were characterized by late spring initiation and continual summer flowering; that genotypes responsive to only vernalization showed a flush of seedheads in spring and no late-season flowering; and, that in genotypes where short-days substitute for vernalization, flowering occurred as a flush of seedheads in spring, no seedheads in summer, and a few seedheads in fall. Vernalization requirements were apparently controlled by only one or two genes. Johnson and White (1994) used four true-breeding *Poa annua* f. *reptans* parents to demonstrate that the genetic control of seasonal flowering (spring-only flowering) was simply inherited and that selection for uniform flowering type should be rapidly achieved.

University of Minnesota *Poa annua* Breeding Program

The most extensive breeding program to date has been that of Dr. Don White, University of Minnesota. In a series of publicly available research reports (White, 1985, 1986, 1987, 1988, 1989, 1990, 1991a, 1991b, 1993, and 1994), Dr. White has demonstrated the following: that ample variation exists in perennial *Poa annua* to warrant a breeder's attention; that the traits of limited flowering, dark green color, and an ability to provide dense turf are desirable and achievable traits for selection when breeding improved cultivars of *Poa annua*; that cultivation of uniform strains of perennial *Poa annua* is possible; that stability across generations is important for a breeder to examine and that variation for stability exists within the species; that uniformity and stability of traits within strains of perennial *Poa annua* is possible to attain; that three selections of *Poa annua* f. *reptans* (Minnesota Nos. 42, 184, and 208) produced sufficient seed to warrant continued seed production; that all selections of perennial *Poa annua* produce excellent turf at 3-mm mowing height and that this low height of cut along with other putting green management practices apparently selects for plants that are all fine-textured, dense, and dark green; and he teaches the expansion of seed increase to enable for commercialization. Dr. White's (1994) release and exclusive licensing of three selections of *Poa annua* var *reptans* (Minnesota Nos. 42, 184, and 208) to Peterson Seed Company, Savage, Minnesota, demonstrated the commercial potential of elite strains of *Poa annua* f. *reptans* that are uniform, stable, have restricted flowering potential, and have potential as turf.

REFERENCES

Adams, W.A. and P.J. Bryan. 1980. Variations in the growth and development of annual bluegrass populations selected from seven different sports turf areas. J. B. Beard, Ed.. Proceedings of the Third International Turfgrass Research Conference, pp. 109–115. American Society of Agronomy, Madison, WI.

Beard, J.B. 1970. An ecological study of annual bluegrass. *United States Golf Association Green Section Record* 8:13–18.

Beard, J.B, P.E. Rieke, A.J. Turgeon, and J.M. Vargas, Jr. 1978. Annual Bluegrass (*Poa annua* L.): Description, Adaptation, Culture, and Control. Michigan State University, Agricultural Experiment Station Research Report. #352. Michigan State University, Agricultural Experiment Station, East Lansing, MI.

Bigelow, C.A. and D.R. Chalmers. 1995. Interseeding establishment of creeping bentgrass (*Agrostis stolonifera* var. *palustris* 'Southshore' (Huds.) Faw.) into an annual bluegrass (*Poa annua* L.) turfgrass system. Proceedings of the 1995 Virginia Turf and Landscape Conference and Tradeshow. (http://sudan/cses.vt.edu/html/Turf/bigelow.htm)

Breuninger, J. 1993. *Poa annua* control in bentgrass greens. *Golf Course Management* 61:68–73.

Briggs, F.N. and P.F. Knowles. 1967. Selection in self-pollinated crops. In *Introduction to Plant Breeding.* pp. 114–121. Reinhold Publishing, New York.

Danneberger, T.K. and J.M. Vargas Jr. 1984. Annual bluegrass seed head emergence as predicted by degree-day accumulation. *Agronomy Journal* 76:756–758.

Darmency, H., A. Berti, J. Gasquez, and A. Matejicek. 1992. Association of esterase isozymes with morphology in F2 progenies of two growth variants in *Poa annua* L. *New Phytologist* 121:657–661.

Dionne, J., Y. Castonguay, P. Nadeau, and Y. Desjardins. 2001. Freezing tolerance and carbohydrate changes during cold acclimation of green-type annual bluegrass (*Poa annua* L.) ecotypes. *Crop Science* 41:443–451.

Duff, D.T. 1978. Disagreements arises over variant of annual bluegrass. *University of Rhode Island Turfgrass Research Review* 3:1–3.

Ellis, W. M. 1973. The breeding system and variation in populations of *Poa annua* L. *Evolution* 27:656–662.

Fenner, M. 1985. *Seed Ecology*. Chapman and Hall, London.

Fermanian, T. W., M. C. Shurtleff, R. Randell, H. T. Wilkinson, and P. L. Nixon. 1997. *Controlling Turfgrass Pests*. Prentice Hall, NJ. p. 655.

Gibeault, V.A. and N.R. Goetze. 1973. Annual meadow-grass. *Journal of Sports Turf Research Institute* 48:9–19.

Hanson, B.D. and C.A. Smith. 2000. Rapid publication Diuron-resistant *Poa annua* is resistant to norflurazon. *Weed Sci.* 48:666–668.

Hayes, H.K., F.R. Immer, and D.C. Smith. 1955. The pure-line method of breeding naturally self-pollinated plants. In *Methods of Plant Breeding*. pp. 94–106.

Hovin, A.W. 1957. Variations in annual bluegrass. *The Golf Course Reporter* 25:18–19

Hovin, A.W. 1958. Meiotic chromosome pairing in amphihaploid *Poa annua* L. *Am. J. Bot.* 45:131–138.

Huff, David. 1996. *Poa annua* for golf course greens. *Grounds Maintenance*. January. pp. G2–G10.

Huff, D.R. 1998. The case for *Poa annua* on golf course greens. *Golf Course Management* October pp. 54–56.

Huff, D.R. 1999. For richer, for *Poa*. *USGA Greens Section Record*. Vol. 37, No. 1, pp. 11–14.

Johnson, B.J. 1999. Weeding out *Poa annua*. *Through The Green*, September/October, pp. 17–40.

Johnson, P.G., B.A. Ruemmele, P. Velguth, D.B. White, and P.D. Ascher. 1993. Chapter 114: An overview of *Poa annua* L. reproductive biology. In *International Turfgrass Society Research Journal*. Proceedings of the Seventh International Turfgrass Research Conference, Palm Beach, Florida. Vol. 7, pp. 798–804. Intertec Publishing Corp., Overland Park, KS.

Johnson, P.G. and D.B. White. 1992. Inheritance of flowering requirements in *Poa annua* L. *Agronomy Abstracts*, p. 171. American Society of Agronomy, Madison, WI.

Johnson, P.G. and D.B. White. 1993. Requirements for flower induction in *Poa annua* L. *Agronomy Abstracts*, p. 159. American Society of Agronomy, Madison, WI.

Johnson, P.G. and D.B. White. 1994. Inheritance of flowering pattern in *Poa annua reptans* (Hausskn) and *Poa annua annua*. *Agronomy Abstracts*, p. 185. American Society of Agronomy, Madison, WI.

Juska, F.V. and A.A. Hanson. 1969. Nutritional requirements of *Poa annua* L. *Agronomy Journal* 61:466–468.

Kane, R.T. 1988. New chemical weapons for fighting *Poa annua* in bentgrass polystands. *The Bull Sheet* 41:8.

Kelly, S.T., G.E. Coats, and D.S. Luthe. 1999. Mode of resistance of triazine-resistant annual bluegrass (*Poa annua*). *Weed Technology* 13:747–752.

Koshy, T. K. 1968. Evolutionary origin of *Poa annua* L. in the light of karyotypic studies. *Can. J. Genet. Cytol.* 10:112–118.

Koyama, T. 1987. *Grasses of Japan and Its Neighboring Regions. An Identification Manual*. p. 523. The New York Botanical Garden, New York.

Law, R. 1977. The turfgrass potential of *Poa annua* ecotypes. *Journal of Sports Turf Research Institute* 53:117.

Law, R. 1979. The cost of reproduction in annual meadow grass. *American Naturalist* 113:3–16.

Law, R. 1981. The dynamics of a colonizing population of *Poa annua*. *Ecology* 62:1267–1277.

Law, R., A.D. Bradshaw, and P.D. Putwain. 1977. Life-history variation of *Poa annua*. *Evolution* 31:233–246.

Lewis, W.H., Ed. 1979. *Polyploidy: Biological Relevance*. Plenum Press, New York, p. 583.

Lush, W.M. 1989. Adaptation and differentiation of golf course populations of annual bluegrass (*Poa annua*). *Weed Science* 37:54–59.

McBurney, S.L. and J.E. Kaufman. 1982. Enhancing the survival and appearance of *Poa annua* with growth regulators. Proceedings of the 52nd Annual Michigan Turfgrass Conference, Vol. 11, pp. 47–51. Michigan State University, Agricultural Experiment Station, East Lansing, MI.

McCarty, B. 1999. Controlling *Poa annua* in bentgrass greens. *Grounds Maintenance* 34:17–20.

McNeilly, T. 1981. Ecotypic differentiation in *Poa annua*: Interpopulation differences in response to competition and cutting. *New Phytologist* 88:539–547.

Mowforth, M.A. and J.P. Grime. 1989. Intra-population variation in nuclear DNA amount, cell size and growth rate in *Poa annua* L. *Functional Ecology* 3:289–295.

Peel, C.H. 1982. A review of the biology of *Poa annua* L.—with special reference to sports turf. *Journal of the Sports Turf Research Institute* 58:28–40.

Piper, C.V. and R.A. Oakley. 1927. Annual bluegrass (*Poa annua*). *Bulletin of the United States Golf Association Green Section* 7(7):128–129.

Till-Buttraud, I., L. Wu, and J. Harding. 1990. Rapid evolution of life history traits in populations of *Poa annua* L. *Journal of Evolutionary Biology* 3:205–224.

Timm, G. 1965. Biology and systematics of *Poa annua*. *Z. Acker-u PflBau*. 122:267–294.

Turgeon, A.J. 2002. *Turfgrass Management*. Prentice Hall, Upper Saddle River, NJ.

Tutin, T.G. 1952. Origin of *Poa annua* L. *Nature* 169:160.

Tutin, T.G. 1957. A contribution to the experimental taxonomy of *Poa annua* L. *Watsonia* 4:1–10.

Vermeulen, P. 1989. Consider *Poa annua* for your new green. *United States Golf Association Green Section Record* 27(5):17.

Warwick, S.I. 1979. The biology of Canadian weeds. 37. *Poa annua* L. *Canadian Journal of Plant Science* 59:1053–1066.

Warwick, S.I. and D. Briggs. 1978. The genecology of lawn weeds. I. Population differentiation in *Poa annua* L. in a mosaic environment of bowling greens lawns and flower beds. *New Phytologist* 81:711–723.

Watschke, T. L., P. H. Dernoeden, and D. J. Shetlar. 1995. *Managing Turfgrass Pests*. Lewis Publishers, Ann Arbor, MI.

Watson, W. 1903. *Poa annua* as a lawn grass. Note: includes quotes from a Treatise on Cultivated Grasses, published by the Lawson Seed and Nursery Company, Edinburgh in 1877. *The Gardeners' Chronicle* 33:380. The Royal Horticultural Society, London, England.

White, D.B. 1985. Breeding of *Poa annua* for improved cultivars. *Annual Turfgrass Research Report*. United States Golf Association Green Section. 1985, p. 6. United States Golf Association, Far Hills, NJ.

White, D.B. 1986. Breeding of *Poa annua* for improved cultivars. *Annual Turfgrass Research Report*. United States Golf Association Green Section. 1986, pp. 7–8. United States Golf Association, Far Hills, NJ.

White, D.B. 1987. Improvement of *Poa annua* and *Poa supina* for golf turf. *Annual Turfgrass Research Report*. United States Golf Association Green Section. 1987, p. 13. United States Golf Association, Far Hills, NJ.

White, D.B. 1988. Improvement of *Poa annua* and *Poa supina* for golf turf. *Annual Turfgrass Research Report*. United States Golf Association Green Section. 1988, pp. 18–19. United States Golf Association, Far Hills, NJ.

White, D.B. 1989. Improvement of *Poa annua* and *Poa supina* for golf turf. *Annual Turfgrass Research Report*. United States Golf Association Green Section. 1989, pp. 19–20. United States Golf Association, Far Hills, NJ.

White, D.B. 1990. Improvement of *Poa annua* for golf turf. *Annual Turfgrass Research Report*. United States Golf Association Green Section. 1990, pp. 13–14. United States Golf Association, Far Hills, NJ.

White, D.B. 1991a. Improvement of *Poa annua* for golf turf. *Turf Craft Australia*, 1991, p. 10. Rural Press Magazines, Australia.

White, D.B. 1991b. Improvement of *Poa annua* var *reptans* for golf turf. *1991 Turfgrass Research Summary*. United States Golf Association Green Section, p. 12. United States Golf Association, Far Hills, NJ.

White, D.B. 1993. Improvement of *Poa annua* for golf turf. *1993 Turfgrass Research Summary*. United States Golf Association Green Section, pp. 22–23. United States Golf Association, Far Hills, NJ.

White, D.B. 1994. Improvement of *Poa annua* var *reptans* for golf turf. *1994 Turfgrass Research Summary*. United States Golf Association Green Section, pp. 18–20. United States Golf Association, Far Hills, NJ.

Wu, L. and A. Harivandi. 1993. Annual bluegrass ecology and management: Understanding key interactions is a prerequisite. *Golf Course Management* 61:100–106.

Younger, V.B. 1959. Ecological studies on *Poa annua* in turfgrasses. *Journal of the British Grasslands Society* 14:233–237.

Zontek, S.J. 1973. A positive approach to *Poa annua* management. *United States Golf Association Green Section Record* 11:1–5.

Chapter 4

Supina Bluegrass (*Poa supina* Schard.)

Suleiman Bughrara

Poa supina Schard., supina bluegrass, has been described as *Poa supena* by Daniel and Freeborg (1977). It is also called *P. sapina* and is currently popular in Germany, Austria, the United Kingdom, and the United States for use on athletic fields, golf courses, and occasionally for lawns. The key characteristics that make it popular for athletic fields include its aggressive stoloniferous growth habit and its tolerance to shade, wear, and cold. In Germany, significant amounts of research have been conducted on the breeding and management of *P. supina* for both athletic fields and golf courses. Research in Germany demonstrated that *P. supina* can be an alternative turfgrass species for athletic fields and other turf areas with heavy traffic. In 1974 it was sprigged into the Munich Olympic Stadium. A survey conducted by Kock (1977) on 12 athletic fields in Germany found no *P. supina* on dry fields, even when irrigated, but it was the dominant species on wet fields. The grass was evenly distributed on heavy traffic areas such as the goal mouth and the center of the soccer field. For an irrigated shady area, *P. supina* might be an ideal turfgrass species.

Nonn (1994) reported that *P. supina*, in a mixture of different grasses on five golf courses, grew dense root systems and maintained good turf quality in heavily shaded areas. Portable plots of *P. supina* and Kentucky bluegrass (*Poa pratensis* L.) were established outdoors, then brought inside a domed stadium for treatment and evaluation under reduced light conditions (5% sunlight). *Poa supina* showed better turf quality than Kentucky bluegrass (Stier and Rogers, 1995). Spatz (1985) recommended *P. supina*, which is cold tolerant, to be used for recultivation of high-altitude skiing grounds. Filiault and Stier (1999) showed that *P. supina* had greater leaf area, leaf number, dry weight, and root mass than other cool-season grasses to assist in overcoming the stress of winter. *Poa supina* is a useful grass in northern areas because it maintains its shoot and root tissues at cold temperatures. *Poa supina* had significantly greater rates of photosynthesis compared to Kentucky bluegrass on a turf area basis but not on a leaf area basis when it was grown in reduced light conditions (Stier et al., 1997). Also, they found that trinexapac-ethyl treatments and nitrogen rate had negligible effects on photosynthesis.

The acceptable mowing height range was defined by Budryte-Aleksandraviciene and Schulz (1999). They found that *P. supina* had the most shade tolerance at a 2.5 cm mowing height when compared with over 15 cultivars from 10 species. The same results were obtained when the grass species were planted under 75% shade (Huber and Schulz, 1997). *Poa supina* exhibited very poor performance on putting greens due to its inability to withstand low mowing heights (Brown and Ross, 1995).

In the United Kingdom, over 25 greens and several Premiership football clubs were seeded with *P. supina* in 1996. Those greens suffered poor drainage, heavy traffic, dense shade, and poor air movement. Unfortunately, no research had been conducted in the United Kingdom up to 2001 (Cheetham, 1997).

At the University of Minnesota, *P. supina* has been researched for breeding and germ plasm enhancement by interspecific crossing with annual bluegrass, *Poa annua* L. No commercial cultivars have resulted from this project. Michigan State University has been conducting shade, traffic, mowing, and management research in *P. supina* for athletic fields. The results of this research indicate *P. supina* is adapted to cutting heights from 1.3 to 3.8 cm. The grass responded best under moderate to high fertility programs (195 to 290 kg N/ha/year). These cutting heights make *P. supina* suitable for tees, fairways, and upper level athletic fields.

The aggressive stoloniferous growth habit of *P. supina* allows it to form a dense turf relatively quickly and to recuperate quickly from damage. In a study by Sorochan et al. (1998), they found aggressiveness of *P. supina* is enhanced with increased fertility and traffic. A seeding mixture of 10% *P. supina* and 90% Kentucky bluegrass, after one season of traffic, was sufficient to increase *P. supina* density to over 50%. Under higher fertility levels, *P. supina* density increased to over 75%. Disadvantages of *P. supina*, due to its stoloniferous spreading ability, includes its ability to be spread by stolon across different turfgrass areas by machinery or runoff. *Poa supina* is a dominant species and should often be planted alone (Brede, 1997).

Christians et al. (2000) evaluated *P. supina* and 34 other turfgrass species established in the fall of 1987 for their adaptability to natural tree shade. They found that mean turf quality of *P. supina* during the season was 1.3 where other species ranged from 1.4 to 6.8 (9 = best quality, 6 = lowest acceptable quality, and 1 = worst quality). In the Chicago Botanical Garden, Bilow et al. (1994) found that *P. supina* initially grew aggressively and formed a dense turf with excessive thatch. Later the turf was thinned because of infestation with sod webworm (*Parapediasia teterrella*), followed by weed invasion.

Climatic (rainfall and temperature) and edaphic (soil texture and moisture) factors are primarily instrumental in determining growth of *P. supina*. It grows best in wet and cold temperatures, often in shaded conditions. It can be established in full sun as long as sufficient moisture is available. It has very poor drought tolerance and limited adaptation to high summer temperatures. Compacted, heavy clay soils with sufficient nutrients and moisture are an ideal environment to establish *P. supina*.

MORPHOLOGY

The morphological traits of *P. supina* are similar to other species in the *Poa* genus. Traits include a smooth leaf blade with a prominent midrib that ends in a boat-shape at the leaf tip. Unlike most *Poa* species, *P. supina* has a stoloniferous growth habit, which produces aboveground runners with short internodes ranging from 1.3 to 5.5 cm reminiscent of rough bluegrass, *Poa trivialis* L.

Comparative Traits

Poa supina is a perennial, while *P. annua* is predominately an annual. The leaves of *P. supina* are coarse and have a light green color. *Poa supina* panicles are purple to red, compacted, and produce flowers once a year (light green grass with colorful panicle). It flowers in early spring, earlier than other cool-season grasses. Its panicles are more upright than those of *P. annua*, which allows the panicle to be cut cleanly off when mowed. On the other hand, the panicles of *P. annua* are often prostrate and never totally removed with mowing.

Steiner and Fuchs (1991) demonstrated similar floret size and seed mass for *P. supina* and *P. annua*. *Poa supina* had a mean floret length of 2.35 mm, width of 0.76 mm, and thousand-seed mass of 0.24 g, while *P. annua* had a mean floret length of 2.36 mm, width of 0.71 mm, and thousand-seed mass of 0.26 g. *Poa supina* also has short membranous ligules that are bright green in color, have no hair visible from the top, and are approximately 0.3 mm in length. *Poa annua* ligules are around 2.0 mm in length and visible from the side, when both are vegetatively growing. However, during panicle formation, *P. supina* ligules lengthen similarly to those of *P. annua* (Entrup, 1975). *Poa supina* has better overall winter hardiness than *P. annua*. Also, *P. supina* has early spring greenup compared to *P. annua*. Most of all, *P. supina* has less disease incidence.

Physiological limitations of *P. supina* are poor drought tolerance and poor adaptability to the southern transition zone. Fertility requirements, proper management practices, and herbicide resistance need additional research. *Poa supina* can tolerate prolonged freezing temperatures and ice cover but is not persistent under heat stress. Light green color, poor high-temperature color retention, and poor seed production are the most limiting characteristics of *P. supina*.

DISTRIBUTION AND CYTOTAXONOMY

Classification

The taxonomic classification of *P. supina* is:

Family	Gramineae (Poaceae)
Subfamily	Pooideae
Tribe	Poeae
Genus	*Poa*
Species	*supina*
Authority	Schard

Poa supina is native to the European Alps and other mountainous areas in Central Europe (European supina bluegrass). It was found growing naturally on cattle paths where no other grass would grow due to the compaction from hooves and a combination of both shade and traffic. It was found in similar environments in France, Switzerland, and Russia. Its common name in Germany is Lägerrispe which means "where the cows lay" (Klapp and Optiz von Boberfeld, 1990). In Germany, it has been used as a turfgrass species since 1960.

Asian supina bluegrass, *P. supina ustulata*, is a subspecies of *P. supina* as classified by Fröhner in 1968. It is common in Asia and has the same chromosome number as *P. supina* except it has no stolons. The spikelets and the anthers are dark violet in color. Date of flowering is late and flowering occurs for a longer period than European *P. supina*. It has not been explored as a turfgrass species.

Genetics and Cytotaxonomy

Berman et al. (1988) separated *P. annua* and *P. supina* by using isozyme electrophoresis. Leaf esterases of 66 *P. annua* and *P. supina* biotypes showed more variability between biotypes than within biotypes. Steiner and Goeritz (1991) used electrophoresis of storage proteins to distinguish between *P. supina* cultivars and *P. annua*. Their tests showed less variation within *P. supina* cultivars and also showed at least one band unique to *P. supina*. Sweeney et al. (1996) used random amplified polymorphic DNA (RAPD) markers from bulk samples and individual seeds to distinguish differences between *P. supina* and *P. annua*. The technique proved to be more useful than amplification of leaf tissue DNA to separate the species.

Poa supina is a diploid ($2n=2x=14$), while *P. annua* is an allotetraploid ($2n=4x=28$). Diploid variants of *P. annua* may also have $2n=2x=14$ and make up approximately 25% of the putting green collections (White, 1988). Low mowing height used in putting green management practices apparently selected for the diploid types.

On the basis of morphological analysis, Tutin (1957) suggested that *P. annua* originated from crosses between *Poa infirma* Kunth and *P. supina*. *Poa supina* is a prostrate perennial species, whereas *P. infirma* is an erect annual. The F_1 hybrids are sterile and similar to *P. annua*. The tetraploid hybrid shows wide morphological variability including annual erect and perennial prostrate ecotypes. Two partially fertile tetraploid plants have been produced from previous crosses and were very similar to *P. annua* plants. Koshy (1968) reported that *P. annua* does not contain the sum of the karyotype of *P. supina* and *P. infirma*, suggesting another species might be involved in the evolution of *P. supina*. Darmency and Gasquez (1997) showed that the three species: *P. infirma*, *P. supina*, and *P. annua* are closely related based on isozymes patterns and morphological analysis. However, their interspecific hybrids are sterile. Unreduced gametes (2n chromosome number) result in polyploidization giving rise to fertile tetraploid hybrids. The two parent diploid species are probably closely related to the allotetraploid species and are perhaps its parental progenitors. Entrup (1975) recognized that *Poa x nannfeldtii* is a hybrid between *P. annua* and *P. supina*. Fertile interspecific F_1 hybrids similar to *P. annua* were reported by Pietsch (1989) in his early works at the Max Planck Institute in Germany, by crossing of *P. supina* with rough bluegrass.

Seed Production

It is generally recognized that breeding-improved cultivars with high seed yield production is the key that can lead to better turf quality, better management practices, and good color retention. Berner (1984) found that *P. supina* reproduces in a panmitic manner, obviously fully self-fertilizing. No apomictic forms were observed over four generations. However, more data are needed to confirm this observation. Although it is difficult to establish *P. supina* from seed, seeded plantings are superior to sodded planting to avoid contamination with *P. annua*. Seed dormancy was also observed (White, 1988). Moist and cold treatment of 7 to 14 days was used to overcome dormancy; otherwise, a few months in cold seed storage is required.

One of the *P. supina* limitations is low seed production. Unmown *P. supina* tops out at 15 cm with panicles extending another 30 to 45 cm. This is considered within the range for mechanical harvesting. Entrup (1979) discussed three management experiments to increase seed yield. The variables were three seeding dates, three planting spacings (15, 30, and 45 cm), and drilling with high and low nitrogen fertilizer application. He suggested that seeding in August will lead to high seed yield. High nitrogen application (67 kg/ha) in the fall depressed seed yield. In the first year of the experiments, 15-cm spaced planting gave the highest yield. Thirty and 45-cm space plantings produced a higher yield in the second year. In the second and subsequent years, *P. supina* stands produced excessive thatch and resulted in reduced flowering and seed yield. In the same study, he also concluded that seed yield depended on the maturity of the flowering shoots and tillers before winter dormancy.

Entrup (1975) showed that *P. supina* had obligatory vernalization requirements for the induction of reproductive growth. Vernalization did not occur in the embryo of the seed or in early stages of vegetative growth. Temperature before vernalization had a great effect on flower formation. High temperature before vernalization decreased floral induction. Maturity of the reproductive tillers was very important in the vernalization process. Older plants required a shorter vernalization period than younger plants. After vernalization, the number of inflorescences increased with the increase in photoperiod from 9 to 21 h.

COLLECTION, SELECTION, AND BREEDING HISTORY

Poa supina was first observed on pastures, cow paths, and around stalls in alpine regions of Central Europe in 1930. After 1960, a turfgrass breeder, P.H. Berner at Steinach, Germany, called attention to *P. supina* which had invaded the turfgrass in high altitudes of alpine habitats. The first cultivar, Supra, was an ecotype cultivar developed from plant material collected on European alpine slopes and trails. It received plant cultivar protection in 1974. The second cultivar, Supernova, was the second ecotype cultivar released from the same breeding program in 1980. These are the only two cultivars of *P. supina* commercially available to date.

The breeding techniques utilized in releasing these two commercial cultivars were based on ecotype selection; no recombinations were involved as described by Berner (1980). The ecotypes and promising individual plants were collected. Plants were vegetatively divided and planted into plots of 1 m and seeds were harvested from high yielding plots after the second winter of phenotype evaluation. Single plants were selected from the progeny test. The plants were divided for replication and were isolated by strips planted to timothy (*Phleum bertolonii* DC). The plots were evaluated for seed production and other turf characteristics. The main objective of Dr. Berner's breeding project was breeding *P. supina* for high seed yield and limited to ecotype selection. The seed were very pure and clean from *P. annua*.

Breeding *P. supina* for dark green color can be an objective of turfgrass breeding projects, but the potential of genetically increasing darkness of color would be slight at best without high seed yield production. Developing *P. supina* cultivars resistant to disease and biotic stresses would be most important, especially drought tolerance for wider adaptation ranges to southern regions. Subsequent breeding efforts for improving *P. supina* performance most likely will require programs based on recombination and selection. In general, improvement has been slow due to the lack of research, low market demand, contamination with *P. annua*, and limited if no international cooperation in germ plasm exchange.

Poa supina has been used to improve turf quality of *P. annua* species. Dr. Donald White at the University of Minnesota made interspecific crosses of *P. supina* x *P. annua*. These crosses produced dark green, vigorous, and rugged progenies that were fertile. The progeny were evaluated and stolons were found to spread on average one-third more than *P. annua*. The plants had one-half the chromosome number (14 chromosomes) of *P. annua*. The success of interspecific hybridization could open the door to much easier inheritance research with *P. annua* and *P. supina*. The excised stem mist emasculation technique and sucrose media for floral pick culture were utilized for making crosses (White, 1988).

In summary, *P. supina* is a relatively new turfgrass in temperate regions of the world. It is well adapted to northern regions with good tolerance to wear and shade. The light green color and low seed yield are disadvantages of *P. supina* due to lack of improved cultivars. Breeding for high seed yield production, wider adaptation ranges, and to drought tolerance are important. Breeding efforts should be enhanced by studies of molecular markers in order to facilitate marker-assisted selection. In vitro studies, DNA manipulations, and improved conventional breeding are required for future development of this species.

REFERENCES

Berman, S.A., D.B. White, and B.A. Ruemmele. 1988. Separation of *Poa annua* L. and *Poa supina* Schrad. Biotypes by Isozyme Electrophoresis. *Agronomy Abstracts*, p. 148.

Berner, P.H. 1980. Characteristics, breeding methods, and seed production of *Poa supina* Schrad. pp. 409–418. In J.R. Beurd, Ed. *Proc. of 3rd Intl. Turfgrass Research Conf., Munich, Germany. 11–13 July 1977.* Int. Turfgrass Soc. and ASA, CSSA, and SSSA, Madison, WI.

Berner, P.H. 1984. Entwicklung der Lägerrispe (*Poa supina* Schrad) zum Rasengrans. *Rasen-Turf-Gazon* 15(1):3–6.

Bilow, K., T. Voigt, and R. Hawké. 1994. Turfgrass evaluations at the Chicago Botanic Garden. *1994 Illinois Turfgrass Research Report*, pp. 46–47.

Brede, D. 1997. Unsung turfgrass species bring diversity to golf courses. *Golf Course Management* 64:54–57.

Brown, W.A. and J.B. Ross. 1995. Evaluation of various low growing turfgrasses for use on putting greens. Prairie Turfgrass Research Center Annual Report 1995. pp. 35–37.

Budryte-Aleksandraviciene, E. and H. Schulz. 1999. Wirkung unterschiedlicher Beschattungsintensität auf die Entwicklung einiger Rasengräserarten und sorten. *Rasen-Turf-Gazon*. 30:89–94.

Cheetham, D. 1997. Opportunity knocks. *Turf Management*. August. pp. 25–26.

Christians, N.E., B.R. Bingaman, L.A. Benning, and G.M. Peterson. 2000. Shade Adaptation Study. 2000 Iowa Turfgrass Research Report pp. 1–4.

Daniel, W.H. and R.P. Freeborg. 1977. *Turf Manager's Handbook.* Business Publications, Harcourt Brace Jovanovich. Cleveland, OH.

Darmency, H. and J. Gasquez. 1997. Spontaneous hybridization of the putative ancestors of the allotetraploid *Poa annua*. *New Phytologist* 136:497–501.

Entrup, N.L. 1975. Vernalization requirement and photoperiod demand for reproductive development in *Poa supina* Schrad. *Zeitschrift-fur-Acker-und-Pflanzenbrun* 141:4, 300–316.

Entrup, N.L. 1979. Die beeinflussung des samenerfrages von *Poa supina* (Schrad.) durch pflanzenbauliche massenahmen. *Zeitschrift-fur-Acker-und-Pflanzenban*. 148:5, 378–392.

Filiault, D.L. and J.C. Stier. 1999. Photosynthetic efficiency and carbohydrate status of overwintering turfgrasses. *Agronomy Abstracts* p. 123. ASA-CSSA-SSA, Madison, WI.

Fröhner, S. 1968. Die asiatischen verwandten von *Poa supina* Schrad. *Bot. Jahrb.* 88:411–442.

Huber, A. and H. Schultz. 1997. Einfluss von Belastung und Beschattung auf einige Rasengraserarten und sorten. *Rasen-Turf-Gazon*. 28:2, 36–40.

Klapp, E. and W. Opitz von Boberfeld. 1990. *Taschanbuch der graser. 12. auflage*, Verlag Paul Parey, p. 282.

Köck, L. 1977. Natürliches Vorkommen von *Poa supina* auf sportplatz-rasen in Tirol. *Rasen, Grunflachen Begrunungen*. Vol. 8, No. 2, pp. 44–46.

Koshy, T.K. 1968. Evolutionary origin of *Poa annua* L. in the light of karyotypic studies. *Can. J. Gen. Cytol*. 10:112–118.

Nonn, H. 1994. Conclusions from the use of seed mixtures and sods with *Poa supina* on golf courses. *Rasen-Turf-Gazon* 25(14):101–102, 104.

Pietsch, R. 1989. *Poa supina* and its value for sports and amenity turf. *Zeitschrift fur vegetationstechnick* 12:21–24.

Sorachan, J.C., J.N. Rogers III, and J.C. Stier. 1998. Seed mixtures of *Supina* bluegrass and Kentucky bluegrass for athletic fields. *Agronomy Abstracts* p. 125.

Spatz, G. Freising-Weihenstaphan 1985. The persistence of ski-run turf at high altitudes. *Rasen. Grunflachen Begrunungen*, 16:15–19.

Steiner, A.M. and H. Fuchs. 1991. Größe und Tausendkornmasse der Spelzfrüchte von lägerrispe (*Poa supina* Schrad.) und Jähriger Rispe (*Poa annua* L.). *Rasen-Turf-Gazon* 22(4):94.

Steiner, A.M. and A. Goeritz. 1991. Die unterscheidung des saatguts von Sorten der Lägerrispe (*Poa supina* Schrad.) und von Järige Rispe (*Poa annua* L.) mittels Electrophorese der speicherproteine. *Rasen-Turf-Gazon* 22(4):90–94.

Stier, J.C. and J.N. Rogers. 1995. Response of *Poa supina* and *Poa pratensis* to plant growth retardant and iron treatments under reduced light conditions. *Agronomy Abstracts*. p. 145.

Stier, J.C., J.N. Rogers III, and J.A. Flore. 1997. Nitrogen and trinexapac-ethyl effects on photosynthesis of *Supina* bluegrass and Kentucky bluegrass in reduced light conditions. *Agronomy Abstracts*. p. 126.

Sweeney, P., R. Golembierski, and K. Danneberger. 1996. Random amplified polymorphic DNA analysis of dry turfgrass seed. *HortScience* 31(3):400–401.

Tutin, T.G. 1957. A contribution to the experimental taxonomy of *Poa annua* L. *Watsonia* 4:1–10.

White, D.B. 1988. Improvement of *Poa annua* and *Poa supina* for golf turf. *1988 Turfgrass Research Summary* (USGA/GCSAA) pp. 18–19.

Chapter 5

Texas Bluegrass (*Poa arachnifera* Torr.)

James C. Read and Sharon J. Anderson

Texas bluegrass (*Poa arachnifera* Torr.) has limited use worldwide, and its major use has been as a range grass in Texas and Oklahoma. It was introduced into North Carolina and Florida but little can be found in the literature concerning its performance (Pitman and Read, 1998; Pitman et al., 2000). There are no released cultivars of Texas bluegrass but synthetic lines have been developed and tested for use as both forage (Read et al., 1997) and turfgrass. One cultivar (Reveille Hybrid bluegrass) resulting from hybridization of Texas bluegrass and Kentucky bluegrass (*P. pratensis* L.) has been released.

DISTRIBUTION AND CYTOTAXONOMY

Texas bluegrass (*Poa arachnifera* Torr.) is a tufted dioecious perennial cool-season grass with long, slender rhizomes that is native to south central United States. It occurs throughout Texas except for the extreme southern and western sections and is most common in the north central region where annual rainfall is between 50 and 82 cm. This corresponds to the Blackland Prairies, Cross Timbers and Prairies, Edwards Plateau, and Rolling Plains areas. It also occurs in most of Oklahoma, southern Kansas, and western Arkansas (Gould, 1975). Populations occur on all soil types, including the acid sandy soils of east Texas and the calcareous clay soils of the Blackland prairies. The grass can be found in river bottoms and upland sites. Some of the more interesting populations have been found in the sand hills of western Oklahoma extending southward into Texas. Texas bluegrass is the predominant grass species on many of these sites.

Texas bluegrass plants vary in height from 20 to 92 cm. Leaf width is highly variable and can be up to 6.5 mm. The pistillate spikelets are covered with fine cobweb-like hairs, and the staminate spikelets are glabrous (Hoover, et al., 1948; Hitchcock, 1951). Habit, inflorescence and spikelet morphology are illustrated in Figure 5.1 (Hatch and Pluhar, 1993).

The base number for the genus *Poa* is $x=7$ (Gould, 1968). Chromosome number of Texas bluegrass was reported to be $2n=84$ (Gould, 1975) but counts by Read (unpublished data) indicated the most common count was $2n=56$ (sample size =5), with considerable number of aneuploids. Counts from root tip smears were very difficult due to

Figure 5.1. Habit and spikelet morphology of *Poa arachnifera* (Texas bluegrass). reproduced from Hatch and Pluhar (1993).

large number and small size of the chromosomes, presence of aneuploidy, and non-uniform counts in different cells of the same root tip.

No apomixis has been observed in *P. arachnifera*. *Poa arachnifera* is dioecious, with normal embryo sac development of sexually reproducing grasses. Facultative apomixis of 15% was estimated for *P. arachnifera* x *P. pratensis* hybrid, 'Reveille' (Read et al., 1999).

The taxonomic classification of Texas bluegrass is (Gould, 1968):

Family	Gramineae (Poaceae)
Subfamily	Pooideae
Tribe	Poeae
Genus	*Poa*
Species	*arachnifera*
Authority	John Torrey

There are eight synonyms listed for *Poa arachnifera* Torr. (Missouri Botanical Garden, 2001). The synonyms reported are *P. arachnifera* var. *glabrata* Vasey ex Beal, *P. arachnifera* var. *glabrata* Vasey ex L. H. Dewey, *P. arachnifera* var. *glabrata* Vasey, *P.*

densiflora Buckley, *P. glabrescens* Nash, *P. nemoralis* L., *P. nevadensis* Vasey ex Scribn., and *P. pratensis* subsp. *agassizensis* (B. Boivin & D. Love) Roy L. Taylor & MacBryde.

COLLECTION, SELECTION, AND BREEDING HISTORY

The first cross between *Poa arachnifera* and *P. pratensis* was by Oliver (Vinall and Hein, 1937). Later crosses between *P. arachnifera* and *P. pratensis*, made by Brown, resulted in first-generation plants with greater heat and drought tolerance and greater herbage yields than the *Poa pratensis* parent (Vinall and Hein, 1937). Interest in these original hybrids was lost until research performed by Read.

Read began collection of wild accessions in 1986. This one collection was screened primarily for forage quality but some plants had turf quality. In 1988, collections from 40 different sites in Texas and Oklahoma were made by Read. Much of his collection has been deposited at the GRIN germ plasm repository in Pullman, WA.

Existing collections have not adequately sampled the natural germ plasm diversity. Genetic diversity of 28 genotypes from the Texas bluegrass germ plasm collection at TAES-Dallas was estimated using amplified fragment length polymorphism (AFLP) and randomly amplified polymorphic DNA (RAPD) markers (Renganayaki et al. 2001). Variability was greater between populations than within. In addition, some populations had higher intrapopulational genetic variation than others.

The largest native populations are most common from Brownwood to Ft. Worth, Texas, and northward to near Woodward, Oklahoma. Populations are infrequent outside of this region. Based upon the frequency of occurrence of Texas bluegrass and the geographic distribution, the center of origin for the species is probably in the Jack county and Palo Pinto county region of Texas. The northwest Oklahoma populations have smaller plants that are unique morphological ecotypes. *Poa arachnifera* has been reported for western Arkansas and southern Kansas, but the author has not made collections in these regions.

Currently breeding efforts are in the early generation selection cycles. Phenotypic selection for leaf and stem rust (*Puccinia* spp.), powdery mildew (*Erysiphe graminis*, Leveille), brown patch (*Rhizoctonia solani*, Kuhn), and fall armyworm (*Spodoptera frugiperda*, J. E. Smith) resistance have been made. Polycross nurseries of elite lines have been developed, and seed has been collected. The next step includes inheritance studies to establish genotype of resistance and to identify inheritance of the resistance traits.

Poa arachnifera is genetically vulnerable, like other native prairie grasses. Wild accessions collected in 1988 were transplanted to a common nursery and grown for one year. In 1989 Read selected 25 plants based upon forage characteristics and 8 plants based upon turfgrass characteristics from the 1988 transplants. The 25 plants selected for forage characteristics were used to produce Syn-1. To produce Syn-1 an isolation nursery was established using 8 replications of all 25 plants using randomized block design. Syn-2 was produced from an isolation block in which eight replications were used and each plant appeared in every row and column. Seeds from Syn-1 and Syn-2 were harvested from all plants and bulked to establish a seed increase block that was used as breeder seed block. Seed from the breeder seed block was used for testing. Syn-2 produced plants inferior to the selected plants in the first generation, whereas Syn-1 produced plants equal to the selected plants. This decrease in plant quality was most likely due to inbreeding depression caused by the small number of plants used to produce the line. This demonstrated the need to maintain high variability in parental accessions to achieve vigorous plants in succeeding generations. It also illustrated the genetic vulnerability inherent within this species. Because of the genetic vulnerability observed with the limited collections, wild germ plasm is still useful.

In 1990, 14 new accessions were selected from the 1989 planting, which were established using plant material from the 1988 collection trips. These 14 plants and five plants from the 1986 collection were combined to create Syn-3. Three plants that were included in Syn-1 were also used in Syn-3. Syn-3 was equivalent in performance to Syn-1 for desired forage characteristics. A need exists to determine the number of parental plants needed to produce synthetic cultivars for turf. All studies were conducted in dry-land conditions, with no supplemental irrigation. Plantings were fertilized at 90 kg N ha^{-1} per year.

As the course of the breeding program progressed, an interspecific hybrid program was initiated in 1989. Early reports provided evidence that the potential for viable crosses between *P. arachnifera* and *P. pratensis* (Vinall and Hein, 1937) was possible. Interspecific crosses included *P. arachnifera* x *P. compressa* and *P. arachnifera* x *P. pratensis*. Crosses of *P. arachnifera* x *P. compressa* were made only one year, and based upon approximately 150 hybrids they were inferior to hybrids produced from the crosses with *P. pratensis*. Approximately 3,300 hybrids with *P. pratensis* have been produced and evaluated. Hybrid progeny from *Poa arachnifera* x *P. compressa* and *P. arachnifera* x *P. pratensis* both exhibited considerable variation. These hybrids can be any combination of dioecious versus perfect flower and apomict versus sexual. This is consistent with other hybrid populations produced by crossing several species with an apomictic species (Read and Bashaw, 1969). Individuals with desirable forage and turf qualities were selected for further development.

Potential for using *P. arachnifera* x *P.* spp. is unlimited, especially for apomictic species. Because *P. arachnifera* is dioecious, it is easy to make large numbers of crosses and in many cases fertility is restored in the apomictic hybrids.

Cultivars

Reveille was developed from the hybrid program. The cross was made in 1990 and Reveille is an F_1 progeny selection from F_1 of *Poa arachnifera* x *P. pratensis*. It is a facultative apomict. It has performed well in turf trials at Colorado State University; University of Arizona; El Paso, TX; Dallas, TX; and GAES at Griffin, Georgia. The trials included minimal maintenance, which suggests that Reveille is adapted to these sites.

Selection Criteria

Only synthetic cultivars have been produced for *P. arachnifera* but none have been released. Syn-1 has done well in Oklahoma and Texas. *P. arachnifera* was introduced to North Carolina and Florida early in the twentieth century (Vinall and Hein, 1937).

Forage yield has been a major selection criterion for Texas bluegrass, for forage applications (Read et al., 1997; Pitman and Read, 1998; Pitman et al., 2000). Insect and disease resistance, and the turf quality parameters of density, color, smoothness were major parameters in selecting Reveille hybrid bluegrass.

Variable levels of resistance to fall armyworm and white grubs (Phyllophaga and Cyctocephala) were observed among selected genotypes of Kentucky bluegrass, Texas bluegrass, and *P. arachnifera* x *P. pratensis* hybrid genotypes (Reinert et al., 1998; Reinert and Read, 1998) Reveille was among the accessions with resistance to fall armyworm and white grubs. Metz and Read (unpublished) identified Texas bluegrass accessions and *P. arachnifera* x *P. pratensis* hybrids resistant to powdery mildew (*Erysiphe graminis*). Resistance to stem rust (*Puccinia* spp.) have also been evaluated for the *P. arachnifera* x

P. pratensis hybrids (Colbaugh et al., 1997). Metz, Read, and Colbaugh (unpublished) identified variation among accessions for resistance to brown patch (*Rhizoctonia solani*). Research is being conducted to increase seed production and determine if these traits are heritable.

Texas bluegrass is drought tolerant in native stands, utilizing its summer dormancy as the primary tolerance mechanism. It is a heat-tolerant cool-season turfgrass. In management trials to maintain good green color, supplemental irrigation was required up to one-third of pan evaporation, indicating irrigation requirements more like those of bermudagrass (*Cynodon* spp.), and less like Kentucky bluegrass (Read, 1998).

Management trials at Dallas, TX of Kentucky bluegrass and Texas bluegrass accessions and *P. arachnifera* x *P. pratensis* hybrids have evaluated turf performance, nitrogen fertilizer requirements, and desired mowing height (Read, 1998).

Flowering

Most *P. arachnifera* will not flower in the greenhouse at TAES-Dallas, and those that do flower will have a reduced number of inflorescences. Vernalization is required for flowering. However, duration of treatment and temperature requirements are unknown. *Poa arachnifera* flowers in the spring, with increasing daylength. Daylength requirements for flowering are similar to those of *Poa pratensis* (KBG), but *P. pratensis* will flower later than *P. arachnifera*. Flower production is determinate, producing only one inflorescence per year. Once the inflorescence is removed, no new inflorescences will be produced during that growing season. New inflorescences will be produced on the spring growth during the following growing season.

At Dallas, *P. arachnifera* flowers earlier than *P. pratensis*. Some *P. pratensis* genotypes have flowered as early as one week after the *P. arachnifera*, thus allowing a limited number of *P. pratensis* x *P. arachnifera* crosses to be made. Successful induction of early flowering in *P. pratensis* genotypes by increasing daylength and temperature has enabled the production of a large number of interspecific crosses with *P. arachnifera*, leading to the development of Reville hybrid bluegrass.

Molecular Genetics and Tissue Culture

Poa arachnifera has transformation potential. Genovesi et al. (2001) reported production of whole plants from callus. Renganayaki et al. (2001) associated AFLP and RAPD fingerprint patterns with county of origin of 28 *Poa arachnifera* genotypes. PCA analysis of AFLP fingerprint profile patterns were associated with gender among the genotypes from two locations.

REFERENCES

Colbaugh, P.F., J.C. Read, and S.P. Metz. 1997. Susceptibility of Texas bluegrass x Kentucky bluegrass hybrids to Puccinia spp. stem rust damage. In *Texas Turfgrass Research—1997*, Consolidated Progress Reports TURF-97-4: 3p. dallas.tamu.edu/pub/(Dec. 10, 1997).

Genovesi, A.D., J.C. Read, and M.C. Engelke. 2001. Plantlet regeneration of zoysiagrass and hybrid bluegrass from meristemoid cultures. *American Society of Agronomy Abstracts*. Title summary number C05-genovesi094438-P.

Gould, F.W. 1968. *Grass Systematics*. McGraw-Hill, Inc. New York.

Gould, F.W. 1975. *The Grasses of Texas*. Texas A&M University, College Station, TX.

Hatch, S.L. and J. Pluhar. 1993. Texas Range Plants. W.L. Moody, Jr. Natural History Series, No. 13. Texas A&M University, College Station, TX.

Hitchcock, A.S. 1951. *Manual of the Grasses of the United States*, 2nd ed, revised by A. Chase. United States Department of Agriculture, Miscellaneous Publication No. 200. Washington, DC.

Hoover, M.M., M.A. Hein, W.A. Dayton, and C.O. Erlanson. 1948. The main grasses for farm and home. pp. 639–700. In *Yearbook of Agriculture*. Kellogg, E.A. 1990. Variation and species limits in agamospermous grasses. *Systematic Botany* 15(1): 11–123.

Missouri Botanical Garden. "w³TROPICOS." Available at mobot.mobot.org/W3T/Search/vast.html (verified May 2, 2001).

Pitman, W.D. and J.C. Read. 1998. Effects of initial plant spacing and applied nutrients on stand development and productivity of Texas bluegrass in the Louisiana coastal plain. *Journal of Plant Nutrition* 21(6): 1093–1102.

Pitman, W.D., D.D. Redfearn, and J.C. Read. 2000. Response of Texas bluegrass to season of nitrogen fertilization on the Louisiana coastal plain. *Journal of Plant Nutrition* 23(4):423–429.

Read, J.C. 1998. TXKY 16-1. A heat resistant bluegrass adapted to the southern United States. Report of Technical Committee on Seed Release and Increase, Texas Agricultural Experiment Station, College Station, TX.

Read, J.C. and E.C. Bashaw. 1969. Cytotaxonomic relationship and the role of apomixis in speciation in buffelgrass and birdwoodgrass. *Crop Sci.* 9: 805–806.

Read, J.C., J.A. Reinert, P.F. Colbaugh, and W.E. Knoop. 1999. Registration of 'Reveille' hybrid bluegrass. *Crop Sci.* 39:590.

Read, J.C., M.A. Sanderson, G.W. Evers, P.W. Voigt, and J.A. Reinert. 1997. Forage production potential of *Poa arachnifera* Torr. in semi-arid climates. pp. 104 In Proceedings XVIII International Grassland Congress. Volume I. Calgary, Alberta, Canada.

Reinert, J.A. and J.C. Read. 1998. Host resistance to white grubs among genotypes of *Poa arachnifera* x *P. pratensis* hybrids. In *Texas Turfgrass Research Reports* 98.

Reinert, J.A., J.C. Read, S.J. Maranz, and B.R. Wiseman. 1998. Resistance to fall armyworm, *Spodoptera frugiperda* in bluegrasses *Poa* spp. and *P. Arachnifera* x *P. pratensis* hybrids. In *Texas Turfgrass Research Reports* 98.

Renganayaki, K., J.C. Read, and A.K. Fritz. 2001. Genetic diversity among Texas bluegrass genotypes (*Poa arachnifera* Torr.) revealed by AFLP and RAPD markers. *Theor. Appl. Genet.* (in press).

Vinall, H.N. and M.A. Hein. 1937. Breeding miscellaneous grasses. pp. 1032–1102. In *Yearbook of Agriculture*.

Chapter 6

Rough Bluegrass (*Poa trivialis* L.)

Richard Hurley

The *Poa* genus comprises over 200 species that are widely distributed throughout the cool, humid climatic regions of the world. The growth habit of the *Poa* genus ranges from an erect, bunch type to a prostrate, stoloniferous type to an extremely dense sod-forming type having a vigorous rhizome system (Hitchcock, 1950).

Studies of regional and global distribution of the *Poa* genus indicates its highest relative differentiation in regions of high latitude or high altitude (Hartley, 1961). *Poa* species are absent or rare in the tropics, except in mountainous regions. However, *Poa* constitutes more than 15% of the grass flora of Alaska, Iceland, Kamchatka, and the high Pamir Mountains. A close association exists between the occurrence of high percentage frequencies of *Poa* species and cool summer temperatures. In the United States the 24°C (75°F) midsummer (July) isotherm effectively demarcates those regions in which *Poa* spp. form more than 5% of the total grass species. A similar relationship between percentage frequency and midsummer temperature is found in most other parts of the world. Climatic factors other than temperature seem to have little influence on the distribution of *Poa* species (Hartley, 1961).

TAXONOMY AND MORPHOLOGY

Poa trivialis has the following taxonomy (Hitchcock, 1950).

Family	Poaceae
Subfamily	Pooideae
Tribe	Poeae
Genus	*Poa*
Species	*trivialis*
Authority	Linneaus

Poa trivialis is commonly known by its scientific name, but is also referred to as rough bluegrass, roughstalk bluegrass, shade bluegrass, roughstalked meadowgrass, and rough meadowgrass (Hubbard, 1954; Beard, 1973). *P. trivialis* produces a moderately fine-textured, light green, medium dense turf. It is a cool-season sod-forming perennial that spreads by creeping leafy stolons, and may be found in soils with a pH range from

5 to 8, with best growth occurring between pH 6 and 7. Besides being well adapted to damp, shaded locations, it is also found growing in wet meadows, as a component of high fertility grasslands, and along ditch banks. It has the ability to germinate and grow at low temperatures, displays good color retention in the fall, produces early spring greenup, germinates rapidly with good seedling vigor, and has excellent winter hardiness (Vartha, 1969; Grime, 1980).

Cytology

P. trivialis is a diploid with a reported chromosome number of $2n = 2x = 14$ (Ahmed et at., 1972). *P. trivialis* is a completely sexual, cross-pollinated species and flowers at approximately 6:00 p.m. under New Jersey conditions at latitude 40° and longitude 74° (Hurley, 1982). Cooper and Calder (1964) reported a floral induction requirement of short days and low temperatures for *P. trivialis*. Budd (1970) showed that seedlings germinating in the fall produced more and earlier seedheads the following spring than seeds germinating during the winter and spring. This species has the ability to produce significant quantities of seed. However, seed shattering provides seed yield inconsistencies and increases cost of seed production (Hurley, 1982; Hurley and Funk, 1985).

Morphology

The morphological description of *P. trivialis* seedheads as provided by Hitchcock (1950) is described as:

"Culms erect from a decumbent base, often rather lax, scabrous below the panicle, 30 to 100 cm tall; sheaths retrorsely scabrous or scaberulous at least toward the summit; ligule 4 to 6 mm long; blades scabrous, 2 to 4 mm wide; panicle oblong, 6 to 15 mm long, the lower branches about 5 in a whorl; spikelets usually 2 or 3 flowered, about 3 mm long; lemma 2.5 to 3 mm long, glabrous except the slightly pubescent keel or lateral nerves rarely pubescent, the web at base conspicuous, the nerves prominent."

The vegetative description of *P. trivialis* provided by Beard (1973):

"Vernation folded; sheaths compressed, retrorsely scabrous, keeled below, frequently purplish split part way; ligule membranous, 4–6 mm long, acute, entire, or sometimes ciliate; collar conspicuous, broad, glabrous, divided; auricles absent; blades flat or V-shaped, 1–4 mm wide, yellowish-green, soft, glossy and keeled below, strongly scabrous, tapering from base to narrowly boat-shaped apex, transparent lines on each side of midrib, margins scabrous; stems compressed, erect to somewhat decumbent at base with creeping, thin, leafy stolons rooting at the nodes."

Geographic Adaptation

P. trivialis is native to all of northern Europe, temperate Asia, and North Africa and was introduced to North and South America and Australia (Hubbard, 1954). Brought to North America from Europe during the Colonial Period, it is best adapted for growth in moist, shaded areas from Newfoundland and Ontario, Canada to North Carolina and west to Minnesota and South Dakota. It has been reported growing in Colorado,

Utah, and as far south as Louisiana. *Poa trivialis* can be readily found on the West Coast from southern Alaska to California (Hitchcock, 1950; Hanson et al., 1969).

Growth Characteristics

P. trivialis often shows as yellow-green clonal patches in Kentucky bluegrass (*P. pratensis* L.), fine or tall fescue (*Festuca* spp.), perennial ryegrass (*Lolium perenne* L.) or bentgrass (*Agrostis* spp.) turfs. Its range of leaf width is similar to that of *P. pratensis*. In contrasting the plant tillers of the two species, *P. trivialis* has a larger ligule, a more scabrous leaf sheath than *P. pratensis*, but shares the general characteristics of *P. pratensis* leaves.

P. trivialis does not tolerate drought and is likely to be short-lived on dry sites (Welton and Carroll, 1940; Carroll, 1943; Beard, 1973). The root system is fibrous, relatively shallow, and annual in nature (Stuckey, 1941). It may be severely damaged or killed during periods of moisture stress, especially in sandy soils. *P. trivialis* also has poor wear tolerance and will not persist under heavy traffic (Beard, 1973; Shearman and Beard, 1975).

There are approximately 5.06 million seeds per kilogram. Seeds germinate under a wide temperature range, with peak germination occurring at approximately 9° C, and reported base temperature of 4°C. Base temperature refers to that temperature below which 50% of potential germination would not occur (Beard and Almodares, 1980). *P. trivialis* seed requires light and alternating temperatures for germination (Chippindale, 1949; Gill and Vear, 1958; Budd, 1970). Milton (1936) showed that some seeds remained viable after being buried in soil for five years. Champness and Morris (1948) found up to 79 million *P. trivialis* seeds per hectare in British grasslands.

In New Zealand, *P. trivialis* invades pastures two to three years after the pasture is sown; the source of *P. trivialis* was probably from buried seed (Vartha, 1969). Invasion of *P. trivialis* is dependent on having suitable conditions for its establishment. Milton and Davies (1947) suggested that a source of *P. trivialis* seed is in the dung of sheep brought from lowlands for grazing. After initial invasion, it probably spreads primarily by vegetative means (Vartha, 1969).

Brown patch (*Rhizoctonia solani* Kuhn), leafspot (*Drechslera poae* [Baudys] Shoemaker and *Bipolaris sorokiniana* [Sacc.] Shoemaker) and dollarspot (*Sclerotinia homoeocarpa* F.T. Bennett) are the most common diseases associated with *P. trivialis*. Gray snow mold (*Typhula incarnata* Lasch ex Fr.), pink snow mold (*Microdochium nivale* [Fr.] Samuels and Hallett), Ophiobolus patch (*Gaeumannomyces graminis* [Sacc.] Arx & Olivier var. avenae [Turner] Dennis) Pythium blight (*Pythium* spp.), Fusarium blight (*Fusarium culmorum* [Sm.] Sacc., F. poae [Peck] Wollenweb.), rust (*Puccinia spp.*), stripe smut (*Ustilago striiformis* [Westend.] Niessl) and powdery mildew (*Erysphe graminis* DC.) have been reported on this species (Hurley, 1982; Hurley and Funk, 1985).

COMMERCIAL USE

P. trivialis has been recommended for use in shaded lawn environments in climates where cool-season grasses are well adapted (Sprague, 1930). In cool, humid climatic regions, *P. trivialis* is best used as a monospecies shade lawn grass on moist sites or areas which are commonly wet. This species does not blend well in seed mixtures with other cool-season turfgrasses. Poor mixing qualities may be attributed to a characteristic light green color, inability to perform well in sunny areas, susceptibility to heat and drought stress and a growth pattern that provides a patchwork appearance when mixed with other turfgrass species.

Naturalization of *Poa trivialis*

Based on early recommendations by agronomists, *P. trivialis* has been intentionally seeded in shaded sites for many years. Additionally, millions of kilograms of *P. pratensis*, commonly known as Kentucky bluegrass, was imported from European seed production during the last 60 to 70 years of the twentieth century. Much of this seed carried with it trace amounts up to significant percentages of *P. trivialis* as a crop contaminant. This provided a large seed pool that has dispersed *P. trivialis* across North America.

P. trivialis can be easily found in pastures, along roadsides, in fields, on golf courses, home lawns, parks, around ponds, streams, and waste areas. Although this species is best adapted to damp shaded areas, it is frequently found growing in full sun (Hurley and Funk, 1985).

P. trivialis, with its shallow root system and poor heat tolerance, will remain green and healthy throughout the summer if kept moist with sufficient irrigation or poor drainage. Surprisingly, if adequate soil moisture is provided throughout a hot summer, this species has been known to survive the summer in Florida (with plants persisting after a winter overseeding on bermudagrass putting greens or tees). More typically in cool-season areas, where the summer heat has dried the soil, *P. trivialis* will turn brown and become dormant during the summer. By late fall, *P. trivialis* turf may have completely recovered from summer dormancy.

Recently, *P. trivialis* has become a problem on northern cool-season golf courses, sod farms and home lawns. More specifically, *P. trivialis* has contaminated sod farms and creeping bentgrass or ryegrass fairways, especially on newly constructed golf courses. Along rivers, streams, ponds, lakes, marshes, sinkholes, poorly drained areas, damp shade, or wetlands on a cool-season grass site *P. trivialis* can be found already growing on the property with viable seed in the soil (Hurley, 1982; Hitchcock, 1950; Hanson et al., 1969; Champness and Morris, 1948). Presence of *P. trivialis* on new golf course construction sites or sod farms creates a high probability that soil or sod movement will contaminate other turf areas with *P. trivialis*.

An additional use for *P. trivialis* has been for winter overseeding of dormant warm-season turfs in the southern United States. For this purpose, it is commonly seeded alone or in combination with perennial ryegrass, with mixtures consisting of 10 to 25% *P. trivialis* by weight (Batten et al., 1980). Recently, *P. trivialis* has gained a reputation for providing an excellent putting surface on greens during the cool winter tourist season. Winter overseeding is typically practiced on bermudagrass (*Cynodon spp.*) putting greens in locations from the Carolinas to Florida, and the Gulf states to the southwest. After providing a desirable winter playing surface on greens and tees, rising spring temperatures cause the *P. trivialis* to thin, the bermudagrass emerges from winter dormancy and, by early summer, the turf has transitioned back to 100% bermudagrass.

Seed Sources and Cultivar Development

Prior to the late 1970s most of the *P. trivialis* grown from seed in North America came from Europe. These common types were often tall-growing, light in color, formed loose-growing sod, and were commonly referred to as 'Danish Common,' which was a source of unimproved seed. Seed of *P. trivialis* cultivars, developed for pasture use and imported into the United States, include the cultivars Polis, Ino, and Dasas. The performance of these European-produced seed sources of *P. trivialis* has been poor. Pasture and common types are normally rather tall-growing, light in color, and form a loose-growing sod. These common types have limited value for closely cut (0.6 to 1.8 cm) and highly maintained turf (Hurley, 1982; Hurley and Funk, 1985).

The cultivar Sabre was released in 1977 by the New Jersey Agriculture Experiment Station as the first improved turf-type cultivar of this species. The parental germ plasm of Sabre was selected from old turfs in New Jersey and surrounding states. This cultivar was developed using three cycles of recurrent selection for darker color, greater density, a lower growth habit, and improved turf performance (Dickson et al., 1980). Sabre was quickly accepted for use in winter overseeding of bermudagrass greens and tees, and for use in seed mixtures adapted for use on shaded lawns of cool moist temperate regions.

The development of turf type cultivars that have a lower growth habit, darker green color, ability to form a denser sod, improved disease resistance, and reduced seed shattering would be helpful in expanding the potential use of this species. The range of genetic variability in a species is important in assessing anticipated gains to be expected in breeding improved cultivars (Burton and DeVane, 1953; Baltensperger and Kalton, 1959). Prior to initiating a breeding program, it is useful to evaluate the kinds and amount of variation within a species for traits that are agronomically important.

In the early 1980s an extensive collection of *P. trivialis* was used in a breeding program that led to the development and commercial release of the cultivar Laser in 1998. Plants selected from golf courses, parks, lawns, cemeteries, and roadsides, primarily in New Jersey, Pennsylvania, New York, and California, provided the original germ plasm source (Hurley et al., 1990). Hurley and Funk (1985) found that significant gains may be anticipated in breeding cultivars of *P. trivialis* to provide a denser sod, dark green color, desirable leaf texture, and improved disease resistance. Additive gene action is the fraction of the total genetic variation most easily exploited in the development of synthetic cultivars in cross-pollinated species such as *P. trivialis*. Additive genetic variance appeared to be the most common form of genetic variance found in *P. trivialis* for the traits studied (Hurley, 1982; Hurley and Funk, 1985).

Following clonal evaluation, selected plants were intercrossed using both single cross and polycross mating schemes. Broad sense heritability estimates show significant genetic variability of *P. trivialis* collected from old turfs in the United States and evaluated for turf quality, leaf color, leaf texture, plant density, and susceptibility to dollar spot, powdery mildew, and stripe smut (Hurley, 1982; Hurley and Funk, 1985). Hurley and Funk (1985) reported that germ plasm is available to develop cultivars superior to Sabre in most characteristics of agronomic interest, especially in resistance to dollar spot and powdery mildew.

In the development of the cultivar Laser, progenies used for breeding were subjected to five cycles of recurrent restricted phenotypic selection for lower-growing turf-type growth profile, a bright darker green color, uniform maturity, freedom from disease, upright growth during reproductive development, and improved seed yield. Laser is a leafy, moderately low-growing, turf-type rough bluegrass capable of producing a moderately compact, fine-textured turf of medium-high density. It has a slower rate of vertical growth than common rough bluegrass and a bright, medium-dark green color. It showed the poor heat and drought tolerance characteristics of this species. Research indicated that selection and breeding for types adapted for use in full sun may prove difficult due to the species' shallow and surface rooted nature (Hurley et al., 1990).

Other cultivars of *P. trivialis* that have been recently developed include Laser II, Winterplay, Cypress, Colt, Stardust, and Darkhorse. The cultivars Laser II and Winterplay were developed using the same germ plasm pool used to breed the cultivar Laser. The above-listed cultivars represent a significant improvement in turf performance and darker green leaf color compared to Danish Common or the European cultivars Dasas, Polis, and Ino.

Priorities for future breeding projects may include qualities that may be useful when using *P. trivialis* for winter overseeding dormant bermudagrass in warm-season climates or for use on shaded sites in cool-season climates: reduction in seed shattering that will assist in providing more consistent and economical commercial seed production, darker green leaf color, and improved resistance to dollar spot.

REFERENCES

Ahmed, M.K., G. Jelenkovic, W.K. Dickson, and C.R. Funk 1972. Chromosome morphology of *Poa trivialis* L. *Can. J. Genetics Cytology* 14: 287–291.

Baltensperger, A.A. and R.R. Kalton, 1959. Variability in reed canarygrass *Phalaris arundinacea* L. II. Seed shattering. *Agronomy Journal* 51:37–38.

Batten, S.M., J.B. Beard, D. Johns, and G. Pitman, 1980. Perennial ryegrass-rough bluegrass polystand composition study for winter overseeding, *Texas Turfgrass Research*. pp. 47–49.

Beard, J.B. 1973. *Turfgrass Science and Culture*. Prentice Hall & Co., Englewood Cliffs, NJ. pp. 64–65, 197–199.

Beard, J.B. and A. Almodares, 1980. Minimum temperature requirements for seed germination of turfgrass. *Texas Turfgrass Research. 1978/79*. Texas Agric. Exp. Stn., College Station, TX. pp.13–15.

Budd, E. 1970. Seasonal germination pattern of *Poa trivialis* L. and subsequent plant behavior. *Weeds Res*. 10:243–249.

Burton, G.W. and E.H DeVane. 1953. Estimating heritability in tall fescue (*Festuca arundinacea*) from replicated clonal material. *Agronomy Journal* 45:478–481.

Carroll, J.C. 1943. Effects of drought, temperature, and nitrogen on turfgrasses. *Plant Physiol*. 18:19–36.

Champness, S. and K. Morris, 1948. The population of buried viable seeds in relation to contrasting pasture and soil types. *J. Ecol*. 36:149.

Chippindale, H.G. 1949. Environment and germination in grass seeds. *J. Gr. Grassld Soc*. 4:57–61.

Cooper, J.P. and D.M Calder, 1964. The Inductive Requirements for Flowering of Some Temperate Grasses. Welsh Plant Breeding Station, Aberystwyth, UK.

Dickson, W.K., G.W. Pepin, R.E. Engel, and C.R. Funk, 1980. Registration of Sabre roughstalk bluegrass. *Crop Sci*. 20:668.

Gill, N.T. and K.C. Vear. 1958. *Agriculture Botany*, Buckworth. p. 636.

Grime, J.P. 1980. An ecological approach to management. pp. 13–17. In I.H. Rorison and R. Hunt, Eds. *Amenity Grassland, and Ecological Perspective*. John Wiley & Sons, New York.

Hanson, A.A., F.V. Juska, and G.W. Burton, 1969. Species and varieties. pp. 406–407. In A.A. Hanson and F.V. Juska, Eds. *Turfgrass Science*, No. 14. Amer. Soc. of Agron., Madison, WI.

Hartley, W., 1961. Studies on the origin, evolution and distribution of the Gramineae. IV The Genus Poa. *Aust. J. Bot*. 9:151–161.

Hitchcock, A.S. 1950. *Manual of the Grasses of the United States*. Vol. 1, Dover Pub., NY. p. 117.

Hubbard, C.E. 1954. *Grasses*. Penguin Books, Inc. Baltimore, MD. p. 428.

Hurley, R.H. 1982. Rough Bluegrass: Genetic Variability, Disease Susceptibility, and Response to Shade. Rutgers University Ph.D. Thesis.

Hurley, R.H. and C.R. Funk, 1985. Genetic variability in disease reaction, turf quality, leaf color, leaf texture, plant density and seed shattering of selected genotypes of *Poa trivialis*. pp. 221–226. In F. Lemaire, Ed. *Proc. 5th Int. Turfgrass Res. Conf., Avigonon, France, July 1–5 1985*. INRA Publ., Versailles, France.

Hurley R.H., M.E. Pompei,. M.B. Clark-Ruh, R Bara, K.W. Dickson, and C.R Funk,. Registration of 'Laser' rough bluegrass. 1990. *Crop Sci*. 30:1357–1358.

Milton, W.E.J. 1936. The buried viable sees of enclosed and unenclosed hill land. pp. 58–84. *Rep. Welsh Pl. Breed Stn*., Series H, No. 14.

Milton, W.E.J. and R.O. Davies, 1947. The yield, botanical and chemical composition of natural hill herbage under manuring, controlled grazing, and hay conditions. *J. Ecol*. 35:65–95.

Shearman, R.C. and J.B Beard. 1975. Turfgrass wear tolerance mechanisms. I. Wear tolerance of seven turfgrass species and quantitative methods for determining turfgrass wear injury. *Agron. J.* 67:215–19.

Sprague, H.B. 1930. Lawn Grass Mixtures for New Jersey Agriculture.

Stuckey, I.H. 1941. Seasonal Growth of Grassroots. *Am. J. Bot.* 28:486–491.

Vartha, E.W. 1969. Aspects of the Agronomy and Ecology of *Poa trivialis* in Pastures.

Welton, F.A. and J.C. Carroll. 1940. Lawn experiments. *Ohio Agric. Exp. Stn. Bull. 613.* pp. 1–43.

Chapter 7

Perennial Ryegrass (*Lolium perenne* L.)

Daniel Thorogood

Perennial ryegrass (*Lolium perenne* L.) is found and used for forage and turf purposes throughout the temperate world but most extensively in the United States and Europe, with significant use, mainly for forage, in Japan, Australia, and New Zealand. Turf-type perennial ryegrass is used in temperate Europe, where it is naturalized, on winter-games pitches and also for heavy-duty lawns, landscaping, tennis courts, cricket fields, golf tees and fairways. It is sometimes used on its own but often in mixture with red fescues (*Festuca rubra* L.) and bentgrasses (*Agrostis* spp.). In North America it is usually mixed with Kentucky bluegrass (*Poa pratensis* L.) for lawns and sports pitches. It is also used at lower latitudes to overseed winter dormant warm-season turfgrass areas. In New Zealand and Australia it is used extensively as both a permanent turf for sports pitches and racecourses and for winter overseeding of warm-season turf areas.

Table 7.1 shows comparisons of perennial ryegrass markets throughout the world. Nearly 90% of seed is used in the United States or Europe. Europe uses just over 75% of the total forage perennial ryegrass with significant quantities being used in Australasia. The turf perennial ryegrass market pattern is different. Nearly all perennial ryegrass for turf is used in the United States and Europe with the United States using 64% of total ryegrass produced. Incredibly, the vast majority of this seed is grown in the Willamette Valley, Oregon, with less than 1% being imported into the United States, mainly from Canada, France, and the Netherlands.

ADAPTATION AND MORPHOLOGY

Climatic and Edaphic Adaptation

Perennial ryegrass is best adapted to cool, moist climates where winterkill is not a problem. Perennial ryegrass grows best on fertile, well-drained soils but has a wide range of soil adaptability. It will tolerate extended periods of flooding (up to 25 days) when temperatures are below 80°F (27°C). Minimum annual rainfall requirement is 18 to 25 inches (457 to 635 mm). Perennial ryegrass tolerates both acidic and alkaline soils, with a pH range of 5.1 to 8.4, with an optimum pH around 6.5 (Beard, 1973). During hot summers, perennial ryegrass becomes dormant and it will not tolerate

Table 7.1. World perennial ryegrass markets, 1999 (figures in tonnes).

	Turf	Percent of Total	Forage	Percent of Total	Turf + Forage	Percent of Total
Europe	21,755	33.6	41,009	75.3	62,798	52.7
USA	41,400	64.0	220	0.4	41,684	35.0
Canada	1,000	1.5	100	0.2	1,102	0.9
Japan	0	0.0	4,250	7.8	4,250	3.6
New Zealand	0	0.0	5,000	9.2	5,000	4.2
Australia	300	0.5	2,500	4.6	2,800	2.3
S. America	250	0.4	1,200	2.2	1,450	1.2
S. Africa	0	0.0	200	0.4	200	0.2
Total	64,705		54,479		119,184	

Source: Mr. Neville Bark, Germinal Holdings Ltd, Banbridge, County Down, N. Ireland, U.K.

climatic extremes of cold, heat, or drought (van Dersal, 1936). Optimum growth occurs between 68 and 77°F (20 to 25°C). (Beard, 1973).

Perennial ryegrass is more sensitive to temperature extremes and drought than annual ryegrass since forage production suffers when daytime temperatures exceed 87°F (31°C) and nighttime temperatures exceed 77°F (25°C) regardless of moisture availability.

Turf Management

Ryegrass produces a reasonably high shoot density at an optimum cutting height of around two centimeters. Nitrogen requirement for optimum growth is about 50 kg ha^{-1} per month. Some modern cultivars produce very good ground cover and shoot density at cutting heights as low as eight millimeters. Ryegrass is the preferred grass for cricket pitches, which are clay-based constructions that are mown at even lower heights and then rolled with a heavy roller to give the desired characteristics of ball speed and bounce. Under these conditions, very little green leaf remains and the root and crown properties of the ryegrass plant are important for producing an appropriate surface. Mowing quality of ryegrass is generally poor because of the tough fibrous vascular bundles in the leaves, although modern cultivars have improved mowing characteristics.

Ryegrass is used on home lawns, parks, fairways, roughs, roadsides, and general landscaping areas. Its major attributes include its speed of germination and establishment, making it a good choice for soil stabilization projects. These characteristics along with its good wear tolerance mean that it is used for creating, reseeding and overseeding sports pitches in Northern Europe where a long playing season precludes the use of slower establishing species.

Most ryegrass cultivars grown for turf in the United States and Australia and New Zealand are deliberately infected with the endophytic fungus *Neotyphodium lolii* Latch. This is a seedborne fungus, often present at high levels in natural ryegrass populations, that is transmitted almost completely effectively through seed generations. It remains viable within the seed during extended seed storage if cool, low humidity conditions are maintained. The alkaloids within the fungus act as a deterrent for a range of shoot-feeding insect pests such as the Argentine stem weevil (*Listronotus bonariensis* Kuschel), sod webworms (*Crambus* spp.), billbug (*Spenophorus* spp.), chinch bug (*Blissus* spp.), a number of aphid species, and fall armyworms (*Spodoptera frugiperda* J.E. Smith) (see review of Fraser and Breen, 1994). This makes the turf much more persistent where insect pests are a problem.

Morphology (see Hubbard, 1984)

Lolium perenne L is a low- to high-tillered tufted perennial grass that grows between 10 and 90 cm high. Stems can be erect or prostrate, are 2–4 noded and smooth. Leaves are mid-green, hairless, sheaths being mid-green with sometimes pinkish-red bases and are folded rather than rolled in the shoot. Ligules are membranous, about 2 mm long. Leaf blades are smooth and waxy on the abaxial side but rougher and ridged on the adaxial side. Leaf tips are blunt or pointed and two small claw-like auricles are located at the blade base.

Inflorescences (Figure 7.1) are single green to reddish spikes consisting of 10–30 stalkless spikelets that alternate on opposite sides of the central rachis. The edge of each spikelet fits into a flat-edged hollow in the rachis, and spikelets consist of 4–14 alternating florets, enclosed by an upper glume that is shorter than the spikelet length. The lower glume is only present in the terminal spikelet. The flower is enclosed by the palea and an awnless lemma that fuses with the seed as it develops to form a caryopsis.

DISTRIBUTION, CYTOTAXONOMY, AND BREEDING SYSTEM

Distribution

The *Lolium* genus is indigenous to Europe, temperate Asia, and North Africa (Terrell, 1968) but the distribution and evolution of the genus has been complicated by the involvement of man, who has encouraged its spread as a crop (*L. perenne* L. and *L. multiflorum* Lam.) and also as a weed in primitive agriculture (*L. remotum* L. Schrank., associated with flax, *Linum usitatissimum* L. and *L. temulentum* L., with wheat, *Triticum aestivum* L.) (Jenkin, 1936). *Lolium perenne* has been introduced throughout the world and is cultivated as a forage and lawn grass on every continent except Antarctica. In Britain it is recorded as being deliberately sown for pasture as early as the seventeenth century (Plot, 1677).

Taxonomy

The taxonomic classification of perennial ryegrass is as follows:

Family	Gramineae (Poaceae)
Subfamily	Pooideae
Tribe	Poeae
Genus	*Lolium*
Species	*perenne*
Authority	Linnaeus

Within the *Lolium* genus Essad (1954) recognized five *Lolium* species, which he separated into an allogamous (self-incompatible) group including *L. perenne* L., *L. multiflorum* Lam., and *L. rigidum* Gaud. and an autogamous (self-compatible) group including *L. temulentum* L. and *L. remotum* Schrank. Terrell (1968) reexamined the groupings of the *Lolium* genus based on morphologies of live plant material and herbaria specimens of European and North American origin. He recognized the same groups but added two other derivative species to the former, namely *L. subulatum* Vis. and *L. canariense* Steud., the latter being relatively recently isolated on the Canary Islands. He also classified *L. persicum* Boiss & Hohen as a separate species which, restricted to Southwest Asia, could well be, at least in part, a prototype stock from which *L. temulentum* and *L. remotum* are derived.

Figure 7.1. Morphological features of the perennial ryegrass inflorescence. Drawing courtesy of Jenny Smith, Aberystwyth, Wales, U.K.

Research over the last 15 years using multivariate methods for describing species relationships based on morphological, protein, and DNA markers has served primarily to confirm these earlier workers' taxonomic arrangements placing the allogamous and autogamous species into distinct groups (Bulinska-Radomska and Lester, 1985; Loos, 1993a,b; Charmet and Balfourier, 1994; Stammers et al., 1995; Charmet et al., 1996; Bennett, 1997; Charmet et al., 1997; Siffelova et al., 1997, Bennett et al., 2000). More specific studies of ribosomal DNA (rDNA) profiles have also been made in *Lolium* (Warpeha et al., 1998b), which showed clear differentiation of fragment sizes between the autogamous and allogamous species. Thomas et al. (1996b) used rDNA clones from wheat and hybridized them onto mitotic chromosome preparations. The chromosome positions of the rDNA probes were easily visualized by appropriate fluorescent staining, since the rDNA genes are large repetitive sequences. In this study, the autogamous species hybridized in the same six chromosome positions. More copies of the 18S-5.8S-26S clone were found in the outbreeders, with *L. rigidum* var. *rigidum* Gaud. having the most. Two copies of the 5S clone were always found on chromosome 2 in the inbreeders and on chromosome 3 in *L. perenne* and *L. multiflorum*. In *L. rigidum* var. *rigidum*, they could be in either of the two positions. The identification of such gene rearrangements can be used to infer species relationships and, in this case, confirms the separation of the inbreeders and outbreeders. The greater variability of the *L. rigidum* accession over *L. perenne* and *L. multiflorum* also suggested that the former is a more ancient species and the isozyme diversity study of Balfourier et al. (1998) confirms this observation.

Cytogenetics

Lolium perenne is diploid (2n = 14) with normal disomic inheritance (Evans, 1926) in common with all other *Lolium* species that are more or less interfertile (Jenkin, 1933;

1935; 1953; 1954a,b; Jenkin and Thomas, 1939; Charmet et al., 1996. Colchicine-induced autotetraploids have been developed for forage purposes.

Lolium perenne also shows degrees of fertility with species of the genus *Festuca* and successful hybridization has occurred with further genera. Natural hybrids in the wild have been formed between *L. perenne* and the *Bovinae* section of *Festuceae*, *F. pratensis* Huds., *F. arundinacea* and *F. gigantea* L. (Ascherson and Groebner, 1902) and such hybrids were also produced in controlled pair-crosses by Jenkin (1933, 1955c). Hybrids between *L. perenne* and *F. pratensis* showed complete pairing between homoeologous chromosomes, indicating genetic information exchange between the species (Peto, 1933) and multivalent formation between chromosomes of *Lolium* and *F. arundinacea* also proved genetic exchange between these two species (Peto, 1933). However, recombination between the species does not occur at random. In *F. pratensis* x *L. multiflorum* hybrids backcrossed to *L. multiflorum*, gametes containing isozyme marker alleles from *L. multiflorum* were favored over those from *F. pratensis* (Humphreys and Thorogood, 1993). Humphreys (1989) used isozyme polymorphism at the PGI/2 locus to differentiate between the three genomes of *F. arundinacea* and the *L. multiflorum* genome, and demonstrated that alleles from all three genomes could be introgressed into *L. multiflorum* by backcrossing. In this case, the isozyme allele from one of the *F. arundinacea* genomes recombined with *L. multiflorum* at a lower rate than from the other two.

The *Lolium* genus was thought to belong to the *Triticeae* because of its spicate inflorescence but Hubbard (1948) declared that the *Festuca* and *Lolium* genera were closely related because they are able to hybridize, they have compound starch grains in their seeds, and their spikelet structures are similar. Stebbins (1956) confirmed the close relationship between *F. pratensis* and *L. perenne*, noting that variation between the caryopses of the two species was no greater than that found between species of the same genus.

The phylogenetic position of *Lolium* with *Festuca* section *Bovinae* has been confirmed by comparison of species DNA patterns. Chloroplast DNA is often used in determining phylogenetic relationships between species and genera because it is a highly conserved genome with few internal rearrangements (inversions and transpositions) and has a low rate of mutation at the nucleotide level (Palmer, 1987). Restriction Fragment Length Polymorphism (RFLP) analysis of chloroplast DNA was made by Lehväslaiho et al. (1987) which indicated a close relationship between *F. pratensis* and *L. multiflorum*, with both species related to *F. arundinacea*. *F. rubra* L., however, was found to be relatively distantly related and the *Festuca* genus was concluded to be an artificial grouping. Subsequent studies have confirmed the close relationship of *Lolium* with the *Bovinae* section of *Festuca* using protein electrophoresis (Bulinska-Radomska and Lester, 1988; Aiken et al., 1998), and nuclear DNA profiles; RFLPs (Xu and Sleper, 1994); RAPDs (Stammers et al., 1995; Siffelova et al., 1997); and a combination of RFLPs and RAPDs (Charmet et al., 1997).

Hybrids have been reported between *Lolium* and members of the Ovinae section of the Festuceae, *F. ovina* L. and *F. rubra* L. (2n = 42) (Jenkin, 1933), *F. capillata* Lam (syn. *F. tenuifolia* Sibth.) (Jenkin, 1955a) and *F. heterophylla* Lam. (Jenkin, 1955b). Jenkin (1955c) also reported hybrids between *L. perenne* and *Glyceria fluitans* L. The widest successful cross reported involving *Lolium* is that between *L. multiflorum* and orchardgrass (*Dactylis glomerata* L.) (Oertel et al., 1996).

Mode of Reproduction

Perennial ryegrass is a self-incompatible wind-pollinated species. Although it can be maintained vegetatively in a sward for many years, its main mode of reproduction is by seed. Short days and cool (< 7°C) temperatures are required for flowering to occur, followed by long days of 12 hours or more. The flowers are typical of the *Gramineae*, each floret possessing a single ovule, from the top of which protrude two feathery stigmas. Three stamens emerge from near the base of the ovule. These floral parts are enclosed by modified leaf structures, the palea and lemma. Once the florets become mature, structures at the base of the floret called lodicules swell with cell sap and force open the lemma and palea. The anthers of the stamens are exserted, extended on long versatile filaments. At the same time, the feathery stigmas project on either side of the floret ready to receive pollen. The anthers split lengthwise from the tip to release clouds of pollen, which is windblown onto the stigmas to effect pollination. Anthesis occurs once daily around midday and is more profuse on warm, bright days. The basal, older florets of the midspike spikelets flower first and flowering linearly progresses toward the outermost floret and basal and apical spikelets.

Perennial ryegrass has a two-locus (SZ) multiallelic gametophytic incompatibility system (Cornish et al., 1980) in common with the grasses, *Secale cereale* L. (Lundqvist, 1956) and *Phalaris coerulescens* L. (Hayman, 1956). This system prevents self seed setting and subsequent inbreeding depression (Jenkin, 1931a; Bean and Yok-Hwa, 1972). Incompatible pollen grains are arrested as the pollen germ-tube makes contact with the stigma surface. Tube growth is then arrested and callose deposits build up in the pollen grain and the tube cell wall (Heslop-Harrison, 1982). The S and Z incompatibility loci act in a complementary manner such that a pollen grain is rejected if it possesses both an S and Z allele in common with the female parent. The pollen gametophyte alone determines the incompatibility reaction, which means that the reaction is localized and, in pairwise crosses, a mixture of compatible and incompatible pollen may occur on the stigma. Up to four different classes of pollination can be expected: fully compatible, if no SZ allele pairs are matched in both parents; three-quarters compatible if three alleles are matched; half-compatible if one SZ allele pair is matched, and fully incompatible if all four alleles are matched (Figure 7.2).

The S and Z loci have recently been mapped in perennial ryegrass to linkage groups 1 and 2 as classified according to the Triticeae consensus map. (Thorogood et al., 2002). They are highly polymorphic and 17 different S and 17 different Z alleles were found in a natural ryegrass population of 39 plants collected from an old permanent pasture (Fearon et al., 1994). It is therefore unlikely that seed production will be reduced in populations with a broad genetic base. However, breeders should be aware that any narrowing of the genetic base by inbreeding (for example by half-sib family selection) may result in a restricted population of S and Z alleles which may result in a degree of within-population incompatibility. Selection for traits controlled by genes linked to both S and Z could have similar effects.

Despite the presence of an effective self-incompatibility system, perennial ryegrass will set seed when selfed. Jenkin (1931b) found seed set as high as 32.3% when obligately selfed and seed setting was genotype-dependent. In general, the number of seeds set on selfing plants is considerably lower than in crosses. Beddows et al. (1962) found self-seed setting to be 3.0% of that obtained by crossing unrelated plants in similar environmental conditions. Foster and Wright (1970) found a similarly low figure (2.2%), although one genotype set as much as 31.0% of that obtained on crossing. Although self-fertility is under polygenic control, a single gene, completely independent of the S and Z loci, producing 100% self-fertile plants, was identified in one inbred line (Thorogood and Hayward, 1991). In a self-fertile relative, *L. temulentum* L., self-fertility was found to be controlled by a mutation of one of the incompatibility loci that was transferred into *L. perenne* and *L. multiflorum* populations by backcrossing (Thorogood and Hayward, 1992). Hayman and Richter (1992) in *Phalaris coerulescens* Desf. and

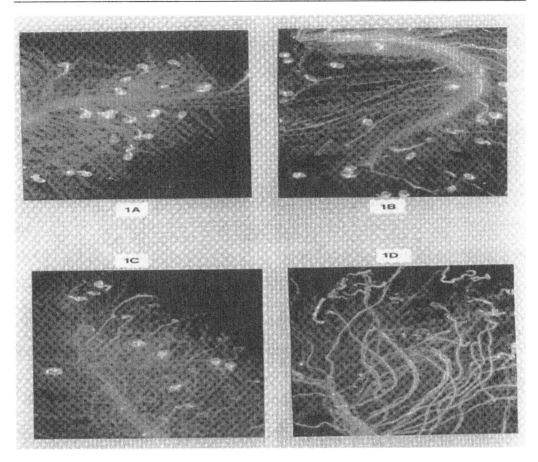

Figure 7.2. Four classes of pollen-stigma response viewed under UV light after staining with aniline blue (×300). 1A = incompatible, 1B = half-compatible, 1C = three-quarters compatible, 1D = fully compatible. (Cornish, 1979).

Egorova and Voylokov (1998) in rye found several self-fertility mutations. In the latter case, about half of these were at a separate locus, designated T. The wide occurrence of self-fertility may explain why Spoor (1976) in his *Lolium* pollinations, which expressed a reasonable degree of self-compatibility, was unable to establish a two-locus system of control.

COLLECTION, SELECTION, AND BREEDING HISTORY

Sources of Germ Plasm

Historically, the grass cultivars traditionally used for turf were developed for agricultural production. 'Aberystwyth S.23' was one of the first ryegrass cultivars developed at the Welsh Plant Breeding Station for use in reseeding the natural fescue/bent upland sheep pastures of mid-Wales and was released for sale in 1933 (Jenkin, 1943). The cultivar was used extensively in sports areas because of its leafy growth, relatively high shoot density, and late-flowering nature. The cultivar was based on a wide base of germ plasm collected from highly productive lowland pastures in England.

A number of cultivars were developed in the United States during the 1960s including 'Linn' (1961), 'NK100' (1962), 'Manhattan' (1967), and 'Pennfine' (1969). At the same time, European cultivars such as 'Pelo' (1962) were released (Beard, 1973). All of these cultivars were based on collections made from natural populations found in old pastures, turfs, and recreational areas. Current breeding efforts rely on existing elite germ plasm that can be combined with native ecotype material collected for specific traits evolved for adaptation to specific sites.

Genetic Diversity/Erosion

Clearly, from both the purely ecological viewpoint and as a basic requirement for cultivar improvement, the genetic diversity of the species must be maintained in current cultivars or in natural populations. Several components of genetic diversity need to be considered, but there is little information on these factors for perennial ryegrass.

Reduction in diversity of new turfgrass cultivars is perhaps inevitable because of the necessity to comply with cultivar registration requirements that demand cultivar uniformity. However, the outcrossing nature of ryegrass and the fact that virtually all cultivars of ryegrass are based on several mother plants that are intercrossed to produce a synthetic cultivar, indicates that genetic diversity is maintained to a certain extent. Perennial ryegrass cultivars are genetically diverse and show considerable polymorphism for AFLP, RAPD, and protein markers, and morphological traits (Köllicker et al., 1999; Roldan-Ruiz et al., 2000). Warpeha et al. (1998a) also found considerable variation for ribosomal DNA and also showed that variability among turfgrass cultivars was higher than for forage. There are two conflicting reports in the literature comparing variation of native populations of ryegrass with cultivars. Sackville Hamilton (1999) analyzed the variation in leaf width of native ryegrass populations held in the UK-based European Core Collection of ryegrasses. Twenty plants from populations of 18 European countries were analyzed and compared with the results obtained for forage cultivars. He found that the pooled within-population variance for each country was always higher than that found in the cultivars and in some cases was more than doubled. However, Casler (1995), who studied accessions from the National Genetic Resources Program perennial ryegrass collection, found that genotypic and phenotypic variance estimates for native populations and cultivars were similar for leaf width, seedling vigor, crown rust (*Puccinia coronata* Corda) incidence, and forage yield.

Another related concern is that the majority of turf ryegrass seed production takes place in a restricted area in Oregon and genetic diversity could be compromised through interpollination of the different ryegrass crops grown there.

Genetic diversity may also be compromised by the fact that old cultivars are being rapidly replaced by new ones. The number of new cultivars being developed is increasing (Table 7.2) but there is no evidence that genetic diversity is also increasing. Twenty-three new ryegrass cultivars have been registered and published in *Crop Science* in the United States since 1990. Although these have been developed and released by different seed companies, they all originate from the New Jersey Agriculture Experiment Station turf breeding program. The cultivars are mostly derived from recently produced elite germ plasm and collections from old turfs in the mid-Atlantic states. A small amount of elite European germ plasm (ex 'Loretta' and 'Elka') and two further European collections (ex Finland and Greece) were included in three cultivars. Effectively there has been only one breeding program in the United States based on a small subsample of the potential ryegrass diversity available worldwide.

The overwhelming worldwide success of the Oregon ryegrass seed industry based on a fairly narrow genetic base suggests a potential for genetic erosion of ryegrass genetic diversity. However, less than 8% of this seed is exported (Young, 1997), so the poten-

Table 7.2. Number of perennial ryegrass cultivars listed by the Sports Turf Research Institute (STRI) in the U.K. and tested by the National Turfgrass Evaluation Program (NTEP) in North America between 1996 and 2001.

Year	Testing Authority	
	STRI	NTEP
1986	27	—
1990	—	90
1991	39	—
1996	64	96
1999	—	134
2001	83	—

tial for genetic erosion of native perennial ryegrass communities in Europe, temperate Asia, and North Africa, from this source, is currently very low.

Whereas Monestiez et al. (1994) found a strong association between geographical distance and genetic distance in natural ryegrass populations in France, Warren et al. (1998) found no such relationship when comparing isozyme patterns of native populations distributed throughout the United Kingdom. This strongly suggested that in the latter case, the genetic structure of wild ryegrassess has been disrupted by artificial introduction of improved cultivars. The genetic variation in native populations may be at risk by pastures being resown with new forage cultivars that may be less diverse. Although this trend is difficult to quantify, grassland improvement programs in western Europe have been extensive, and are now being implemented in eastern Europe. Of far greater risk is the complete and irreversible loss of permanent pasture through urbanization, deforestation, and conversion to arable farming.

Many old cultivars are no longer available and may represent a further loss of useful genetic variation to the plant breeder. *Ex-situ* conservation of genetic diversity addresses the problem of genetic erosion and there are several *Lolium perenne* gene banks worldwide. The United States National Genetic Resources Program (NGRP) currently holds 590 *L. perenne* accessions that include known cultivars and native populations from the old world. The Genetic Resources Unit at the Institute of Grassland and Environmental Research, Aberystwyth, in the United Kingdom holds the most extensive collection. This research group holds the collections for all *Lolium* species and white clover, *Trifolium repens* L. and is responsible for the European Central Crop Databases for these two species. Currently 1,879 accessions of *L. perenne* are available that have been accrued from seed exchange with other research institutes and botanic gardens (26), known cultivars (463), breeders' lines (49), old land races (33), but mainly from collections made throughout Europe by the Unit's staff (1,308). The almost random collection of populations is now being replaced by a more methodical approach. Critical core collections are being conserved with appropriate selected populations being identified that account for most of the genetic diversity available. Charmet and Balfourier (1995) found that by using geographical clustering information they could maintain 92% of the variation present in their French collection in a 5% core collection.

For *ex-situ* collections to be effective, appropriate collecting and maintenance is necessary. Sackville Hamilton (1999) gives an example of genetic erosion caused by collecting ryegrass seed as compared to vegetative units. It would be expected that a seed collection would be heavily biased toward those plants setting seed at the time of collection. He found that the mean within-population variance from seed collections was significantly lower than that of vegetative-unit collections and was similar to the variance found in cultivars deliberately selected for uniformity. Genetic erosion may also occur at any of the stages of preparation, subsampling, exchange, storage, and regeneration of seed because of differences in seed viability and fecundity of parent plants (Sackville Hamilton, 1998).

Breeding Techniques

All commercial turf ryegrass cultivars have been bred as synthetic cultivars produced by the interpollination, usually in isolation, of several selected plants. The method (reviewed by Corkill, 1956) is ideal for a self-incompatible, allogamous, wind-pollinated species, since seed is readily obtained and the amount of inbreeding depression encountered due to selfing is very low. The high degree of heterozygosity of individuals and heterogeneity within populations means that recessive lethals and semilethals are rarely encountered using the method. Simple phenotypic selection of individual plants for appropriate characteristics is probably the most common method of breeding perennial ryegrass. However, more efficient breeding methods are used by which progeny of parental clones are tested by either a top-cross to a standard cultivar or by a poly-cross between individuals. Performance or general combining ability is then assessed in the progeny as small replicated family plots or spaced plant rows. Parents whose progeny have the best general combining ability are then selected to be used as parents in a new synthetic. Several cycles of selection can be made to obtain yet further improvements in additive traits. Figure 7.3 shows a generalized example of a typical poly-cross breeding cycle used in the Institute of Grassland and Environmental Research (IGER) breeding program. The number of plants used to make a synthetic can vary considerably from population to population. Too few may cause inbreeding depression; too many may result in a highly heterogeneous population that would segregate for many characters in subsequent generations of seed multiplication. Although Corkill (1956) found that six plants was the optimum number to obtain, and retain in subsequent generations, maximum heterosis for yield, this number would be dependent on the heterozygosity of the parents and the heterogeneity of the population for any desired character.

A number of attempts have been made to justify F1 hybrid production in perennial ryegrass in order to exploit heterosis fully in forage breeding programs. If heterosis for turfgrass traits such as shoot density or wear tolerance could be found, these methods could be applied to the production of turfgrass varieties. F1 hybrid production requires either mass emasculation of the female parent in the cross or the use of a male sterility system. The former is not feasible in ryegrass because of the flower structure. Male sterile ryegrass plants have been identified by many authors and have also been genetically characterized (Wit, 1974; Gaue, 1977; Connolly and Wright-Turner, 1984). Male sterility is genetically controlled by cytoplasmic genes with modifier genes present in the nucleus. Maintainer lines have been identified that would serve as male parents for the multiplication of the male sterile lines. So far, no commercial examples of F1 hybrid cultivars have been documented. The success of the much simpler method of producing heterotic combinations through synthetic production, and the fact that seed yields would be lower when only the seed from the male-sterile parent were to be harvested, would seem to mitigate against using a complex F1 hybrid system.

England (1974) devised a scheme exploiting the self-incompatibility system of ryegrass to increase the proportion of hybrid progeny when intercrossing two parental popula-

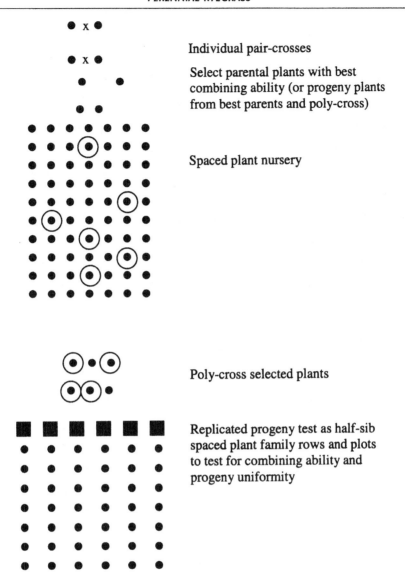

Figure 7.3. One cycle of selection using the poly-cross method to produce a synthetic turf ryegrass cultivar.

tions of plants. His scheme depends on the ability to produce self-seed from an initial parent plant. Ryegrass plants will often set a small amount of seed and this can be improved by heat treatment (Wilkins and Thorogood, 1992). The progeny from the self would then be interpollinated to produce a population in which all of the double homozygous plants ($S_{11}Z_{11}$ and $S_{22}Z_{22}$) would be eliminated. The remaining genotypes would be largely intra-incompatible. By intercrossing this population with another similarly produced population with good specific combining ability, an 83% hybrid population exhibiting heterosis would be achieved. This has the advantage of not having to go to the expense of producing male sterile lines and it also means that seed could be harvested from both populations and therefore potential seed yields would not be reduced compared with those produced from a synthetic.

Further Tools Available to the Breeder

Genetic Markers

The rediscovery of Gregor Mendel's plant hybridization experiments by William Bateson at the start of the twentieth century (Olby, 1987) heralded an explosion of knowledge of the nature of genes and how they interact with each other and the environment to give rise to the characteristics of living organisms. Vast research effort has been expended on determining exactly where the genes are located on the chromosomes or in the cytoplasm in relation to each other in a range of crop species. This was first done for genes encoding for distinct, simply inherited morphological traits and the information gained could be used to determine linkage relationships between the limited set of genes to build up rudimentary genetic maps. In *Lolium*, a number of traits controlled by a single gene have been identified and in some cases linkage relationships determined. T.J. Jenkin identified a number of such morphological variants; brittle (recessive) vs. normal stem (Jenkin, 1955d); two lethal seedling mutants (Ll and LlII) (Jenkin, 1928a); another type of chlorophyll deficient seedling (Yy) (Jenkin, 1928b); two complementary genes (C and R) in which the double recessive lacks anthocyanin at all stages of growth (Jenkin, 1930). Further work by Jenkin (1954c) has shown that Yy and Rr are linked with each other and a further dwarf mutant (Dd). Gentner (1929) discovered that the roots of *L. multiflorum* seedlings fluoresce under UV light. In segregating crosses between *L. multiflorum* and *L. perenne*, this trait was found to be dominant, controlled by a single gene (Corkill, 1933) Other genes that have been characterized are the S and Z loci, the former also being found to be linked to the isozyme locus PGI/2 (Cornish et al., 1980). More recently, Thorogood et al. (in press) mapped the S and Z loci to linkage groups 1 and 2 in accordance with the Triticeae map that also showed direct synteny with the S and Z loci of rye. They also identified a third locus (or loci) on LG3 that interacts with the S locus (or a tightly linked locus) causing disturbed ratios of marker loci on LG3. Thorogood and Hayward (1991) identified a self-compatibility locus in an inbred line of *L. perenne* that was a mutation of an independent locus and Thorogood and Hayward (1992) identified a mutation of one of the incompatibility loci in *L. perenne* and *L. multiflorum* backcross progeny of hybrids between *L. perenne/L. multiflorum* and the self-fertile *L. temulentum*, now thought to be the S locus (Thorogood et al., in press).

Apparently neutral genetic markers can be visualized as isoenzymes. These are allelic forms of the same enzyme, encoded for by the same gene, which can be separated and visualized, using appropriate staining techniques, as allelic bands on a gel subjected to an electrical charge. Enough polymorphism was found by Hayward and McAdam (1977) at five isozyme loci to distinguish a number of commercial ryegrass cultivars. Isozyme markers have the advantage over morphological markers in that they are genetically neutral in their effects, are codominant, and can quite often be highly polymorphic in a range of different populations, whereas recessive alleles that determine easily distinguishable morphological traits tend to be detrimental and occur at relatively low frequencies.

The advent of DNA technology has resulted in a number of easily screened polymorphic markers. Restriction Fragment Length Polymorphisms, RFLPs (see Hayward et al.,

1998) Randomly Amplified Polymorphic DNA, RAPDs (see Stammers et al., 1995), Amplified Fragment Length Polymorphisms, AFLPs (Bert et al., 1999), Simple Sequence Repeats, SSRs (Jones et al., 2001a), and gene sequences or part-sequences themselves called Expressed Sequence Tags, ESTs (Bert et al., 1999) have all been found to be highly polymorphic in *L. perenne* and therefore are potentially useful markers in segregating populations.

If easily screened markers are associated with traits that plant breeders are interested in, opportunities to supplement, and maybe reduce the reliance on, long-term field selection trials by indirectly selecting for the marker instead become available. The association of markers with some traits of agronomic interest in *Lolium* has only been determined in the last 15 years or so. Hayward and McAdam (1988) found that selecting for a particular allele at the *PGI/2* locus in some populations affected spaced plant yield. Furthermore, different alleles were associated with the higher yield in different populations, suggesting that the locus is linked to a quantitative trait locus (QTL) influencing herbage yield rather than the isozyme locus itself having a direct effect. Associations between PGI/2 and herbage water-soluble carbohydrate and ACP/1 and date of ear emergence are reported by Humphreys (1992).

Although associations between simply inherited markers and agronomically important traits are useful, selection for complex traits will require many polymorphic markers covering the whole genome so that any QTL having a significant effect on the trait of interest can be identified and selected for indirectly on the basis of linked marker genes. Such complete maps can also be used to identify the number, mode of action, and interaction of QTL involved in determining the expression of a trait, giving the breeder useful information to make marker-assisted selection decisions.

The first published *Lolium perenne* map was based on 101 segregating RFLP, RAPD, and isozyme markers (Hayward et al., 1994) in a population derived from a pair-cross between a heterozygous *L. perenne* plant crossed onto a doubled haploid (homozygous) *L. perenne* plant. This family only consisted of 89 plants and large errors could be associated with estimates of linkage, possibly explaining the large number of linkage groups (13) found, some of which gave linkage distances well in excess of 50cM. Despite these shortcomings, the principle of marker-assisted selection was established for a number of traits, mainly associated with flowering response, which could be associated with certain linkage groups. Of particular note, three regions were found to account for 40% of the phenotypic variation for ear emergence date and one of these regions was also associated with inflorescence production in the year of sowing. Selection for *L. perenne* derived markers associated with this trait would have the effect of virtually eliminating inflorescence production in the establishment year. This map was refined (Hayward et al., 1998) by the addition of more RFLP and RAPD markers and the use of the linkage calculation program, JOINMAPÔ (Stam and van Ooijen, 1995). This gave a linkage map consisting of seven linkage groups, agreeing with the haploid chromosome number of *Lolium*, based on 17 isozyme markers, 48 RAPDs and 41 RFLPs.

More recently, Bert et al. (1999) produced a marker map based on the same family as Hayward et al. (1994, 1998) using a large number of AFLPs from 17 primer pairs, three isozymes, and 5 EST markers. Extensive coverage was achieved over seven linkage groups but an uneven distribution of markers was found with a number of significant gaps in five of the linkage groups. There was also strong clustering of AFLP markers, which may be caused by suppression of recombination in the centromeric regions of the chromosomes.

The mapping population described above, known as P150/112, has been used to provide a framework of markers and is currently being used to construct a definitive map consisting of RFLP, AFLP, isozymes, ESTs, and most recently, SSRs (Jones et al., 2001a). The number of plants in the framework map has been increased to around 150 plants, allowing linkage distances between markers to be more accurately determined. SSRs are extremely valuable especially for trait mapping and for doing marker-assisted selection in breeding programs because they are codominant PCR-based markers, are highly polymorphic, making them applicable to a wide range of populations. They are also known to be abundant, occurring every 21–65kb in plant genomes (Morgante and Olivieri, 1993; Wang et al., 1994). Although a limited number of SSRs were identified by Kubik et al. (1999), Jones et al. (2001a) identified nearly 60 out of a possible 100 SSR amplification products that were polymorphic in the P150/112 family. Many of the products were also polymorphic in a range of grass species and as many as seven allelic forms of some SSRs were found.

A number of research groups throughout the world have access to the P150/112 mapping family through the International *Lolium* Genome Initiative, ILGI (Forster, 1999), which was established to coordinate the development of molecular markers and their exploitation. The map aligns remarkably well to those of other members of the *Triticeae* (Jones et al., 2001b), which means that biochemical, physiological, agronomic, and genetic information from the major Graminaceous crop species can be exploited in *Lolium*. Several trait-specific families are being developed that can be broadly mapped using a subset of markers from the ILGI mapping family. Of particular note is an F2 family that has been characterized for water-soluble carbohydrate content, which is an important digestibility criterion for forage grasses (Turner et al., 2000) and crown rust (*Puccinia coronata* Corda) resistance (Thorogood et al., 2001). Identification of QTL will lead to a greater understanding of the genetic basis of important agronomic traits, and to the use of marker-based selection methods in practical breeding programs.

Intergeneric Introgression to Increase Adaptability

Festuca pratensis provides an excellent source of genetic variation, particularly for a range of stress tolerances (cold and drought) and disease resistance that can potentially be incorporated into *L. perenne* by introgression. Significant improvements have been made in ryegrass in terms of growth characteristics and cultivars with high shoot density, slow growth, and good winter and summer wear tolerance have been produced. Further improvements in increasing the adaptability and sustainability of ryegrass are necessary. The Bovinae section of the *Festuca* genus has been demonstrated to be closely related to ryegrass. *F. pratensis, F arundinacea,* and *F. glaucescens,* in particular have been shown to be particularly tolerant of environmental stresses such as cold, drought, and disease. Almost unlimited recombination can occur between the two genera pointing to the potential for introgressing traits to improve the performance of perennial ryegrass turf. For example, Humphreys and Thomas (1993) identified ryegrass plants with introgressed chromosome segments derived from *F. arundinacea* that were as drought tolerant as the original *F. arundinacea* parent. These plants were later found to have the same *Festuca*-derived introgressed chromosome fragment as revealed by genomic *in-situ* hybridization (GISH) of mitotic chromosome preparations with fluorescent dyes able to differentiate DNA from the two species. By inference, these fragments were deemed to contain the genes for drought tolerance (Humphreys and Pasakinskiene, 1996). Temperature-stable crown rust resistance in *F. pratensis* has also been transferred into *L. perenne* lines from *F. pratensis* (Adomako et al., 1997).

An alternative approach to trait mapping has been initiated by King et al. (1998) that will potentially aid the introgression of stress tolerance traits into ryegrass. They have developed a number of *L. perenne* substitution lines with single *F. pratensis* chromosome arms. Each line can then be crossed to *L. perenne* to generate a series of plants that contain different amounts of the substituted *F. pratensis* chromosome arm, identified by genomic *in-situ* hybridization (GISH) and subsequent differential staining of *Festuca*- and *Lolium*-derived DNA of metaphase root-tip cells. It is then possible to map these segments using markers that are shown to have *Festuca* specific alleles. QTL specific to the introgressed segments could then be identified and mapped.

Genetic Transformation

Significant variation waits to be exploited in naturally occurring populations within the species and in closely related species that can hybridize with *Lolium*. But it is also possible to overcome the natural breeding barriers and insert useful genes into ryegrass from unrelated species. Although gene transfer mediated through vectors such as viruses and *Agrobacterium* has been possible in the dicotyledonous species this method of transfer is not as readily applicable to *Lolium*. Instead, physical methods of transfer have been effective but have required (1) the culture of undifferentiated tissues and cells devoid of cell walls, (2) the stable, constitutive incorporation of the new gene into the donor plant genome, and (3) the ability to regenerate green fertile plants from these cultures. The various techniques are now in place and, although only reporter genes such as that encoding for Glucuronidase (GUS) and marker genes such as herbicide and antibiotic resistance genes have been transformed (Spangenberg et al., 1995), the results demonstrate the possibilities for transferring simply inherited agronomically important genes, such as those encoding for disease and pest resistance.

Regeneration of *Lolium* protoplasts is possible (Dalton, 1993) and protoplast transformation of *Lolium* has been achieved by Wang et al. (1997) by direct gene transfer achieved through electroporation of the protoplast membrane. The use of microscopic silica carbide whiskers onto which the DNA is coated as a vector for direct gene transfer into *Lolium* protoplasts (Dalton et al., 1998) and biolistic bombardment of calli with DNA-coated gold particles (Altpeter et al., 2000) have been successful. In the latter case, evidence for the sexual transmission of the transgene was reported, whereas in previous reports, stability of the transgene in transformed tissue has not been demonstrated. The success of transformation has always been shown to be genotype dependent (Olesen et al., 1995; Altpeter, et al., 2000). However, by optimizing culture media and regeneration treatments and conditions, Altpeter and Posselt (2000) were able to increase regeneration ability, managing to produce green shoots from four commercial cultivars and four inbred lines of ryegrass.

The techniques of tissue culture and gene transfer across the breeding barrier clearly offer opportunities to extend the variation within the species available to breeders with possibilities for incorporating herbicide tolerance to allow selective control of weeds in turfgrass and seed production areas. Pest and disease resistance, and abiotic stress resistance, and a range of turfgrass quality traits could be targeted in the future. However, these traits are often under complex genetic control and a clearer understanding of the number of genes involved in controlling these traits needs to be gained before gene transfer techniques can be applied. Many issues still need to be addressed if the technology is to be routinely used as a part of the turfgrass breeder's tool kit, issues such as gene stability in transformed plants, the effect of the transgene on the expression of other genes within the plant, and the removal of agronomically unimportant selectable marker genes. Of particular importance in the wind-pollinated outcrossing crops is the issue of how specific transgenes might behave when incorporated into wild populations.

Breeding Objectives and Progress

Breeding objectives depend on where and how the grass is to be grown. The North American continent covers a vast area with climatic extremes. Perennial ryegrasses are tested by the National Turfgrass Evaluation Program (NTEP) coordinated by the United States Department of Agriculture from Beltsville, Maryland, since 1982. In the most recently completed set of trials (1994–1998), a total of 96 cultivars were tested at 30 sites across North America, over a period of four years in statistically robust, replicated, and randomized designs. Three standard entries are included in the most recent complete set of trials: 'Linn' (Funk and Engel, 1967) was selected at Corvallis, Oregon, from a 1928 introduction from New Zealand and released in 1961. 'Pennfine,' the first ryegrass cultivar to be selected specifically for turfgrass use at the Pennsylvania Agriculture Experiment Station, was released in 1969. 'Saturn' was selected and released relatively recently (1986) from a number of crosses of good turfgrass cultivars available at that time and represents the best ryegrass cultivar in the previous set of NTEP trials overall (Dr. Kevin Morris, personal communication).

A number of traits were measured over a period of four years, generating a large data set. This data set can be divided into three main performance criteria: (1) turf growth characteristics and appearance, (2) disease resistance, and (3) resistance or tolerance to environmental factors.

Growth Characteristics

Over 30 sites, visual quality, shoot density, percentage ground cover, and leaf texture were measured for each variety being tested. These measurements were taken at different times of the year over four years at all sites, giving a total of more than 300 scores for each cultivar entered for trial. When all of these scores are analyzed using Principal Components, most of the variation (47.3%) can be accounted for in the first component. The second component accounts for only 6.7%. This shows that most of the growth traits, regardless of where and when they were assessed, are correlated and have similar principal component vector loadings (Figure 7.4). By plotting the first two principal component's scores for growth characters of all of the cultivars in the trial (Figure 7.5), it is possible to get an overall impression of how cultivar improvements have been made over the standard check cultivars. In this case, where the first component accounts for the vast majority of the variation observed, the old standard cultivars 'Linn' and 'Pennfine' have high positive scores indicating poor performance. Most of the other cultivars with the exception of some European-bred cultivars group toward the other end of the scale, indicate good performance. The differences between these cultivars are relatively small, but the new entry, Palmer III is apparently the best of the group.

Turf Color

So-called 'genetic color' is measured in NTEP trials as a visual assessment of green color under nonstressed conditions, with low scores given for light green and high scores for dark green turf. Correlation of cultivar color between the 30 sites is, in nearly all cases, high (Figure 7.6). Darker colors are perceived in the United States as being preferable and increases in modern cultivars over Linn, Pennfine, and European-bred entries are highly significant. Some of the newest cultivars such as Brightstar II, Radiant, Prelude III, and Wind Dancer have an extremely dark green color.

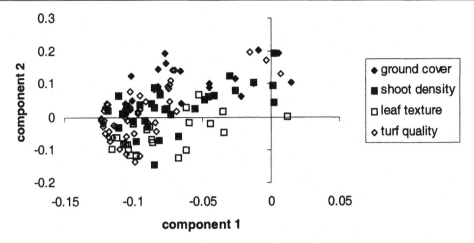

Figure 7.4. Bi-plot of first two principal components for major turf growth characters measured at NTEP sites. Component 1 accounts for 47.3% of the variation and component 2, 6.7%.

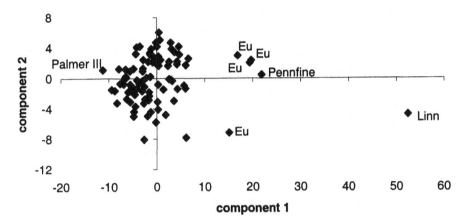

Figure 7.5. Bi-plot of first two principal component scores for cultivars in NTEP trials. Eu = Cultivar submitted by European breeder.

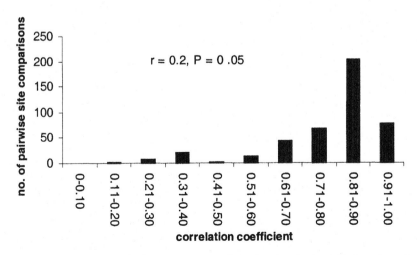

Figure 7.6. Frequency distribution of correlation coefficients of paired NTEP sites, based on genetic color ratings of perennial ryegrass entries.

Spring Greenup

This character is a measure of new green leaf growth in the spring. Each site has its own unique environmental and edaphic variables (or measuring criteria), such that correlation coefficients between sites vary between 0.82 ($P<0.001$) and -0.73 ($P<0.001$). Although Linn and Pennfine on average have the worst spring greenup ratings, at some sites they are the best performers. Although clear varietal differences were identified under site-specific conditions, determination of whether progress has been made in improving this character is difficult when large cultivar x site interaction occurs. Targeted improvements are more likely to be made if the various physiological and genetic components of the character are identified, dissected, mapped, and reassembled into breeding populations.

Turf Diseases

A number of diseases are recorded in NTEP trials at varying numbers of sites. Diseases scored at the greatest number of sites are brown patch (caused by *Rhizoctonia solani* Kuehn.), dollar spot (caused by *Sclerotinia homoecarpa* Bennett), and red thread (caused by *Laetisaria fuciformis* McAlpine).

Considerable site x cultivar interaction occurs for incidence of brown patch that may be caused by any number of unidentified environmental or pathotype variables or may be due to limited genetic variation for resistance. This latter possibility seems to be backed up by the fact that Pennfine and/or Linn are not significantly worse than the best-ranked cultivar at most sites.

Dollar spot is a relatively new disease on perennial ryegrass that is thought to have become prevalent through constant and prolonged exposure to inoculum. There is little site x cultivar interaction in NTEP trials and Linn is apparently as resistant as the best cultivars overall.

The old standard cultivars, Linn and Pennfine, appear to have good resistance to red thread over all sites, relative to many of the new cultivars. This may be explained by the fact that the recently developed, slower growing, or dwarf cultivars are more prone to attack (Baldwin, 1987). Cultivar rankings in general show little relationship between NTEP sites, a situation that is mirrored in Europe.

Other diseases recorded at NTEP sites include *Pythium* blight, which is a disease that affects young establishing swards and prefers warm humid conditions. For this reason it is particularly problematic on ryegrass used for annual overseeding of warm-season turfs. Pink snow mold (*Microdochium nivale* Samuels and Hallet) is also recorded at two sites and occurs in mild wet winters or under snow cover, which induces a similarly appropriate microclimate. For both of these diseases the cultivar rank correlations between the two sites where measurements were recorded are not significant. New cultivars are apparently more resistant to *Pythium* but there is no improvement for pink snow mold resistance, with Linn being one of the most resistant cultivars. This may be due to the fact that the disease is more prevalent on cultivars that have greater thatch, producing a humid microenvironment which encourages the development of the fungus.

Gray leaf spot (*Pyricularia grisea* Sacc.) is a highly pathogenic and potentially devastating disease becoming prevalent in ryegrass turf, especially at high temperature and humidity. A potential source of resistant germ plasm has been found in populations deriving from Bulgarian collections tested in Rutgers turfgrass trials (Clarke and Meyer, personal communication).

Drought Stress

Although several breeders have claimed drought tolerant cultivars of ryegrass, the current NTEP trials list the results from two sites that show no improvements over Pennfine. However some cultivars with good turf quality traits are equally drought tolerant as older cultivars. No correlation exists between cultivar scores at either of the two NTEP sites where apparent drought tolerance was recorded. Clearly it may be possible to identify drought tolerant cultivars under a specific set of drought inducing conditions, but as the range of drought inducing variables vary from site to site, general drought tolerance cannot be inferred across a range of environmental and climatic conditions.

Winter Hardiness

The NTEP data strongly suggest that no improvement in winter hardiness has been achieved by breeders. Both Linn and Pennfine have very good winter hardiness scores unsurpassed by any other cultivar at both sites where it was measured. However, modern cultivars combine winter hardiness with good turf quality traits. Correlation between winter hardiness scores for the two NTEP sites at which the trait is recorded is completely lacking despite significant cultivar differences being identified at each site, which makes interpretation of results difficult.

The European Perspective

Perennial ryegrass is ideally suited to the cool, wet climate of temperate Europe and emphasis has been placed on improving plant growth characteristics such as sward density, short growth, and leaf fineness. These characteristics have been greatly improved in European breeding programs since the early 1970s. No significant improvements have been made for durable resistance to diseases and resistance to environmental factors (van Wijk, 1993), which would make for a significantly more persistent turf in suboptimal conditions. Diseases and pests are generally not a big problem in many parts of Europe, although resistance to crown rust, black stem rust (*Puccinia graminis* Fuckel) snow mold, leaf spot *(Drechslera* spp.), and red thread are useful additional breeding objectives for some areas.

Ryegrass is used alone or as a major component of mixtures for low- to medium-quality lawns that may be mown as close as 8 mm. In Europe and New Zealand the species is also used as a surface for winter-games pitches (primarily soccer and rugby) where resistance to treading and wear, over a prolonged period of extensive play, often under cold and waterlogged conditions, is important.

Breeding and selection for growth characteristics is relatively easy, as the major component characteristics such as tolerance of close mowing, shoot density, leaf-fineness, slow regrowth under both regular and infrequent mowing regimes are highly correlated. In lawn and artificial wear trials conducted at the STRI (Sports Turf Research Institute) in the United Kingdom, the correlation between quality growth characteristics, such as shoot density, and abrasive summer wear at an 8 mm cutting height is high ($r=0.64$, $P<0.05$). Yet there is no correlation between shoot density and winter wear at a 20 mm cutting height and even some indication that the correlation may be significantly negative if a larger sample of cultivars were to be tested ($r=-0.57$; data from Anon., 1997 and Newell and Jones, 1995). This suggests that it may be possible to breed cultivars with high winter wear tolerance combined with good lawn growth characteristics, but, in practice, this rarely happens. Figure 7.7 shows the latent vectors

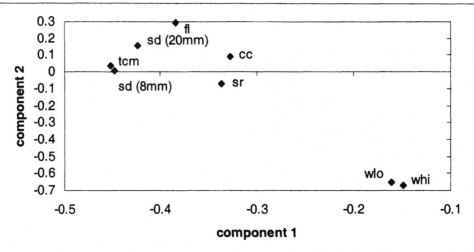

Figure 7.7. Latent vectors of turf ryegrass characters listed in STRI's 'Turfgrass Seed, 2001'. cc = cleanness of cut, fl = fineness of leaf, sd (8 mm) = shoot density at 8 mm cutting height, sd (20 mm) = shoot density at 20 mm cutting height, sr = short regrowth between cutting intervals, tcm = visual merit under 8 mm cutting regime, whi = ground cover after winter wear under high nitrogen regime, wlo = ground cover after winter wear under low nitrogen regime.

of the first two principal components of variation of measurements taken from the STRI's 'Turfgrass Seed' booklet which lists the performance of cultivars for a range of characters. A range of lawn performance characters is correlated, as indicated by the fact that they produce vectors with similar direction. Winter wear under two nitrogen regimes, on the other hand, give completely different vector scores because they are not correlated with lawn performance scores. A plot of the first two cultivar principal component scores illustrates differences in cultivar performance (Figure 7.8). Although cultivars like Darius and Verdi perform well under lawn and winter wear conditions respectively, only AberElf and AberImp combine both lawn performance and winter wear tolerance (highly negative component 1 and 2 scores).

Both summer and winter turf color are important aesthetic considerations in breeding and selection programs, possibly relating to environmental or pest and disease stresses and turf persistency under such conditions. Genetic variation exists for both intrinsic turfgrass color and the stability of that color during stress conditions, causing varying degrees of color loss through senescence (Thorogood, 1996). One encouraging advance in ryegrass turf breeding is the use of stay-green grass genotypes. There are several types of stay-green identified in a number of plant species including *Arabidopsis thalliana* (L.) Heynh., maize (*Zea mays* ssp. mays L), sorghum (*Sorghum bicolor* [L.] Moench), soybean (*Glycine max* Merr.), and pea (*Pisum sativum* L.) (Thomas and Smart, 1993).

The physiological processes involved in delaying the chlorophyll catabolic process, producing stay green phenotypes are diverse (Thomas and Howarth, 2000). A stay-green mutant has been found in *Festuca pratensis* in which one of the enzymes (
-oxygenase) involved in chlorophyll catabolism is disabled. This enzyme breaks the ring structure of chlorophyll to produce a colorless linear molecule (Matile et al., 1996). When this enzyme is disabled, chlorophyll and its immediate breakdown products, chlorophyllide and phaeophorbide, both of which are green pigments, increase. The effect on grass leaves is dramatic under conditions in which leaf senescence outstrips growth. The stay-green allele has been introgressed into turf-type ryegrasses by back-

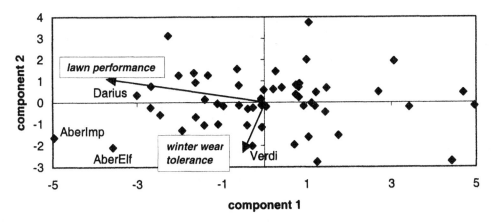

Figure 7.8. First two principal component scores for ryegrass cultivars listed in 'Turfgrass Seed 2001.' Variation accounted for = 69.4%.

crossing and the resultant populations show distinct color retention advantages over nonstay-green populations under conditions in which turf greenness is normally reduced (Figure 7.9). A stay-green ryegrass cultivar, AberNile, was added to the European Community Common Catalogue in 2001 and is currently being commercially produced.

Crown rust resistance, derived from meadow fescue, appears to be linked to the stay-green locus (Adomako et al., 1997). This resistance is markedly different from that expressed in ryegrass cultivars. Although ryegrass cultivars are often resistant when tested at 10°C, this resistance is not temperature stable and breaks down at higher temperatures (25°C) (Roderick et al., 2000). Resistance in meadow fescue, including stay-green types, is expressed at both temperatures and stay-green introgression lines have been shown to express the same temperature stable resistance (Figure 7.10).

The New Zealand Perspective

There is a significant turf ryegrass breeding effort in Oceania and particularly in New Zealand (Dr. Alan Stewart, personal communication). The New Zealand environment is very different from that where turf ryegrass has traditionally been bred in northern Europe and temperate North America. Grass growth occurs during the mild New Zealand winters, whereas the summers are dry with the additional problem of the grass being attacked by the Argentine stem weevil and black beetle (*Heteronychus arator* Fabricius). A small number of turf perennial ryegrass cultivars have been developed in New Zealand including 'Grasslands Trophy' and 'Grasslands Coronet.' These cultivars are quite coarse textured compared to European and U.S. bred cultivars but perform well under the mild conditions encountered where winter growth is required. High endophyte cultivars from the United States have been evaluated and the incorporation of Mediterranean-derived germ plasm in breeding programs has produced cultivars such as 'Endurance' and 'Arena' with increased winter growth and recovery from winter wear.

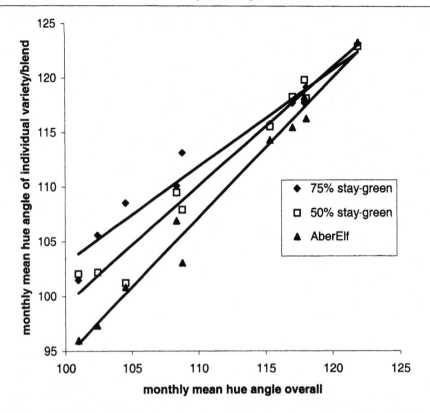

Figure 7.9. Regression of monthly hue angle scores (higher score = greener turf) of turf plots of the cultivar, AberElf and blends with stay-green ryegrass with mean monthly hue angle score (Thorogood, unpublished data).

Future Breeding Objectives

Growth and Appearance

Significant advances in plant growth characteristics such as shoot density, leaf texture, tolerance of close mowing, and short growth have been achieved over the last two decades both in United States and Europe. Darker green colors have also been produced in North America, largely to achieve color compatibility with Kentucky Bluegrass (*Poa pratensis* L.). The future challenge to ryegrass turf breeders is to be able to maintain these characteristics under a range of different pathogen and environmental stresses in order to increase the range of use and reliability of the species.

Wear Tolerance

Ryegrass is one of the most wear tolerant temperate turfgrasses available and improvements in summer wear tolerance have been achieved indirectly by increasing shoot density. Winter wear on European sports pitches has been reduced partly by empirical selection of wear-resistant ryegrass germ plasm using artificial wear machines with studded rollers (Canaway, 1981). Little is known about the mechanisms and genetic control of resistance and tolerance on which more directed selection could be practiced.

Biomass before wear commences has been found to be a good indicator of winter wear tolerance (Ellis, 1981). Such material will have an equivalent higher biomass after

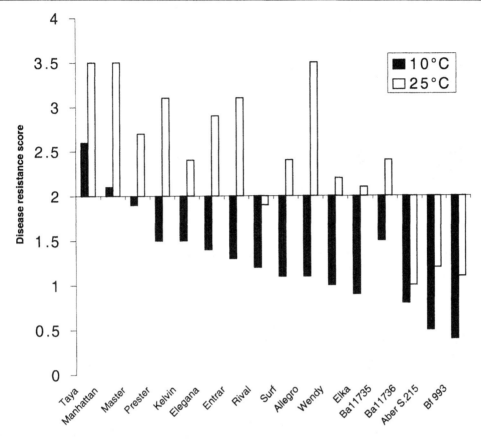

Figure 7.10. Crown rust resistance score for 12 cultivars of turf perennial ryegrass, two ryegrass/meadow fescue stay-green introgression lines, Ba11735 (temperature sensitive) and Ba11736 (temperature stable) and two meadow fescue cultivars, Aber S215, nonstay-green and Bf993, stay green (both temperature stable). 0 = resistant, 4 = susceptible (Roderick et al., 2000).

wear and will also maintain a greater leaf area index with a greater photosynthetic capacity for recovery regrowth. This will be more important in winter when light and temperature may well limit photosynthetic activity.

Lush and Rogers (1992) found that turf follows the general ecological principle of the self-thinning rule, where biomass is inversely proportional to shoot density (White, 1981). For a given turf population, there is a ceiling (self-thinning line) where biomass can only be increased with a reduction in shoot density (Figure 7.11). In breeding for combined winter wear tolerance and textural playing quality characteristics, it is important to select for high shoot density and high biomass, thus raising the self-thinning line (dotted line on Figure 7.11). Breeding and selection is the only way of concurrently improving both biomass and shoot density, as any management strategy will only increase one of the traits. For example, lowering the cutting height will increase shoot density, while raising the cutting height will increase biomass. The correlation between shoot density and biomass in ryegrass cultivars is weak (Shildrick, 1981) and hence many cultivars listed by the STRI as having good wear tolerance have poor shoot density and visual appeal (Anon., 2001). The lack of association between wear tolerance and visual appeal (shoot density) was also reported by Bonos et al. (2001).

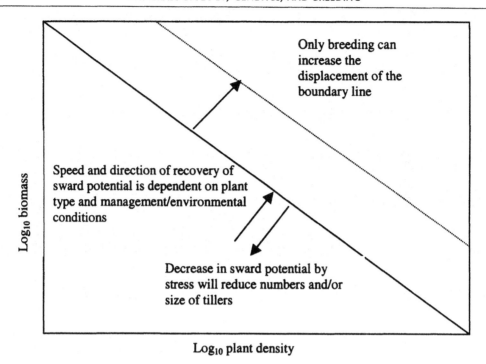

Figure 7.11. Relationship between shoot density and biomass according to the −1/2 self-thinning rule.

Disease Resistance

Resistance to the many diseases of ryegrass will remain an important breeding objective, especially as many of the improvements in other turfgrass traits such as shoot density and slow vertical growth have increased the success of certain pathogens. Breeding for durable resistance will rely on a number of factors: (1) identification of germ plasm from within the species or from other related genera or even from unrelated plants or bacteria through the techniques of transgenics (see Meyer and Belanger, 1997); (2) understanding the interaction of these resistances to environmental and pathotype variables; and (3) understanding how selection for resistance may affect other important turfgrass traits. The ability to map genes that encode for resistance to a wide range of pathotypes under a range of different environments will be a useful tool for identifying durable resistance to important ryegrass turf diseases. Once identified, these genes can be pyramided together in marker-assisted selection programs.

Drought Tolerance

Drought tolerance is a difficult trait to measure since it is a complex character with many components. Drought strategies can be divided into two types (May and Milthorpe, 1962). The first of these is escape, where a plant avoids drought by completing its life-cycle before the onset of drought. This strategy is not relevant to perennial grasses except perhaps in seed production, where early flowering cultivars may be harvestable before the onset of drought. The other type is drought endurance where either high water content is maintained during drought, perhaps by producing a deep or dense rooting system, or low internal water content is tolerated and the plant has an ability to recover after drought. In grasses, this latter strategy might involve tolerance of stem and root primordia to dehydration, leaf drop, and low cuticular conductance. Volaire et al. (1998) found that high molecular weight fructans were remobilized from leaf primordia of a drought tolerant line of orchard grass and in two cultivars of perennial ryegrass which enabled rapid recovery after drought.

Single- or multisite field trials will also add to the complexity of breeding for drought tolerance with variation for soil, climate, year, and management factors all having an influence on drought response. Unsurprisingly, the drought tolerance ratings for the two NTEP sites at which data were collected were not correlated with each other. Thomas et al. (1996a) assessed the response of a range of *Festuca pratensis* populations to drought under two glasshouse regimes and two experiments using mannitol as an osmoticum. Although highly significant population differences were found in each experiment, there was no consistency between experiments. Although individual experiments and trials can identify plants, populations, and cultivars with positive components which can then be used in breeding programs, no one set of experiments can fully explain the variation for drought tolerance. A genetic mapping approach to drought tolerance components may help to target functional genes, QTL, and associated marker loci, which can be used as selection criteria in a breeding program.

Winter Hardiness

Waldron et al. (1998) top-crossed a number of turf cultivars onto NK200, a ryegrass cultivar with reasonable winter hardiness. They identified a number of half-sib families with greater winter hardiness than NK200 and they obtained high narrow-sense heritability estimates (59%) suggesting that improvement using single trait selection for winter hardiness could be made. However, gains in winter hardiness were negatively correlated with crown rust resistance and leaf texture, suggesting that multitrait selection would be desirable.

Winter hardiness, like drought tolerance is a complex trait and gains made in breeding and selection programs are limited by the restricted conditions under which the selections take place. Components of winter hardiness include freezing tolerance, snow mold resistance, waterlogging, desiccation, ice-encasement, and starvation caused by negative net assimilation of photosynthate under low light conditions (Dr. Mike Humphreys, personal communication). Variability for a combination of these components would need to be identified and selected, perhaps with the aid of genetic mapping and marker selection tools, if future advances are to be made.

CONCLUSIONS

Lolium perenne is a considerably successful species, much of this success being due to its artificial spread by man who has exploited it as a valuable forage and turfgrass species. This success is likely to continue because of its genetic diversity that is maintained by its outcrossing nature and its ability to grow in a wide range of climatic and edaphic conditions. Active conservation of the species both *in-situ* and in *ex-situ* gene banks will enhance its success. Conventional breeding of the species has not appeared to reduce genetic diversity, while improvements in a number of characteristics have been made. However, the possibility of genetic erosion through breeding and management activities still exists. Genetic marker technology will enhance breeders' abilities to identify and use genetic variation more efficiently both within the species and in intergeneric hybrids and introgressions with species mainly from the *Festuca* genus. Ryegrass, by virtue of its worldwide adaptation and economic importance, and relatively simple genetic structure, has become the model perennial grass for genomic study through collaborations such as the International *Lolium* Genome Initiative (ILGI). The production of trait-specific mapping families will increase our knowledge of the genetic control of many agronomically important traits that can be efficiently manipulated in marker-assisted breeding and selection programs. If appropriate variation cannot be found even in this material, genetic transformation technology offers the possibilities of extending variation even further.

REFERENCES

Adomako, B., D. Thorogood, and B.C. Clifford, 1997. Plant reaction types to crown rust (*Puccinia coronata* Corda) disease inoculations in meadow fescue (*Festuca Pratensis* L.), perennial ryegrass (*Lolium Perenne* L.) and *L. Perenne* L. introgression lines. *International Turfgrass Society Research Journal* 8: 823–831.

Aiken, S.G., S.E. Gardiner, H.C.M. Bassett, B.L Wilson, and L.L. Consaul. 1998. Implications from SDS-PAGE analyses of seed proteins in the classification of taxa of *Festuca* and *Lolium* (*Poaceae*). *Biochem. Syst. Ecol.* 26, 5: 511–533.

Altpeter, F. and U.K. Posselt. 2000. Improved plant regeneration from cell suspensions of commercial cultivars, breeding- and inbred lines of perennial ryegrass (*Lolium perenne* L.). *J. Plant Physiol.* 156:790–796.

Altpeter, F., J.P Xu, and S. Ahmed, 2000. Generation of large numbers of independently transformed fertile perennial ryegrass (*Lolium perenne* L.) plants of forage- and turf-type cultivars. *Molecular Breeding* 6: 519–528.

Anon. 2001. *Turfgrass Seed 2001*. Sports Turf Research Institute, Bingley, England.

Anon. 1997. *Turfgrass Seed 1997*. Sports Turf Research Institute, Bingley, England.

Ascherson, P. and P. Groebner. 1902. *Synopsis der Mitteleuropäischen Flora 2*. Leipzig: Wilhelm Engelmann.

Baldwin, N.A. 1987. *Turfgrass Diseases*. The Sports Turf Research Institute, Bingley, West Yorkshire, England.

Balfourier. F., G. Charmet, and C. Ravel. 1998. Genetic differentiation within and between natural populations of perennial and annual ryegrass (*Lolium perenne* and *L-rigidum*). *Heredity* 81: 100–110.

Bean, E.W. and C. Yok-Hwa, 1972. An analysis of the growth of inbred progeny of *Lolium*. *J. Agric. Sci.* 79: 147–153.

Beard, J.B. 1973. *Turfgrass: Science and Culture*. Prentice-Hall, Upper Saddle River, NJ.

Beddows, A.R., E.L. Breese, and B. Lewis, 1962. The genetic assessment of heterozygous breeding material by means of a diallel cross. *Heredity* 17: 501–512.

Bennett, S.J. 1997. A phenetic analysis and lateral key of the genus *Lolium* (*Gramineae*). *Gen. Res. Crop Evol.* 44: 63–72.

Bennett, S.J., M.D. Hayward, and D.F. Marshall, 2000. Morphological differentiation in four species of the genus *Lolium*. *Gen. Res. Crop Evol.* 47: 247–255.

Bert, P.F., G. Charmet, P. Sourdille, M.D. Hayward, and F. Balfourier, 1999. A high density map for ryegrass (*Lolium perenne*) using AFLP markers. *Theor. Appl. Genet.* 99: 445–452.

Bonos, S.A., E. Watkins, J.A. Honig, M. Sosa, J.A. Molnar, and W.A. Meyer, 2001. Breeding cool-season turfgrasses for wear tolerance using a wear simulator. *International Turfgrass Society Research Journal* 9: 137–145.

Bulinska-Radomska, Z. and Lester, R.N. 1985. Relationships between five species of *Lolium* (*Poaceae*). *Pl. Syst. Evol.* 148: 169–175.

Bulinska-Radomska, Z. and R.N. Lester, 1988. Intergeneric relationships of *Lolium*, *Festuca*, and *Vulpia* (*Poaceae*) and their phylogeny. *Pl. Syst. Evol.* 159: 217–227.

Canaway, P.M. 1981. Wear tolerance of turfgrass species. *J. Sports Turf Res. Inst.* 57: 108–121.

Casler, M.D. 1995. Patterns of variation in a collection of perennial ryegrass accessions. *Crop Sci.* 35: 1169–1177.

Charmet, G. and F. Balfourier, 1994. Isozyme variation and species relationships in the genus *Lolium* L (Ryegrasses, Graminaceae). *Theor. Appl. Genet.* 87, 6: 641–649.

Charmet, G. and F. Balfourier, 1995. The use of geostatistics for sampling a core collection of perennial ryegrass populations. *Genet. Resour. Crop Ev.* 42: 303–309.

Charmet, G., F. Balfourier, and V. Chatard, 1996. Taxonomic relationships and interspecific hybridization in the genus *Lolium* (grasses). *Gen. Res. Crop Evol.* 43: 319–327.

Charmet, G., C. Ravel, and F. Balfourier, 1997. Phylogenetic analysis in the *Festuca-Lolium* complex using molecular markers and ITS rDNA. *Theor. Appl. Genet.* 94: 1038–1046.

Connolly, V. and R. Wright-Turner, 1984. Induction of cytoplasmic male-sterility into ryegrass (*Lolium perenne*). *Theor. Appl. Genet.* 68: 449–453.

Corkill, L. 1933. Inheritance of fluorescence in ryegrass. *Nature* 130: 134.

Corkill, L. 1956. The basis of synthetic strains of cross-pollinated grasses. *Proceedings of the 7th International Grassland Congress*, pp. 427–438. Massey Agricultural College, Palmerston North, New Zealand.

Cornish, M.A. 1979. The Genetics of Self-Incompatibility in *Lolium perenne*. Ph.D. thesis, University of Birmingham, United Kingdom.

Cornish, M.A., M.D. Hayward, and M.J. Lawrence, 1980. Self-incompatibility in ryegrasses. V. Genetic control, linkage and seed set in diploid *Lolium perenne* L. *Heredity* 44: 333–340.

Dalton, S.J. 1993. Regeneration of plants from protoplasts of *Lolium* (Ryegrasses) and *Festuca* (Fescues). In Y.P.S. Bajaj, Ed., *Biotechnology in Agriculture and Forestry vol.* 22. Plant Protoplasts and Genetic Engineering III. Springer-Verlag, 1993, pp. 46–68.

Dalton, S.J., A.J.E. Bettany, E. Timms, and P. Morris, 1998. Transgenic plants of *Lolium multiflorum, Lolium perenne, Festuca arundinacea* and *Agrostis stolonifera* by silicon carbide fibre-mediated transformation of cell suspension cultures. *Plant Science* 132: 31–43.

Egorova, I.A. and A.V. Voylokov, 1998. Localization of self-fertility mutations at the S locus in inbred rye lines from the Peterhoff genetic collection. *Genetika* 34: 1094–1099.

Ellis, C.J. 1981. An Experimental Approach to Wear Tolerance in *Lolium perenne*. Ph.D. thesis, University of Liverpool, United Kingdom.

England, F.J.W. 1974. The use of incompatibility for the production of F1 hybrids in forage grasses. *Heredity* 32: 183–188.

Essad, S. 1954. Contribution à la sytèmatique du genre *Lolium*. Ministère de l'Agriculture. Annales de l'Institute Natlionale Recherche Agronomie, Paris. *Amelioration des Plantes* 4: 325–351.

Evans, G. 1926. Chromosome complements in grasses. *Nature* 118: 841.

Fearon, C.H., M.A. Cornish, M.D Hayward, and M.J. Lawrence, 1994. Self-incompatibility in ryegrass. X. Number and frequency of alleles in a natural population of *Lolium perenne* L. *Heredity* 73: 254–261

Forster, J.W. 1999. The International *Lolium* Genome Initiative. *Plant and Animal Genome IX Conference* p.W100, p.53. San Diego.

Foster, C.A. and C.E. Wright, 1970. Variation in the expression of self-fertility in *Lolium perenne* L. *Euphytica* 19:61–70.

Fraser, M. and J.P. Breen, 1994. The role of endophytes in integrated pest management for turf. In *Handbook of Integrated Pest Management*, pp. 521–528. A.R. Leslie, Ed., Lewis Publishers, Boca Raton, FL.

Funk, C.R., and R.E. Engel, 1967. Performance of Kentucky bluegrass, perennial ryegrass and tall fescue varieties. 1967 Report on Turfgrass Research at Rutgers University. Bulletin No. 818. pp. 67–71.

Gaue, I. 1977. Ms-based breeding in *Lolium perenne* L. *Proceedings of the 13th International Grassland Congress*, pp. 435–437. Leipzig, Germany.

Gentner, G. 1929. Über die Verwedbarkeit von ultravioletten Strahlen bei der Samenprüfung. *Praktische Blätter für Pflanzenbau und Planzenscutz.* 6: 166–172.

Hayman, D.L. 1956. The genetical control of incompatibility in *Phalaris coerulescens* Desf. *Austral. J. Biol. Sci.* 9: 321–331

Hayman D.L. and J. Richter, 1992. Mutations affecting self-incompatibility in *Phalaris Coerulescens* Desf (*Poaceae*). *Heredity* 68: 495–503.

Hayward, M.D. and N.J. McAdam, 1977. Isozyme polymorphism as a measure of distinctiveness and stability in cultivars of *Lolium perenne* L. *Z. Pflanzenzüchtg* 81: 228–234.

Hayward, M.D. and N.J. McAdam, 1988. The effect of isozyme selection on yield and flowering time in *Lolium perenne*. *Plant Breeding* 101: 24–29.

Hayward, M.D., J.W. Forster, J.G. Jones, O. Dolstra, C. Evans, N.J. McAdam, K.G. Hossain,. M. Stammers, J. Will, M.O. Humphreys, and G.M. Evans, 1998. Genetic analysis of *Lolium*. I.

Identification of linkage groups and the establishment of a genetic map. *Plant Breeding* 117: 451–455.

Hayward, M.D., N.J. McAdam, J.G. Jones, C. Evans, G.M. Evans, J.W. Forster, A. Ustin, K.G. Hossain, B. Quader, M. Stammers, and J.K. Will, 1994. Genetic markers and the selection of quantitative traits in forage grasses. *Euphytica* 77: 269–275.

Heslop-Harrison, J. 1982. Pollen-stigma interaction and cross-incompatibility in the grasses. *Science* 215: 1358–1364

Hubbard, C.E. 1948. Gramineae. In *British Flowering Plants*. pp. 284–348. Hutchinson, John. B. Gawthorn Ltd, London.

Hubbard, C.E. 1984. *Grasses*. 3rd ed. Revised by J.C.E. Hubbard. Penguin Books, London, United Kingdom.

Humphreys, M.O. 1992. Association of agronomic traits with isozyme loci in perennial ryegrass (*Lolium perenne* L.). *Euphytica* 59: 141–150.

Humphreys, M.W. 1989. The controlled introgression of *Festuca arundinacea* genes into *Lolium multiflorum*. *Euphytica* 42: 105–116.

Humphreys M.W. and I. Pasakinskiene, 1996. Chromosome painting to locate genes for drought resistance transferred from *Festuca arundinacea* into *Lolium multiflorum*. *Heredity* 77: 530–534.

Humphreys M.W. and H. Thomas 1993. Improved drought resistance in introgression lines derived from *Lolium multiflorum* x *Festuca arundinacea* hybrids. *Plant Breeding* 111: 155–161.

Humphreys, M.W. and D. Thorogood, 1993. Disturbed mendelian segregations at isozyme marker loci in early backcrosses of *Lolium multiflorum* x *Festuca pratensis* hybrids to L multiflorum. *Euphytica* 66:11–18.

Jenkin, T.J. 1928a. Inheritance in *Lolium perenne*. I. Seedling characters, lethal and yellow-tipped albino. *J. Genet.* 19: 391–402.

Jenkin, T.J. 1928b. Inheritance in *Lolium perenne*. II. A second pair of lethal factors. *J. Genet.* 19: 403–417.

Jenkin, T.J. 1930. Inheritance in *Lolium perenne*. III. Base-color factors, C and R. *J. Genet.* 22: 389–394.

Jenkin, T.J. 1931a. The method and technique of selection, breeding and strain building in grasses. *Imp. Bureau Plant Genet. Herbage Plants Bull.* No. 3: 5–34.

Jenkin, T.J. 1931b. Self-Fertility in *Lolium perenne* L. Welsh Plant Breeding Station Bulletin Series H. No. 12, pp. 100–119.

Jenkin, T.J. 1933. Interspecific and intergeneric hybrids in herbage grasses. Initial crosses. *J. Genet.* 28: 205–264.

Jenkin, T.J. 1935. Interspecific and intergeneric hybrids in herbage grasses. II. *Lolium perenne* x *L. temulentum*. *J. Genet.* 31: 379–411.

Jenkin, T.J. 1936. Natural selection in relation to the grasses. In A Discussion on the Present State of the Theory of Natural Selection. *Proc. Roy Soc,* Ser. B. 121:52–56.

Jenkin, T.J. 1943. Aberystwyth strains of grasses and clovers. *J. Min. Agric.* 50:343–349. London.

Jenkin, T.J. 1953. Interspecific and intergeneric hybrids in herbage grasses. V. *Lolium rigidum* sens. Ampl with other *Lolium* species. *J. Genet.* 52: 252–281.

Jenkin, T.J. 1954a. Interspecific and intergeneric hybrids in herbage grasses. VI. *Lolium italicum* Abr. intercrossed with other *Lolium* types. *J. Genet.* 52: 282–299.

Jenkin, T.J. 1954b. Interspecific and intergeneric hybrids in herbage grasses. VII. *Lolium perenne* with other *Lolium* species. *J. Genet.* 53: 105–111

Jenkin, T.J. 1954c. Interspecific and intergeneric hybrids in herbage grasses. VIII. *Lolium loliaceum, Lolium remotum* and *Lolium temulentum,* with references to '*Lolium canadense.*' *J. Genet.* 52: 318–331.

Jenkin, T.J. 1955a. Interspecific and intergeneric hybrids in herbage grasses. XII. *Festuca capillata* in crosses. *J. Genet.* 53: 105–111.

Jenkin, T.J. 1955b. Interspecific and intergeneric hybrids in herbage grasses. XIII. The breeding affinities of *Festuca heterophylla*. *J. Genet.* 53: 112–117.

Jenkin, T.J. 1955c. Interspecific and intergeneric hybrids in herbage grasses. XVII. Further crosses with *Lolium perenne*. *J. Genet.* 53: 442–466.

Jenkin, T.J. 1955d. Interspecific and intergeneric hybrids in herbage grasses. XVIII. Various crosses including *Lolium rigidum* sens. Ampl. with *L. temulentum* and *L. loliaceum* with *Festuca pratensis* and with *F. arundinacea. J. Genet.* 53: 467–486

Jenkin, T.J. and P.T. Thomas, 1939. Interspecific and intergeneric hybrids in herbage grasses. III. *Lolium loliaceum* x *L. rigidum. J. Genet.* 37: 255–286.

Jones, E.S., M.P. Dupal, R. Kölliker, M.C. Drayton, and J.W. Forster, 2001a. Development and characterisation of simple sequence repeat (SSR) markers for perennial ryegrass (*Lolium perenne* L). *Theor. Appl. Genet.* 102: 405–415.

Jones, E.S., M.D Hayward, T. Yamada, I.P Armstead,. and J.W. Forster, 2001b. Comparative Mapping of *Lolium perenne* with Other Members of the Poaceae. Plant and Animal Genome IX Conference W31_02. San Diego.

King, I.P., W.G. Morgan, I.P. Armstead, J.A. Harper, M.D. Hayward, A. Bollard, J.V. Nash, J.W. Forster, and H.M. Thomas, 1998. Introgression mapping in the grasses. I. Introgression of *Festuca pratensis* chromosomes and chromosome segments into *Lolium perenne. Heredity* 81: 462–467.

Kölliker, R., F.J. Stadelmann, B. Reidy, and J. Nosberger, 1999. Genetic variability of forage grass cultivars: A comparison of *Festuca pratensis* Huds., *Lolium perenne* L., and *Dactylis glomerata* L. *Euphytica* 106: 261–270.

Kubik, C., W.A. Meyer, and B.S. Gaut, 1999. Assessing the abundance and polymorphism of simple sequence repeats in perennial ryegrass. *Crop Sci.* 39: 1136–1141.

Lehväslaiho, H., A. Saura, and J. Lokki, 1987. Chloroplast DNA variation in the grass tribe *Festuceae. Theor. Appl. Genet.* 74: 298–302.

Loos, B.P. 1993a. Allozyme variation within and between populations in *Lolium* (*Poaceae*). *Plant Syst. Evol.* 188: 101–113.

Loos, B.P. 1993b. Morphological variation in *Lolium* (*Poaceae*) as a measure of species relationships. *Plant Syst. Evol.* 188: 87–99.

Lundqvist, A. 1956. Self-incompatibility in rye. I. Genetic control in the diploid. *Hereditas* 42: 293–348.

Lush, W.M. and M.E. Rogers 1992. Cutting height and the biomass and tiller density of *Lolium perenne* amenity turfs. *J. Appl. Ecol.* 29: 611–618.

Matile, P., S. Hörtensteiner, H. Thomas, and B. Kräutler, 1996 Chlorophyll breakdown in senescent leaves. *Plant Physiology* 112: 1403–1409.

May, L.H. and F.L Milthorpe, 1962. Drought resistance in crop plants. *Field Crop Abs.* 15: 171–179.

Meyer, W.A. and F.C. Belanger. 1997. The role of conventional breeding and biotechnical approaches to improve disease resistance in cool-season grasses. *International Turfgrass Society Research Journal* 8: 777–790.

Monestiez, P., M. Goulard, and G. Charmet, 1994. Geostatistics for spatial genetic structures: Study of wild populations of perennial ryegrass. *Theor. Appl. Genet.* 88: 33–41.

Morgante, M. and A.M. Olivieri, 1993. PCR-amplified microsatellites as markers in plant genetics. *Plant J.* 3: 175–182.

Newell, A.J. and A.C. Jones, 1995 Comparison of grass species and cultivars for use in lawn tennis courts. *J. Sports Turf Res. Inst.* 71: 99–106.

Oertel, C., J. Fuchs, and F. Matzk, 1996. Successful hybridization between *Lolium* and *Dactylis*. *Plant Breeding* 115: 101–105.

Olby, R.C. 1987. William Bateson's Introduction of Mendelism to England. *British Journal of History of Science* 20: 399–420.

Olesen, A., M. Storgaard, M. Folling, S. Madsen, and S.B. Andersen, 1995. Protoplast, callus and suspension culture of perennial ryegrass—effect of genotype and culture system. Current issues in plant molecular and cellular biology. pp. 69–74. *Proceedings of the 8th International Congress on Plant Tissue and Cell Culture,* Florence, Italy, 12–17 June, 1994. Kluwer Academic Publishers, Dordrecht, Netherlands. 1995.

Palmer, J.D. 1987. Chloroplast DNA evolution and biosystematic uses of chloroplast DNA variation. *Amer. Natur.* 130: S6–S29.

Peto, F.H. 1933. The cytology of certain intergeneric hybrids between *Festuca* and *Lolium. J. Genet.* 28: 113–137.

Plot, R. 1677. *The Natural History of Oxfordshire.* Oxford University Press.

Roderick, H.W., D. Thorogood, and B. Adomako, 2000. Temperature dependent resistance to crown rust infection in perennial ryegrass, Lolium perenne L. *Plant Breeding* 119: 93–95.

Roldan-Ruiz, I., J. Dendauw, E. Van Bockstaele, A. Depicker, and M. De Loose, 2000. AFLP markers reveal high polymorphic rates in ryegrasses (*Lolium* spp.). *Molecular Breeding* 6: 125–134.

Sackville Hamilton, N.R. 1998. The regeneration of accessions in seed collections of the main perennial forage grasses and legumes of temperate grasslands: Background considerations. pp. 103–108. In L. Maggioni, P. Marum, N.R. Sackville Hamilton, I. Thomas, T. Gass and E.Lipman, Eds. *Report of a Working Group on Forages.* 6th meeting, 6–8 March 1997, Beitstolen, Norway. International Plant Genetic Resources Institute, Rome, Italy.

Sackville Hamilton, N.R. 1999. Genetic erosion issues in temperate grasslands. pp. 48–55. *Proceedings of the Technical Meeting on the Methodology of the FAO World Information and Early Warning System on Plant Genetic Resources.* Prague, Czech Republic. J. Serwinski and I. Faberova, Eds. International Plant Genetic Resources Institute, Rome, Italy.

Shildrick, J.P. 1981. Shoot numbers, stem bases and persistence in artificially worn perennial ryegrass cultivars. *J. Sports Turf Research Institute* 57: 84–107.

Siffelova, G., M. Pavelkova, A. Klabouchova, I. Wiesner, and V. Nasinec, 1997. Computer-aided RAPD fingerprinting of accessions from the ryegrass-fescue complex. *J. Agr. Sci.* 129: 257–265.

Spangenberg, G., Z.Y. Wang, X.L. Wu, J. Nagel, and I. Potrykus, 1995. Transgenic perennial ryegrass (*Lolium perenne*) plants from microprojectile bombardment of embryogenic suspension cells. *Plant Science* 108: 209–217.

Spoor. W. 1976. Self-incompatibility in *Lolium perenne* L. *Heredity* 37:417–421.

Stam, P. and J.W. van Ooijen, 1995. JOINMAPÔ version 2.0. Software for the calculation of genetic linkage maps. CPRO-DLO, Wageningen.

Stammers, M., J. Harris, G.M. Evans, M.D. Hayward, and J.W. Forster, 1995. Use of random PCR (RAPD) technology to analyze phylogenetic relationships in the *Lolium-Festuca* complex. *Heredity* 74: 19–27.

Stebbins, G.L. 1956. Taxonomy and evolution of genera with special reference to family Gramineae. *Evolution* 10: 235–245.

Terrell, E.E. 1968. A Taxonomic Revision of the Genus Lolium. Technical Bulletin No. 1392. United States Department of Agriculture, Washington, D.C.

Thomas, H., S.J. Dalton, C. Evans, K.H. Chorlton, and I.D. Thomas. 1996a. Evaluating drought resistance in germplasm of meadow fescue. *Euphytica* 92: 401–411

Thomas H.M., J.A. Harper, M.R. Meredith, W.G. Morgan, I.D. Thomas, E. Timms, and I.P. King, 1996b. Comparison of ribosomal DNA sites in Lolium species by S.J. fluorescence *in situ* hybridization. *Chromosome Research* 4: 486–490

Thomas, H. and C.J. Howarth, 2000. Five ways to stay green. *J Exp. Bot.* 51: 329–337

Thomas, H. and C.M. Smart, 1993. Crops that stay green. *Ann. of Appl. Biol.* 123: 193–219.

Thorogood, D. 1996. Varietal colour of *Lolium perenne* L. turfgrass and its interaction with environmental conditions. *Plant Varieties and Seed* 9: 15–20.

Thorogood, D. and M.D. Hayward, 1991 The genetic control of self compatibility in an inbred line of *Lolium perenne. Heredity* 67: 175–181.

Thorogood, D. and M.D. Hayward, 1992. Self-fertility in *Lolium temulentum*—Its genetic control and transfer into *L. perenne* and *L. multiflorum. Heredity* 68: 71–78.

Thorogood, D., M. Paget, M.O. Humphreys, L. Turner, I. Armstead, and H. Roderick, 2001. QTL analysis of crown rust resistance in perennial ryegrass—implications for breeding. *International Turfgrass Society Research Journal*: 9: 218–223.

Thorogood, D., W.J. Kaiser, J.G. Jones, and I. Armstead, 2002. Self-incompatibility in ryegrass 12. Genotyping and mapping the S and Z loci in *Lolium perenne* L. *Heredity* in press.

Turner L.B., M.O. Humphreys, A.J Cairns, and C.J. Pollock, 2000. Resource allocation in *Lolium perenne*—A mapping approach. *J. Exp. Bot.* 51: 49.

Van Dersal, W.R. 1936. The ecology of a lawn. *Ecology* 17: 515–527.

Van Wijk, A.J.P. 1993 Turfgrasses in Europe: Cultivar evaluation and advances in breeding. *International Turfgrass Society Research Journal* 7: 26–38.

Volaire F., H. Thomas, N. Bertagne, E. Bourgeois, M.F. Gautier, and F. Lelievre, 1998. Survival and recovery of perennial forage grasses under prolonged Mediterranean drought—II. Water status, solute accumulation, abscisic acid concentration and accumulation of dehydrin transcripts in bases of immature leaves. *New Phytol.* 140: 451–460.

Waldron, B.L., N.J. Ehlke, D.L. Wyse, and D.J Vellekson, 1998. Genetic variation and predicted gain from selection for winter hardiness and turf quality in a perennial ryegrass top-cross population. *Crop Sci.* 38: 817–822.

Wang, G.R., H. Binding, and U.K. Posselt, 1997. Fertile transgenic plants from direct gene transfer to protoplasts of *Lolium perenne* L. and *Lolium multiflorum* Lam. *J. Plant Physiol.* 151: 83–90.

Wang, Z., J.L. Weber, G. Zhong, and S.D. Tanksley, 1994. Survey of plant short tandem repeats. *Theor. Appl. Genet.* 88: 1–6.

Warpeha, K.M.F., I. Capesius, and T.J. Gilliland, 1998a. Genetic diversity in perennial ryegrass (*Lolium perenne*) evaluated by hybridization with ribosomal DNA: Implications for cultivar identification and breeding. *J. Agr. Sci.* 131: 23–30.

Warpeha, K.M.F., T.J. Gilliland, and I. Capesius, 1998b. An evaluation of rDNA variation in *Lolium* species (ryegrass). *Genome* 41: 307–311.

Warren, J.M., A.F. Raybould, T. Ball, A.J. Gray, and M.D. Hayward, 1998. Genetic structure in the perennial grasses *Lolium perenne* and *Agrostis curtsii*. *Heredity* 81: 556–562.

White, J. 1981. The allometric interpretation of the self-thinning rule. *J. Theor. Biol.* 89: 475–500.

Wilkins, P.W. and D. Thorogood, 1992. Breakdown of self-incompatibility in perennial ryegrass at high temperature and its use in breeding. *Euphytica* 64: 65–69.

Wit, F. 1974. Cytoplasmic male-sterility in ryegrasses (*Lolium* spp.) detected after intergeneric hybridization. *Euphytica* 23: 31–38.

Xu, W.W. and D.A. Sleper, 1994. Phylogeny of tall fescue and related species using RFLPs. *Theor. Appl. Genet.* 88: 685–690.

Young, W.C. 1997. Grass Seed Exports...Where Does It All Go? Department of Crop and Soil Science, Oregon State University, Corvallis, OR.

Chapter 8

Tall Fescue
(*Festuca arundinacea*)

William A. Meyer and Eric Watkins

Tall fescue (*Festuca arundinacea* Schreb.) is adapted for use as a turfgrass in many areas of the world, including many parts of North America, South America, Europe, the cooler parts of Asia, Africa, Australia, and New Zealand. In the United States, tall fescue is mainly adapted to the eastern half of the country, California, and parts of the Pacific Northwest. The species is most widely used as a turfgrass in the transition zone of the southeastern United States that stretches from Missouri to Virginia and Maryland. The primary reason that it does well in this region is its ability to survive under the drought conditions that are common to this part of the country.

DISTRIBUTION, PHYSIOLOGY, AND CYTOTAXONOMY

History and Worldwide Use

According to Borrill (1976), European settlers introduced tall fescue into North and South America. Tall fescue can be found in damp pastures throughout Europe and North Africa and in the mountains of east Africa and Madagascar (Buckner et al., 1979). Western Europe is the main center of origin of Festuceae.

In U.S. trials in the 1800s tall fescue was recognized as being more resistant to crown rust (*Puccinia coronata* Cda.), more vigorous and more drought tolerant than meadow fescue (*F. pratensis* Huds.) (Ten Eyck, 1903; Buckner et al., 1979). By the early 1900s the superior qualities and persistence of tall fescue were being recognized.

With the release of the cultivars Alta and Kentucky-31, tall fescue became a prominent species. In 1940 Alta was released jointly by the Oregon Agriculture Experiment Station and the Forage and Range Section of the United States Department of Agriculture. It was registered by the American Society of Agronomy in 1944 as a productive forage grass. Alta had better forage quality than turf quality. Two of the progenitors of Alta were from Germany and one was from a Missouri collection (Cowan, 1956).

Kentucky-31 tall fescue was derived from an ecotype growing before 1890 on a Menifee County farm in Kentucky. Dr. E.N. Fergus collected seed from this farm during 1931 (Fergus and Buckner, 1972). After small plot evaluations at the Kentucky Agricultural Experiment Station and other sites in the state, Kentucky-31 was released as a cultivar in 1943 (Fergus, 1952; Fergus and Buckner, 1972). This cultivar was described as having a very tough root system and good sod-forming ability on the rolling hills of western Kentucky. It also had the ability to persist at low soil fertility and had improved tolerance to many insects compared to other cool-season species (Murray and Powell, 1979). This grass became very popular as a forage, conservation, and turfgrass. Over 50 million kg of Kentucky-31 seed are still produced each year, mainly in Missouri (personal communication, Ronnie Stapp, Pennington Seed Company, 2001).

Large amounts of Kentucky-31 are sold each year, primarily in the transition zone of the United States, to annually overseed tall fescue lawns in September and October. The quality of Kentucky-31 tall fescue seed is lower than tall fescue seed grown in the northwestern United States because of contamination from undesirable grasses such as orchard grass (*Dactylis glomerata* L).

By the 1960s states in the transition zone were recommending Kentucky-31 because of its wide adaptation to variable soil pH, rainfall, and sunlight (Murray and Powell, 1979). It was also used widely in the southwestern U.S. under irrigation and in the higher elevations of Arizona and Nevada (Meyer and Funk, 1989).

Climatic and Physiological Limitations

Hartley (1950) mentioned that temperature extremes limit the distribution of tall fescue. In colder regions such as central and northern Canada and at high elevations, tall fescue occurrence is restricted (Borrill et al., 1976). Burns and Chamblee (1979) reported data from Saskatoon, Saskatchewan, that showed complete winterkill of tall fescue in the first year after sowing, while survival was good only 175 miles to the south. Survival under cold temperatures can be greatly modified by snow cover.

Observations over 20 years in New Jersey indicate that extended periods of ice sheeting can cause winter injury to tall fescue; however, damage is not as severe as on perennial ryegrass (*Lolium perenne* L.), which can be completely killed (personal communication, C.R. Funk). Compared to perennial ryegrass, tall fescue has greater winter dormancy in New Jersey.

The only report of winter injury to tall fescue in the National Turfgrass Evaluation Program showed up to 14.5% winterkill in 1994–95 in a trial seeded in 1992 (Shearman and Morris, 1996). Tall fescue requires higher soil temperatures for seed germination than either Kentucky bluegrass (*Poa pratensis* L.) or bentgrass (*Agrostis* spp.) (Beard, 1982). In areas like Nebraska, spring seeding of tall fescue has become popular. If tall fescue seedings are made late in the fall, the limited development of seedlings can result in desiccation losses due to frost heaving.

Tall fescue develops the deepest root system of any of the cool-season turfgrass species. Rather than being a drought-tolerant grass, it is better described as a drought-avoidant grass because of its deep root system and its ability to use water deep in the soil profile. Observations in a tall fescue turf mowed at 5 cm in southern California showed excavated roots at depths of 140 cm (Rector J., personal communication). Supplemental irrigation will allow tall fescue to perform well as turf in arid areas. In southern California and parts of southern Europe, it has become the most popular cool-season turfgrass for lawns mowed at 4 cm or higher.

Tall fescue will thrive in a wide variety of soil types but prefers heavy to medium textured soils with considerable humus (Buckner and Cowan, 1973). It will grow in alkaline and saline soils better than most other cool-season species. Optimum growth occurs at a pH of 5.5 to 6.5, but the species will tolerate pH as low as 4.7 to as high as 8.5 (Beard, 1973). Tall fescue tolerates wet soils and survives in compacted soils better that other cool-season turfgrasses (Younger, 1965).

This species is best adapted to the transition zone between the cool-humid and warm-humid regions where neither warm nor cool-season species are well adapted (Juska et al., 1969; Beard, 1973). Due to its good shade tolerance, tall fescue fills a special need in the southern United States where commonly used bermudagrass has poor shade tolerance.

Brown patch (*Rhizoctonia solani* Kühn) and leaf spot (*Dreschlera dictyoides* Drechsl.) are the most serious diseases of tall fescue (Couch, 1962). Brown patch is a serious problem in most areas and presents the greatest challenge to turfgrass breeders. Leaf spot is more severe during establishment and under wet, shaded conditions. Much progress has been made in developing cultivars with resistance to leaf spot (Shearman and Morris, 1996). In areas like the Pacific Northwest and the northeastern United States, *Microdochium* patch (*Microdochium nivale* Fr.) can seriously limit the use of tall fescue, especially the highly dense dwarf-types (Morris and Shearman, 2000).

Tall fescue has good seedling vigor but has a reduced rate of secondary tiller formation compared to perennial ryegrass. The basic principles of turf establishment (liming and fertilization, seed placement and irrigation) are similar for all seeded cool-season turfgrass species (Murray and Powell, 1979). The one unique requirement of tall fescue is the need to get an area well established before applying traffic pressure (this is not the case with perennial ryegrass and improved Kentucky bluegrass). Under traffic from mowing, leaf spot can be a severe disease on new seedlings.

Turfgrass quality of tall fescue is best with mowing heights of 3.8 to 7.5 cm (Davis, 1958; Murray and Powell, 1979). When tall fescue is maintained at cutting heights below 3.3 cm, competition from annual bluegrass (*Poa annua* L.) can be severe. Compared to other cool-season species, tall fescue has a medium fertility requirement. Tall fescue is typically seeded at rates of 200–400 kg/ha, which is greater than other cool-season species (Younger, 1965; Murray and Powell, 1979). The only species that can be effectively mixed with tall fescue is Kentucky bluegrass. In this mixture, only 5 to 10% Kentucky bluegrass, by seed weight, should be used. This type of mixture helps the turf grower by adding sod strength and reducing the amount of brown patch disease.

Taxonomy and Cytology

The taxonomic classification of tall fescue is:

Family	Poaceae
Subfamily	Festucoideae
Tribe	Festuceae
Genus	*Festuca*
Species	*arundinacea*
Authority	Schreb.

Tall fescue belongs to the Bovinae section of the genus and is classified as belonging to the tribe Festuceae (Buckner et al., 1979); however, it shares many characteristics with its close relative meadow fescue. Originally, some confusion developed about the classification of tall fescue and meadow fescue pertaining to whether they were separate species. In 1771 the German botanist Schreber described tall fescue as being more robust than meadow fescue (*F. elatior* L.). In 1935, Hitchcock regarded tall fescue as *F. elatior* var *arundinacea*, and in 1950, he described them as separate species, *F. elatior* L. (meadow fescue) and *F. arundinacea* Schreb. (tall fescue) (Buckner et al., 1979; Hitchcock, 1935; Hitchcock, 1950). The two grasses have even been considered as the same species. Many researchers have studied the relationship between these two species. The

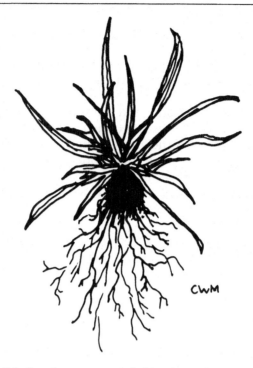

Figure 8.1. Bunch-type growth habit of a turf-type tall fescue.

allohexaploid genome (2n=6x=42) of tall fescue is given the designation $PPG_1G_1G_2G_2$. The P genome appears to have come from meadow fescue (2n=2x=14) while the two G genomes come from the tetraploid *Festuca arundinacea* var. *glaucescens* Boiss. (Eizenga et al., 1998). This relationship was confirmed by restriction fragment length polymorphism (RFLP) mapping of meadow fescue, although the P genome was shown to have diverged substantially during evolution (Chen et al., 1998).

Morphological Description

Tall fescue is a tufted bunchgrass that may or may not have short rhizomes (Figure 8.1) (Terrell, 1979). Rhizomes are more commonly seen on spaced plants in sandy soils but are rarely seen in turf.

Culms are erect, stout, smooth (or rough below panicle), to 2 m tall; ligules are membranous and up to 2 mm long on common types, and are usually absent on turf-types. Tall fescue has a rolled vernation and pointed leaf tip. It has a coarse medium-fine leaf texture with leaves having prominent veins on the upper side and lacking a mid rib (Figure 8.2) (Christians, 1998).

Leaf blades are minutely scaborous or smooth, scaborous on the margin, up to 60 cm long and 3–12 mm wide. Auricles are absent or short, and auricles and collars are ciliate or sometimes glaborous (Terrell, 1979). The collar is broad and may have a light-green shiny appearance (Christians, 1998). Panicles are typically 10–50 cm long, broad, and loosely branched (Figure 8.3), although they can be narrow with short branches.

Spikelets are elliptical to oblong, each with 3 to 10 florets. Rachilla internodes are scabrous. The lower glumes are narrow-lanceolate, 3–6 mm long with one nerve. The upper glumes are lanceolate to narrow-oblong, 4.5 to 7 mm long and three-nerved. The lemmas are narrowly elliptic or lanceolate, 6 to 10 mm long and fine-nerved, often scaborous with minute prickles or rarely glabrate with awns up to 4 mm long or

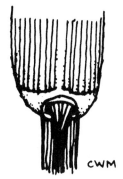

Figure 8.2. Collar region of a tall fescue leaf.

Figure 8.3. Typical panicle of a tall fescue plant.

absent. The paleas are as long as the lemmas. The oblong caryopsis is tightly enclosed by the lemma and palea (Terrell, 1979). On many fine-textured, dwarf tall fescue cultivars, the above measurements can be reduced, especially for leaf width, leaf length, and plant height.

Control of Flowering

Tall fescue is primarily an outcrossing species that has disomic inheritance of chromosomes, and the species shows almost complete self-incompatibility. A cross-pollinating species offers both advantages and disadvantages to plant breeders. Creating new genetic combinations in the breeding populations of tall fescue is quite easy; however, maintaining genetic combinations in a population during cultivar development can be a challenge. Due to its extensive adaptation in many parts of the United States, tall fescue plants can be found growing in many natural areas and roadside ditches. This presents a problem for plant breeders because foreign pollen can fertilize breeding nurseries and delay progress in cultivar development; therefore, tall fescue breeders must remove or mow down these naturally occurring tall fescue plants whenever possible.

The factors that affect the flowering of tall fescue have not been extensively studied. The research that has been done in this area shows that many factors are probably involved in floral induction in tall fescue.

Tall fescue floral development is enhanced by exposure to cold temperatures during the winter months. In Kentucky, Templeton et al. (1961) found that all tall fescue shoots that were exposed to an entire winter (appeared aboveground by December) produced panicles, while only about two-thirds of shoots that appeared during late February or early March formed panicles. The lack of an absolute vernalization requirement for flowering can speed up cultivar development in situations where large amounts of seed are not needed.

Templeton et al. (1961) also found that tall fescue will flower under a range of photoperiods; a short photoperiod was not an absolute requirement for floral induction (when vernalization had occurred during the winter) and a long photoperiod resulted in earlier flowering but fewer panicles per plant (Templeton et al., 1961). Tall fescue appears to prefer short days and/or low temperature for induction of flowering, and long days (16 h) for initiation and development of flowers (Hicks and Mitchell, 1968).

Tall fescue plants vary substantially in floral fertility, with some cultivars producing extremely high yields. Cultivars such as Kentucky-31 that produce large amounts of seed are used more widely due to their low cost. When developing tall fescue cultivars, breeders try to identify and select plants that appear to have high floral fertility. Tall fescue genotypes are diverse in the amount of panicles they produce, as well as in the amount of pollen they shed. In New Jersey, it has been observed that tall fescue plants shed their pollen in the early to midafternoon hours.

Tall fescue (turf-type) seed producers prefer cultivars that mature early in the growing season. Earlier maturing cultivars typically have higher seed production because they are able to develop before heat and disease can negatively affect seed set. As a result, tall fescue breeders try to select plants with early anthesis dates. The most effective way to select for anthesis date is to select across multiple environments (to reduce genotype X environment interaction), and to only select in the years after establishment (Watson Jr. and McLean, 1991). Because of the time required for this type of evaluation, turfgrass breeders rarely use it. Current turf-type cultivars have a wide range of maturity dates. Some of the earlier maturing cultivars include Kentucky-31, Falcon, and Endeavor; while the later maturing cultivars include Matador, Finesse, and Apache II (Watkins and Meyer, 2000). One of the more difficult challenges that plant breeders face is being able to develop an early maturing cultivar that has superior turf characteristics such as dark green color, fine leaf texture, and disease resistance.

The identification of genes that control flowering in tall fescue has yet to occur. Understanding the genetic control of flowering in the species would be beneficial, in that it would allow for improved selection protocols.

TURF-TYPE CULTIVAR DEVELOPMENT

Early Cultivars

In 1962, Dr. C. Reed Funk at the New Jersey Agricultural Experiment Station started to make collections of persistent tall fescue clones (0.25 to 1 meter in diameter) from old turf areas. In naturalized stands throughout the country, a small percentage of plants were found to have a lower growth habit, finer leaves, and a reduced rate of vertical growth. These clones, along with germ plasm from the Plant Introduction Service, were used to initiate cycles of population improvement of tall fescue under closely mowed single plant progeny turf trials. After 19 years of testing and 5 cycles of recombination, the turf-type cultivar Rebel was released in 1981 (Funk et al., 1981b).

This release was closely followed by the release of the turf-type cultivars Falcon and Olympic (Funk et al., 1981a; Meyer et al., 1982).

These new turf-type cultivars had leaves that were 30–40% finer and tiller densities almost twice that of Kentucky-31. They also had a darker-green color, improved persistence under close mowing, and better overall disease resistance (Meyer and Funk, 1989). These turf-type cultivars (especially Falcon) had good seed yields in the Pacific Northwest. Beginning in the mid-1980s, many new improved turf-type cultivars were released. Some of the prominent new cultivars were Adventure, Apache, Arid, Bonanza, Cimarron, Finelawn, Finelawn 5GL, Houndog, Jaguar, Mesa, Mustang, Rebel II, Tribute, and Wrangler (Meyer and Funk, 1989).

Lower-growing turfgrasses are important because of decreased mowing costs and improved performance. The first turf-type cultivars Rebel and Falcon had mature heights ranging from 130 to 145 cm as individual spaced plants. In breeding populations of the new turf-type cultivars, a small percentage of the plants were found that were 100 cm or less in plant height. These shorter plants also showed a reduced rate of vertical growth and higher tiller densities than Rebel or Falcon when grown as a mowed turf (Meyer and Funk, 1989).

The first dwarf-type cultivar was Trailblazer. Other prominent dwarf-type cultivars that were released soon after were Bonsai, Eldorado, Murietta, and Silverado (Meyer and Funk, 1989). These new dwarf-type cultivars have higher tiller densities, which can result in increased disease pressure from brown patch, especially under hot and humid conditions. They have performed exceptionally well under irrigation in arid climates, such as southern California, where disease pressure is reduced. These shorter tall fescues also have an advantage in seed production because of reduced lodging.

In the year 2000, over 50 million kg of turf-type and dwarf-type tall fescue seed were produced in western Oregon. The average seed yields of turf-type tall fescue have increased from 900 kg/ha (Rebel) to well over 1800 kg/ha on recently developed cultivars.

Tremendous progress has been made since the first turf-type and dwarf-type cultivars were released. In recent years, the cultivars that are referred to as intermediate types (between turf-types and dwarf-types) have been shown to have the best adaptation throughout the United States. Many of these cultivars also have high seed production potential.

Major Sources of Germ Plasm

As described previously, tall fescue was introduced into the United States in the late 1800s. Cowan (1956) noted that the earliest collections of tall fescue in the United States were made in 1886. The first official plant introduction of *F. arundinacea* by the USDA was PI 5835 from Sweden in 1901 (Terrell, 1979).

Lawns and pastures were seeded in the United States by the settlers with seed they brought to the United States or with seed from USDA sources. Over a period of 150 years, much selection has occurred under grazing and mowing, which has lead to the development of the germ plasm presently used in breeding programs. The earliest cultivars were Alta (Cowan, 1956), Kentucky-31 (Fergus, 1952), and Rebel (Funk et al., 1981b).

In 1962, Dr. C. Reed Funk started a 5,000 hour trek to examine old, naturalized turf and pasture areas throughout the United States. He was able to find tall fescue clones that persisted in a wide range of environments including Alabama, Georgia, Idaho, Kansas, Kentucky, Maryland, Missouri, Mississippi, New Jersey, North Carolina, South Carolina, Ohio, Kansas, Pennsylvania, Tennessee, Texas, and Virginia. These collections were made from 1962 to 1982 (Funk and Meyer, 2001). The senior author has also made similar collections in the mid-Atlantic, Midwest, and western U.S. from 1975 to 2000.

Whether many of these collections originated from the widely used Kentucky-31 or from other introductions of tall fescue, is difficult to determine. The important turf collections represented a few rare plants that had finer leaves and a low growth habit. These persistent clones were grown as spaced plants and interpollinated with other similar plants. Many plants had to be discarded prior to anthesis. Seed from these plants was used to seed single plant progeny turf trials. Interplant competition within the turf, and evaluations after a number of years, resulted in selecting plants with which to initiate cycles of selection as spaced plants. Rebel tall fescue resulted from 19 years of turf trials and five cycles of recombination (Funk et al., 1981b).

The senior author has taken part in, and coordinated collection trips in eastern Europe. After many days of collecting, only a single low-growing, dense clone was found. A recent trip to central and northern Portugal indicated that a higher percentage of collections from this region will have a lower turf-type growth habit.

Obviously, the turf-type cultivars presently being released are vastly different than most of their wild sources of parental germ plasm. This has resulted from 38 years of selection and breeding work on these cultivars. Considering that Kentucky-31 is most likely one of the original parental sources, many morphological changes have obviously occurred to this species.

Wild Germ Plasm

The search for wild germ plasm of tall fescue is needed to add to the present breeding programs. The importance of this can be seen in recent experiences with collections of perennial ryegrass from eastern Europe, which have provided excellent sources of resistance to gray leaf spot (*Pyricularia grisea* [Cooke] Sacc.) (Meyer et al., 2001a). With additional cycles of recurrent phenotypic selection in conjunction with turf plot progeny evaluations, the need for new sources of germ plasm continues to be a high priority.

Breeding Techniques

Classical Approach

The main traits of tall fescue that breeders exploit in cultivar development are cross-pollination and self-incompatibility. Allowing a large number of tall fescue plants to intercross creates a myriad of genetic combinations in the progeny. These progeny are then evaluated in either turf plots or as spaced plants in a nursery, and superior genotypes are selected and evaluated further. Eventually, a group of plants with similar maturity and phenotypic traits are intercrossed in a polycross. This final cross produces the breeder seed of a new cultivar.

At Rutgers University, evaluation of single plant progenies under close mowing has proven to be an effective method to select superior plant material. These trials are usually conducted by sowing a 1.0 x 1.3 m plot and maintaining it under a 3.3 cm mowing height for 1 to 4 years. After germination occurs in the fall, a large roller is applied to the turf each week to enhance the development of leaf spot. In most years, sufficient segregation occurs within the progenies for resistance to this disease, and up to 200 resistant seedlings per plot can be selected in December. These seedlings are then screened in the greenhouse under short day conditions at a temperature of about 7°C for 3 to 4 months. This technique has worked very effectively to make dramatic improvements in the level of leaf spot resistance in tall fescue. Coronado (Rose-Fricker et al., 1999a) and Barlexus are tall fescue cultivars with improved resistance to leaf spot that were developed using the above technique (Shearman and Morris, 1996).

After the tall fescue progeny trials have gone through two summers with heat and drought stress and brown patch damage, selections are made in late summer, based on turf performance and disease incidence. These selections involve the removal of 1–200 individual plants/plot which are then increased, after screening, for another cycle of selection as spaced plants. When brown patch disease pressure is low, inoculations are made by growing *R. solani* on sterilized Kentucky bluegrass seed and applying the seed to the turf with a drop spreader in order to create uniform epidemics.

If a large percentage of plants in an isolated nursery of 2,000–3,000 plants have a similar growth habit and maturity, off-type plants can be removed prior to anthesis, while the remaining plants are allowed to interpollinate. The cultivars Millennium (Bara et al., 2000) and Rembrandt (Ford et al., 2000) were developed using this strategy.

In many cases, the most effective approach to use when selecting from a large spaced plant nursery is to select plants with similar phenotypes, and move them to an isolated crossing block. These crossing blocks typically contain 50 or more plants each. Selections are based on color, basal tiller density, growth habit, leaf texture, absence of foliar diseases, plant height and maturity. These plants are then flood-irrigated until the seed begins to mature. Prior to seed harvest, each plant is examined for the quality and quantity of seed. Approximately 30% of all plants that were allowed to pollinate are discarded before harvest. In many cases, seed from individual plants harvested from these crossing blocks is sent to the Pacific Northwest for further evaluation. The seed is grown out as spaced plants in a replicated nursery, while maintaining the identity of each maternal line. Concurrently, single progeny turf plots are established in New Jersey. Turf performance records, before and after anthesis, are used to guide selection in the Pacific Northwest. These selections are based on seed production and disease resistance with special emphasis on stem rust resistance (*Puccinia graminis* subsp. *graminicola*) (Welty and Mellbye, 1989). Usually the turfgrass breeders in the northwestern United States go through one or two cycles of phenotypic recurrent selection before releasing breeder seed of a new cultivar.

In cases where individual tall fescue clones have shown exceptional seed yield (over 250 g/plant), vegetative clones are kept and planted as replicated spaced plants. These are maintained until turf performance can be determined in single progeny turf plots. After one year of progeny turf evaluations, the top performing clones are allowed to interpollinate in isolation. The cultivar Plantation was developed using this approach (Meyer et al., 2001b).

Another technique used in breeding tall fescue and many other turfgrass species includes the evaluation of single clones as tiller plots under frequent mowing. This is usually done by killing an old Kentucky bluegrass turf with a nonselective herbicide and then planting individual plants at 0.3 x 0.3 m spacings into the old turf. The old thatch serves to prevent soil from being carried by the mowing equipment and prevents erosion until the individual plants grow to cover the area. After 1–2 years of frequent mowing at 5 cm, selections are made based on turf quality and absence of disease.

Molecular Approach

Although the conventional breeding techniques that have been used to develop turf-type tall fescue have been effective, they are quite slow. Alternative ways to create new gene combinations in tall fescue germ plasm would be advantageous to plant breeders. The simplest way that new gene combinations could be found without going through the typical breeding process is through somaclonal variation. Although this method relies heavily on chance, somaclonal variation has been found in tissue culture regenerants of tall fescue (Roylance et al., 1994).

Another method that has been studied is the use of doubled haploids. The first step would be to generate haploid plants via anther culture. These haploids would usually be cultured from either F_1 or F_2 plants so that they would possess a high amount of genetic diversity (Poehlman and Sleper, 1995). The plants would be propagated, by tillering (Kasperbauer and Eizenga, 1985), and evaluated in the field in a replicated study and superior plants would be selected. Identifying different genotypes in a population of tall fescue haploids is possible (Kasperbauer and Eizenga, 1985). The chromosomes in these plants would be doubled (procedure in Kasperbauer and Eizenga, 1985), resulting in plants that are homozygous at all loci. Doubled haploid plants could then be integrated into a breeding program.

One of the ways that researchers could accelerate the breeding process is through the use of genetic transformation. A number of methods are being studied that could eventually lead to transgenic turf-type tall fescue cultivars. The primary methods that have been studied to this point are direct gene transfer and *Agrobacterium tumefaciens*-mediated gene transfer.

Direct gene transfer involves the uptake of foreign DNA by protoplasts. The primary techniques used in direct gene transfer include incubation in polyethylene glycol (PEG), insertion of DNA using a particle gun, and electroporation (Poehlman and Sleper, 1995). One of the first steps toward transformed tall fescue was the refinement of a procedure to develop plants from cell suspension protoplasts (Dalton, 1988). Without the development of this procedure, direct gene transfer of tall fescue would not have been possible. The performance of nontransformed cell suspension-derived tall fescue regenerants and their progenies was similar to seed-grown plants in a field trial; thus, plants that are regenerated from protoplasts could be incorporated into breeding programs without major restrictions (Stadelmann et al., 1999). Currently, no reports of transformed tall fescue plants carrying a foreign gene of agronomic significance have been documented; however, the insertion of reporter genes has been done successfully using electroporation (Ha et al., 1992), PEG (Wang et al., 1992; Kuai and Morris, 1996; Bettany et al., 1998; Kuai et al., 1999), and particle gun bombardment (Spangenberg et al., 1995). One of the concerns with these transformed plants is whether or not they will grow properly under field conditions. The primary concern would be the stability of transgene expression during tillering of the tall fescue plant. Bettany et al. (1998) found that a reporter gene in tall fescue was unstable during early generations of tillering. In order for these methods to be effective, stable transgene expression is necessary. *Agrobacterium tumefaciens*-mediated transformation may prove to be an effective method of gene transfer in grasses such as tall fescue, and is currently being studied as an option for transformation of the species (Lakkaraju et al., 2001).

A breeding method that has not yet been used in tall fescue, but may have great potential, is marker-assisted selection. The first step in this method of breeding is to develop a map of the tall fescue genome. Once a gene is mapped, it might be associated with a RFLP band; thus, the breeder could search for plants that have the band that is associated with the desirable gene, and the gene could be transferred into a proper genetic background (Poehlman and Sleper, 1995).

Very little work has been done in the area of gene mapping in tall fescue. Monosomic lines, which can aid in gene location, have been developed (Eizenga and Phillips, 1998). RFLPs have been used to make a tall fescue genomic library (Xu et al., 1991), and a tall fescue genetic map (Xu et al., 1995). Future research will focus on developing more extensive maps and identifying genes of agronomic importance.

Interspecific Hybridization

Interspecific hybridization has been used in the past to increase the forage quality of tall fescue. Many examples of research dealing with the crossing of tall fescue with annual ryegrass (*Lolium multiforum* L.) (Buckner et al., 1961; Hill and Buckner, 1962; Buckner et al., 1965; Webster and Buckner, 1971; King et al., 1987; Eizenga et al., 1991), perennial ryegrass (Buckner et al., 1961; Hill and Buckner, 1962; Webster and Buckner, 1971), and giant fescue [*Festuca gigantea* (L.) Vill.] (Buckner et al., 1976; King et al., 1987; Burner et al., 1989; Eizenga et al., 1990; Burner et al., 1991; Eizenga et al., 1991) can be found. Although some of these species, with the exception of perennial ryegrass, lack superior turfgrass characteristics, they may contain useful genes for traits such as disease resistance. Interspecific hybridization is a useful tool to transfer genes between related species. The cultivar Kenhy was derived from annual ryegrass X tall fescue hybridization (Asay et al., 1979).

Major Selection Criteria

Tall fescue breeders select for a number of traits when evaluating germ plasm, including dark green color, good winter color, fine leaf texture, density, insect and disease resistance, drought tolerance, seedling vigor and establishment, seed yield, and maturity. The major selection criteria for turf-type tall fescue are resistance to brown patch, turf density, endophyte infection, and acid soil tolerance.

Brown Patch

Brown patch is a disease of many cool-season turfgrass species. Brown patch will produce varying symptoms on tall fescue depending on height of cut and weather conditions. In high-cut turf, the disease first appears as a light brown, circular patch that can eventually grow to 0.3 to 0.9 m in diameter (Smiley et al., 1992). The distinctive, irregularly shaped lesions have a tan center surrounded by an outer black margin (Smiley et al., 1992). Disease development is favored by 10 hours of consecutive leaf wetness, high nitrogen, and minimum nighttime temperatures above 20°C (Smiley et al., 1992). The pathogen is able to survive unfavorable conditions in the form of bulbils or thick-walled mycelia in plant debris (Smiley et al., 1992). Chemical control is available for the disease. In some parts of the country, brown patch is only a severe problem for a few weeks in mid to late summer. In these areas, as soon as favorable weather conditions for the disease subside, the tall fescue turf will often recover completely.

Although some management practices can reduce brown patch severity in tall fescue, the ideal solution to the problem would be genetic resistance. Green II et al. (1999) showed that the greater levels of resistance to brown patch in Kentucky-31 compared to 'Mojave' were due to wider leaves and slower lesion expansion. They proposed that future research look at the development of genetic markers of biochemical factors that might be associated with slower lesion expansion. Identification of resistance genes in tall fescue and avirulence genes in *R. solani* would also be very helpful to plant breeders.

Unfortunately no complete genetic resistance to brown patch has been found in any available turf-type tall fescue cultivars. Researchers must continue to look for resistance genes within the species. One way that genetic resistance could be identified is through the use of a clonal nursery of tiller plots (see description of procedure in 'Breeding Techniques'). Superior tall fescue germ plasm could be planted into tiller plots and inoculated with *R. solani*. Individual plants that showed no disease could then be cloned and replicated (to remove any genotype X environment complications) in another tiller plot study the following year. These plants could then be inoculated with the disease and plants with resistance could be identified. These plants could then be crossed with one another and with susceptible genotypes, and their progeny evaluated in the same manner. This method should result in the identification of any resistance genes, if they exist.

Because direct genetic resistance to the pathogen has been difficult to obtain, the development of cultivars with reduced density may be necessary. Giesler et al. (1996a) showed that cultivar stature affects canopy density and disease development. They looked at three different stature types of tall fescue (dwarf, medium, tall), and found that disease levels were highly correlated with blade density and verdure measurements. They also showed that the tall types (Kentucky-31 and Fawn) exhibited less disease than both the medium and dwarf types. In a separate study (Giesler et al., 1996b), they seeded Fawn tall fescue in the greenhouse at two different rates, and later inoculated the seedlings with *R. solani*. The fungal hyphae in the high-density treatment were shown to have spread at a faster rate than the hyphae in the low-density treatment. This study sheds light on the reason that cultivars with reduced density show less disease severity; the fungus can still grow and spread, but at a reduced rate. This reduced spread helps reduce rapid disease buildup during favorable weather conditions.

Density

Because turf-type tall fescue was developed from forage tall fescue, many years of intensive breeding were needed to develop acceptable turf density. Compared to earlier developed cultivars such as Falcon and Rebel, newer cultivars have much superior density (Morris and Murray, 1992). This increased density has been primarily due to the selection of plants that exhibit narrow leaves in spaced-plant nurseries. The major drawback to this increased density is that many of these newer cultivars are more susceptible to brown patch than the earlier developed cultivars.

Recently, current tall fescue cultivars have been classified into groups based on morphological and physiological traits (Watkins and Meyer, 2000). These traits include leaf texture, plant height, and maturity. In order to select tall fescue genotypes with the proper density, the different types of tall fescue and how they relate to turf performance must be understood. Based on recent data, current turf-type tall fescue cultivars have been classified into six categories: forage-type, early standard, standard, early semidwarf, semidwarf, and dwarf (Watkins and Meyer, 2000).

The 'Forage' group consists of the cultivar Kentucky-31, and other forage-types with similar characteristics. This group is defined by having very low turf quality due to reduced density and light green color. One advantage that this group has is high seed production, which allows Kentucky-31 to remain the top selling tall fescue cultivar in the country. Another reason that Kentucky-31 seed is inexpensive is that it is often produced as part of a pasture rotation; therefore, the grower is able to utilize the field for more than one use. This group shows less susceptibility to brown patch, which may be due partly to its wider leaves (Green II et al., 1999). The decreased density associated with these tall fescue types leads to increased invasion of weed species.

The 'Early Standard' group includes cultivars such as Falcon and Olympic. The turf quality of these grasses is improved over the previous group; however, the density is not acceptable and these cultivars exhibit an unattractive color. These cultivars also show a fairly early maturity.

'Standard' cultivars include Rebel Jr., Falcon II, and Coronado Gold (Figure 8.4). This type is characterized by a later maturity, average turf quality, and average turf density. These cultivars are quite successful in areas that require low maintenance. In a

Figure 8.4. Growth habit of a standard, turf-type tall fescue spaced plant.

space-planted nursery, these plants are quite large, much like the plants in the forage and early standard groups.

The 'semidwarf' group consists of most of the cultivars that have been developed in more recent years. These plants have a medium plant height in the nursery, have a medium maturity, increased turf density, and very good turf quality (Figure 8.5). Some examples of this group include Millennium, Rembrandt, and Plantation. The 'early semidwarf' group has the same characteristics as the semidwarf group with the exception of having earlier maturity. 'Prospect,' 'Empress,' and 'Endeavor' are members of this group.

'Dwarf-type' tall fescue cultivars have very narrow leaves and exhibit the highest turf density of any group (Figure 8.6). This increased turf density often leads to increased susceptibility to brown patch and other diseases. Many cultivars in this group are also subject to heat stress. The advantage of this group is that they display very good turf quality when they are not affected by disease or other summer stresses. This group includes the cultivar Matador and several experimental lines that will soon be released.

In New Jersey turf plot evaluations, these groups differed significantly in overall turf quality (semidwarf, early semidwarf, dwarf > standard > early standard and forage). Although statistically insignificant, noticeable differences were also observed in resistance to brown patch with the dwarf type consistently ranking last.

Based on nursery and turf plot observations, in the absence of genetic resistance, breeding efforts should be focused on developing cultivars that exhibit above-average density, but not the high density that is found in the dwarf group. By slightly reducing turf density while retaining other favorable turf characteristics, cultivars with reduced susceptibility to brown patch can be developed.

Endophyte Enhancement

Since the mid-late 1980s, another procedure used to evaluate individual clones of tall fescue has been the microscopic examination of each plant for the presence of

Figure 8.5. Growth habit of a semidwarf, turf-type tall fescue spaced plant.

Figure 8.6. Growth habit of a dwarf, turf-type tall fescue spaced plant.

Acremonium coenophialum Morgan-Jones and Gams endophytes (Saha et al., 1988; Johnson-Cicalese and Funk, 1990; Murphy et al., 1993). Before harvest, each plant is declared E+ or E− based on examinations. These endophytes enhance resistance to aboveground feeding insects such as billbugs (*Sphenophorus* spp.) (Murphy et al., 1993), and can also enhance heat and drought tolerance (Breen, 1994; Funk and White, 1997). The percentage of endophyte in a new cultivar can be calculated and adjusted according to these records. Because of the toxic effects that endophytes have on livestock that consume infected foliage and straw, the percentage of endophyte in a new tall fescue should be kept to approximately 50% (Bacon et al., 1977). In the 1996 National Tall Fescue Test, the endophyte content, in seed, varied greatly: Coronodo Gold-97%,

Rembrandt-95%, Jaguar 3-72%, Plantation-51%, Falcon II-16%, Genesis-8%. Twenty-eight cultivars in the test contained no endophyte (Johnson-Cicalese et al., 1998).

Acid Tolerance

An additional selection criterion that might be considered in some breeding programs is aluminum-tolerance for acid soils. Tall fescue is occasionally grown in acidic soils, and therefore cultivars with superior aluminum-tolerance would be desirable. Foy and Murray (1998) used recurrent selection to obtain aluminum-tolerant tall fescue populations. The two methods that they used were (1) growing the plants in soil with varying pH, and (2) growing the plants in nutrient solutions with varying amounts of aluminum. Both methods were effective in increasing acid tolerance; however, the first method was more successful. They found that tall fescue germ plasm contains "considerable genetic variability with respect to acid soil tolerance..." (Foy and Murray, 1998, p. 1321).

Genetic Vulnerability

The native distribution of tall fescue is quite extensive, and includes most of Europe, North Africa, and western and central Asia and Siberia (Terrell, 1979). Collection efforts should be made in these areas in an attempt to find beneficial genes. Unfortunately, collecting tall fescue for use as turf has not been as successful as collections of other cool-season turfgrass species from these areas. As a result, the genetic base for turf-type tall fescue appears to be quite narrow. RFLP analysis of both forage and turf-type tall fescue shows that the turf-type cultivars are probably more related to each other than are the forage cultivars (Xu et al., 1994).

To date, conditions such as white and yellow seedling disease, which can be an indication of inbreeding, have not been noted in tall fescue breeding populations, while they have been seen often in perennial ryegrass breeding populations. Observations in New Jersey support other research (Fraser et al., 1996) showing Kentucky-31 and Torpedo to be quite susceptible to gray leaf spot while many of the newer commercial cultivars and experimentals have excellent resistance to this disease. All of the commercial perennial ryegrasses were found to be susceptible to gray leaf spot while some new sources of germ plasm from eastern Europe were found to be resistant (Meyer et al., 2001a).

Cultivar Compositions

All of the present commercial tall fescue cultivars are synthetics composed of varying numbers of clones. Examples include: three clones in the case of 'Bonanza' (Meyer and Rose-Fricker, 1990), 36 for 'Tomahawk' (Rose-Fricker et al., 1999b) and 162 clones for 'Millennium' (Bara et al., 2000).

Maintaining many different populations in a breeding project that are going through cycles of recurrent phenotypic and genotypic selection is normal. This approach includes evaluations in single turf plot progeny tests. The potential for inbreeding is reduced by selecting new parental clones from the different populations when selecting the final parents for new cultivars.

Breeder seed is maintained by one of two ways. The seed is kept in storage under proper conditions until needed, or it is maintained vegetatively. Maintaining the breeding material in a vegetative state is usually most effective when a cultivar contains an endophyte, because after a number of years in storage, endophyte content can be lost, even under ideal storage conditions.

BREEDING PROGRESS

Geographic Distribution of Breeding Efforts

Over 15 turf-type tall fescue breeding programs are located in the Pacific Northwest, mainly in the Willamette Valley of Oregon. This is an excellent area to make improvements in seed yield and resistance to stem rust, *Microdochium* patch, pink patch (*Limonomyces roseipellis* Stalpers & Loerakker) and leaf spot (*Dreschlera dictyoides* [Drechs] Shoemaker). At times, severe selection pressure for stem rust has resulted in plants with reduced tolerance to other summer stresses. The most effective programs use an integrated approach of progeny testing in the northeastern or southeastern United States with a concurrent effort of evaluating material for seed production in the northwest.

Turf-type tall fescue breeding programs can be found at Rutgers University; Clemson University; University of Georgia, Griffin; University of Illinois, Urbana; Pure Seed Testing, Inc., Rolesville, NC; and other locations in the east and southeast. Some of these stations are maintained by private companies to evaluate tall fescue breeding material. These locations are favorable for tall fescue breeding because the weather conditions allow breeders to evaluate germ plasm for resistance to many different stresses including, heat, drought, and disease. These programs have been able to make final selections of cultivars (without inputs from the Northwest) that have performed very well in the NTEP trials, some examples being Rembrandt, Millennium, and Plantation that were originally developed in New Jersey and are now being produced in the northwestern United States (Morris and Shearman, 2000). One advantage of selecting for seed production in warmer areas in the eastern United States includes selection for excellent seed set under drought and heat stress. Those lines selected under these conditions have yielded well in the Northwest. If a cultivar is highly susceptible to stem rust, seed production may be severely reduced.

Cultivar Evaluation

Since the release of Rebel (a tremendous improvement over Kentucky-31 and Alta) in 1981, improvement in the turf quality of tall fescue cultivars available to the public has continued. The National Turfgrass Evaluation Program started in 1983 for tall fescue and additional trials were initiated in 1987, 1992, and 1996.

The NTEP trials always include the top ranked cultivars from the previous trial. In most cases, the top cultivar from one trial is ranked well below many (20–30) other cultivars in the next trial. In the 1983 trial, Arid was the top ranked cultivar, with Jaguar ranking second (Murray and Morris, 1988). In the 1987 trial, Arid ranked 47th and Jaguar 49th (Morris and Murray, 1992). Shenandoah was the second-ranked cultivar in the 1987 trial (Morris and Murray, 1992) and ranked 64th in the 1992 trial (Shearman and Morris, 1996).

In the 1992 NTEP trial, 'Jaguar 3' was the top entry (Shearman and Morris, 1996), while Falcon II was ranked 2nd. Jaguar 3 ranked 10th in the 1996 trial while Falcon II dropped to 86th out of 129 entries (Morris and Shearman, 2000). A new NTEP trial will be seeded in 2001 and recent results, from trials seeded since 1996, indicate that a similar pattern of improvement will continue.

In all of these trials, Kentucky-31, with or without endophyte, has ranked at the bottom or near the bottom of the trial. The one characteristic where Kentucky-31 has been competitive, especially in the early trials, is brown patch resistance. Kentucky-31 is an open cultivar with much less density than most cultivars and, as a result, has usually had less brown patch. In the 1987 NTEP trial, no significant differences were found between Kentucky-31 and the top dense, turf-type cultivars for brown patch resistance (Morris and Murray, 1992). This was again true in the 1992 NTEP, in which Jaguar 3 was the most brown patch resistant; however, Kentucky-31 was not significantly different for this trait. The fact that Jaguar 3, which has much greater density, and Kentucky-31 have similar brown patch ratings, shows significant progress for brown patch resistance breeding (Shearman and Morris, 1996).

The brown patch ratings in the 1996 NTEP trial show significant improvements for many of the denser turf-type cultivars (Morris and Shearman, 2000). Many of the top performing entries such as Plantation, Jaguar 3, Millennium, and Rembrandt showed higher levels of resistance than Kentucky-31. One trial from Kentucky in these results did not show a significant percent brown patch difference between Kentucky-31 and the top cultivars.

Leaf spot resistance in turf-type tall fescue has continued to improve relative to Kentucky-31 and older turf-type cultivars. This can be seen from the data in the above-referenced NTEP results. Many of the present improved turf-type tall fescue cultivars have shown improved resistance to gray leaf spot (Fraser et al., 1996; Meyer et al., 2001a), when compared to Kentucky-31 and cultivars such as Falcon and Rebel. This disease can cause severe devastation on stands of tall fescue.

Data have been taken in tall fescue NTEP trials on drought tolerance and summer performance (Morris and Shearman, 1999). Many of the top performing cultivars in the 1996 NTEP are rating significantly better than Kentucky-31 in these evaluations. In greenhouse studies (Huang and Fry, 1998; Huang et al., 1998; Huang and Gao, 2000), 'Houndog V,' Kentucky-31 and Mustang had better drought stress tolerance than the dwarf-type 'MIC 18,' Observations at Rutgers University would indicate that some of the very dense low-growing cultivars are more susceptible to severe drought stress compared to semidwarf or conventional turf-type tall fescue. The dense, compact cultivars such as Bonsai and Matador can also experience more damage from brown patch under severe drought and heat stress conditions (Watkins et al., 2000).

Current Challenges and Future Prospects

During the past 20 years, tremendous progress has been made in the development of tall fescue cultivars with improved turf quality (color, density, texture, etc.), persistence, and disease resistance. Sources of resistance to *Pythium* blight, which can be a devastating disease under flooded, hot conditions, are still needed. There is also a need to continue to improve the resistance of tall fescue to brown patch disease.

An additional challenge is to find improved resistance to feeding by soil-inhabiting grubs (Coleoptera:Scarabaeidae). It appears that current endophytes in tall fescue have little or no effect in controlling this insect. Observations by the senior author suggest that those cultivars with improved heat and drought tolerance have also shown tolerance to feeding by grubs.

Another weakness of tall fescue has to do with its slow rate of establishment under traffic. This can result in severe competition from weeds such as *Poa annua* in low-cut turf. If a tall fescue were developed that had better competitive abilities in mixtures with other cool-season species, it would expand the usage of this species. At this time, tall fescue is not as effective as perennial ryegrass for renovating old, thin turf areas.

Tall fescue has become one of the top three cool-season turfgrass species produced and used in the United States and Europe. The potential to further improve the species and expand its use is great.

REFERENCES

Asay, K.H., R.V. Frakes, and R.C. Buckner. 1979. Breeding and cultivars. pp. 111–139. In R.C. Buckner and L.P. Bush, Eds. Tall fescue. *Agron. Monogr. 20.* ASA, CSSA, and SSSA, Madison, WI.

Bacon, C.W., J.K. Porter, J.D. Robbins, and E.S. Lutrell. 1977. *Epichloe typhina* from toxic tall fescue. *Appl. Environ. Microbiology* 34:576–581.

Bara, R.F., M. Richardson, W.A. Meyer, R. Bara, D.A. Smith, S. Tubbs, and C.R. Funk. 2000. Registration of 'Millennium' tall fescue. *Crop Science* 40:1503–1504.

Beard, J.B. 1973. *Turfgrass: Science and Culture.* Prentice Hall, Englewood Cliffs, NJ.

Beard, J.B. 1982. *Turf Management for Golf Courses.* Burgess Publishing Co., Minneapolis.

Bettany, A.J.E., S.J. Dalton, E. Timms, and P. Morris. 1998. Stability of transgene expression during vegetative propagation of protoplast-derived tall fescue (*Festuca arundinacea* Schreb.) plants. *Journal of Experimental Botany* 49(328):1797–1804.

Borrill, M. 1976. Temperate grasses. pp. 137–142. In N.W. Simmonds, Ed. *Evolution of Crop Plants.* Longman, London and New York.

Borrill, M., B.F. Tyler, and W.G. Morgan. 1976. Studies in Festuca 7. Chromosome atlas (part 2). An appraisal of chromosome race distribution and ecology, including *F. pratensis* var. *apennia* (De Not.) Hack.-tetroploid. *Cytologia* 41:219–236.

Breen, J.P. 1994. Acremonium endophyte interactions with enhanced plant resistance to insects. *Ann. Rev. Entomol.* 39:401–423.

Buckner, R.C. and J.R. Cowan. 1973. The fescues. pp. 297–306. In M.E. Heath, D.S. Metcalfe, and R.F. Barnes. *Forages.* 3rd ed. The Iowa State University Press, Ames, IA.

Buckner, R.C., H.D. Hill, and P.B. Burrus Jr. 1961. Some characteristics of perennial and annual ryegrass X tall fescue hybrids and of the amphidiploid progenies of annual ryegrass X tall fescue. *Crop Science* 1:75–80.

Buckner, R.C., H.D. Hill, A.W. Hovin, and P.B. Burrus II. 1965. Fertility of annual ryegrass X tall fescue amphiploids and their derivatives. *Crop Science* 5:395–397.

Buckner, R.C., J.B. Powell, and R.V. Frakes. 1979. Historical development. pp. 31–39. In R.C. Buckner and L.P. Bush, Eds. Tall fescue. *Agron. Monogr. 20.* ASA, CSSA, and SSSA, Madison, WI.

Buckner, R.C., G.T. Webster, P.B. Burrus II, and L.P. Bush. 1976. Cytological, morphological, and agronomic characteristics of tall x giant fescue hybrids and their amphiploid progenies. *Crop Science* 16:811–816.

Burner, D.M., G.C. Eizenga, R.C. Buckner, and P.B. Burrus Jr. 1989. Meiotic instability of tall fescue X giant fescue amphiploids. *Crop Science* 29:1484–1486.

Burner, D.M., G.C. Eizenga, R.C. Buckner, and P.B. Burrus Jr. 1991. Genetic variability of seed yield and agronomic characters in *Festuca* hybrids and amphiploids. *Crop Science* 31:56–60.

Burns, J.C. and D.S. Chamblee. 1979. Adaptation. pp. 9–30. In R.C. Buckner and L.P. Bush, Eds. Tall fescue. *Agron. Monogr. 20.* ASA, CSSA, and SSSA, Madison, WI.

Chen, C., D.A. Sleper, and G.S. Johal. 1998. Comparative RFLP mapping of meadow and tall fescue. *Theor. Appl. Genet.* 97:255–260.

Christians, N. 1998. *Fundamentals of Turfgrass Management.* Ann Arbor Press, Chelsea, MI.

Couch, H.B. 1962. *Diseases of Turfgrasses.* Reinhold Publ. Corp., New York.

Cowan, J.R. 1956. Tall fescue. *Adv. Agron.* 8:283–320.

Dalton, S.J. 1988. Plant regeneration from cell suspension protoplasts on *Festuca arundinacea* Schreb. (tall fescue) and *Lolium perenne* L. (perennial ryegrass). *J. Plant Physiol.* 132:170–175.

Davis, R.R. 1958. The effect of other species and mowing height on persistence of lawn grasses. *Agron. J.* 50:671–673.

Eizenga, G.C., D.M. Burner, and R.C. Buckner. 1990. Meiotic and isozymic analyses of tall fescue x giant fescue hybrids and amphiploids. *Plant Breeding* 104:202–211.

Eizenga, G.C., P.B. Burrus Jr., J.F. Pedersen, and P.L. Cornelius. 1991. Meiotic stability of 56-chromosome tall fescue hybrid derivatives. *Crop Science* 31:1532–1535.

Eizenga, G.C. and T.D. Phillips. 1998. Registration of 13 monosomic tall fescue genetic stocks. *Crop Science* 38:555.

Eizenga, G.C., C.L. Schardl, T.D. Phillips, and D.A. Sleper. 1998. Differentiation of tall fescue monosomic lines using RFLP markers and double monosomic analysis. *Crop Science* 38:221–225.

Fergus, E.N. 1952. Kentucky 31 Fescue—Culture and Use. Kentucky Agric. Ext. Circ. 497.

Fergus, E.N. and R.C. Buckner. 1972. Registration of 'Kentucky-31' tall fescue. (Reg No. 7). *Crop Science* 12:714.

Ford, T.M., D.A. Smith, R.F. Bara, W.A. Meyer, and C.R. Funk. 2000. Registration of 'Rembrandt' tall fescue. *Crop Science* 40:1506–1507.

Foy, C.D., and J.J. Murray. 1998. Developing aluminum-tolerant strains of tall fescue for acid soils. *Journal of Plant Nutrition* 21(6):1301–1325.

Fraser, M.L., W.A. Meyer, and C.A. Rose-Fricker. 1996. Susceptibility of tall fescue to gray leaf spot. p. 139. In *Agronomy Abstracts*. ASA, Madison, WI.

Funk, C.R., W.K. Dickson, W.A. Meyer, and R.J. Pedersen. 1981a. Registration of 'Falcon' tall fescue. *Crop Science* 21:632.

Funk, C.R., W.K. Dickson, and R.H. Hurley. 1981b. Registration of 'Rebel' tall fescue. *Crop Science* 21:632.

Funk, C.R. and W.A. Meyer. 2001. Seventy years of turfgrass improvement at the New Jersey Agriculture Experiment Station. *Proceedings of the Tenth Annual Rutgers Turfgrass Symposium.* pp. 11–19.

Funk, C.R. and J.F. White Jr. 1997. Use of natural and transformed endophytes for turf improvement. In Bacon and Hill, Eds. *Neotyphodium/Grass Interactions*. Phenum Press, New York.

Giesler, L.J., G.Y. Yuen, and G.L. Horst. 1996a. Tall fescue canopy density effects on brown patch disease. *Plant Disease* 80:384–388.

Giesler, L.J., G.Y. Yuen, and G.L. Horst. 1996b. The microclimate in tall fescue turf as affected by canopy density and its influence on brown patch disease. *Plant Disease* 80:389–394.

Green II, D.E., L.L. Burpee, and K.L. Stevenson. 1999. Components of resistance to *Rhizoctonia solani* associated with two tall fescue cultivars. *Plant Disease* 83:834–838.

Ha, S-B, F-S Wu, and T.K. Thorne. 1992. Transgenic turf-type tall fescue (*Festuca arundinacea* Schreb.) plants regenerated from protoplasts. *Plant Cell Reports* 11:601–604.

Hartley, W. 1950. The global distribution of tribes of the Graminae in relation to historical and environmental factors. *Aust. J. Agric. Res.* 1:355–373.

Hicks, D.H. and K.J. Mitchell. 1968. Flowering in pasture grasses. 1. Interactions of day length and temperature on inflorescence emergence for *Festuca arundinacea* Schreb. variety Manade. *N.Z. J. Bot.* 6:86–93.

Hill, H.D. and R.C. Buckner. 1962. Fertility of *Lolium-Festuca* hybrids as related to chromosome number and meiosis. *Crop Science* 2:484–486.

Hitchcock, A.S. 1935. Manual of the Grasses of the United States. USDA Misc. Publ. 200.

Hitchcock, A.S. 1950. Manual of the Grasses of the United States. 2nd Ed. USDA Misc. Publ. 200.

Huang, B. and J.D. Fry. 1998. Root anatomical, physiological, and morphological response to drought stress for tall fescue cultivars. *Crop Science* 38:1017–1022.

Huang, B. and H. Gao. 2000 Root physiological characteristics associated with drought resistance in tall fescue cultivars. *Crop Science* 40:196–203.

Huang, B., J.D. Fry, and B. Wang. 1998. Water relations and canopy characteristics of tall fescue cultivars during and after drought stress. *HortScience* 35(5):837–840.

Johnson-Cicalese, J.M. and C.R. Funk. 1990. Additional host plants of four species of billbug found on New Jersey turfgrasses. *J. Am. Soc. Hort. Sci.* 115(4):608–611.

Johnson-Cicalese, J., N. Haas, T. Masters, B. Bhandari, G. Mansueand, and W.A. Meyer. 1998. Incidence of Neotyphodium endophyte in seed lots of cultivars and selections in the 1996 National Tall Fescue Test. *Rutgers Turfgrass Proceedings* 29:151–156.

Juska, F.V., A.A. Hansen, and A.W. Hovin. 1969. Evaluation of tall fescue, *Festuca arundinacea* Schreb., for turf in the transition zone of the United States. *Agronomy Journal* 61:625–628.

Kasperbauer, M.J. and G.C. Eizenga. 1985. Tall fescue doubled haploids via tissue culture and plant regeneration. *Crop Science* 25:1091–1095.

King, M.J., L.P. Bush, R.C. Buckner, and P.B. Burrus II. 1987. Effect of ploidy on quality of tall fescue, Italian ryegrass x tall fescue and tall fescue x giant fescue hybrids. *Annals of Botany* 60:127–132.

Kuai, B., S.J. Dalton, A.J.E. Bettany, and P. Morris. 1999. Regeneration of fertile transgenic tall fescue plants with a stable highly expressed foreign gene. *Plant Cell, Tissue and Organ Culture* 58:149–154.

Kuai, B. and P. Morris. 1996. Screening for stable transformants and stability of b-glucuronidase gene expression in suspension cultured cells of tall fescue (*Festuca arundinacea*). *Plant Cell Reports* 15:804–808.

Lakkaraju, S., L.H. Pitcher, X. Wang, and B.A. Zalinskas. 2001. *Agrobacterium*-mediated transfer of turfgrasses. *Proceedings of the Tenth Annual Rutgers Turfgrass Symposium.* p. 20.

Meyer, W.A., S.A. Bonos, K.A. Plumley, R.F. Bara, D.A. Smith, C.R. Funk, M.M. Mohr, E. Watkins, J.A. Murphy, and W.K. Dickson. 2001a. Progress in breeding for disease resistance in open-pollinated cool-season turfgrasses. *Proceedings of the Tenth Annual Rutgers Turfgrass Symposium.* pp. 22–23.

Meyer, W.A. and C.R. Funk. 1989. Progress and benefits to humanity from breeding cool-season grasses for turf. pp. 31–48. In Sleper, D.A., K.H. Asay, and J.F. Pedersen, Eds. Contributions from Breeding Forage and Turf Grasses. CSSA special publication number 15. CSSA, Madison, WI.

Meyer, W.A. and C.A. Rose-Fricker. 1990. Registration of 'Bonanza' tall fescue. *Crop Science* 31:1089.

Meyer, W.A., B.L. Rose, C.A. Rose-Fricker, and C.R. Funk. 1982. Registration of 'Olympic' tall fescue. *Crop Science* 22:1260–1261.

Meyer, W.A., R. Stapp, K. Hignight, D.A. Smith, R.F. Bara, and C.R. Funk. 2001b. Registration of 'Plantation' tall fescue. Accepted for publication in *Crop Science*.

Morris, K.N. and J.J. Murray. 1992. National Tall Fescue Test—1987. Final report 1988–1991. NTEP No. 92-11. USDA-ARA. Beltsville, MD.

Morris, K.N. and R. Shearman. 1999. National Tall Fescue Test—1996. Progress report 1998. NTEP No. 99-2. USDA-ARS. Beltsville, MD.

Morris, K.N. and R. Shearman. 2000. National Tall Fescue Test—1996. Progress report 1999. NTEP No. 00-5. USDA-ARS. Beltsville, MD.

Murphy, J.A., S. Sun, and L.L. Betts. 1993. Endophyte-enhanced resistance to billbug (Coleoptera:Curculionidae), sod webworm (Lepidoptera:Pyralidae), and white grub (Coleoptera:Scarabeidae) in tall fescue. *Environmental Entomology* 22(3):699–703.

Murray, J.J. and K. Morris. 1988. National Tall Fescue Test—1983. Final report 1984–1987. PSI No. 8. USDA-ARS. Beltsville, MD.

Murray, J.J. and J.B. Powell. Turf. 1979. In R.C. Buckner and L.P. Bush, Eds. Tall fescue. *Agron. Monogr. 20.* ASA, CSSA, and SSSA, Madison, WI.

Poehlman, J.M. and D.A. Sleper. 1995. *Breeding Field Crops.* 4th ed. Iowa State University Press, Ames.

Rose-Fricker, C.A., M.L. Fraser, W.A. Meyer, and C.R. Funk. 1999a. Registration of 'Coronado' tall fescue. *Crop Science* 39:288.

Rose-Fricker, C.A., M.L. Fraser, W.A. Meyer, and C.R. Funk. 1999b. Registration of 'Tomahawk' tall fescue. *Crop Science* 39:288–289.

Roylance, J.T., N.S. Hill, and W.A. Parrott. 1994. Detection of somaclonal variation in tissue culture regenerants of tall fescue. *Crop Science* 34:1369–1372.

Saha, D.C., M.A. Jackson, and J.M. Johnson-Cicalese. 1988. A rapid staining method for detection of endophytic fungi in turf and forage grasses. *Phytopathology* 78:237–239.

Shearman, R. and K. Morris. 1996. National Tall Fescue Test—1992. Final report 1993–1995. NTEP No. 96-13. USDA-ARS, Beltsville, MD.

Smiley, R.W., P.H. Dernoeden, and B.B. Clarke. 1992. *Compendium of Turfgrass Diseases.* 2nd ed. APS Press, St. Paul, MN.

Spangenberg, G., Z.Y. Wang, X.L. Wu, J. Nagel, V.A. Iglesias, and I. Potrykus. 1995. Transgenic tall fescue (*Festuca arundinacea*) and red fescue (*F. rubra*) plants from microprojectile bombardment of embryogenic suspension cells. *J. Plant Physiology* 145:693–701.

Stadelmann, F.J., R. Kölliker, B. Boller, G. Spangenberg, and J. Nösberger. 1999. Field performance of cell suspension-derived tall fescue regenerants and their half-sib families. *Crop Science* 39:375–381.

Templeton Jr., W.C., G.O. Mott, and R.J. Bula. 1961. Some effects of temperature and light on growth and flowering of tall fescue, *Festuca arundinacea* Schreb. II. Floral development. *Crop Science* 1:283–286.

Ten Eyck, A.M. 1903. Meadow Fescue. Agric. Dept. Kansas. Agric. Exp. Stn. Press Bulletin. 125.

Terrell, E.E. 1979. Taxonomy, morphology, and phylogeny. pp. 31–39. In R.C. Buckner and L.P. Bush, Eds. Tall fescue. *Agron. Monogr. 20.* ASA, CSSA, and SSSA, Madison, WI.

Wang, Z-Y, T. Takamizo, V.A. Iglesias, M. Osusky, J. Nagel, I. Potrykus, and G. Spangenberg. 1992. Transgenic plants of tall fescue (*Festuca arundinacea* Schreb.) obtained by direct gene transfer to protoplasts. *Bio/Technology* 10:691–696.

Watkins, E. and W.A. Meyer. 2000. Morphological characterization of turf-type tall fescue genotypes. p. 157. *Agronomy Abstracts.* ASA, Madison, WI.

Watkins, E., W.A. Meyer, J.A. Murphy, S.A. Bonos, R.F. Bara, D.A. Smith, and W.K. Dickson. 2000. Performance of tall fescue cultivars and selections in New Jersey turf trials. *Rutgers Turfgrass Proceedings* 31:181–205.

Watson Jr., C.E. and S.D. McLean. 1991. Response to divergent selection for anthesis date in tall fescue. *Crop Science* 31:422–424.

Webster, G.T. and R.C. Buckner. 1971. Cytology and agronomic performance of *Lolium-Festuca* hybrid derivatives. *Crop Science* 11:109–112.

Welty, R.E. and M.E. Mellbye. 1989. *Puccinia graminis* subsp. *graminicola* on tall fescue in Oregon. *Plant Disease* 73:775.

Xu, W.W., D.A. Sleper, and S. Chao. 1995. Genome mapping of polyploid tall fescue (*Festuca arundinacea* Schreb.) with RFLP markers. *Theor. Appl. Genet.* 91:947–955.

Xu, W.W., D.A. Sleper, and D.A. Hoisington. 1991. A survey of restriction fragment length polymorphisms in tall fescue and its relatives. *Genome* 34:686–692.

Xu, W.W., D.A. Sleper, and G.F. Krause. 1994. Genetic diversity of tall fescue germplasm based on RFLPs. *Crop Science* 34:246–252.

Younger, V.B. 1965. A report on tall fescue for turf. *West. Landscape News* 5:6,20.

Chapter 9

Fine-Leaved *Festuca* Species

Bridget A. Ruemmele, Joseph K. Wipff,
Leah Brilman, and Kenneth W. Hignight

The genus *Festuca* contains approximately 450 species found in temperate regions throughout the world, extending through the tropics and on mountaintops (Clayton and Renvoize, 1986). Fine-leaved *Festuca* spp. were used as early as the sixteenth century for golf turfs (Beard, 1973). Since its naming as *Festuca*, an old Latin name for a weedy grass (Hitchcock, 1950), breeders diligently transformed several species within this genus into vital turfgrass cultivars used throughout the world (Ruemmele et al., 1995).

Fine-leaved *Festuca* spp. have been grown on greens in Scotland for centuries (Beard, 1998). Use on fairways will likely increase with the desire for more environmentally-sustainable golf courses (Christians, 2000; Stier, 1999). Some fine-leaved *Festuca* spp. may also be used on golf course roughs and other natural areas (Dernoeden, 1998; Riordan, 1997). Cultivars vary in seedling vigor, seasonal quality, spring greenup, and color (Landschoot et al., 2000). Perdomo et al. (1999) also noted variability in establishment rates among specimens.

This large and variable genus includes nine subgenera, organized along the main lines of variation, although Darbyshire and Warwick (1992) combined *Subulatae* and *Subuliflorea* to make eight. *Festuca* subgenus *Festuca* is the largest, most taxonomically difficult subgenus in *Festuca*, in which identification of species often depends on the disposition of sclerenchyma strands in the transverse section of a leaf blade (Clayton and Renvoize, 1986). This subgenus also contains the collection of agronomically-important turf species known as the "fine fescues." Within this grouping, further subdivision forms two categories known as the *F. rubra* (red fescue) and *F. ovina* (sheep fescue) complexes, or aggregates. Cytological analyses have resulted in conflicting conclusions regarding species delineation (Vinall and Hein, 1937).

F. rubra L. *sensu lato* is a morphologically diverse polyploid, outcrossing complex (Bowden, 1960; Löve and Löve, 1961, 1975; Calder and Taylor, 1968; Taylor and Mulligan, 1968; Welsh, 1974; Markgraf-Dannenberg, 1980; Pavlick, 1985), native to Europe (Beard, 1973), with widespread distribution in arctic and temperate regions of Europe (Auquier, 1971b; Markgraf-Dannenberg, 1980), Asia (Kreczetovich and Bobrov, 1934; Tzvelev, 1976) and North America (Piper, 1906; Saint-Yves, 1925; Hultén, 1942, 1968; Pavlick, 1985). It is a complex containing morphologically distinct taxa (e.g.,

Hackel, 1882; Beal, 1896; Auquier, 1968, 1971a, 1971b; Alexeev, 1982), although plants morphologically intermediate to the recognized taxa were reported (e.g., Piper, 1906; Saint-Ives, 1925; Hultén, 1942). Taxa within this complex also occupy a wide array of habitats [e.g., beaches, sand dunes, coastal rock, cliffs, salt marshes, riverine gravel, moist meadows, boreal grasslands, and disturbed roadsides (Pavlick, 1985)], with very distinct ecotypes in these different habitats. *F. rubra* ssp. *litoralis* G.F.W. Meyer (Auquier) (slender creeping red fescue), for example, grows in salt marshes periodically inundated with sea water (Duyvendak et al., 1981). Ecotypes are often separated by spatial isolation or crossing barriers, preventing them from hybridizing in nature.

Species within the *F. ovina* complex form noncreeping, fine-leaved turf particularly suited to poor growing conditions. Their colors range from dark green to blue green and powdery blue-green hues. Species within this complex exhibit a diverse range of ploidy levels, as well as morphology.

INTRODUCTION

Environmental Adaptation

Fine-leaved *Festuca* spp. are generally characterized by their fine to very narrow leaves, usually less than 1 mm (Beard, 1973); adaptation to cool-humid regions of the world; drought, but not heat tolerance; general shade tolerance; adaptation to infertile, acid soils with a pH of 5.5 to 6.5 (Hanson et al., 1969; Beard, 1973; Roberts, 1987, 1990; Ruemmele et al., 1995); low tolerance to poorly-drained soils and high nitrogen fertilization (Davis, 1967; Beard, 1973; Meyer and Funk, 1989); and good adaptation in low-maintenance situations (Roberts, 1990; Meier, 1992). Meyer and Funk (1989) noted fine fescues perform well under both low and high management practices. Intra- and interspecific genetic variation associated with nitrogen use demonstrated the possibility to develop cultivars tolerant to low maintenance (Bourgoin, 1997). Wojcik and Skogley (1991) documented adaptation of fine-leaved *Festuca* spp. to limited moisture conditions. Fine-leaved *Festuca* spp. are valued for erosion control situations (Meyer, 1994). Watschke (1990) noted their superior advantage for use as low-maintenance roadside turf, while maintaining an attractive appearance. Many of these species tolerate tree root competition (Meyer, 1982). These species withstand northern cold, as well as heat of the upper south (Roberts, 1990). They are commonly used in cool-humid regions of the world, as well as cool, semiarid areas, when the latter are irrigated (Beard, 1973). Rossi and Sausen (1995) observed *F. rubra* ssp. *commutata* (Thuill.) Nyman (Chewings fescue) tolerated heat better than most of the other fine-leaved *Festuca* spp. in a 60-cultivar trial. They grow well in sun or shade, adapting to mowing heights of 3.75 cm or less (Roberts, 1990).

Tolerance to physical wear and playability need improvement for many potential uses. McNitt and Waddington (1992) observed the poorest traction on fine-leaved *Festuca* spp. compared to *Poa pratensis* L. (Kentucky bluegrass), *F. arundinacea* Schreb. (tall fescue), and *Lolium perenne* L. (perennial ryegrass). While fine-leaved *Festuca* spp. are well-adapted to many park and school ground areas, with persistence under moderate foot traffic, they do poorly in areas continually used for intense athletic activities (Meyer, 1986). Meyer et al. (1983) and Baker and Hunt (1997) also reported poor tolerance to traffic. Fifteen cultivars of five species and subspecies of *Festuca*, grown at four localities, were each subjected to fine turf management with abrasive or golf-spike wear and to sports field management with football-stud wear (Shildrick et al., 1983). Data on ground cover, appearance, height of growth between cuts under sports field management and resistance to *Fusarium* spp., *Laetisaria fuciformis* (McAlpine) Burdsall (red thread), and *Sclerotinia homeocarpa* F. T. Bennett syn. *Lanzia* or *Moellerodiscus* spp. (dollar spot) were recorded. Under all treatments, cultivars of *F. rubra* ssp. *commutata* and *F. rubra* ssp. *litoralis* were superior for wear and required less frequent cutting than *F. rubra* ssp. *rubra* Gaudin (strong creeping red fescue).

Heavy metal tolerance varies among fine-leaved *Festuca* spp. germ plasm. Harrington et al. (1996) demonstrated one cultivar of *F. rubra* ssp. *litoralis* ('Merlin') exhibited higher zinc and cadmium tolerance compared to the *F. rubra* ssp. *commutata* cultivar ('Cascade'). Brown and Brinkmann (1992) observed zinc and lead tolerances for *F. ovina*. Aluminum toxicity from soils naturally high in this element may affect turfgrasses (Murray et al., 1976). Liu et al. (1997) determined fine-leaved *Festuca* spp. were more tolerant of high aluminum concentrations compared with *Poa pratensis, Lolium perenne*, and *Festuca arundinacea*. Endophyte-containing cultivars exhibited greater aluminum tolerance compared to nonendophyte-containing fine-leaved *Festuca* cultivars (Liu et al., 1995).

Disease susceptibility has been considered a major weakness among fine-leaved *Festuca* spp. *Laetisaria fuciformis, Sclerotinia homeocarpa, Gaeumannomyces incrustans* (take-all patch) (Kemp and Clarke, 1991), *Microdochium nivale* (Fr.) Samuels & I.C. Hallett; syn. *Gerlachia nivale* (Ces. ex Sacc.) W. Gams & E. Muller (telemorph *Monographella nivalis* [Schaffnit]) (fusarium patch/pink snow mold) (Nilsson and Weibull, 1981), *Drechslera dictyoides* (Drechs.) Shoemaker f. sp. *Dictyoides*; syn. *Helminthosporium dictyoides* Drechs. (net blotch), *Drechslera* spp. (leaf spot) (Ruemmele et al., 1995), *Leptosphaeria korrae* J.C. Walker & A.M. Sm. (necrotic ring spot) (Sann, 1993), *Fusarium roseum* Link; Fr. (fusarium blight), *Erysiphye graminis* D.C. ex Merat, syn. *Blumeria graminis* (DC) E.O. Speer, (powdery mildew), *Colletotrichum graminicola* (Ces.) G.W. Wilson (anthracnose), *Puccinia crandallii* Pammel & H. Hume in H. Hume (leaf rust), and *Ustilago striiformis* (Westend. Niessl) (stripe smut) (Meyer and Funk, 1989; Ruemmele et al., 1995) adversely affect fine-leaved *Festuca* spp. *Magnaporthe poae* Landschoot and Jackson (summer patch) (Fraser et al., 1998) is another serious disease of fine-leaved *Festuca* spp. Susceptibility to this disease severely limits the use of *F. trachyphylla* (Hackel) Krajina (hard fescue) in compacted areas. *Puccinia graminia* ssp. *graminicola* (stem rust) has been found on *F. rubra* ssp. *commutata* in seed production areas of the Willamette Valley of Oregon (Welty and Azevedo, 1995). Occasionally stem rust has also been reported in *F. rubra* ssp. *rubra* and *F. trachyphylla*, with differential sensitivity to different strains of the stem rust fungus (Pfender, 2001). *Sclerophthora macrospora* (Sacc.) Thirum, Shaw, and Naras. (downy mildew or yellow tuft) was found on fine-leaved *Festuca* spp. growing in golf course fairways (Jackson, 1977).

Potential for genetic resistance has been documented. Casler et al. (2001) observed genetic resistance to snow mold (various species) in fine-leaved *Festuca* spp. growing on golf course fairways in Wisconsin. Mass selection for 1–2 generations in *F. rubra* led to noticeable improvement in resistance to *Fusarium nivale* (*Microdochium nivale*) (Nilsson and Weibull, 1981). Leaf spot resistance (species not stated) varied among species and cultivars of fine-leaved *Festuca* spp. (Fraser and Rose-Fricker, 1999a, 1999b). Hodges et al. (1975) reported greatest *Sclerotinia homeocarpa* resistance in *F. trachyphylla* compared to *F. rubra* ssp. *rubra* and *F. rubra* ssp. *commutata*.

Insect susceptibilities among *Festuca* spp. include *Blissus leucopterus hirtus* (hairy chinch bug) (Davis et al., 1989; Heller and Walker, 1997; Watschke, 1990), and *Parapediasia* spp. (sod webworm) (Sargent, 1982). Some resistance may be attributed to the presence or absence of endophytes. A major advancement in germ plasm improvement of fine-leaved *Festuca* spp. resulted from the discovery and introduction of *Epichloe* spp./*Neotyphodium* spp. (previously *Acremonium* spp.) into commercial cultivars. Funk et al. (1993) reported the discovery of *Neotyphodium coenophialum* in naturalized *F. rubra* ssp. *commutata* led to the incorporation of this beneficial fungus into 'Jamestown

II,' 'SR 5000,' and 'Banner II.' These fungi enhance insect resistance, disease resistance, and drought resistance (Funk and White, 1997). Aboveground insects deterred by these fungi include *Blissus* spp. (chinch bugs), *Sphenophorus parvulus* Gyllenhal (bluegrass billbugs), *Schizaphis graminum* Rodani (greenbugs), *Parapediasia* spp., and *Spodoptera mauritia* Boisduval (armyworms) (Saha et al., 1987). *Spodoptera eridania* Cramer (fall armyworm) feeding was deterred on endophyte-enhanced *F. rubra* ssp. *commutata* and *F. trachyphylla* (Breen, 1993). *Sclerotinia homeocarpa* resistance associated with endophytes was reported by Funk (1993) and Clarke et al. (1994). Funk et al. (1989, 1994) described use of endophytes in grasses for turf and soil conservation. *Epichloe typhina* (Fr.) Tul. (choke or cattail disease), leading to low seed yield concerns must be addressed when introducing endophytes into selected germ plasm (Funk and White, 1997). A thorough review of the literature, as well as current and potential progress using endophytes in fine-leaved *Festuca* spp. breeding is provided by Funk and White (1997).

Acremonium-type (now *Epichloe* spp./*Neotyphodium* spp.) endophytic fungi were detected in 19 of 83 seed lots of fine-leaved *Festuca* cultivars and selections by Saha et al. (1987). Most infected seed lots were of European origin, with endophyte levels at least 50%. Endophytes were also detected in 10 of 328 fine-leaved *Festuca* plants collected from old established turfs throughout the United States. Most endophytes were nonchoke-inducing (NCI). Saha et al. (1987) found NCI endophytes occurred in *F. glauca* Lamarck (blue fescue), *F. ovina* ssp. *hirtula* (Hackel *ex* Travis) M. Wilkinson (sheep fescue), and *F. valesiaca* Schleicher *ex* Gaudin (false sheep fescue). Nonchoke-inducing endophytic fungi were also found in *F. rubra* ssp. *rubra*, *F. rubra* ssp. *litoralis*, *F. rubra* ssp. *commutata*, and *F. trachyphylla*. Endophytes appeared to be associated with enhanced host plant resistance to *Blissus* spp. Useful NCI endophytes are being incorporated into leading cultivars and elite germ plasm collections of fine-leaved *Festuca* spp. DaCosta et al. (1998) documented endophyte levels in cultivars entered in the 1998 National Fine Fescue test.

Breeding, Genetics, and Taxonomy

Fine-leaved *Festuca* ssp. breeding has been conducted at several universities throughout the United States, as well as at numerous private companies. Included among universities are Michigan State University (Vargas et al., 1980), The Pennsylvania State University (Anonymous, 1995), The Rutgers University (Duell et al., 1978), and The University of Rhode Island. Material developed originally by Queen's University in Northern Ireland is also undergoing commercialization. Private companies conducting fine fescue breeding include Advanta Seeds, in the Netherlands and the United States; Barenbrug, Inc., with programs in the Netherlands, France, and the United States; Cebeco International, with programs in the United States and the Netherlands (Guthrie, 1989); Deutsche Saatveredlung Lippstadt-Bremen GmbH, Lippstadt, Germany; Pickseed West, Inc. in Tangent, Oregon; Pure Seed Testing, Inc. in Hubbard, Oregon; and Seed Research of Oregon, Inc. in Corvallis, Oregon.

Breeding objectives for fine-leaved *Festuca* spp. include improved wear tolerance, insect and disease resistance, and better sod-forming ability (Ruemmele et al., 1995), as well as reduced fertilizer, water, and mowing requirements (Anonymous, 1995). Heat and cold tolerances could also be improved (Beard, 1973). Improved transfer of endophytes from plant to seed, as well as improved seed storage to retain endophyte-viability, are desired (C.R. Funk, 1993, personal communication). Germ plasm breeding pools have been expanded with collections from areas of origin (Europe and Asia) and naturalized stands, such as along the eastern United States where initial colonization occurred. Specific breeding objectives for fine-leaved *Festuca* spp. at The Rutgers University in New Brunswick, New Jersey, include collection and evaluation of potentially useful germ plasm, collection and evaluation of endophytes associated with these species, continued breeding and development of new cultivars by conventional methods, development and application of several new tools to improve discrimination among endophyte isolates from nature, and synthesis of new grass-endophyte combinations for experimental testing and possible commercial use (Funk, 1998). Evaluation criteria include establishment vigor, turf quality, pest resistance, stress tolerance, texture, density, vertical growth rate, and persistence.

Molecular technology may enhance conventional breeding. Plant regeneration from suspension cells and protoplasts, symmetric and asymmetric somatic hybridization, and gene transfer methods and recovery of transgenic plants were developed by Spangenberg et al. (1998). Biolistic bombardment of cell suspension cultures resulted in transgenic *F. rubra* (Spangenberg et al., 1995b). *F. rubra* was successfully regenerated from tissue culture callus originally induced from mature seed (Torello et al., 1984; Zaghmont and Torello, 1988; Spangenberg et al., 1994; Wang et al., 1995). Regeneration of *F. rubra* also succeeded via cell suspension and from protoplasts (Spangenberg et al., 1994; Wang et al., 1995). Dale (1977) regenerated *F. rubra* plants from shoot apex culture.

Improved cultivars will produce attractive, durable, and persistent turfs with reduced establishment and maintenance costs (Funk et al., 1989). Grasses adapted to poor soils, shade, heavy wear, close mowing, and other specialized uses and environments are also essential (Funk et al., 1989). Bourgoin (1997) demonstrated the possibility of breeding cultivars better adapted to low maintenance conditions for turf. Breeding efforts have already produced improved turf-type fine-leaved *Festuca* spp. cultivars with better establishment than older cultivars (Perdomo et al., 1999). The Germplasm Resource Information Network (GRIN) maintains accessions of many fine-leaved *Festuca* spp., including 337 listed only as *F. rubra*, 164 listed only as *F. ovina*, 2 *F. rubra* ssp. *litoralis*, 20 *F. rubra* ssp. *commutata*, 6 *F. rubra* ssp. *rubra*, 12 *F. trachyphylla* (under *F. longifolia*), 4 *F. filiformis* Pourret (hair or fine-leaved fescue), 61 *F. idahoensis* Elmer (Idaho fescue), 11 *F. valesiaca* and 11 *F. heterophylla* Lamarck (shade fescue) (USDA, 2001).

Taxonomic classification has changed through the years. Figure 9.1 contains the current classification of relevant *Festuca* spp. Hackel (1882) divided the hundreds of *Festuca* spp. into six sections, which remain to date. Within these groupings, the taxonomy and nomenclature has evolved. The *Festuca* section (originally section *Ovinae*) contains 129 of 170 species recognized in Europe (Markgraf-Dannenberg, 1980). Ninety-one species belong to *F. ovina* sensu Hackel, while 21 species are grouped under *F. rubra* sensu Hackel. Stace and Ainscough (1984) placed 14 of the latter species group into the *F. rubra* aggregate. Sodium dodecylsulphate-polyacrylamide gel electrophoresis (SDS-PAGE) banding patterns of seed storage proteins supported recognition of *F. rubra* and *F. ovina* as species complexes (Aiken and Gardiner, 1991). Both infrageneric and species levels could be distinguished.

Extensive documentation regarding current taxonomic classification of North American *Festuca* spp. is maintained by Aiken et al. (1996 onward). Both complexes vary in morphology and adaptation, with introduced cultivars important both ecologically and economically (Stace et al., 1992). Stace et al. (1992) found two characters to successfully distinguish *F. rubra* and *F. ovina* complexes in the British Isles. Other identification characters varied in usefulness, from being helpful after considerable experience to being highly misleading. The two complexes are distinguished in Figure 9.2.

> The fine-leaved *Festuca* spp. classification hierarchy includes:
>
> Family: Poaceae
> Subfamily: Festucoideae
> Tribe: Festuceae
> Genus: ***Festuca*** L.
> Subgenus: ***Festuca***
> Section: ***Festuca***
>
> ***Festuca rubra*** aggregate
>   *F. heterophylla* Lamarck
>   *F. rubra* L.
>   ssp. ***commutata*** (Thuill.) Nyman
>   ssp. ***litoralis*** (G.F.W. Meyer) Auquier
>   ssp. ***rubra*** Gaudin
>
> ***Festuca ovina*** aggregate
>   *F. filiformis* Pourret
>   *F. idahoensis* Elmer
>   *F. ovina* L. ssp. ***hirtula*** (Hackel *ex* Travis) M. Wilkinson
>   *F. trachyphylla* (Hackel) Krajina
>   *F. valesiaca* Schleicher *ex* Gaudin

Figure 9.1. Classification of turf-type fine-leaved *Festuca* species within the Poaceae family.

> Sheath of young tiller-leaves fused into a tube almost to top; some or all tillers extravaginal .... ***F. rubra*** complex
>
> Sheath of young tiller-leaves with at least the upper 40% with free, overlapping margins; all tillers intravaginal .... ***F. ovina*** complex

Figure 9.2. Distinguishing features delineating *Festuca rubra* and *F. ovina* fine fescue complexes, based on Stace et al., 1992.

Morphology, cytology, anthesis, and/or physiological activity have been used to delineate individual species, although controversy exists about utility of some characters. As technology has expanded, taxonomy has been refined, based on new revelations of affinities or discontinuities among fine-leaved *Festuca* spp.

Transverse leaf blade sections of tiller leaves have been used in taxonomy of the genus *Festuca* since 1882, when Hackel (1882) illustrated the characteristics of 37 taxa using block diagrams. Howarth (1924) presented diagrams and descriptions of several members from the *F. rubra* and *F. ovina* complexes in Great Britain. Tzvelev (1976) and Alexeev (1972, 1980), considering *Festuca* in the U.S.S.R. and North America, made considerable use of leaf cross-section characters as a taxonomic tool. Markgraf-Dannenberg (1980) included leaf cross-section characters in keys and descriptions in the Flora Europaea treatment of *Festuca*. Kjellqvist (1961, 1964) and Auquier (1971b) relied on sclerenchyma tissue formations to distinguish fine-leaved *Festuca* spp., although Stace and Cotton (1974) contended their categorizations were not completely accurate. Habitat influenced anatomy, which complicated morphological analyses in Swedish specimens (Kjellqvist, 1961). Aiken et al. (1985) evaluated how reliably Canadian narrow-leaved *Festuca* could be identified using differences in leaf blade anatomy. They reported leaf cross-section data provided information useful in identification, especially when geographic origin and habitat were known. Leaf blade anatomy could be used to separate species into three *Festuca* complexes (Rough Fescue [*F. altaica* Trin., *F. campestris* Rydb., and *F. hallii* (Vase) Piper], *F. rubra*, and *F. ovina*) in Canada. The *F. rubra* complex is distinguished by prominent ribs and relatively narrow strands of sclerenchyma tissue opposite the bundles, often resulting in an angular outline to the leaf. The *F. ovina* complex have well-developed sclerenchyma tissue in an interrupted or continuous band along the abaxial epidermis, resulting in a rounded outline to the leaf.

Morphology, cytology, and anthesis analyses distinguished three species within the *F. rubra* complex from three species within the *F. ovina* complex (Schmit et al., 1974). Of the species studied, one noncreeping and two creeping types were assigned to the *F. rubra* complex, while the remaining noncreeping types fit the *F. ovina* complex. Leaf width placed the sole noncreeping species into the *F. rubra* complex, rather than the *F. ovina* complex with the remaining noncreeping species.

Burg and Vierbergen (1979) described a root fluorescence method to distinguish between *F. rubra* and *F. ovina*. Seedlings were sprayed with ammonia and observed under fluorescent light. Reddish-yellow fluorescence occurred around *F. rubra* roots, while bluish-green fluorescence occurred around the *F. ovina* roots.

Restriction fragment length polymorphisms (RFLP) of cpDNA and nuclear rDNA were used to identify species and cultivars within species of turfgrasses, including fine-leaved *Festuca* spp. (Ohmura et al., 1993). Phylogenetic relationships among turfgrasses could be studied by this method. Forty-eight accessions of 10 *Festuca* spp. were analyzed using laser flow cytometry as an adjunct to morphological separation of the species (Huff and Palazzo, 1998). Ploidy level determinations, based on laser flow cytometry results, compared favorably to current taxonomic treatments of *Festuca* spp., including *F. trachyphylla* and *F. ovina* ssp. *hirtula*.

Esterase banding patterns on electrophoresis gels were successful in distinguishing fine-leaved *Festuca* spp. in blends or in mixes with *P. pratensis* (Bell et al., 1995). Bands unique to cultivars or species could distinguish concentration within a blend or mix, based on the intensity of the band. Reversed-phase high-performance liquid chromatography (RP-HPLC) analysis of seed proteins determined the amount of each component in an *F. trachyphylla*/*F. rubra* ssp. *commutata* mix (Freeman and Mileva, 1998). This proved a rapid and effective method to analyze seed mixes for these species.

Wilson (1999) studied isozyme variation to establish the relationship between *F. rubra* ssp. *commutata*, *F. idahoensis*, *F. roemeri* var. *roemeri*, and *F. roemeri* var. *klamathensis*. The *F. rubra* var. *commutata* was most distinct, followed by the *F. roemeri* var. *roemeri*, with the *F. idahoensis* and *F. roemeri* var. *klamathensis* difficult to separate based on isozymes, although they differ in leaf shape and sclerenchyma. Wilson (1999) also looked at flow cytometry of the latter species to determine ploidy level. She utilized 'Covar,' *F. valesiaca* as a diploid and 'Azay,' *F. trachyphylla* (cited as *F. huonii*), as a hexaploid control and determined that these species, *F. idahoensis*, *F. roemeri* var. *roemeri*, and *F. roemeri* var. *klamathensis* are all tetraploid, unlike *F. rubra* ssp. *commutata*, which is a hexaploid.

Figure 9.3 provides a key to the fine-leaved *Festuca* taxa currently used as turf. Individual descriptions are included in the separate section for each species.

THE RED FESCUE COMPLEX

Taxonomy

Confusion and debate continues regarding classification within this complex. The first consideration is how to delineate a species. Without a universally-accepted definition, most taxonomists agree species are groups of individuals sharing morphological characters in common, with different groups separable by specialized traits, excluding geographical distribution; interbreeding populations; and reproductively isolated from other populations. Subspecies have been widely accepted as a considerable segment of a species with a distinct geographical or ecological area, more or less morphologically distinct, and often showing some intergradation (Davis and Heywood, 1973).

The term 'variety' (= cultivar) in the agronomic sense is not the same as used in the taxonomic sense. Cultivars are man-made groups in which only a limited number of individuals are selected, maintained, and given a cultivar (unique, commercial) name. These genetically-restricted materials do not encompass the total taxonomic variation. Therefore, a cultivar cannot be considered equal to a taxonomic variety.

Hackel (1882) used leaf tips and sclerenchyma to segregate the *F. rubra* aggregate into six subspecies. He considered two of these subspecies of importance to domestication, *F. rubra* ssp. *eu-rubra* and *F. rubra* ssp. *dumetorum* (L.) Hackel. Synonyms for the latter included *F. arenaria* Gren & Godr., *non* Osbeck, *F. juncifolia* St. Amans, and *F. sabulicola* Duf. Within *F. rubra* ssp. *eu-rubra*, Hackel (1882) listed seven varieties, including *F. rubra* ssp. *eu-rubra* var. *genuifolia* and *F. rubra* ssp. *eu-rubra* var. *fallax* (Thuill.) Hackel. *F. rubra* ssp. *eu-rubra* var. *genuifolia* was further divided into seven subvarieties, including *F. rubra* ssp. *eu-rubra* var. *genuifolia* subvar. *arenaria* (Osbeck) Hackel (synonyms *F. dumetorum* Rafr. *non* L., *F. arenaria*, and *F. oraria* Dumetorum. Stace and Cotton (1974), urging against use of sclerenchyma for species delineation, disagreed with Hackel's taxonomic classifications.

Minor changes were made to Hackel's classification by Howarth (1924). *F. rubra* ssp. *eu-rubra* was grouped simply as *F. rubra*. *F. rubra* ssp. *dumetorum* was reclassified as *F. juncifolia*. Most subvarieties of *F. rubra* ssp. *eu-rubra* var. *genuina* became varieties of *F. rubra* ssp. *genuina*. This subspecies is considered synonymous with *F. rubra* ssp. *duriuscula* and *F. rubra* ssp. *rubra* (Stace and Cotton, 1974). Another variety of *F. rubra* ssp. *eu-rubra*, var. *fallax* is considered synonymous with the current *F. rubra* ssp. *commutata*.

Schmit et al. (1974) asserted *F. rubra* should be classified into three distinct noninterbreeding species, using comparative morphology, turf characteristics, chromosome numbers, reproductive isolation of flowering time, and pollen longevity. They proposed *F. rubra* ssp. *rubra* (spreading fescue) and *F. rubra* ssp. *tricophylla* (creeping fescue, now *F. rubra* ssp. *litoralis*) for the two rhizomatous groups, while retaining *F. rubra* ssp. *commutata* for nonrhizomatous forms. *F. rubra* ssp. *rubra* and *F. rubra* ssp. *litoralis* differed in chromosome number ($2n = 56$ and $2n = 42$, respectively). While *F. rubra* ssp. *commutata* had $2n = 42$ (like *F. rubra* ssp. *litoralis*), this species shed pollen in early morning, rather than late afternoon, as occurred with both *F. rubra* ssp. *rubra* and *F. rubra* ssp. *litoralis*. Ruemmele et al. (1995) noted heavy dew or rain may limit pollen shed at these times, acting as a species barrier.

1. Sheath margins of young tillers fused for over three-fourths of their length; rhizomes present or absent; basal sheaths often shredding into fibers
    2. Plants cespitose, without rhizomes
        3. Ovary and caryopsis apex glabrous .... *F. rubra* ssp. *commutata*
        3'. Ovary apex and caryopsis apex pubescent .... *F. heterophylla*
    2. Plants with rhizomes
        4. Plants strongly rhizomatous with long creeping rhizomes; 2n = 42 or 56 .... *F. rubra* ssp. *rubra*
        4. Plants cespitose, with short, slender, rhizomes; 2n = 42 .... *F. rubra* ssp. *litoralis*
1'. Sheath margins of young tillers fused of less than three-fourths of their length; rhizomes absent; basal sheaths persistent not shredding into fibers
    5. Lemma awnless, or with a mucro to 0.4 mm .... *F. filiformis*
    5'. Lemma awns 0.5 mm long or longer
        6. Lemmas 5.0 mm or longer
            7. Spikelets 7.5 mm or longer; awns half as long as lemma or longer; lemmas 5.0 to 8.5 mm long; $2n = 28$ .... *F. idahoensis*
            7'. Spikelets less than 7.5 mm long; awns less than half as long as lemma; lemmas 3.4 to 5.2 mm long; $2n = 14$ .... *F. valesiaca*
        6'. Lemmas less than 5.0 mm long
            8. Leaf blades with sclerenchyma in an even and continuous band, uniform in thickness, with 1 (-3) well developed adaxial ribs; $2n = 28$ .... *F. ovina* ssp. *hirtula*
            8'. Leaf blades with sclerenchyma uneven in thickness, interrupted, or forming distinct bundles at midrib and margins, with 1–7 well developed adaxial ribs; $2n = 14$ or 42.
                9. Leaf blades with three sclerenchyma fascicles, found opposite the midvein and at each margin; with 1–5 adaxial rib; $2n = 14$ .... *F. valesiaca*
                9'. Leaf blades with sclerenchyma fascicles present between the margins and midvein or rarely a continuous, uneven band, with 5–7 well-developed adaxial ribs; $2n = 42$ .... *F. trachyphylla*

Figure 9.3. Morphological key to the fine fescue taxa, derived from descriptions provided by Aiken and Darbyshire (1990), Markgraf-Dannenberg (1980), and Wilkinson and Stace (1991).

Morphologically, *F. rubra* ssp. had wider leaves than species in the *F. ovina* complex, ranging from 0.8 to 1.8 mm (Schmit et al., 1974). These species also typically had five or more epidermal ridges, with five being most common. *F. rubra* ssp. *commutata* lacked rhizomes, while *F. rubra* ssp. *rubra* had strong, long rhizomes and *F. rubra* ssp. *litoralis* contained finer, shorter rhizomes. Shildrick (1976a) concurred with the classification proposed by Schmit et al. (1974). Rhizome formation, chromosome number, and heading date were used to group 80 cultivars into one of the three *F. rubra* subspecies. Boeker (1974) observed *F. rubra* ssp. *rubra* cultivars had higher root masses in upper soil layers compared to *F. rubra* ssp. *commutata* cultivars.

F. rubra specimens collected from 76 sites in Poland were subjected to chromosome analyses (Konarska, 1974). Three subspecies were determined: *F. rubra* ssp. *rubra* (2n = 42 and 2n = 56), *F. rubra* ssp. *commutata* (2n = 42), and *F. rubra* ssp. *arenaria* (2n = 56). Chromosome analyses of *F. rubra* cultivars by Luesink and van Hardeveld (1975) confirmed 2n = 56 for specimens with strong rhizome development and 2n = 42 for samples without rhizomes or fine rhizomes. No hybridization between the two chromosome-differentiated groups was observed. Chromosome analyses by Dreven (1976) determined *F. rubra* ssp. *rubra*, a creeping form, contained 2n = 56, while *F. rubra* ssp. *commutata*, a noncreeping type had 2n = 42. They also documented *F. tenuifolia* (now *F. filiformis*) was 2n = 28, while *F. trachyphylla* was 2n = 42. Others (Hubbard, 1984; Aiken et al., 1996 onward; and Wilson, 1999) determined *F. filiformis* as 2n = 14, although Aiken also cites 2n = 28 for this species.

Chromosome number, growth habit, rhizome development, presence of anthocyanin in leaf sheath, dimension of culm and leaves, heading date, and disease resistance were used by Duyvendak and Vos (1974) to distinguish nearly 30 varieties of *F. rubra* spp. Duyvendak et al. (1981) also studied 59 *F. rubra* cultivars to delineate taxa and cultivars. *F. rubra* cultivars were classified into hexaploid noncreeping, hexaploid creeping, and octoploid creeping species. Classification of individual plants as creeping or noncreeping proved less definitive than had been theorized. This resulted not only because of the gradual transition in proportion of intra- and extravaginal shoots, but also due to occurrence of prostrate, rooted, vegetative, elongated, intravaginal shoots. This finding complicates determination of this character, as well as the whole classification. With the exception of chromosome number, none of the investigated characters showed a discontinuity in expression, while the ranges of variation of the groups overlapped. The hexaploid noncreeping group would be named *F. rubra* var. *commutata*. Most creeping hexaploid and some octoploid cultivars had folded leaves. Duyvendak et al. (1981) concluded hexaploids with folded basal leaf blades were either *F. rubra* ssp. *litoralis* or *F. rubra* ssp. *rubra*. Octoploids were *F. rubra* ssp. *rubra*. Further research was urged to more accurately define the complex.

Dubé et al. (1985) reported two *Festuca* cytotypes in eastern Quebec, 2n = 42 and 56. Dubé and Morisset (1987) also reported two cytotypes when they analyzed variation of 32 *F. rubra* s.l. populations from salt marshes, coastal rocks, coastal sands, and anthropogenic sites in eastern Quebec, Canada. The more widespread chromosome number was 2n = 21, with this group found mainly in natural habitats. The 2n = 28 types were localized in non-natural habitats, areas assumed planted with cultivars. Variation of characters was essentially continuous within and between populations. Some populations were much more variable than others. Generally, character variation patterns were mainly related to ecological, rather than geographical factors. They believed it would be difficult to assess effects of massive introduction of various lawn and pasture cultivars on variability of native populations of *F. rubra*. Dubé and Morisset (1987) concluded, if differentiation is mainly ecologically controlled, one could predict variation patterns should also be influenced by the degree of habitat heterogeneity. Dubé and Morisset (1987) also asserted the weak structure revealed by similarity analysis of populations was the source of classification problems within species. Although a continuum in morphological characters existed between populations, ecological differences between taxa suggested these taxa were best recognized as subspecies of one taxon (i.e., *F. rubra*), rather than as distinct species.

Morphoanatomical differentiation within *F. rubra*, due mainly to ecological factors, was reported by Huon (1972) from work in western France. Huon (1972) studied reproductive biology of some infraspecific taxa with $2n = 42$ from different habitats (e.g., cliffs, shores, dunes), concluding infraspecific taxa were mainly ecological. Ecological differentiation with respect to salts (Rozema et al., 1978; Humphreys, 1982) or heavy metals has also been reported (Karantanglis, 1980). Anderson and Taylor (1979) studied two distinct morphological types of *F. rubra* growing in proximity, one on stabilized sand dunes and the other on unstable dunes. Physiological differences in response to sand accretion, salinity, and solar radiation between the two populations were found, but no barriers to cross-pollination or dispersal of seeds between their respective habitats were apparent.

Chemical analyses successfully distinguished species and/or cultivars within species. Villamil et al. (1979, 1982) consistently characterized *F. rubra* ssp. *commutata* cultivars using esterase patterns obtained by isoelectric focusing extracts of ground seeds in polyacrylamide gels. Clapham and Almgard (1978) also examined fine-leaved *Festuca* spp. by this method. Weibull et al. (1986) developed a starch gel electrophoretic method to study isoenzyme variation in *F. ovina*. Of twelve loci examined, eight were considered polymorphic. Weibull et al. (1986) concluded the greater genetic variation was representative of species from large ranges, found in habitats representing later stages of succession, and with high fecundities, outcrossing reproductive mode, wind pollination, and long generation time. Fredriksson (1999) demonstrated some success in separating *F. rubra* spp. and *F. trachyphylla* using thin layer chromatography. Wilson (1999) studied isozyme variation for *F. roemerii* and *F. rubra* ssp. *commutata*, comparing it to the enzymatic variation reported earlier in the fine-leaved *Fesctuca* fescues. Wilson showed similar levels of genetic diversity among species she studied compared to other species studied, and concluded the species she studied did not appear to have gone through a genetic bottleneck.

Pavlick (1985) concluded the native North American subtaxa of *F. rubra* from the Pacific Northwest were, for the most part, morphologically separable from those of Europe, except for *F. rubra* ssp. *pruinosa*. Complicating taxonomic matters in the North American *F. rubra* complex is the widespread introduction into North America of European *F. rubra* subspecies (Pavlick, 1985). Despite problems with classifying members of this complex, most recent researchers (e.g., Piper, 1906; Saint-Ives, 1925; Auquier, 1968, 1971a, 1971b; Taylor and MacBryde, 1977; Alexeev, 1982; Pavlick, 1985) suggested subtaxa of *F. rubra* should be recognized, a view accepted and followed here.

The following list includes the subspecies of *F. rubra* and closely related species found in North America and Europe. Although not comprehensive, this list provides the starting point for researchers seeking potential gene pools for future breeding efforts. Pavlick (1985) reported the following North American native taxa and subspecies of *F. rubra*: *F. rubra* ssp. *aucta* (Krecz. & Bobr.) Hult.; *F. rubra* ssp. *mediana* (Pavlick) Pavlick (as var. *mediana* in Pavlick, 1985); *F. rubra* ssp. *pruinosa* (Hack.) Piper; *F. rubra* ssp. *secunda* (J.S. Presl) Pavlick; *F. rubra* ssp. *vallicola* (Rydb.) Pavlick, *F. ammobia* Pavlick, and *F. richardsonii* Hook. Pavlick (1985) also provided a key to distinguish the native North American subspecies of *F. rubra* and related species. *F. rubra* ssp. *aucta* occurs along the Pacific coast from the Kamchatka Peninsula through the Aleutian Islands to southern British Columbia in moist areas, often areas of high annual rainfall, on sand (stabilized sand dunes, beaches, etc.) or silt deposits, from just above high tide line upward. *F. rubra* ssp. *mediana* occurs along the Pacific coast from Vancouver Island to Oregon on sand beaches and dunes. *F. rubra* ssp. *pruinosa* is a seashore taxon which occurs along the Atlantic coasts of Europe from Iceland through the British Isles to Portugal, and along the Pacific coast of North America from Alaska to California. It grows mainly in soil pockets and crevices of rocks and rock cliffs, from the upper littoral zone and above, being occasionally found on pebble or sand beaches. *F. rubra* ssp. *secunda* occurs along the Pacific coast of North America from Alaska southward to Oregon along the seashore, growing on exposed coasts having high annual rainfall, from the upper part of the intertidal zone (pebble beaches) and above (in soil pockets on rocks, in meadows, and on cliffs, bank, and stabilized sand dunes). *F. rubra* ssp. *vallicola* occurs in the mountains above 1000 m elevation in southern British Columbia from the Yukon border area to Wyoming in moist situations such as wet meadows, lake margins, etc., and is also found in disturbed soil such as roadsides. *F. ammobia* (synonym = *F. rubra* ssp. *arenicola* Alexeev) occurs along the Pacific coast from eastern Vancouver Island to California above the high-tide line on sandy beaches and spits, but in British Columbia it appears to be restricted to the summer-dry coastal Douglas fir zone.

Festuca richardsonii is a circumpolar plant. In North America it occurs in Alaska, northwestern British Columbia, Yukon, Northwest Territories and Greenland, south to northwestern British Columbia, along the coast of Hudson Bay, James Bay, Québec and Labrador, and southward in the Rocky Mountains of Alberta. Also in arctic and subarctic Europe and Asia, and the Ural Mountains. This species grows in a variety of habitats; sands, gravels, rocky soil, and silts of river banks, bars and flats, glacial outwash (near glaciers), and beaches, sand dunes, muskegs, and in dry open areas in the mountains. *F. richardsonii* is often regarded as a subtaxon of *F. rubra*. It is morphologically different from typical *F. rubra sensu stricto* from Europe in leaf sclerenchyma pattern, leaf blade vestiture, panicle size and shape, lemma vesititure, awn length, habitat, geography, and other characters. *F. rubra* ssp. *arenaria* (Osbeck) Syme. has been misapplied in North America to *F. richardsonii*.

Hubbard (1984) provided a key to distinguish subspecies of *F. rubra*, including *arctica, arenaria, commutata, litoralis, megastachys, multiflora, pruinosa,* and *rubra*. Three of these subspecies (*commutata, litoralis,* and *rubra*) have undergone germ plasm improvement and are presented in greater detail under separate headings after this section. *F. rubra* ssp. *arctica* (Hack.) Govor. grows in compact tufts with long ascending rhizomes, many ending in shoots. This species thrives at high altitudes on mountains above 600 m. It has also been classified as *F. richardsonii* Hooker. *F. rubra* ssp. *arenaria* (Osbeck) Syme. is commonly known as sand fescue. This perennial forms loosely-tufted patches with solitary culms near the sea on loose and consolidated sand dunes, shingle-sand ridges, and waste ground. *F. rubra* ssp. *megastachys* Gaudin forms loosely-tufted turf with slender rhizomes. Sometimes sown on grassland, it is found on grass banks, waste ground, roadsides, and pastures. The much larger spikelets separate this subspecies from *F. rubra* ssp. *rubra*. *F. rubra* ssp. *multiflora* (Hoffm.) Wallr. grows as a compactly-tufted plant with short, spreading, upturned rhizomes. Found on grassy banks and margins of recently sown grassland, this subspecies differs from other rhizomatous *F. rubra* by its wider leaf blades, larger inflorescence, and spikelets with numerous flowers. *F. rubra* ssp. *pruinose* (Hack.) Piper is commonly named bloomed fescue. Its growth habit is loosely-tufted, with short, slender, wiry, upturned rhizomes. Found on sea cliffs, stony seashores, and salt marshes, this subspecies is distinctive based on its dense habit, very pruinose leaf blades (waxy 'bloom'), very rigid bristle-like leaf blades, and tough, short lemmas.

Use, Management, and Environmental Limitations

F. rubra has become important agronomically, with 17,000 metric tons produced in the European Union in 1996/1997, second to only *L. perenne* and *L. multiflorum* L. (annual ryegrass), among turf-type grasses (Bondesen, 1997). In 1994 Oregon produced over 4.8 million kg of *F. rubra* seed, the largest area of fine fescue production in the northwestern United States (Young et al., 1997). Prior to 1960, limited cultivars were available for turf use (Youngner, 1957; Hanson, 1965), including 'Clatsop,' 'Duraturf,' 'Illahee,' 'Olds,' 'Pennlawn,' and 'Rainier.' Pennlawn was the most widely used of the listed cultivars. In the United States, 43 cultivars have been granted Plant Variety Protection. The 2000 OECD list of Cultivars Eligible for Certification contained 259 cultivars.

The versatility of species within this complex make it suitable for dry, shaded conditions, including lawns, roadsides, airfields, and other general use situations (Beard, 1973). Among the varied uses for *F. rubra* spp. is closely-mown sports fields in combination with *Agrostis capillaris* L. (Colonial bentgrass) (Shildrick, 1977). Wear tolerance was considered moderate, comparable to *A. capillaris*, although less than *L. perenne* and *P. pratensis*. Slow recovery deters its use for some sports turf uses (Beard, 1973). Putting and bowling greens have also been seeded with *F. rubra* (Beard, 1973). Overseeding dormant warm-season turfgrass has been successful with these species (Schmidt and Shoulders, 1977).

F. rubra commonly grows in Nova Scotia, Canada, along roadsides, in pastures (especially those exposed and close to the coast), in sand and gravel along beaches, the upper zone of salt marshes and, sometimes even in boggy soils (Roland and Smith, 1969). Poor submersion tolerance has been reported (Beard and Martin, 1968). Salt tolerance is generally good with these subspecies (Skirde, 1970), although Ahti et al. (1977) reported a wide range among 19 cultivars examined. *F. rubra* ssp. *litoralis* cultivars showed the highest salt tolerance, while *F. rubra* ssp. *rubra* salt tolerance was intermediate, and *F. rubra* ssp. *commutata*, *F. ovina* ssp. *hirtula*, and *F. trachyphylla* salt tolerances were rated low (Ahti et al., 1977). Cold acclimation and deacclimation is good (–45 to –32 C), although less than for *A. stolonifera* L. (creeping bentgrass) and *P. pratensis*, but superior to *L. perenne* (White and Smithberg, 1977).

Originating in Europe, Asia, and/or North America, *F. rubra* utilization led to distribution throughout the cool-humid portions of much of the world, including North America, Eurasia, and northern Africa (Hitchcock, 1950), as well as Australia (Beard, 1973). Within the United States, this genus extends from coast to coast in cooler parts of the country, including mountainous regions (Hitchcock, 1950). Aiken and Darbyshire (1990) reported *F. rubra* has both native and introduced forms in Canada.

Cytology, Cytogenetics, and Breeding

The chromosome number of cultivated *F. rubra* is $2n = 28$, 42 or 56, with 56 being the most frequent count. Within *F. rubra* ssp., $2n = 70$ has also been reported (Bolkhovskikh et al., 1969). Huff and Palazzo (1998) determined counts of $2n = 56$ for seven cultivars of *Festuca rubra* ssp. *rubra*. Delineations along the two most common chromosome counts show $2n = 42$ representative of *F. rubra* ssp. *litoralis*, while $2n = 56$ is found with *F. rubra* ssp. *rubra* (Schmit et al., 1974). Like *F. rubra* ssp. *litoralis*, *F. rubra* ssp. *commutata* also has $2n = 42$ (Schmit et al., 1974).

Nilsson (1933) reported hexaploid *F. rubra* as an autopolyploid. Jauhar (1975) hypothesized allopolyploidy was also possible. The diploid-like chromosome pairing was deemed under genetic control, with polyploidy likely resulting from the integration of three genomes (Jauhar, 1975). *F. rubra* spp. were considered segmental allopolyploids with genetically-controlled diploid-like meiosis by Jauhar (1975). Meiosis was proposed under the control of multivalent suppressor genes.

The remaining cultivated species in the *F. rubra* complex, *F. heterophylla*, has $2n = 28$ (Brandberg, 1948), the lowest ploidy level reported in the *F. rubra* complex. *F. ovina* ssp. *hirtula*, *F. filiformis*, and *F. idahoensis* in the *F. ovina* complex also contain this chromosome complement. Stahlin (1929) and Hubbard (1984) reported this species also may be 2n = 42. Jenkin (1955c) questioned Stahlin's conclusion, noting the plants in question more closely resembled an *F. rubra* type.

The presence of B chromosomes, which may influence recombination of A chromosomes at meiosis, were reported in *F. rubra* by Flovik (1938, 1940). Rees (1974) and Jones (1975) speculated the B chromosome influence is important in adaptive variability of diploid species. Önder and Jong (1977) noted their importance in polyploid species was not yet investigated.

Most *F. rubra* plants are hexaploids and octoploids, with all taxa behaving as outbreeding diploids (Stace and Ainscough, 1984). Auquier (1977) reported up to 49.6% self-fertility in some genotypes, with self-fertilities above 10% considered very exceptional (Auquier, 1977; Barker and Stace, 1982).

Schmit el al (1974) suggested interspecific hybrids among fine-leaved *Festuca* spp. in both *F. rubra* and *F. ovina* complexes may be limited by anthesis date, hour of pollen shed, and/or chromosome number. In the *F. rubra* complex, they determined *F. rubra* ssp. *rubra* (2n = 8x = 56) and *F. rubra* ssp. *litoralis* (2n = 6x = 42) both shed pollen in late afternoon from 15:00 to 17:00 and 14:00 to 16:00, respectively. *F. rubra* ssp. *commutata* (2n = 6x = 42), in the *F. rubra* complex, and *F. trachyphylla* (2n = 6x = 42), in the *F. ovina* complex, shed pollen in early morning, prior to 6:00 and 8:00, respectively. Although these two species from different complexes shed pollen at similar times, anthesis occurred several days later for *F. rubra* ssp. *commutata* compared to *F. trachyphylla*. Schmit et al. (1974) also observed pollen viability dropped rapidly after shed. For example, pollen shed in late afternoon was largely not viable by late evening. This served to effectively isolate *F. rubra* ssp. *commutata* from rhizomatous types in the *F. rubra* complex, even when they shed pollen on the same dates. They noted all fine-leaved *Festuca* spp. appeared to be highly crossbred, with the possibility to select for varied degrees of self-compatibility.

Since original ecological/geographical distribution patterns are disturbed, hybridization between cultivars and ecotypes may occur. Intergeneric hybrids occur naturally between the *F. rubra* aggregate and species of annual *Vulpia* (Stace and Cotton, 1974; Willis, 1975; Hubbard, 1984; Ainscough et al., 1986; Stace and Al-Bermani, 1989). Hybrids between hexaploid *F. rubra* and *V. fasciculata* (Forsskal) Fritsch exist in many localities on the coasts of southern Britain and the Channel Isles (Stace and Cotton, 1974; Willis, 1975). Hybrids have also been found between hexaploid and octoploid *F. rubra* aggregates and tetraploid *V. membranacea* (L.) Dumetorum (dune fescue) in southern England (Stace and Cotton, 1974). The authors argued the octoploids to actually be *F. juncifolia* St. Amans (formerly *F. arenaria* Osbeck, sec. Kjellqvist). Both hybrids were highly sterile. These hybrids were placed in the hybrid genus x *Festulpia* Melderis *ex* Stace & Cotton as x *F. hubbardii* Stace & Cotton *hybr. nov.* (2n = 35) and x *F. melderisii* Stace & Cotton *hybr. nov.* (2n = 42) (suggested to be *F. juncifolia* x *V. membranacea*). The rhizomatous pentaploid appeared more like *F. rubra* vegetatively, but was intermediate between the two species in floral characteristics. Artificial hybrids using *F. rubra* as the female parent were less successful than hybrids using *V. membranacea* as the female parent (Stace and Cotton, 1974). Hexaploid hybrids between diploid *V. bromoides* (L.) S.F. Gray (squirrel-tail fescue) and hexaploid *F. rubra* have been recorded from England (Hubbard, 1984; Stace and Ainscough, 1984) and the Netherlands. Natural hybrids have also been reported between hexaploid *V. myuros* (L.) C.C. Gmelin. (rat's-tail fescue) and *F. rubra* agg. (Hubbard, 1984; Stace and Ainscough, 1984), as well as *V. fasciculata* (tetraploid; sect. *Monachne* Dumort.) (Stace and Ainscough, 1984).

Heptaploid hybrids (2n = 49) were reported from crosses between hexaploid and octoploid specimens of *F. rubra s.l.* in Poland (Skalinski et al., 1971). Artificial hybrids were also produced between hexaploid *F. rubra* ssp. *rubra* and *V. fasciculata* (Forsskal) Fritsch (2n = 28), but only when *V. fasciculata* was used as the female parent (Stace and Cotton, 1974). Barker and Stace (1982) produced intergeneric hybrids of diploid *V. sicula* x hexaploid *F. rubra* aggregate, hexaploid *F. rubra* aggregate x diploid *V. sicula*, diploid *V. sicula* x octoploid *F. rubra* aggregate, tetraploid *V. fasciculata* x hexaploid *F. rubra* aggregate, and hexaploid *V. myuros* x hexaploid *F. rubra* aggregate. In each case, the female parent was listed first.

Wild pentaploid specimens of x *Festulpia hubbardii* were determined to be hybrids of *F. rubra* (2n = 42) and *V. fasciculata* (2n = 28) (Stace and Cotton, 1974). Examination of the highly sterile hybrids revealed two sets of chromosomes from *Vulpia* and three sets of chromosomes from *Festuca* (Bailey and Stace, 1992). Most chromosome pairings in the hybrids were homogenetic, with a smaller level of heterogenetic pairing of up to nine bivalents out of a maximum possible 14 bivalents (Bailey and Stace, 1992). A chance backcrossing of a natural pentaploid hybrid (*F. rubra* x *V. fasciculata*) to hexaploid *F. rubra* produced a hexaploid plant with reasonable fertility and morphologically similar to *F. rubra* (Stace and Ainscough, 1984). This was cited as evidence of a mechanism for continued introgression of *V. fasciculata* into *F. rubra*, thereby increasing the variability of the latter.

L. perenne, crossed as the female with *F. rubra*, produced a hybrid plant successfully grown in vitro by embryo culture (Nitzsche and Hennig, 1976). The resulting plant was confirmed as a hybrid based on its resemblance to the male parent, *F. rubra*. A report of *L. perenne* x *F. rubra* by Holmberg (1926) was later retracted (Jenkin, 1933). Jenkin (1924, 1933, 1955a, 1955b, 1955c, 1955d, 1955e, 1955f) attempted numerous hybridizations using *F. rubra* (2n = 42 or 56) as one of the parents, as summarized in Table 9.1. Jenkin (1924, 1933) crossed *L. perenne* (2n =14) with *F. rubra* (2n = 42), producing F_1 hybrids that resembled *F. rubra*, but were male sterile. Reciprocal crosses were unsuccessful. Jenkin (1924, 1933) suggested natural hybrids between these two species were possible, but lacked detection due to their morphological similarity to *F. rubra*. *L. loliacea* (2n = 21) x *F. rubra* (2n = 56) produced one seed, which did not successfully germinate (Jenkin, 1955f). The reciprocal cross could not be completed, due to nondehiscent anthers of *L. loliacea*. *L. temulentum* (darnel) x *F. rubra* produced seed, but not plants. Crossing *F. rubra* by the hybrid *L. rigidum* Gaudin x *L. perenne* failed to create seed. The remaining crosses using *F. rubra* are reported in the sections with the other parents utilized in those crosses. Jauhar (1975) also successfully crossed *L. perenne* x hexaploid *F. rubra*. Meiotic analysis of the hybrid (2n = 28) revealed extensive pairing of homeologues. Jenkin (1933) also crossed *F. pratensis* (2n = 14) with *F. rubra* (2n = 42 or 56). Low seed set, with no viability resulted. The reciprocal cross yielded no seed from two attempts, but success with three other parents, still with no viable progeny. *F. arundinacea* (2n = 42) x *F. rubra* (2n = 42) yielded seed, none of which germinated (Jenkin, 1933). The reciprocal cross resulted in seed that germinated to produce viable, but weak plants. One plant survived the winter season, but failed to develop inflorescences. Jenkin (1933) suggested natural hybrids were unlikely to survive.

Yet another set of crosses used *F. ovina* (2n = 28) x *F. rubra* (2n = 42 or 56, but not verified) (Jenkin, 1933). The resulting viable plants most resembled *F. rubra*. Seed, but no viable plants were produced from the reciprocal cross. Jenkin (1933) speculated natural hybrids were possible, since parental species occupied contiguous, though disparate habitats. Hybrids were likely to be found in intermediate zones between the two habitats, if they existed.

Table 9.1. *Festuca* spp. inter- and intrageneric hybridization experiments conducted at the Welsh Plant Breeding Station by T. J. Jenkin and associates (Jenkin 1924, 1933, 1955a, 1955b, 1955c, 1955d, 1955e, 1955f, and Jenkin and Thomas, 1949).

Female Parent	Male Parent	Result
F. filiformis	F. rubra	No seedling survival[a]
F. filiformis	L. perenne	No seed produced
F. heterophylla	F. filiformis	Established F1 hybrid
F. pratensis	F. filiformis	Seed did not germinate
F. ovina	F. filiformis	Seed did not germinate
L. loliaceum	F. filiformis	Established F1 hybrid[b]
L. loliaceum x F. filiformis hybrid	L. loliaceum	No seed produced
L. perenne	F. filiformis	No seed produced
L. rigidum	F. filiformis	Seed did not germinate
F. heterophylla	F. arundinacea	Seed did not germinate
F. heterophylla	F. pratensis	No seedling survival
F. pratensis	F. heterophylla	No seed produced
F. heterophylla	F. rubra	No seedling survival
F. rubra	F. heterophylla	Established F1 hybrid
F. heterophylla	F. vaginata (?)	Seed did not germinate
L. loliaceum	F. heterophylla	Seed did not germinate
L. perenne	F. heterophylla	Established F1 hybrid
F. ovina	F. rubra	Established F1 hybrid
F. rubra	F. ovina	Seed did not germinate
F. ovina x F. rubra hybrid	F. rubra	Established F1 hybrid
F. ovina x F. rubra hybrid	F. ovina	Established F1 hybrid
F. ovina	L. loliaceum	No seed produced
L. loliaceum	F. ovina	Seed did not germinate
F. ovina	L. perenne	Seed did not germinate
L. perenne	F. ovina	No seedling survival
F. rubra	F. arundinacea	Established F1 hybrid
F. arundinacea	F. rubra	Seed did not germinate
F. rubra	F. capillata	Seed not produced
F. loliacea	F. rubra	Seed did not germinate
F. rubra	F. pratensis	Seed did not germinate
F. pratensis	F. rubra	Seed did not germinate
F. rubra	L. perenne	Seed did not germinate
L. perenne	F. rubra	Established F1 hybrid
L. perenne x F. rubra hybrid	F. rubra	No seed produced
L. perenne x F. rubra hybrid	L. perenne	No seed produced
F. rubra	L. rigidum x L. perenne hybrid	No seed produced
L. temulentum	F. rubra	Seed did not germinate

[a] Hybrid status was considered questionable.
[b] The hybrid was pollen sterile.

Several attempts at hybridization of *Festuca* spp. used *F. pratensis* Huds. (meadow fescue) as one parent and a fine-leaved *Festuca* spp. as the other parent (Jenkin, 1955b). *F. pratensis* (2n = 14) crossed with *F. filiformis* (2n = 14) produced seeds, but no plants. Although both are diploids, Jenkin (1955b) noted they differed greatly in morphology, as well as habitat preference. Natural hybrids were considered unlikely. Crossing *F. pratensis* (2n = 14) with *F. heterophylla* (2n = 14) produced no seeds. The reciprocal cross produced seeds, although no viable plants resulted. *F. pratensis* (2n = 14) crossed with *F. rubra* (chromosome number not stated, although both hexaploid and octoploid plants were used) produced opposite results from a previous report (Jenkin, 1933). Less seed was produced this direction than the reciprocal cross the second time this hybridization was attempted. Both types of crosses produced seed, but no viable plants from the latest pollinations.

Several hybridizations were attempted using *F. filiformis* as one parent (Jenkin, 1955c). *F. ovina* (2n = 28) x *F. filiformis* (2n = 14) produced seed, but no viable plants. *F. heterophylla* (2n = 14) x *F. filiformis* resulted in seeds that produced one viable plant out of two germinations. The hybrid had smaller spikelets and florets than *F. heterophylla*. Its fertility was not studied. *F. filiformis* x *F. rubra* produced seed that may have actually been the result of self-pollination. The seed germinated, but the plant failed to survive. The chromosome number of *F. rubra* was unknown, but was hypothesized as 2n = 42. *F. filiformis* x *L. perenne* showed evidence of ovary stimulation, but no viable seeds resulted. The reciprocal cross also produced no viable seed. *L. rigidum* x *F. filiformis* yielded no viable seeds. *L. loliaceum* x *F. filiformis* produced seeds that grew initially as weak plants. Surviving hybrids resembled neither parent, and were weak, with sterile pollen. Backcrossing attempts to *L. loliaceum* were unsuccessful.

Numerous hybridization experiments were conducted using *F. heterophylla* (2n = 28, most likely) as one of the parents (Jenkin, 1955d). *F. heterophylla* x *F. arundinacea* (2n = 42) produced seeds, but no plants. *F. pratensis* (2n = 14) x *F. heterophylla* yielded no seeds, while the reciprocal cross produced seed, but no plants. *F. rubra* (2n = 56) x *F. heterophylla* produced some plants, although their hybrid status was not confirmed. The same type of cross with *F. rubra* (hypothesized as 2n = 42) also produced a hybrid, but its fertility was not tested. The reciprocal cross using supposed hexaploid *F. rubra* as the male parent produced germinable seed, but no surviving plants. *F. heterophylla* x *F. vaginata* (undetermined ploidy level) and *L. loliaceum* (2n = 14) x *F. heterophylla* both produced seed, although none germinated. *L. perenne* (2n = 14) x *F. heterophylla* produced weak seedlings. The two hybrids reaching maturity were more *Festuca* in character and male sterile. Female fertility was not examined.

Hybrid attempts also utilized *F. ovina* as one of the parents (Jenkin and Thomas, 1949; Jenkin, 1955b, 1955e). In addition to crosses reported earlier, *F. ovina* (2n =28) x *F. rubra* (2n=42) produced hybrid plants, while the reciprocal cross yielded seed that did not germinate (Jenkin and Thomas, 1949; Jenkin, 1955e). The hybrid that survived appeared more like *F. rubra* in morphology and had nondehiscent anthers. This hybrid was crossed with its female parent (*F. ovina*), producing weak seedlings, with only one plant surviving to maturity. This backcross hybrid appeared intermediate between its parents and had nondehiscent anthers, although its fertility as a female was not tested. The same hybrid was backcrossed to an unrelated *F. rubra*, yielding stronger hybrid plants, although its fertility was also not studied.

Jenkin (1955e) also crossed *F. ovina* with two *Lolium* species. *F. ovina* x *L. loliaceum* produced no seed, while the reciprocal cross yielded seed, but no plants. *F. ovina* x *L perenne* and the reciprocal cross both succeeded in seed development. Only *L. perenne* x *F. ovina* resulted in weak seedlings, which did not survive long term.

Stebbins (1971) theorized the *F. rubra* aggregate is a "declining polyploid complex." Stace and Ainscough (1984) concurred with Stebbins, noting diploids were more rare than polyploids and they could account for the amount of variation found in polyploids. Since diploids had restricted geographic ranges, continued hybridization to form new polyploid derivatives had apparently ceased. Stace and Ainscough (1984) also hypothesized hybridizations between existent polyploids could lead to new combinations and modes of variation. Introgression of relatively unrelated taxa (e.g., *Vulpia* spp.) also could provide a means of increasing genetic variation and of initiating new lines that would not be available to most "declining polyploids."

Genetic transformation in *F. rubra* may circumvent the need for conventional hybridization as a means to transfer desired genes into *F. rubra*. Altpeter et al. (2001) produced a stable genetic transformation using a turf-type *F. rubra* cultivar, 'Borfesta.' Spangenberg et al. (1995a) successfully electrofused metabolically-inactive protoplasts isolated from *F. rubra* with protoplasts of *L. perenne*. This procedure showed potential for directed one-step partial and alien gene transfer between sexually incompatible species.

Floral induction experiments varied in optimal recommendations. Extended cold periods for primary induction of flowering were indicated by observations from Bommer (1959), Cooper and Calder (1964), and Bean (1970). Bean (1970) noted only 20% flowering after 19 weeks exposure to winter conditions in Wales, UK, in one experiment, while no flowering occurred after 26 weeks in another experiment. Primary and secondary inductions for inflorescence development and anthesis required long day conditions (Cooper and Calder, 1964; Bean, 1970; Murray et al., 1973). Murray et al. (1973) and Heide (1990) observed variation among biotypes of *F. rubra* clones for floral induction. Floral induction was optimized under normal daylight with supplementary lighting for one hour during the night, rather than continuous lighting (Murray et al., 1973). After natural cool temperature floral induction in the field to 24 January 1971, followed by 28 days at 10–13°C at normal daylength, optimal floral development occurred after 21–24°C for 28 days under normal daylight supplemented with 1 h light at midnight, resulting in earlier heading, increased panicle size, and greater seed number compared to continuous lighting (Murray et al., 1973). Heide (1990) observed floral induction and development differences among three high-latitude populations studied from the *F. rubra* complex. Optimum flowering occurred using short day (8 h) and 9°C for 15–21 weeks, followed by 12–16 long day (24 h) cycles (Heide, 1990), although one population responded equally well to short and long day cycles at 6 and 9°C. Maximum floral induction for *F. rubra* was also reported to occur under short photoperiods of less than 12 h and cool temperatures between 4 and 10°C (Gardner and Loomis, 1953; Cooper and Calder, 1964). *F. rubra* does not respond to floral induction during its juvenile stage of development (Cooper and Calder, 1962). Klebesadel (1971) observed differential response to floral induction between *F. rubra* spp., based on cultivar adaptation. The subarctic-adapted cultivar 'Arctared' responded to treatments simulating normal subarctic nyctoperiods (lengthening 9 to 15 h), while the temperate-adapted cultivar Illahee suffered winter injury, with poor inflorescence production. Flowering intervals vary somewhat among species. In Great Britain, *F. heterophylla* flowers during June and July, while *F. rubra* ssp. *commutata* flowers during June (Hubbard, 1984).

Native and naturalized locations in temperate regions serve as germ plasm resources for fine-leaved *Festuca* spp. Rutgers University in New Brunswick, New Jersey, contains one of the most extensive collections of specimens used for breeding (Meyer, 2000). Enhanced saline tolerance for *F. rubra* ssp. *litoralis* may be obtained by collections from coastal sites. Plants with heavy metal tolerances have been collected from mining sites. Considerable collection and germ plasm evaluation is needed for this subspecies (Ruemmele et al., 1995). Coastal areas and old mine sites particularly support growth of *F. rubra* ssp. *litoralis*. Greenhouse screening has been used extensively to select *F. rubra* ssp. *litoralis* germ plasm with saline and/or heavy metal tolerances.

Selection criteria for all fine-leaved *Festuca* spp. include increased seed yield, improved stress tolerances (heat, drought, disease), and improved turf quality of high seed-yielding common types (Meyer and Funk, 1989; Anonymous, 1995). Genetic potential exists to produce more vigorous, rhizomatous, sod-forming cultivars better adapted to sod production use (Beard, 1973).

For *F. rubra* ssp. *commutata*, heat tolerance, disease, and insect resistances have already been introduced in recent cultivars. *Puccinia graminis* ssp. *graminicola* resistance could be enhanced in this subspecies (Welty and Azevedo, 1995), since this disease has been found throughout seed production areas of the Willamette Valley of Oregon. Leaf spot resistance (genus and species not stated) appears moderate for some cultivars of this species (Fraser and Rose-Fricker, 1999a; Morris and Shearman, 1994, 1996). *Laetisaria fuciformis* resistance has improved in recent cultivars, rating intermediate between *F. trachyphylla* and *F. rubra* ssp. *rubra* (Morris and Shearman, 1994, 1996). For low-maintenance situations, *Laetisaria fuciformis* and *Sclerotinia homeocarpa* genetic resistance could be improved (Meyer, 1982; Meyer and Belanger, 1997). Nonchoke-inducing endophytes have been detected in a number of cultivars and selections of *F. rubra* ssp. *commutata* (Hurley et al., 1984). Field observations suggested improved summer performance and enhanced resistance to *Blissus leucopterus* might be associated with the presence of these nonchoke-inducing endophytes in 'Longfellow.' Morris and Shearman (1994, 1996) noted cultivars containing the *Neotyphodium* spp. endophytes were associated with *Sclerotinia homeocarpa* resistance.

New cultivars of *F. rubra* ssp. *litoralis* need improved *Laetisaria fuciformis* and *Sclerotinia homeocarpa* resistances (Meyer and Funk, 1989). Endophytes in germ plasm sources have been absent (Ruemmele et al., 1995). New germ plasm has been found with natural endophytes in this species (L. Brilman, 2001, personal communication) or inserted into this species (K.W. Hignight, 2001, personal communication). Salt resistance separates this species from others in this genus. Rose-Fricker and Wipff (2001) observed better seed germination for two cultivars of *F. rubra* ssp. *litoralis* compared to one cultivar of *F. trachyphylla* when salts were added to the germination medium. Disease resistance continues as a major selection criterion for germ plasm improvement. Leaf spot resistance (species not stated) has been observed in improved cultivars (Fraser and Rose-Fricker, 1999a).

Recent *F. rubra* ssp. *rubra* cultivars have shown improved seed yield, lower turf-type growth, and good *Erisyphye graminis* resistance. *Drechslera dictyoides*, *Laetisaria fuciformis* and *Sclerotinia homeocarpa* resistances have been moderately improved (Meyer and Funk, 1989), although Morris and Shearman (1994, 1996) reported moderate to severe susceptibility to *Laetisaria fuciformis* among improved cultivars. Increased resistance of *Laetisaria fuciformis* and *Sclerotinia homeocarpa* will permit better growth in low-maintenance situations (Meyer and Belanger, 1997). Nonchoke-inducing endophytes are desired. *Neotyphodium* spp. endophytes are associated with increased *Sclerotinia homeocarpa* resistance (Morris and Shearman, 1994, 1996). Sources of these endophytes have been identified (Ruemmele et al., 1995) and incorporated into cultivars such as 'Jasper II' and 'SR 5200E' (L. Brilman, 2001, personal communication). *Magnaporthe poae* resistance has been observed in *F. rubra* ssp. *rubra* germ plasm (Kemp et al., 1990). Leaf spot resistance (genus and species not stated) appears moderate for this group (Fraser and Rose-Fricker, 1999a, 1999b), but is not considered a major need for improvement by Meyer and Belanger (1997).

Johnson-Cicalese et al. (2000) developed a technique to inoculate mature *F. rubra* L. ssp. *commutata* and *F. rubra* L. ssp. *rubra* tillers with fungal endophytes (*Epichloe festucae* Leuchtm., Schardl, and Siegel and *Neotyphodium* spp.) from fine-leaved *Festuca* spp. and *Poa ampla* Merr. (big bluegrass). Inoculated plants transmitted endophytes to their offspring, indicating this method could be used as an alternative to cross-pollination to introduce endophytes into endophyte-free germ plasm. Endophytes have been primarily introduced to nonendophyte-containing germ plasm using a modified backcross method with phenotypic selection of progeny (Meyer and Funk, 1989). The initial cross uses the endophyte-containing germ plasm as the maternal parent. Endophyte-containing progeny are then backcrossed to the recurrent parent for one or more cycles, with examination for the presence of endophytes in the progeny after each backcross.

Phenotypic recurrent selection is popular in conventional breeding programs working with outbreeding species, such as the fine-leaved *Festuca* spp. Parental lines vary from a few to more than 50. Aminotriazole herbicide tolerance increased in *F. rubra* ssp. *commutata* after three cycles of selection (Lee and Wright, 1981). Resistance approximately doubled with each selection cycle. Tolerance to the nonselective herbicide, glyphosate, was significantly increased after each of four cycles of recurrent selection using the *F. rubra* ssp. *litoralis* cultivar 'Manoir' (Johnston et al., 1989).

Festuca heterophylla Lamarck, Shade Fescue

F. rubra L. var. *heterophylla* (Lamarck) Mutel is another synonym by which this species has been known (Hitchcock, 1950; Hubbard, 1984). Another common name used for this species is various-leaved fescue (Hubbard, 1984).

F. heterophylla originates from southern England and Poland south to northwest Spain and Greece. It is considered native to temperate Asia and Europe (USDA, 2001). Clapham et al. (1952) stated this species was either native or, more likely, introduced to Great Britain. Hubbard (1984) contended introduction to the British Isles occurred during the early nineteenth century. It has become naturalized as far north as Northumberland and in Scotland. This species is also thinly distributed through central and southern Europe and southwest Asia (Hubbard, 1984). *F. heterophylla* was reported introduced into the United States as a turfgrass. It is found mainly in forests.

F. heterophylla has had limited use for turf purposes. Where utilized, it has been sown in woodland areas, as well as being an early fodder plant (Hubbard, 1984). Low fertility sites favor this species, as with other *Festuca* spp. Specimens have been found in woods and wood margins (Hubbard, 1984). Dry, sandy or gravelly soils favor this species (Hubbard, 1984). *F. heterophylla* has naturalized readily to dry, shady places (Jenkin, 1955d).

Plants are perennial, densely tufted, and without rhizomes. Culms range from 60 to 150 cm tall, with sheaths closed for more than three-quarters of their length. Dead basal sheaths scarcely shred into fibers at the leaf base. Ligules are 0.1 to 0.3 mm long. Leaves grow to 25 cm long, and 2.0 to 4.0 mm wide, being flat to conduplicate, 3–5 (7) veined, with 1(3) ribs; and sclerenchyma tissue in 3–5 very slender strands. The inflorescence grows 6–17 cm long, with the panicle contracted or open, and branches scabrid. Spikelets are 8.0 to 11.5 mm, with 3–6 florets. The first glume ranges from 3.0 to 5.0 mm long, while the second glume spans 4.0 to 7.0 mm. Lemmas may be 5.0 to 8.0 mm long. Awns grow to 6.0 mm. Anthers are half as long as the palea. The ovary apex is pubescent. Hubbard (1984) also noted leaves are very hairy at their base. This species may be distinguished from *F. ovina* by longer three-angled basal leaflets, closed sheaths, wider, flat culm blades, larger spikelets, and longer lemmas (Hubbard, 1984).

Two turf cultivars are available in Europe. 'Liget' is sold in Hungary, while 'Sawa' is distributed in Poland.

Festuca rubra L. ssp. *commutata* (Thuill.) Nyman, Chewings (Chewing's) Fescue

Taxonomy of Chewings fescue has fluctuated in recent years. Prior to the current classification, *F. nigrescens* Lamarck and *F. rubra* var. *commutata* Gaudin were also used in reference to this subspecies (Hubbard, 1984). Hitchcock (1950) noted the *F. rubra* var. *commutata* name, as well as *F. commutata* Thuill. (USDA, 2001), *F. rubra* var. *fallax*

Hackel, and *F. rubra* ssp. *eu-rubra* var. *commutata* St.-Yves. Although sometimes containing an apostrophe in the common name (Chewing's), this grass was named after George Chewings, who called the grass "Chewings fescue" (Morgan, 1998). The common name appears both capitalized and uncapitalized in literature.

The origin of *F. rubra* ssp. *commutata* ranges from southern Sweden southward in Europe, throughout most parts of Europe (Hubbard, 1984). This subspecies has been widely introduced elsewhere in the world. *F. rubra* ssp. *commutata* was first cultivated, harvested, and sold as seed in New Zealand by George Chewings (Aldous and Semos, 1999; Hubbard, 1974; Morgan, 1998). This grass was originally imported from England and sold to the previous farm's owner (William Tarlton) as "hard fescue" (Morgan, 1998). Mr. Chewings observed the unusual patch of grass on his newly-purchased farm in Mossburn, New Zealand and began seed production, selling the grass as "Chewings fescue" (Morgan, 1998). Specimens have been located on open grasslands, road verges, and waste ground (Hubbard, 1984).

Utilization of this subspecies is highly varied throughout the world, with wide distribution in the United States of America, Great Britain, and New Zealand (Hubbard, 1984). For lawns, *F. rubra* ssp. *commutata* may be planted alone or with *Agrostis* spp. or *F. rubra* (Hubbard, 1984). It is also blended with other fine fescues and *Poa pratensis* in mixtures for sun or partial shade of home lawns (Erdmann and Harrison, 1947). It has been planted on the grass courts at Wimbledon in England, the greens and fairways at St. Andrews golf course in Scotland, and home lawns in New Zealand (Morgan, 1998). *F. rubra* ssp. *commutata* has been successfully maintained on hard clay fairways (Ermer, 1928). Henensal et al. (1977) contended this species was one of the most suitable in France for roadside embankments. Australian cricket pitches use this subspecies in overseeding mixes (Walker, 1969). In the United States, some *F. rubra* ssp. *commutata* is used for winter overseeding dormant warm-season turf to provide winter color and cover for lawns and sports turf (Schmidt and Shoulders, 1977; Hurley, 1990; Skogley et al., 1993). *F. rubra* ssp. *commutata* cultivars performed best in Japanese management experiments of *F. rubra* spp. (Razmjoo et al., 1993). Cultivar adaptability varied among locations, including utilization for overseeding *Zoysia japonica* L (Japanese lawngrass).

Wear tolerance superior to *F. rubra* ssp. *rubra* has been reported by Shildrick (1976b). Medium-to-low maintenance conditions in full sun to moderate shade permit this species to thrive. *F. rubra* ssp. *commutata* is drought resistant, tolerating drier soils, such as well-drained, chalky, gravelly, or sandy soils (Hubbard, 1984). Poor low temperature color retention has been reported (Jackson, 1962). Excess thatch accumulation may result at pH less than 5.0 (Edmond and Coles, 1958). High nitrogen fertilization has also been associated with excess thatch accumulation in this species.

Plants are perennial and tufted, lacking rhizomes. Culms reach 25–90 cm tall, with sheaths closed for most of their length and glabrous or puberulent at apex. Basal sheaths appear reddish-brown, with dead sheaths shredding into fibers at their base. Living sheaths are usually glabrous and reddish, browning as they age. The ligule is minute. Leaves are 4–15 cm long and 0.3 to 0.7 mm wide. They are conduplicate, appear green or glaucous, and can be glabrous or scabrous on the leaf angles. Leaves also have (3-) 5 adaxial ribs, sclerenchyma tissue usually in (5) 7 strands, and no costal sclerenchyma tissue. The inflorescence ranges from 4 to 13 cm long. It is narrow and often secund, with scabrous branches spreading only at anthesis. Spikelets are 6.5 to 9.5 (11) mm long, with 5–8 florets. The first glume is 2.0 to 4.0 mm long and 1 (-3)-veined, while the second glume is 3.3 to 5.0 mm long and also (1-) 3-veined. Glabrous lemmas may be 4.6 to 6.2 mm long and green, although usually reddish-violet on the upper portion. The awn is 1.0 to 3.0 mm long, less than half as long as the lemma. Anthers are 1.8 to 2.2 mm long, usually more than half as long as the palea. The ovary apex is glabrous. Hubbard (1984) reported *F. rubra* ssp. *commutata* could be distinguished from other *F. rubra* by the lack of rhizomes, and *F. ovina* spp. by its thicker leaf blades, tubular leaf sheaths, smooth culms, and longer-awned lemmas. Extensive tillering (Parks and Henderlong, 1967) enables this species to maintain its density in turf use.

Selected cultivars are described with regard to method of cultivar development and/or improvements to this species. 'Cascade' was released as a noncreeping cultivar developed from New Zealand germ plasm (Frakes, 1974). A questionable classification of another New Zealand cultivar occurred with 'Grasslands Tasman' ('Tasman'). This cultivar was described as a "creeping red fescue" with rhizomes, yet it was reportedly developed as a seven-parent synthetic cultivar (2n = 42) selected for rhizomatous spread from a population composed mainly of lines of 'New Zealand Chewings' fescue (Rumball, 1982b). 'Grasslands Tasman' was distinguished from yet another New Zealand cultivar, 'Grasslands Cook' ('Cook') by its longer, coarser leaves, fewer tillers, larger rhizomatous spread, and larger root base. Grasslands Cook (2n = 42) was described as a tufted, nonrhizomatous cultivar (Rumball, 1982a). This cultivar resulted from selection from a polycross of a population containing mainly lines of New Zealand Chewings fescue. It is brighter green, less diseased, more uniform, and more densely-tillered, with shorter leaves than New Zealand Chewings fescue. Both Tasman and Cook are reported as "Chewings fescue" cultivars developed in New Zealand (Rumball and Rolston, 1988), despite Tasman being described as having a creeping habit, atypical of this subspecies.

The Oregon Agricultural Experiment Station released 'Checker' (Oregon Agricultural Experiment Station, 1978). This cultivar was selected from turf plots established by seed derived from open pollination of 60 cultivars. Its attributes include moderate spread, acceptable dark green color, and fine leaf texture.

Herbicide resistance was selected for in the cultivar 'Countess' by recurrent selection, using increasing concentrations of the herbicide through successive generations of selection (Johnston and Faulkner, 1986). This cultivar tolerates applications of aminotriazole, which may be used to eliminate weedy grass species from lawns and seed production fields. Unfortunately, shortly after this cultivar was developed, along with the *A. capillaris*, 'Duchess,' which was also developed for resistance, this herbicide became a restricted use chemical.

Pickseed, Inc. of Tangent, Oregon, released 'Victory' (Pepin et al., 1988a), an advanced generation synthetic of progeny from 178 clones from the Rutgers University turfgrass breeding program in New Brunswick, New Jersey. Parental material included collections from old turfs in the eastern United States, as well as the cultivars 'Barfalla,' 'Highlight,' 'Koket,' 'Menuet,' and 'Waldorf.' Victory demonstrated improved seed yield, disease resistance, and phenotypic uniformity. Tolerance to close mowing, moderate shade, acidity, and low fertility were additional attributes.

Longfellow was an early endophyte-containing cultivar developed through cooperative efforts between International Seeds, Inc. (now Cebeco, Inc.) and the New Jersey Agricultural Experiment Station (Edminster et al., 1993). This advanced generation synthetic was selected from progeny of 10 clones, based on medium dark-green color, fine texture, improved disease resistance, and drought stress tolerance. Most newer cultivars of *F. rubra* ssp. *commutata* that are endophyte-infected contain the Longfellow endophyte.

'Jamestown' and Jamestown II were introduced from the University of Rhode Island breeding program. Jamestown exhibited improved heat tolerance compared to earlier cultivars (Funk et al., 1980). Jamestown II was enhanced with *Neotyphodium coenophialum* from *F. rubra* ssp. *commutata* germ plasm closely related to Longfellow (Skogley et al., 1993). Jamestown II was developed using a modified backcross procedure with

Jamestown as the recurrent parent and LF-1 (related to Longfellow) as the endophyte donor parent. A high level of endophyte makes this cultivar inappropriate to forage or pasture usage. This dark green, fine-texture cultivar produces a turf of high density. Winter color retention, pest resistance, and cold hardiness are additional attributes. Tolerance to low mowing heights permits use of this cultivar in overseeding dormant bermudagrass greens in warmer climates, as well as other dormant warm-season lawns and sports turfs. Like Jamestown, 'Banner' exhibited improved heat tolerance compared to earlier releases (Funk et al., 1980).

Additional cultivars were released with improved attributes. 'Shadow' demonstrated improved *Erisyphye graminis* resistance (Funk et al., 1980). 'SR 5000' and 'SR 5100' were released by Seed Research of Oregon, Inc. (Lynch, 1996). Both cultivars form dense, fine-textured turf, bright, dark green in color and high in endophyte level. SR 5000 shows improved summer performance and disease resistance. SR 5100 grows very slowly, with dwarf characters. It tolerates shade and low maintenance well. Pickseed West developed 'Silhouette' using the Rutgers University germ plasm. It was released by Roberts Seed and Ampac Seeds for use in low maintenance shade areas (Anonymous, 2000). This dark green cultivar exhibits very fine leaf texture and excellent heat and drought tolerances. It is recommended for hard-to-mow areas. 'Tiffany' was released by Turf-Seed, Inc. of Hubbard, Oregon, in 1992 (Rose-Fricker et al., 1999b). An advanced-generation synthetic, this cultivar exhibits low growth habit and many reproductive tillers. It contains the *Neotyphodium coenophialum* from Longfellow. Recurrent selection for low plant height, uniform maturity, numerous reproductive tillers and turf performance resulted in release of a cultivar that produces a dark green, fine-textured turf, with resistance to several fungal diseases.

Festuca rubra L. ssp. *litoralis* (G.F.W. Meyer) Auquier, Slender Creeping Red Fescue

Prior to separation from *F. rubra* ssp. *rubra*, this subspecies was considered part of the entire complex known as *F. rubra* (Hitchcock, 1950). Stahlin (1929) designated this species as *F. rubra* ssp. *eu-rubra* var. *genuina* Hackel. subvar. *vulgaris* Hackel. It has also been referred to as *F. rubra* ssp. *tricophylla* and *F. rubra* var. *litoralis* Vasey (Huff and Palazzo, 1998).

Native to the coasts of Western Europe and the Baltic region, *F. rubra* ssp. *litoralis* is typically found in salt marshes and damp maritime sands. *F. rubra* ssp. *litoralis* grows in muddy tidal sediment of coastal areas of Great Britain (Hubbard, 1984). Salinity and heavy metal tolerances set this species apart from many others. Some cultivars ('Dawson') display excellent salt tolerance (McHugh et al., 1985). Merlin and 'Sterling' withstand exposure to heavy metals (Shildrick, 1980). Wear tolerance superior to *F. rubra* ssp. *rubra* has been reported by Shildrick (1976b), making this species suitable for some moderate traffic situations.

Plants are perennial and tufted, with short, slender, rhizomes. Culms grow to 60 cm. Sheaths are closed for most of their length, appearing glabrous or short pubescent. Basal sheaths become reddish-brown, with dead sheaths shredding into fibers at their base. Living, glabrous sheaths are usually reddish, browning with age. Ligules are 0.1 to 0.4 mm long. Leaves grow 5–30 cm long and 0.4 to 0.8 mm wide. They may be flat or conduplicate, are abaxially green and smooth, and are scabrous or have short pubescent ribs (3–5) adaxially. Sclerenchyma tissue is usually in (5) 7 strands, with no costal sclerenchyma tissue present. Inflorescences extend 6–8 cm, being erect to slightly nodding, with scabrous branches. Spikelets measure 8–11 mm long, with 5–8 florets. The first glume is 2.0 to 4.5 mm long, (1–3)-veined, while the second glume is 4.6 to 5.2 mm long, and also (1-) 3-veined. Lemmas are 6.0 to 7.0 mm long, bright green to glaucous, and glabrous, except for the scabrid apex (occasionally along upper margins as well). The awn grows to 0.6 to 3.0 (4.0) mm long, less than half as long as the lemma. Anthers are usually more than half as long as the palea. The ovary apex is glabrous. *F. rubra* ssp. *litoralis* may be distinguished from *F. rubra* ssp. *rubra* by its longer lemmas, glabrous outer sheaths and the longer rhizomes of *F. rubra* ssp. *rubra* (Hubbard, 1984).

Numerous cultivars have been successfully developed from evaluation and selection within ecotypes, as well as hybridization between ecotypes. 'Seabreeze' was derived from hybridization of *F. rubra* ssp. *commutata* and *F. rubra* ssp. *litoralis* populations, followed by selection for rhizome development as found in *F. rubra* ssp. *litoralis* (Rose-Fricker et al., 1999c). Maternal *F. rubra* ssp. *litoralis* cultivars were topcrossed to *F. rubra* ssp. *commutata* cv. Highlight in the Netherlands. Plants having 2n = 42 chromosomes, rhizomes, narrow leaves, good spring growth, and *Erisyphye graminis* resistance were selected from space plantings. Further recurrent selection increased reproductive capacity as well as *Drechslera dictyoides* and *Puccinia crandallii* resistance. Although from hybrid origins, Seabreeze was released as *F. rubra* ssp. *litoralis*. This low-growing cultivar has bright green color, good establishment rate, high tiller density, improved turf quality, excellent winter color, cold and shade tolerances. In addition to diseases noted earlier, *Sclerotinia homeocarpa* resistance is also good.

'Dawson' demonstrated more *Drechslera dictyoides* resistance than other fine-leaved *Festuca* spp. cultivars (Meyer, 1982). 'Dawson (E)' was developed by introducing an endophyte into multiple seedlings of Dawson. A single, dwarf plant of Dawson was detected in 1974 (Fisher and Johnston, 1985). The plant had very short leaves and high vegetative tiller density. When crossed with Dawson, F_1 progeny were tall. Three dwarf groups were isolated by the F_3 generation, ultimately resulting in the advancement of this dwarf habit as 'Logro' and 'Elfin.' Further selection for color and uniformity resulted in the cultivar 'Count' developed from this same germ plasm (L. Brilman, 2001, personal communication). European breeders have developed many additional cultivars including 'Barcrown,' Manoir, 'Cezanne,' and 'Mocassin.' Many cultivars of this species have not shown the stress tolerance necessary for many areas in the United States. They also often have very poor seed yields.

Festuca rubra L. ssp. *rubra* Gaudin, Strong Creeping Red Fescue

As with *F. rubra* ssp. *litoralis*, this subspecies was once known simply as *F. rubra* (Hitchcock, 1950). Earlier sources list this species as *F. rubra* ssp. *eu-rubra* var. *genuina* Hackel (Hackel, 1882; Levitsky and Kuzmina, 1927). The USDA (2001) also lists *F. glaucescens* Hegetschw. & Heer, nom. illeg., *F. rubra* ssp. *glaucodea* Piper, *F. rubra* ssp. *vulgaris* (Gaudin) Hayek, *F. rubra* var. *glaucescens* K. Richt, nom. illeg., *F. rubra* var. *glaucescens* Hack. and *F. rubra* var. *lanuginosa* Mert. & W.D.J. Koch as synonyms.

This species is native throughout most of Europe, including coastline areas (Wild-Duyfjes, 1973) and Great Britain (Hubbard, 1984). *F. rubra* ssp. *rubra* grows abundantly on short grassland dunes, moors, and mountain slopes, as well as in mixed woodland, hedge rows, and wasteland (Hubbard, 1984). Huon (1972) reported its presence in eastern France. The USDA (2001) reports this species as native to Asia and North America, in addition to Europe.

Lawns, parks, cemeteries, and golf course roughs are recommended establishment sites. This species shows more compatibility in lawn mixtures with *P. pratensis* and *L. perenne* than other fine-leaved *Festuca* spp. (Meyer and Funk, 1989). Temperate regions throughout the world favor this species. Coastal areas with saline soils are prime locations for germ plasm. Old mine sites often contain specimens from this species.

Plants are perennial and usually loosely tufted, with long rhizomes. Culms extend to 110 cm, with glabrous or pubescent sheaths closed for most of their length. Basal sheaths are reddish-brown, with dead sheaths shredding into fibers at their base. Living sheaths usually appear reddish, browning as they age. Ligules are 0.1 to 0.4 mm long. Leaves reach 5–30 cm in length, and 0.6 to 4.0 mm in width. They are flat or conduplicate, abaxially green or glaucous and smooth. Adaxially, they may be scabrous or have 3–5 pubescent ribs. Sclerenchyma tissue is usually present in 5–7 strands, with no costal sclerenchyma tissue evident. The inflorescence grows to 7–14 (20) cm. It is erect to slightly nodding, with scabrous branches. Spikelets range from 6.0 to 14.0 mm long, with 5–8 florets. The first glume is 2.0 to 4.5 mm long, 1 (-3)-veined, while the second glume is 3.5 to 6.0 mm long, and also (1-) 3-veined. Lemmas are 5.0 to 8.0 mm long and green, with purple to mostly reddish-purple margins. They are glabrous, except for a scabrid apex (occasionally also along upper margins). Awns are 0.6 to 3.0 (4.0) mm long, less than half as long as the lemma. Anthers may be 2.4 to 3.5 mm long and are usually more than half as long as the palea. The ovary apex appears glabrous. Seeds are larger than *F. rubra* ssp. *commutata* (Meyer and Funk, 1989).

Several improved cultivars have been released since the 1960s. The Canada Department of Agriculture Research Station in Beaverlodge, Alberta, Canada released 'Boreal' (Elliott, 1968). Due to its high seed yield and public release, much of the seed currently sold as Canadian red fescue is actually Boreal. Another early cultivar of *F. rubra* ssp. *rubra*, 'Wintergreen,' compared favorably with Pennlawn in Michigan trials (Copeland, 1972). This cultivar was deemed suitable for cool climates on dry sandy soils of moderate acidity and shaded sites.

The University of Alaska Agricultural Experiment Station, SEA, and USDA jointly developed Arctared (Hodgson et al., 1978). This cultivar was derived from seed collected from the Matanuska Valley in Alaska in 1957. It has excellent winter survival, even when mown as closely as 1.25 cm for the entire growing season. Rapid germination and excellent seedling vigor are additional attributes.

'Fortress,' 'Ruby,' and 'Ensylva' displayed less leaf spot (species not stated), *Laetisaria fuciformis*, and *Sclerotinia homeocarpa* than the best fine-leaved *Festuca* cultivars of the time (Funk et al., 1980). Fortress (2n = 56) consists of a six-clone synthetic cultivar (Duell et al., 1976). Parental clones were based on good turf quality and similar maturity of original clones and polycross progenies. Fortress shows moderately good resistance to *Erysiphe graminis*. It forms a moderately dense, leafy turf of medium fine texture and moderately dark green color. Although it has aggressively-spreading rhizomes, Fortress produces a significantly less dense sward than other contemporary improved cultivars of *F. rubra* ssp. *commutata*, which makes it better for blending with other turf species. Seedling vigor is good, with tolerance to low soil pH, low fertility and shade.

'Jasper,' 'Flyer II,' 'Fenway,' 'Pathfinder,' and 'Flyer' cultivars were developed from germ plasm from Rutgers University (Meyer, 2000). 'Flyer' was developed jointly by Pure-Seed Testing, Inc. of Hubbard, Oregon, and the New Jersey Agricultural Experiment Station (Meyer et al., 1990). This advanced generation synthetic cultivar resulted from progenies of 17 clones. Parental material was collected from old turfs in the northwestern and southern United States. Improved *Drechslera dictyoides*, *Puccinia crandallii*, and *Erysyphye graminis* resistance, as well as improved vigor and high seed yields are among attributes for this improved cultivar.

Pure Seed Testing, Inc. released 'Shademaster' in 1987 as an advanced-generation synthetic cultivar composed of progenies from nine plants (Rose-Fricker et al., 2000). Progeny emanated from 'Vista,' with phenotypic recurrent selection based on improved turf performance, dark green color, improved seed yield, and resistance to *Drechslera dictyoides*. The resulting low-growing cultivar produces a dense, fine-textured turf having medium-dark-green color, good shade and low maintenance tolerances.

'SR 5200E' was released by Seed Research of Oregon, Inc. (Lynch, 1996). This dark green, fine-textured, aggressive cultivar produces strong, vigorous rhizomes, making it suitable for erosion control sites. This cultivar was developed from germ plasm selected by Mr. Jack Murray, USDA, for persistence in the mid-Atlantic states in the United States.

THE *FESTUCA OVINA* COMPLEX

Taxonomy

Considerable confusion over classification of species within this complex exists (Ruemmele et al., 1995). Hackel (1882) grouped *F. ovina* into one species, while others have split this complex into a large number of species, based on morphological differences (Markgraf-Dannenberg, 1980). Turesson (1926) described morphology of *F. ovina* spp. in Scandinavia. British diploid and tetraploid races within the *F. ovina* complex were delineated by Watson (1958). Önder and Jong (1977) contended the diploid form was more properly classified as *F. valesiaca*. The formation of leaf sclerenchyma tissue was used by Horanszky et al. (1971) to separate *F. ovina* aggregates. *F. vaginata* and *F. glauca* contained sclerenchyma in a ring-like structure. *F. pseudovina, F. valesiaca*, and *F. sulcata* sclerenchyma tissue was located in three bundles. *F. wagneri* sclerenchyma tissue formed an intermediate arrangement between the first two groups. Pavlick (1984) recognized seven native taxa of the *F. ovina* complex in the mountains of British Columbia, Canada, including *F. idahoensis* as a segregated species from *F. ovina*.

Chromosome counts have limited use in species delineation, with overlapping distribution among species. Soo (1973) reported chromosome numbers for species from Hungary. *F. pallens* included $2n = 14, 28, 42$; *F. vaginata* was $2n = 14$; *F. wagneri* and *F. dalmatica* and *F. pseudodalmatica* contained $2n = 28$; *F. valesiaca* consisted of $2n = 14, 28, 42$; *F. rupicola* was only $2n = 42$; and *F. pseudovina* had both $2n = 14$ and $2n = 28$. Tracey (1977) noted two hexaploid taxa within the *F. ovina* complex, *F. brevipila* and *F. carnuntia*, and one tetraploid, *F. eggleri* as three new species. *F. pseudovina* was suggested better represented as *F. valesiaca* ssp. *parviflora* (Tracey, 1977). *F. valesiaca* and *F. pseudovina* have since been placed in the same taxon under *F. valesiaca* (Huff and Palazzo, 1998). The cultivar 'Covar' has been shown to originate from Turkish populations showing introgression between *F. valesiaca* and *F. callieri* (Wilson, 1999). Bolkhovskikh et al. (1969) reported $2n = 14, 21, 28, 42, 49$, and 70 for this complex.

To distinguish nonrhizomatous fine-leaved *Festuca* spp. included in both *F. rubra* and *F. ovina* complexes (*F. rubra* ssp. *commutata, F. valesiaca, F. ovina* ssp. *hirtula*, and *F. trachyphylla*), Schmit et al. (1974) used leaf morphology, date and hour of anthesis. One species from the *F. rubra* complex, *F. rubra* ssp. *commutata*, had wider leaves than the remaining species from the *F. ovina* complex. For species within this complex, *F. ovina* ssp. *hirtula* pollen shed was observed between 11:00 and 12:00, while *F. trachyphylla* pollen was shed before 8:00.

Morphologically, Schmit et al. (1974) determined *F. filiformis* (cited as *F. tenuifolia*) had no prominent epidermal ridges, forming a roundly infolded leaf. Leaf width ranged from 0.2 to 0.5 mm. *F. ovina* leaves were V-shaped and narrow (0.3 to 0.5 mm), with an indistinct epidermal midrib. *F. trachyphylla* leaves were also V-shaped, but wider (0.4 to 0.8 mm) than *F. ovina* or *F. filiformis*. The midrib of *F. trachyphylla* was well-defined, with two less-distinct side ridges. Hubbard (1984) noted *F. ovina* has smaller spikelets than *F. rubra*, as well as younger leaf sheaths being united in *F. rubra*, but not *F. ovina*. *F. trachyphylla* and *F. glauca* leaf blades are slightly thicker and awns are longer than those of *F. ovina* (Hubbard, 1984). As with species studied in the *F. rubra* complex, Schmit et al. (1974) noted species from the *F. ovina* complex also appeared highly crossbred, with the possibility to select for varied degrees of self-compatibility.

Use, Management, and Environmental Limitations

Weibull et al. (1991) noted the 1990 OECD list contained 33 cultivars of *F. ovina* s.l., with most from *F. trachyphylla*. Other varieties on this list were classified as *F. ovina* ssp. *capillata* Lam. (now *F. filiformis*), *F. ovina* ssp. *vulgaris* (Koch) Sch. & Kell., and *F. ovina* ssp. *vallesiaca* Koch (now *F. valesiaca*) (OECD, 1990). By 1999, the OECD list contained 44 cultivars.

Species within the *F. ovina* complex typically tolerate droughty, infertile soils. They are often used where fertility and irrigation will be minimal, if not absent. Some have provided grazing for sheep in dry areas, as well as for turf purposes (Weibull et al., 1991). Low maintenance usage includes roadsides, reclamation areas, roughs, and railway banks. Lawns and wildflower mixtures may include these species. Some species, like *F. ovina* ssp. *hirtula* and *F. glauca*, may be better suited for ornamental plantings than as turf. Lead tolerance has been documented in diploid *F. ovina* (Urquhart, 1971). Diploid forms display excellent drought tolerance and low requirement for nitrogen (Weibull et al., 1991).

An outbreeding, wind-pollinated and widely distributed grass, *Festuca ovina* grows on the Baltic island of Oland (Prentice et al., 1995). Roland and Smith (1969) noted *F. ovina* was naturalized in Nova Scotia from its origin in Europe. It persists and forms clumps on sterile, sandy, and poor dry soils, even appearing as a lawn weed (Roland and Smith, 1969). In Scandinavia, this species grows on heaths, dry meadows, mountain grassland, and dry, open woodland (Hylander, 1953; Hubbard, 1974).

Cytology, Cytogenetics, and Breeding

F. valesiaca may be $2n = 14, 28, 42$, while *F. pseudovina* had both $2n = 14$ and $2n = 28$ (Soo, 1973). *F. valesiaca* and *F. pseudovina* have since been placed in the same taxon under *F. valesiaca*. *F. filiformis* has two chromosome levels reported, $2n = 14$ (Stahlin, 1929; Hubbard, 1984) and $2n = 28$. These ploidy levels are similar to *F. ovina* ssp. *hirtula* and *F. idahoensis* ($2n = 28$) and *F. valesiaca* ($2n = 14$) in the *F. ovina* complex, as well as *F. heterophylla* in the *F. rubra* complex. *F. idahoensis* was reported with $2n = 28$, comparable to *F. ovina* ssp. *hirtula* (Hubbard, 1984) and *F. filiformis* in the *F. ovina* complex, as well as *F. heterophylla* in the *F. rubra* complex. Turesson (1930) listed sexual *F. ovina* as diploid ($2n = 14$). Weibull et al. (1986) considered *F. ovina* as an allogamous perennial grass, $2n = 14$. SDS-PAGE analyses of seed proteins supported classification of *F. idahoensis* as a separate species, but did not support separation of subspecies within this taxon (Aiken et al., 1992; Aiken et al., 1998). The chromosome number of *F. trachyphylla* was reported as $2n = 42$ (Hubbard, 1984).

The presence of B chromosomes, which may influence recombination of A chromosomes at meiosis, were reported in *F. ovina* by Önder and Jong (1977) and Bidault (1964, 1968). Rees (1974) and Jones (1975) speculated the B chromosome influence is important in adaptive variability of diploid species. Önder and Jong (1977) noted their importance in polyploid species was not yet investigated.

Studies by Prentice et al. (1995) showed the association between allele frequencies and habitat variation may be the result of either direct or indirect selection on enzyme loci studied. The among-habitat component of genetic variation corresponded to niche diversification. Results supported their hypothesis that "niche variation may contribute to maintenance of genetic polymorphism within populations of *F. ovina*." (Prentice et al., 1995).

Hybridization involving species within the *F. ovina* complex has not been as extensive as with the *F. rubra* complex species. No natural hybrids between *L. perenne* (2n = 14) and *F. ovina* (2n = 28) have been reported (Jenkin, 1933). Artificial crosses yielded nonviable seed (Jenkin, 1933). Watson (1958) crossed *F. filiformis* (as *F. tenuifolia*, 2n = 14) with *F. ovina* (2n = 28). When the diploid served as female parent, hybrids were triploid. The reciprocal cross produced tetraploid progeny. Although there was no emasculation while performing the crosses, Watson (1958) asserted the low self-fertility of the parents made self-fertilization unlikely. Hybrids between diploid *F. glauca* and *F. vaginata* varied greatly in male and female sterility, with the two types being uncorrelated (Horanszky et al., 1971). Pollen grain size also varied among the hybrids. Brilman (2001) reported hybrids from *F. filiformis* and *L. perenne* crosses obtained through mutual bagging and no emasculation. Although comparable with *F. rubra* ssp. *commutata*, *F. rubra* ssp. *rubra*, and *F. rubra* ssp. *litoralis*, interspecific hybridization of *F. trachyphylla* in nature is minimized by unique pollen shed periods (Schmit et al., 1974). *F. glauca* has been crossed with *F. trachyphylla* to produce an advanced generation synthetic hybrid released as the cultivar 'Minotaur' (Anonymous, 2000). Minotaur forms a dense, dark green to blue-green turf. This high endophyte-containing cultivar produces short plants. Although they pollinate at different times of the day, many researchers consider *F. glauca* a 'glaucous' or 'blue' form of *F. trachyphylla*, rather than a separate species. These glaucous forms have also been erroneously designated *F. ovina* ssp. *hirtula*. The cultivar 'SR 3200' also originated from a controlled cross of *F. glauca* x *F. trachyphylla* that differed naturally in time of pollen shed.

Short days of cool temperatures induce flowering of *F. ovina* (Wycherley, 1954; Cooper and Calder, 1964; Beard, 1973). In the British Isles, flowering of *F. filiformis*, *F. ovina* ssp. *hirtula*, and *F. trachyphylla* occurs in May and June (Hubbard, 1984). Seed has been difficult to obtain from *F. ovina* ssp. *hirtula* plants. Morning flowering was observed for *F. valesiaca* in the Khakassy area of Siberia, Russia (Anonymous, 1973).

Disease resistance could be improved in new cultivars from species of this complex. Nonchoke-inducing endophytes have been found in germ plasm of *F. glauca* (Anonymous, 1995; Meyer and Funk, 1989). *Phleospora idahoensis* (stem eyespot) has been identified on native *F. idahoensis* (Smith, 1971). Breeding for resistance to this disease would be beneficial. Among fine-leaved *Festuca* spp., *F. ovina* ssp. *hirtula* exhibited poor leaf spot (species not stated) resistance (Fraser and Rose-Fricker, 1999a, 1999b). Enhanced disease tolerance to *Laetisaria fuciformis* (Jackson, 1960), *Microdochium nivale*, *Erisyphye graminis*, *Rhizoctonia solani*, and *Ustilago striiformis* is desired as well for this species (Beard, 1973). Pest resistance and performance in low maintenance settings are key selection goals for *F. trachyphylla*. Among fine-leaved *Festuca* spp., *F. trachyphylla* exhibited poor leaf spot (species not stated) resistance according to Fraser and Rose-Fricker (1999a, 1999b). Results from Morris and Shearman (1994, 1996) showed some increase in leaf spot resistance (species not stated) for some cultivars of *F. trachyphylla*. Superior *Laetisaria fuciformis* resistance was shown in improved *F. trachyphylla* cultivars compared to improved cultivars of *F. rubra* ssp. *commutata* and *F. rubra* ssp. *rubra* (Morris and Shearman, 1994, 1996). Recent cultivars of *F. trachyphylla* demonstrate texture and density comparable to *F. rubra* ssp. *commutata*, reduced rate of growth, lower fertility requirements, and better resistance to *Laetisaria fuciformis*, *Drechslera dictyoides*, *Colletotrichum graminicola* and *Sclerotinia homeocarpa* (Meyer, 1982; Meyer and Funk, 1989). Breeders must continue to improve resistance to *Laetisaria fuciformis* and *Sclerotinia homeocarpa*, enabling use of this species in low-maintenance conditions (Meyer and Belanger, 1997).

Plants containing nonchoke-inducing endophytes were identified in a number of cultivars and selections of *F. trachyphylla* (Hurley et al., 1984). Observations in field settings suggested improved summer performance and enhanced resistance to *Blissus leucopterus* might be associated with the presence of these nonchoke-inducing endophytes in 'Valiant.' The *Neotyphodium* sp. endophyte has been associated with *Sclerotinia homeocarpa* resistance in this species (Morris and Shearman, 1994, 1996).

Collections from sites of origin, as well as naturalized turfs in non-native locations comprise key sources of germ plasm for species within the *F. ovina* complex. Cultivars were successfully developed from materials collected in both native and naturalized sites. Native populations in the Pacific Northwest of the United States and Canada may be utilized for germ plasm refinement of *F. idahoensis* into turf-type cultivars. Native and naturalized germ plasm sources have been used to produce cultivars of *F. ovina* ssp. *hirtula*. European native sites as well as naturalized turfs comprise key sources of germ plasm for *F. trachyphylla*.

Conventional breeding techniques employ a variety of methods to refine germ plasm within the *F. ovina* complex. Recurrent phenotypic selection has been successfully used with *F. trachyphylla* (Johnston et al., 1989). Synthetic cultivars were formed using varied numbers of parents. Seedling screening for disease resistance in *F. trachyphylla* germ plasm was successful (Pepin et al., 1988b). Endophytes were primarily introduced to nonendophyte-containing germ plasm using a modified backcross method with phenotypic selection of progeny. The initial cross uses the endophyte-containing germ plasm as the maternal parent. Endophyte-containing progeny are then backcrossed to the recurrent parent for one or more cycles, with examination for the presence of endophytes in the progeny after each backcross.

F. ovina produces 'light' seed, is wind-pollinated, and largely self-sterile (Urquhart, 1971; Auquier, 1977). Breeding efforts for several species within this complex have been limited (Weibull et al., 1991). Most available cultivars are, therefore, morphologically similar to wild material. Comparing two commercial and two wild populations from the *F. ovina* complex, Weibull et al. (1991) determined a slight difference in allele frequencies for the cultivars compared to wild material. Since this difference was not statistically significant, the authors concluded members of this complex could be considered in the early stages of domestication.

Festuca filiformis Pourret, Hair Fescue or Fine-leaved Fescue

This species has changed taxonomically from *F. tenuifolia* Sibth. (Hubbard, 1984) and *F. capillata* Lamarck (USDA, 2001) to its current designation. Hitchcock (1950) listed this species as the latter, with the additional synonyms, *F. ovina* var. *capillata* Alefeld., *F. ovina* ssp. *capillata* Lamarck, and *F. ovina* var. *tenuifolia* (Sibth.). *F. dumetorum* was also listed as a synonym for this species by Weibull et al., (1991). Hubbard (1984) and Jenkin (1955c) list another common name as fine-leaved sheep's fescue.

The native range for this species extends from western and central Europe to Asia Minor and North Africa. Although widespread in Great Britain, it is less common than *F. ovina* (Hubbard, 1984). Howarth (1924) reported this species native to Scotland. This species grows on heaths, moors, parkland, hill-grassland, and in open woodland (Hubbard, 1984). Jenkin (1955c) suggested *F. filiformis* was probably not native to Wales. It is common in the northeastern United States, as well as southeastern Canada. Roland and Smith (1969) reported its extensive distribution in Nova Scotia, where it has been considered a weedy grass. *F. filiformis* can become weedy in dry, sandy, or rocky areas. Under poor conditions, it may persist for years as nonflowering tufts. This species was introduced into New Zealand (Hubbard, 1984). *F. filiformis* has been used as a lawngrass

(Hubbard, 1984). While preferred to *F. ovina* in England, this species is seldom planted alone (Beard, 1973). It withstands mowing as low as 1.2 cm (Ruemmele et al., 1995).

Roland and Smith (1969) considered *F. filiformis* to have little commercial value, asserting it was too short and wiry and matured early, leaving a reddish-brown appearance. Since this assessment, breeders have obtained germ plasm worthy of consideration for cultivar development, resulting in releases noted below. This species is fairly drought-resistant and tolerant of close mowing (Hubbard, 1984). *F. filiformis* grows well on poor, acid, sandy, gravelly, or peaty soils in dry or damp sites (Hubbard, 1984). Common in Nova Scotia, *F. filiformis* grows on dry or sterile soils of pastures, lawns, along roadsides and fields (Roland and Smith, 1969). Tolerant of moderate shade, this species has been sown in woodlands (Hubbard, 1984). Such conditions may eventually yield isolated tufts. Salt tolerance varies among species, but is generally high. Skirde (1970) reported good salt tolerance for *F. valesiaca* and *F. ovina* (ssp. *hirtula*) cv. 'Biljart.'

Plants lack rhizomes, making them densely tufted. Leaves are a distinctive bluish or yellowish-green. Culms grow to 15–55 cm. Internodes are strongly scabrous or puberulent. Sheaths open to the base, while dead sheaths remain prominent at the base of plants, not splitting with age. Upper sheaths appear glabrous or finely scabrous-hirsute, rarely with any purple coloration. Ciliate ligules are 0.1 to 0.3 mm in length. Blades are 5–30 cm long and 0.2 to 0.4 mm wide. They are setaceous or threadlike, plicate, apically scabrous, with 3 large veins, 0–4 small veins, and one central rib. Sclerenchyma tissue forms a continuous or almost continuous abaxial ring. The inflorescence may be 1–7 cm long. At maturity, it is open or narrowly contracted, with scabrous branches. Yellowish-green spikelets extend 3.0 to 6.5 mm, with 2–6 florets. Glumes appear glabrous or apically scabrous. The first glume is 1.0 to 2.5 mm long and 1-veined, while the second glume is 1.7 to 3.9 mm long and 3-veined. Lemmas are 2.3 to 4.4 mm long and glabrous or apically scabrous. Awns may be 0.0 to 0.4 mm long. Anthers are 1.5 to 2.2 mm long and about half as long as the palea. The ovary apex is glabrous. *F. filiformis* differs from other *F. ovina* by its relatively thinner leaf blades and lack of short-awned, longer lemmas (Hubbard, 1984). Vegetative plants are identified by their light brown, persistent sheaths and fine leaf blades. In cross-section, the blade has no ribs, while the continuous band of sclerenchyma tissue is evident (Aiken and Darbyshire, 1990).

One cultivar, 'Barok,' is available in the United States. This diploid Dutch cultivar was bred by Barenbrug Holland B.V., Arnhem, the Netherlands (Weibull et al., 1991).

Festuca idahoensis Elmer, Idaho Fescue

Numerous synonyms are listed by Hitchcock (1950). These include *F. ovina* var. *ingrata* Hack. ex Beal, *F. ovina* var. *columbiana* Beal, *F. ovina* var. *oregona* Hack. ex Beal, *F. ingrata* Rydb., *F. ingrata nudata* Ryb., *F. amethystina* var. *asperrima* subvar. *idahoensis* St.-Yves, and *F. amethystina* var. *asperrima* subvar. *robusta* St.-Yves. Beal (1896) and Piper (1906) recognized it as a distinct segregate in the *F. ovina* complex. Bluebunch fescue is a common name associated with this species (Huff and Palazzo, 1998).

This fine-leaved *Festuca* spp. is native to the United States and Canada, including the Rocky Mountains and the Pacific coast (Lynch, 1996). It has been observed from southwestern Saskatchewan west to British Columbia in Canada and south to Colorado and central California in the United States (USDA, 2001). *F. idahoensis* is prevalent at higher elevations in Montana, Utah, and Idaho (Hanson, 1965). Belsky (1979) documented its distribution in alpine meadows of the Olympic Mountains in Washington.

F. idahoensis grows best on deep, fertile, silty and clayey soils, but is not uncommon on poorer soils (Lynch, 1996). This species also tolerates weakly alkaline, saline, and acid soils, surviving on as little as 25 cm of rainfall in cool climates. Higher altitudes favor this species, which can be competitive against other turfgrasses once it has matured.

Plants are perennial and densely tufted, lacking rhizomes. Leaves appear blue or yellowish-green. Culms grow to 30–100 cm. Sheaths are open, glabrous or scabrous, and persistent. Ligules are short, 0.3 to 0.6 mm in length. Leaves may be (5) 15–30 cm long and 0.6 to 1.0 mm wide. They are conduplicate, adaxially glabrous or pubescent, scabrous, and often glaucous or pruinose, with 3–5 large veins and 2–5 small veins. Adaxially, there are (1) 3–5 ribs. Sclerenchyma tissue is located in broad irregular strands. The inflorescence grows to (5) 7–16 cm. It is narrow or open, with scabrid branches that are erect or spreading. Spikelets may be 7.5 to 13.5 mm long, with 4–9 florets. The first glume is 2.4 to 4.5 mm long and 1-veined, while the second glume is 3.0 to 6.0 mm long and 3-veined. Lemmas are 5.0 to 8.5 mm long, with scabrid apices. Awns are (1.5) 3.0 to 6.0 (7.0) mm long. Paleas have distinct pubescence between veins. Anthers are 2.5 to 4.0 mm long. The ovary apex is glabrous.

Aiken et al. (1985) reported that, although leaf anatomy of *F. idahoensis* was similar to both the *F. ovina* complex and *F. occidentalis* Hooker [which more closely resembles the *F. rubra* complex (Pavlick, 1983b; Aiken et al., 1985)], it is not clearly related to either, based on leaf anatomy. Pavlick (1983a, 1983b) used a number of characters to distinguish this species, including spikelet size.

Species improvement has been limited. The Plant Materials Center of the Soil Conservation Service at Pullman, Washington, created 'P-6435' using mass selection over several generations from an outstanding selection among 61 accession collected in Idaho (Hanson, 1965). Although vigorous, with excellent seedling vigor, strong roots, dark green color and good seed production, this selection was never released as a cultivar (Hanson, 1965). Presently, 'Trident' is the only cultivar developed for utilization as turf.

Festuca ovina L. ssp. *hirtula* (Hackel *ex* Travis) M. Wilkinson, Sheep Fescue

Seven subspecies have been described within *F. ovina* (Hubbard, 1984). This subspecies' nomenclature has also included *F. ovina* L., with no subspecies designation (Hubbard, 1984). Many synonyms listed by Hitchcock (1950) have since been designated as unique species or subspecies apart from this subspecies. This subspecies is also commonly named sheeps (Bonos et al., 2001) or sheep's fescue (Hubbard, 1984). Throughout the world, this species has received numerous other common names, including ovina; coquiole, fétuque des moutons, and fétuque ovine (French); Schafschwingel (German); festuca-ovelha (Portuguese); and cañuela de oveja (Spanish) (USDA, 2001).

F. ovina ssp. *hirtula* naturally occurs in northern and central Europe. It is considered indigenous to North American and Eurasia (Beard, 1973). The most common and widespread taxon of the *F. ovina* aggregate in the British Isles, *F. ovina* ssp. *hirtula* is found throughout England, Wales, Scotland, and Ireland. It is most common in western Britain, but becomes scarce in the mountains on northern and western Scotland, where it is replaced by *F. vivipara* (L.) Sm. (Wilkinson and Stace, 1991). Hubbard (1984) noted its presence in open situations, such as on heaths, moors, and especially hill and mountain grasslands where it frequently dominates. Huon (1972) noted growth of this taxon on dryland swards in eastern France. *F. ovina* ssp. *hirtula* has been documented in Japanese swards (Ohba et al., 1973). Widespread in northern temperate zones, this subspecies thrives from sea level to 1,220 m (Hubbard, 1984).

F. ovina ssp. *hirtula* is used in difficult-to-mow areas (roadsides, railway banks, roughs and reclamation areas), lawns, and wildflower mixtures. This species is adapted to moderate shade and medium-low fertility in temperate climates. Weibull et al. (1991) noted its excellent tolerance to drought and low fertility levels, requiring minimal nitrogen. Specimens have been located on poor, well-drained, shallow soils, both acidic and basic (Hubbard, 1984), although Beard (1973) stated *F. ovina* prefers acidic, coarse-textured soils of low fertility. Heat tolerance is considered poor (Beard, 1973). This drought-resistant species withstands close mowing and heavy grazing (Hubbard, 1984). Under close mowing, *F. ovina* ssp. *hirtula* develops shorter roots than *F. rubra* ssp. *commutata* (Evans, 1931). Fairly good wear tolerance has been reported (Morrish and Harrison, 1948).

Plants are densely-tufted perennials. Culms reach (6) 10–45 (70) cm in height. Sheaths open for over half their length, appearing pubescent or, occasionally, glabrous. Auricles and ligules are short and minutely ciliate. Leaves may be 2–10 (15) cm long and 0.2 to 0.7 mm wide. They are conduplicate, mostly circular to oval in cross section, and green or slightly glaucous, but usually not pruinose. Their upper half is usually scabrid, while typically pubescent abaxially at their base (occasionally glabrous). Leaves are also pubescent adaxially, where they have 5 obscure veins, and 1 (-3) ribs. Sclerenchyma tissue forms a thin broken ring or, sometimes, an unbroken ring 1–2 cells thick. The inflorescence extends (1.5) 5–10 (12) cm long, but remains contracted, spreading only at anthesis. Branches are scabrid. Spikelets are 4.0 to 6.0 (7.0) mm long, with 3–8 florets. Glumes may be glabrous or scabrid in their distal half or pubescent in their distal third. Distal margins are ciliolate. The first glume is 1.7 to 2.5 mm long, and 1-veined. The second glume is 2.2 to 4.0 mm long and 3-veined. Lemmas measure 3.0 to 4.0 mm in length. They are 5-veined, green, glaucous, or tinged with reddish violet on the upper portion, and pubescent or scabrid in the distal half (rarely glabrous). Awns may be 0.7 to 2.0 mm long. Anthers are 1.6 to 2.5 mm long, yellow or purple, and usually more than half as long as the palea. The ovary apex is glabrous. Harberd (1962) observed a single *F. ovina* ssp. *hirtula* could spread 10 cm in one year by tillering alone. This species is distinguished by its very fine leaf texture, rather tufted growth habit, and blue-green stiff leaves (Beard, 1973).

Limited cultivars have been developed within this species. 'P-274' was selected at the Plant Materials Center of the Soil Conservation Service at Pullman, Washington, from materials originating in Turkey (Hanson, 1965). This dwarf, blue-green, densely-tufted, erect-growing selection had abundant fine stems, dense leaves, and short, stiff, abundantly basal leaves. Although breeder seed has been available, it was never released as a cultivar.

'Career' was released by Gebr. Van Engelen of the Netherlands (Beard, 1973). This cultivar featured more greenish color than common *F. ovina*. It was described as a fine-textured, low-growing cultivar, with excellent drought resistance and shade tolerance.

'Quatro' is sold in the United States for turf. Turf-Seed, Inc. introduced 'Bighorn' in 1987 (Rose-Fricker and Meyer, 1993). This advanced-generation cultivar resulted from three cycles of phenotypic recurrent selection for improved turf quality and color uniformity. Germ plasm originated from old turf areas in New Jersey. Specific selection criteria included overall attractiveness, a powdery blue color, early maturity, leafy (not wiry) types, mowing quality, winter color, and seedhead formation. Improved resistances to *Microdochium nivale*, *Drechslera dictyoides*, *Sclerotinia homeocarpa*, *Laetisaria fuciformis*, and *Fusarium roseum* have been observed in this cultivar.

Jacklin Seed Company (a division of J.R. Simplot) of Post Falls, Idaho, released 'MX-86' in 1988 (Sellmann et al., 1997). Selections for uniformity and stability were employed in screening a seedlot of common *F. ovina* ssp. *hirtula* from Eastern Germany. This low-maintenance turfgrass displays medium blue-green color, enhanced seedling vigor, and improved resistance to *Sclerotinia homeocarpa* and *Drechslera* spp.

'Azay' is available from Seed Research of Oregon, Inc. (Lynch, 1996). Although designated as *F. ovina* ssp. *hirtula*, controversy exists regarding species' status of this cultivar. Wilson (1999) designated this cultivar as *F. huonii*, while Huff and Palazzo (1998) and D. Floyd (2001, personal communication) determined this cultivar most closely resembles *F. trachyphylla*.

Festuca trachyphylla (Hackel) Krajina, Hard Fescue

Nomenclature has changed dramatically with *F. trachyphylla*. *F. ovina* var. *duriuscula* (L.) Koch (Beard, 1973) and *F. duriuscula* L. (Hitchcock, 1950), *F. longifolia* Thuill. (Hubbard, 1984), and *F. brevipila* Tracey (USDA et al., 2001) have also been associated with this species.

F. trachyphylla is native to open forests and forest edge habitats of Central Europe. It was introduced and naturalized throughout many temperate regions, including France, Great Britain, and Scandinavia. Introduction to Great Britain probably occurred in the nineteenth century in seed from Germany (Hubbard, 1984). This species is adapted to a Mediterranean climate (Billot et al., 1982). It was first common in the eastern United States and southeastern Canada, but is now widely established in North America.

Turf use and reclamation sites are recommended uses, including areas where less mowing is preferred. Suitable planting sites include roadsides, railway banks, parks and sports grounds, and home lawns (Hubbard, 1984). Henensal et al. (1977) contended this species was one of the most suitable in France for roadside embankments. Shade or full sun in temperate regions of the world support *F. trachyphylla*. This species tolerates well-drained stony and sandy soils (Hubbard, 1984). Drought tolerance is considered less than *F. ovina*, but greater than *F. rubra* (Beard, 1973).

Plants are perennial and densely tufted, without rhizomes. Culms extend (9) 20–75 cm in length. Sheaths are open, closed only at the base. They are pubescent or occasionally glabrous. Auricles and ligules are short and minutely ciliate. Leaves reach 3.5 to 19.0 cm in length and 0.4 to 1.0 mm in width. They are conduplicate, green or subpruinose, scabrous or puberulent, 5–7 veined with 5–7 usually well-developed adaxial ribs. Sclerenchyma tissue is unevenly thickened, usually forming three tailing islets at the midrib and margins. They are 1–4 cells thick and rarely continuous. The inflorescence may be 3.0 to 9.5 (13) cm long, erect to nodding, and contracted, with branches usually scabrid on the angles. Spikelets are 5.5 to 9.0 mm long and yellow-green, blue-green or purple, with 3–8 florets. The rachilla is commonly visible between florets. Glumes are generally glabrous, although they may be scabrous apically, or even pubescent. The first glume is 2.0 to 3.5 mm long and 1-veined, while the second glume is 3.0 to 5.5 mm long and 3-veined. Lemmas are 3.8 to 5.0 mm long; 5-veined; glabrous, scabrous or pubescent apically; rarely entirely pubescent; and usually with apically-ciliate margins. The awn is 0.5 to 2.5 mm long. Yellow or purple anthers may be (2.3) 2.5 to 3.4 mm long. They are usually more than half as long as the palea. The ovary apex is glabrous. Hubbard (1984) noted leaf blades are slightly thicker and awns are longer on *F. trachyphylla* compared to *F. ovina*. Beard (1973) distinguished this species from *F. ovina* as being grayish-green, with wider and tougher leaves in the former.

Active progress in recent years yielded several improved cultivars using varied breeding techniques. An early cultivar, 'Durar,' was developed by recurrent selection in Pullman, Washington (Schwendiman et al., 1964) from material maintained in an old planting at the Eastern Oregon Branch Experiment Station in Union, Oregon (Hanson, 1965). This cultivar was released in 1949 jointly by the Agricultural Experiment Stations of Idaho, Oregon, and Washington, as well as the Soil Conservation Service in the United States (Beard, 1973). Originally released as 'P-2517,' it was named Durar in

1963 (Hanson, 1965). This cultivar was well-adapted to cool subhumid and semiarid regions where close mowing would be nonessential. Durar was more uniform, drought resistant, and shade tolerant than *F. rubra* ssp. *commutata* of the time (Schwendiman et al., 1964).

Biljart (C-26), 'Scaldis,' and 'Waldina' were additional early cultivars with similar appearance to *F. rubra* ssp. *commutata*, but with slower vertical growth rates and improved heat tolerance (Meyer, 1982). N.V.H. Mommersteeg of the Netherlands introduced Biljart in 1963 (Beard, 1973). This cultivar produced a deep green turf having very fine texture and high shoot density. Additional attributes included medium-low growth habit, somewhat tufted, with stiff leaves, excellent drought resistance, and very good resistance to *Laetisaria fuciformis* and leaf spot (listed under a former name, *Helminthosporium* sp.). Tolerance to the nonselective herbicide, glyphosate, was significantly increased after each of four cycles of recurrent selection using one of these cultivars (Waldina) (Johnston et al., 1989).

The Rutgers University breeding program in New Brunswick, New Jersey, developed 'Reliant,' 'Spartan,' 'Ecostar,' 'SR 3000,' 'Oxford,' and 'Nordic' cultivars (Meyer, 2000). Ecostar exhibits excellent shade tolerance, summer density and color (Anonymous, 2000). Low fertility, drought, heat and cold tolerances are additional features of this cultivar. Ecostar has shown resistance to *Sclerotinia homeocarpa, Rhizoctonia solani*, and leaf spot (species not designated). Oxford was developed for its very dark green color that blends well with darker *P. pratensis* (Anonymous, 2000). This makes it useful in sod production areas where *P. pratensis* is the major species planted.

Reliant is a 43-clone synthetic resulting from screening more than 30,000 seedlings for *Erysiphye graminis* resistance (Duell et al., 1983). It originated from European cultivars and germ plasm collected from old turfs. Numerous enhancements make this a desirable cultivar, including the bright, medium-dark green color, low turf-type habit, uniform maturity and high seed yield potential. Winter hardiness, improved summer performance, and improved resistance to *Erysiphye graminis, Colletrotrichum graminicola, Pyrenophora dictyoides,* and *Laetisaria fuciformis* are additional attributes.

Pickseed, Inc. released Spartan in 1984 (Pepin et al., 1988b) as an advanced generation synthetic from progeny of 142 clones. The clones originated from the cultivars (or related germ plasm) Biljart, Reliant, Scaldis, SR 3200, and Waldina, as well as old turfs in the eastern United States and western Europe. This cultivar was developed for its leafy, low-management, persistent qualities. It produces an attractive, dense, low-growing fine-textured turf with cold tolerance, as well as good disease, heat, and drought tolerance.

'Aurora' was released by Turf-Seed, Inc. after development in a joint effort between Pure-Seed Testing, Inc. and the New Jersey Agricultural Experiment Station (Meyer et al., 1991). Seventeen *E. graminis*-resistant plants from old turfs of the northeastern United States, as well as Biljart and Scaldis cultivars formed the parentage of Aurora. Improved turf density and quality, reduced vertical growth, high seed yield, and overall attractiveness resulted from three cycles of recurrent phenotypic selection. Aurora shows early maturity, medium, dark green color, and improved resistance to *Sclerotinia homeocarpa, Laetisaria fuciiformis, Puccinia crandallii, Microdochium nivale*, and *Drechslera* spp.

Seed Research of Oregon, Inc. released SR 3000, an advanced generation synthetic cultivar developed from progeny produced by a polycross using five endophyte-containing clones from the New Jersey Agricultural Experiment Station (Robinson et al., 1989). This cultivar was developed for its outstanding turf quality, including fine leaf texture and dark green color. Other attributes include shade tolerance, disease and insect resistance, excellent heat and drought tolerance, and very high endophyte level (Lynch, 1996).

Another release of Seed Research of Oregon, Inc. was 'SR 3100' (Lynch, 1996). This cultivar produces a more dwarf growth habit than normal, having dark green color, very high endophyte content, heat and drought tolerance, and high resistance to many common turfgrass diseases. Under minimal irrigation and fertility, this cultivar forms an attractive, dense turf.

'Discovery' was released by Pure Seed Testing, Inc., Hubbard, Oregon, in 1996 (Rose-Fricker et al., 1999a). This advanced-generation synthetic cultivar was produced from progenies of plants selected for low growth habit and high numbers of reproductive tillers. Maternal parents, containing *Neotyphodium coenophialum,* were topcrossed with Aurora and Waldina. Maternal progenies were then topcrossed with Aurora and an experimental line tracing back to Aurora. Recurrent selection for *Puccinia crandallii* resistance, high panicle number, uniform maturity, turf quality, and low growth resulted in release of Discovery as a highly-endophyte-infected, low-growing cultivar with dense, fine-textured, dark green turf.

'Rescue 911,' released by Jacklin Seed Company, Post Falls, Idaho, displays a dark green color and very fine texture (Anonymous, 2000). This cultivar has exhibited resistance to *Drechslera dictyoides, Rhizoctonia solani, Pythium* spp., and *Sclerotinia homeocarpa.*

Festuca valesiaca Schleicher *ex* Gaudin, False Sheep Fescue

Alternate nomenclature has included *F. pseudovina* Hackel *ex* Weisb. (Huff and Palazzo, 1998) and *F. valesiaca* ssp. *pseudovina* (Hackel *ex* Weisb.) Hegi. Hitchcock (1950) also noted the synonym *F. ovina* var. *valesiaca* Link., as well as an optional spelling of 'vallesiaca,' *F. ovina* ssp. *vallesiaca* Koch was listed by Weibull et al. (1991) as another synonym. As with other sheep fescue, the common name sometimes includes sheeps or sheep's in place of 'sheep.'

F. valesiaca originates from central Germany and north central Russia southward to the Pyrenees, central Italy and south central Greece. A few collections have been made in the western United States, most likely deriving from introductions. USDA (2001) reports this species native to temperate and tropical Asia, as well as Europe. Henensal et al. (1977) contended this species was one of the most suitable in France for roadside embankments. *F. valesiaca* is adapted to well-drained soils with a mean annual precipitation of 25 cm. It is found in dry meadows and open rocky and sandy areas.

These blue plants are perennial and densely tufted, but without rhizomes. Culms reach 20–50 cm in height. Sheaths are open to the base and glabrous, smooth or sparsely scabrid. The ligule is short, less than 0.5 mm. Leaves are 0.2 to 0.6 mm wide, conduplicate, glaucous or pruinose, scabrid, 5-veined, and have 1–5 adaxial ribs. Leaf blades are sometimes deciduous. Sclerenchyma tissue occurs in 3 stout strands, rarely with additional small strands. The inflorescence is 3–10 cm long and contracted, with branches sparsely scabrid and erect or spreading. Pruinose spikelets range from (5.5) 6.0 to 6.7 mm long, with 3–5 (8) florets. The first glume is 2.0 to 2.5 mm long, while the second glume is 2.6 to 3.9 mm long. Lemmas may be 3.4 to 5.2 mm long, appearing glabrous or ciliate. Awns are 1.0 to 2.0 mm long. Anthers are 2.2 to 2.6 mm long. The ovary apex is glabrous.

Limited cultivar development has occurred in this species. Covar was developed from Turkish populations (Schwendiman et al., 1980) for soil erosion control and revegetation of disturbed lands. It is shorter, and has a deeper blue color than other sheep or hard fescue cultivars. Though it is somewhat slow to establish, it is very persistent, competitive, winter hardy and drought tolerant once it has established. It could perform well in low maintenance areas. 'Liwally' was released from a German breeding company, Deutsche Saatveredlung Lippstadt-Bremen GmbH, Lippstadt (Weibull et al.,

1991). Nine clones were selected from about 20 Hungarian diploid ecotypes to produce this synthetic cultivar.

REFERENCES

Ahti, K., A. Moustafa, and H. Kaerwer. 1977. Tolerance of turfgrass cultivars to salt. pp. 165–171. J.B. Beard, Ed. In *Proceedings Third International Turfgrass Conference.* International Turfgrass Society and ASA and CSSA, Madison, WI.

Aiken, S.G. M.J. Dallwitz, C.L. McJannet, and L.L. Consaul. 1996 onward. 'Festuca of North America: Descriptions, Illustrations, Identification, and Information Retrieval.' Version: 12th September 2000. biodiversity.uno.edu/delta/.

Aiken, S.G. and S.J. Darbyshire. 1990. Fescue Grasses of Canada. Publication 1844/E. Canadian Government Printing Centre. Ottawa, Ontario, Canada.

Aiken, S.G., S.J. Darbyshire, and L.P. Lefkovitch. 1985. Restricted taxonomic value of leaf sections in Canadian narrow-leaved *Festuca* Poaceae. *Canadian Journal of Botany* 63:995–1007.

Aiken, S.G. and S.E. Gardiner. 1991. SDS-PAGE of seed proteins in *Festuca* (Poaceae): Taxonomic implications. *Canadian Journal Botany* 69:1425–1432.

Aiken, S.G., S.E. Gardiner, H.C.M. Bassett, B.L. Wilson, and L.L. Consaul. 1998. Implications from SDS-PAGE analyses of seed proteins in the classification of taxa of *Festuca* and *Lolium* (Poaceae). *Biochemical Systematics Ecology* 26:511–533.

Aiken, S.G., S.E. Gardiner, and M.B. Forde. 1992. Taxonomic implications of SDS-PAGE analyses of seed proteins in North American taxa *Festuca* subgenus *Festuca* (Poaceae). *Biochemical Systematics Ecology* 20:615–629.

Ainscough, M.M., C.M. Barker, and C.A. Stace. 1986. Natural hybrids between *Festuca* and species of *Vulpia* section *Vulpia*. *Watsonia* 16:143–151.

Aldous, D.E. and P.S. Semos. 1999. Turfgrass identification and selection. pp. 19–47. D.E. Aldous. Ed. In *International Turf Management*. CRC Press. London.

Alexeev, E.B. 1972. The significance of degree of integration of leaf vaginae of vegetative shoots in the systematism of Fescues of the group *Festuca ovina* L. s.l. *Vestn. Mosk. Univ. Ser. 16 Biol.* 5: 48–51.

Alexeev, E.B. 1980. *Festuca* L. Subgenera et sections Novae ex *America boreali* et Mexica. *Nov. Syst. Vyssh. Rast.* 17:42–53.

Alexeev, E.B. 1982. Genus *Festuca* L. (Poaceae) in Oriente Extremo URSS. *Nov. Syst. Pl. Vasc.* 19:6–45.

Altpeter, F., J. Zu, and S. Ahmed. 2001. Stable genetic transformation of commercial cool season turfgrass (*Lolium perenne* L. and *Festuca rubra* L.) cultivars. *International Turfgrass Society Research Journal* 9:123–128.

Anderson, C. and K. Taylor. 1979. Some factors affecting the growth of two populations of *Festuca rubra* var. *arenaria* on the dunes of Blackeney Point, Norfolk. pp. 129–143. R.L. Jefferies and A.J. Davy, Eds. In *Ecological Processes in Coastal Environments*. Blackwell Scientific Publications. Oxford, UK.

Anonymous. 1973. On the diurnal rhythms of pollination of closely related allopatric grass taxa. *Botanicheskii Zhurnal* 58:484–492.

Anonymous. 1995. "Environment" impacts public turf breeding programs. *Seed World* 133(9):9.

Anonymous. 2000. New turfgrass variety review. *Turf News* 24(4):17, 20–25.

Auquier, P. 1968. *Festuca rubra* subsp. *litoralis* (G. F. W. Mey.) Auquier: Morphologie, écologie, taxonomie. *Bull. Jard. Bot. Nat. Belg.* 38:181–192.

Auquier, P. 1971a. *Festuca rubra* L. subsp. *pruinosa* (Hack.) Piper: Morphologie, écologie, taxonomie. Lejeunia, ser. 2. 56:1–16.

Auquier, P. 1971b. Le probleme de *Festuca rubra* L. subsp. *arenaria* (Osb.) Richt. Et de ses relations avec. *F. juncifolia* St-Amans. Lejeunia. 57.

Auquier, P. 1977. Biologie de la reproduction dans le genre *Festuca* L. (Poaceae). 1. Systemes de pollinisation. *Bulletin de la Societe Royale de botanique de Belgique* 110:129–150.

Bailey, J.P. and C.A. Stace. 1992. Chromosome banding and pairing behaviour in *Festuca* and *Vulpia* (*Poaceae, Pooideae*). *Plant Systematics Evolution* 182:21–28.

Baker, S.W. and J.A. Hunt. 1997. Effect of shade by stands on grass species and cultivar selection for football pitches. *International Turfgrass Society Research Journal* 8:593–601.

Barker, C.M. and C.A. Stace. 1982. Hybridization in the genera *Vulpia* and *Festuca*: The production of artificial F_1 plants. *Nord. J. Bot.* 2:435–444.

Beal, W.J. 1896. *Grasses of North America for Farmers and Students*. Vol. 2. Henry Holt & Company. New York.

Bean, E.W. 1970. Short-day and low-temperature control of floral induction in *Festuca*. *Annals Botany* 34:57–66.

Beard, J.B. 1973. *Turfgrass: Science and Culture*. Prentice-Hall, Inc.: Englewood Cliffs, NJ.

Beard, J.B. 1998. Traditional fine-leaf fescue putting greens. *Turfax* 6(6):3.

Beard, J.B. and D.P. Martin. 1968. Submersion tolerance of creeping bentgrass, Kentucky bluegrass, annual bluegrass and red fescue as affected by water temperature. *Agronomy Abstracts* 60:62.

Bell, G.E., M.B. McDonald Jr., and T.K. Danneberger. 1995. Electrophoretic evaluation of esterase isozymes from turfgrass seed blends and mixtures. *Journal Turfgrass Management* 1(3):1–11.

Belsky, A.J. 1979. Determinants of Ecological Amplitude in *Festuca idahoensis* and *Festuca ovina*. Ph.D. Dissertation. Washington University. Seattle, WA.

Bidault, M. 1964. Sur la presence de chromosomes surnemeraires dans divers taxa du *Festuca* gr. *Ovina* L. C.R. hebd. Seanc. *Acadamie Sci.* 259:4779–4782.

Bidault, M. 1968. Essai de taxonomie experimentale et numerique sur *Festuca ovina* L. s.l. dans le sud-est de la France. *Revue Cytol. Biol. Veg.* 31:217–356.

Billot, C., C. Chevallier, and J Peyriere. 1982. Adaptation des especes et cultivars de graminees a gazon au climat mediterraneen. *Rasen, Grunflachen, Begrunungen* 13:79–86.

Boeker, P. 1974. Root development of selected turfgrass species and cultivars. pp. 55–61. E.C. Roberts, Ed. In *Proceedings Second International Turfgrass Conference*. International Turfgrass Society and ASA and CSSA, Madison, WI.

Bolkhovskikh, Z. V. Grif, T. Matvejava, and O. Zakharyeva, Eds. 1969. *Chromosome Numbers of Flowering Plants*. Leningrad, Russia.

Bommer, D. 1959. Über Zeitpunkt und Verlauf der Blütendifferenzierung bei Perennierenden Gräsern. *Z. Acker Pflanzenbau* 109:95–118.

Bondesen, O.B. 1997. Statistics on Turfgrass Seed Production in Europe. Danish Seed Council. Copenhagen, Denmark.

Bonos, S.A., E. Watkins, J.A. Honig, M. Sosa, T. Molnar, J.A. Murphy, and W.A. Meyer. 2001. Breeding cool-season turfgrasses for wear tolerance using a wear simulator. *International Turfgrass Society Research Journal.* 9:137–145.

Bourgoin, B. 1997. Variability of fine-leaved fescues (*Festuca* spp.) grown at low nitrogen levels. *International Turfgrass Society Research Journal* 8:611–620.

Bowden, W. M. 1960. Chromosome numbers and taxonomic notes on northern grasses. II. Tribe Festuceae. *Canadian Journal Botany* 38:117–131.

Brandberg, B. 1948. On the chromosome numbers of some species of *Festuca* sect. Ovinae. *Ark. Bot.* 33B:1–4.

Breen, J.P. 1993. Enhanced resistance to fall armyworm (*Lepidoptera*: Noctudiae) in *Acremonium* endophytic-infected turfgrasses. *Journal Economic Entomology* 86:622–629.

Brilman, L.A. 2001. Utilization of interspecific crosses for turfgrass improvement. *International Turfgrass Society Research Journal* 9:157–161.

Brown, G. and K. Brinkmann. 1992. Heavy metal tolerance in *Festuca ovina* L. from contaminated sites in the Eifel Mountains, Germany. *Plant and Soil* 143:239–247.

Burg, W.J., van der, and G. Vierbergen. 1979. Distinguishing *Festuca rubra* and *F. ovina*. *Seed Science & Technology* 7:569.

Calder, J.A. and R.L. Taylor. 1968. Flora of the Queen Charlotte Islands. Part I. Canada Department Agriculture Monograph No. 4. Ottawa, Ontario, Canada.

Casler, M., J. Gregos, and J. Stier, 2001. Seeking snow mold-tolerant turfgrasses: A tough winter disease isn't so hard on some grasses. *Golf Course Management.* 69(5):49–52.

Christians, N. 2000. Fairway grasses for Midwest golf courses: Diseases and tough climatic conditions make choices tricky. *Golf Course Management* 68(4):49–56.

Clapham, D. and G. Almgard. 1978. Biochemical identification of cultivars leads to award of breeders' rights. *Agri. Hort. Gen.* 36:88–94.

Clapham, A.R., T.G. Tutin, and E.F. Warberg. 1952. *Flora of the British Isles*. Cambridge University Press. London.

Clarke, B.B., D.R. Huff, D.A. Smith, and C.R. Funk. 1994. Enhanced resistance to dollar spot in endophyte-infected fine fescues. *Agronomy Abstracts* 86:187.

Clayton, W.D. and S.A. Renvoize. 1986. Genera *Graminum*. Grasses of the World. Kew Bulletin, Additional Series 13:1–389.

Cooper, J.P. and D.M. Calder. 1962. Flowering Responses of Herbage Grasses. Report of the Welsh Plant Breeding Station for 1961. pp. 20–22.

Cooper, J.P. and D.M. Calder. 1964. The inductive requirements for flowering of some temperate grasses. *Journal British Grassland Society* 19:6–14.

Copeland, L.O. 1972. Wintergreen. A new fine-leaved red fescue for Michigan lawns. *Michigan Agricultural Experiment Station Research Report No. 173*. p. 2.

DaCosta, M., B. Bhandari, J. Carson, J. Johnson-Cicalese, and W.A. Meyer. 1998. Incidence of endophytic fungi in seed of cultivars and selections in the 1998 National Fine Fescue Test. *1998 Rutgers Turfgrass Proceedings* 30:185–190.

Dale, P.J. 1977. Meristem tip culture in *Lolium, Festuca, Phleum*, and *Dactylis*. *Plant Science Letters* 9:333–338.

Darbyshire, S.J. and S.I. Warwick. 1992. Phylogeny of North American *Festuca* (Poaceae) and related genera using chloroplast DNA restriction site variation. *Canadian Journal Botany* 70:2415–2429.

Davis, M., G. Kortier, and D.R. Smitley. 1989. Feeding preferences of chinch bugs for fine fescue and Kentucky bluegrass. *Proceedings of the 59th Annual Michigan Turfgrass Conference* 18:115–118.

Davis, P.H. and V.H. Heywood. 1973. *Principles of Angiosperm Taxonomy*. Krieger Publishing Company. New York.

Davis, R.R. 1967. Population changes in Kentucky bluegrass-red fescue mixtures. *Agronomy Abstracts* 59:51.

Dernoeden, P.H. 1998. Fine fescues on golf courses: Around the edges of courses, fine-leaf fescues offer low-maintenance alternatives. *Golf Course Management* 66(4):56–60.

Dreven, Z., van. 1976. Chromosomen Önderzoek door het R.P.v.Z. ter bepaling van de ploidiegraad en ten behoeve van de Önderscheiding van soorten en varieteiten. *Mededelingen* 33:19–22.

Dubé, M. and P. Morisset. 1987. Morphological and leaf anatomical variation in *Festuca rubra sensu lato* (Poaceae) from eastern Quebec. *Canadian Journal Botany* 65:1065–1077.

Dubé, M., P. Morisset, and J. Murdock. 1985. Chromosome numbers in populations of *Festuca rubra sensu lato* (Poaceae) from eastern Quebec. *Canadian Journal Botany* 63:227–231.

Duell, R.W., C.R. Funk, and S.W. Cosky. 1978. A step in developing superior fine fescues. *1978 Rutgers Turfgrass Proceedings* 9:109–114.

Duell, R.W., C.R. Funk, R.H. Hurley, and F.B. Ledeboer. 1983. Reliant hard fescue. *Crop Science* 23:1011–1012.

Duell, R.W., R.M. Schmit, C.R. Funk, R. Bailey, and B.L. Rose. 1976. Registration of Fortress fine fescue. *Crop Science* 16:123–124.

Duyvendak, R., B. Luesink, and H. Vos. 1981. Delimitation of taxa and cultivars of red fescue (*Festuca rubra* L. sensu lato). *Rasen, Grunflachen, Begrunungen* 12(3):53–63.

Duyvendak, R. and H. Vos. 1974. Registration and evaluation of turfgrasses in the Netherlands. pp. 62–73. E.C. Roberts, Ed. In *Proceedings Second International Turfgrass Conference*. International Turfgrass Society and ASA and CSSA. Madison, WI.

Edminster, C.W., G. W. Pepin, M.C. Engelke, M.L. Fraser, and C.R. Funk. 1993. Registration of 'Longfellow' chewings fescue. *Crop Science* 33:1415–1416.

Edmond, D.B. and S.T.J. Coles. 1958. Some long-term effects of fertilizers on a mown turf of browntop and chewing's fescue. *New Zealand Journal of Agricultural Research* 1:665–674.

Elliott, C.R. 1968. Registration of Boreal red fescue. *Crop Science* 8:398.

Erdmann, M.H. and C.M. Harrison. 1947. The influence of domestic ryegrass and redtop upon the growth of Kentucky bluegrass and chewing's fescue in lawn and turf mixtures. *Journal American Society Agronomy* 39:682–689.

Ermer, F. 1928. Growing Chewing's fescue on hard clay fairways. *The National Greenkeeper.* 2(2):11.
Evans, T.W. 1931. The root development of New Zealand browntop, chewing's fescue and fine-leaved sheep's fescue under putting green conditions. *Journal Board Greenkeeping Research* 2(5):119–124.
Fisher, R. and D.T. Johnston. 1985. The breeding of dwarf *Festuca rubra* cultivars. pp. 147–157. F. Lemaire, Ed. In *Proceedings Fifth International Research Conference.* International Turfgrass Society and Institute National de la Recherche Agronomy, Paris, France.
Flovik, K. 1938. Cytological studies of arctic grasses. *Hereditas* 24:265–375.
Flovik, K. 1940. Chromosome numbers and polyploidy within the flora of Spitzbergen. *Hereditas* 26:430–440.
Frakes, R.V. 1974. Registration of Cascade chewings fescue (Reg. No. 9). *Crop Science* 14:338.
Fraser, M.L., B.B. Clarke, P.J. Landschoot, and C.R. Funk. 1998. Susceptibility of fine fescue species, cultivars and selections to *Magnaporthe poae. Proceedings of the Seventh Annual Rutgers Turfgrass Symposium.* p. 32.
Fraser, M.L. and C.A. Rose-Fricker. 1999a. Reaction of fine fescues to leaf spot, 1997. *Biological and Cultural Tests for Control of Plant Diseases* 14:143.
Fraser, M.L. and C.A. Rose-Fricker. 1999b. Reaction of fine fescues to leaf spot, 1998. *Biological and Cultural Tests for Control of Plant Diseases* 14:144.
Fredricksson, P. 1999. Identifying seeds of hard and red fescue, based on phenolic compounds, by means of thin layer chromatography. *Sveriges Utsadesforenings Tidskrift* 109:160–166.
Freeman, G.W. and M.M. Mileva. 1998. Separation of fine fescues using reversed-phase high performance liquid chromatography. *Proceedings of the Seventh Annual Rutgers Turfgrass Symposium.* p. 33.
Funk, C.R. 1993. Breeding and evaluation of Kentucky bluegrass, tall fescue, fine fescue, perennial ryegrass, and bentgrass for turf. *1993 Turfgrass Research Summary (USGA).* United States Golf Association. Far Hills, NJ. pp. 5–6.
Funk, C.R. 1998. Breeding and evaluation of Kentucky bluegrass, perennial ryegrass, tall fescue, fine fescues, and bentgrass for turf. *1998 Turfgrass and Environmental Research Summary (USGA).* United States Golf Association. Far Hills, NJ. pp. 34–35.
Funk, C.R., F.C. Belanger, and J.A. Murphy. 1994. Role of endophytes in grasses used for turf and soil conservation. pp. 201–209. C.W. Bacon and J.F. White Jr. Eds. In *Biotechnology of Endophytic Fungi of Grasses.* CRC Press. Boca Raton, FL.
Funk, C.R., B.B. Clarke, and J.M. Johnson-Cicalese. 1989. Role of endophytes in enhancing the performance of grasses used for conservation and turf. pp. 203–210. A.R. Leslie and M.P. Kenna, Eds. In *Integrated Pest Management for Turfgrass and Ornamentals.* United States Environmental Protection Agency, Office of Pesticide Programs, Field Operations Division. Washington, DC.
Funk, C.R., R.W. Duell, and W.K. Dickson. 1980. Performance of fine fescue cultivars in New Jersey. *Rutgers Turfgrass Proceedings* 11:98–104.
Funk, C.R. and J.F. White, Jr. 1997. Use of natural and transformed endophytes for turf improvement. pp. 229–239. C.W. Bacon and N.S. Hill, Eds. In *Neotyphodium/Grass Interactions.* Plenum Press. New York.
Funk, C.R., R.H. White, and J.P. Breen. 1993. Importance of *Acremonium* endophytes in turfgrass breeding and management. *Agriculture, Ecosystems and Environment* 44:215–232.
Gardner, F.P. and W.E. Loomis. 1953. Floral induction and development in orchard grass. *Plant Physiology* 28:201–217.
Guthrie, A. 1989. Cebeco: Breeding tomorrow's turfgrass cultivars. *Parks, Golf Courses & Sports Grounds* 54(11):12–14.
Hackel, E. 1882. Monographia *festucarum* europaearum. Verlag von Theodor Fischer. Kassel and Berlin.
Hanson, A.A. 1965. Grass Varieties in the United States. United States Department of Agriculture Agricultural Handbook No. 170. USDA-ARS. U.S. Government Printing Office. Washington, DC.

Hanson, A.A., F.V. Juska, and G.W. Burton. 1969. Species and varieties. pp. 370–409. A.A. Hanson and F.V. Juska, Eds. In *Turfgrass Science. Agronomy Monograph 14*. American Society of Agronomy. Madison, WI.

Harrington, C.F., D.J. Roberts, and G. Nickless. 1996. The effect of cadmium, zinc, and copper on the growth, tolerance index, metal uptake, and production of malic acid in two strains of the grass *Festuca rubra*. *Canadian Journal Botany* 74:1742–1752.

Harberd, D.J. 1962. Some observations on natural clones in *Festuca ovina*. *New Phytologist* 61: 85–100.

Heide, O.M. 1990. Primary and secondary induction requirements for flowering of *Festuca rubra*. *Physiologia plantarum* 79:51–56.

Heller, P.R. and R. Walker. 1997. Summer management of hairy chinch bug with dursban pro, dursban turf, and experimental formulations on established fescue turfgrass, 1996. *Arthropod Management Tests* 22:378–379.

Henensal, P., G. Arnal, and J. Puig. 1977. Research into the establishment of roadside embankments. pp. 391–400. J.B Beard, Ed. In *Proceedings Third International Turfgrass Conference*. International Turfgrass Society and ASA and CSSA, Madison, WI.

Hitchcock, A.S. 1950. *Manual of the Grasses of the United States*. 2nd ed. Revised by A. Chase. United States Department of Agriculture Miscellaneous Publication 200. United States Government Printing Office. Washington, DC.

Hodges, C.F., W.M. Blaine, and P.W. Robinson. 1975. Severity of *Sclerotinea homeocarpa* blight on various cultivars of fine-leaved fescues. *Plant Disease*. 59: 12–14.

Hodgson, H.J., R.L. Taylor, L.J. Klebesadel, and A.C. Wilton. 1978. Registration of Arctared red fescue (Reg. No. 13). *Crop Science* 18:524.

Holmberg, O.R. 1926. *Hartmans Handbok: Skandinaviens Flora*. Haftel. Norstedt and Soner. Stockholm, Sweden.

Horanszky, A., B. Janko, and G. Vida. 1971. Zur biosystematik der *Festuca ovina*-gruppe in Ungarn. *Hungary Annales Universitatis Scientiarum Budapestinensis de Rolando* 13:95–101.

Howarth, W.O. 1924. On the occurrence and distribution of *Festuca rubra*, Hack. in Great Britain. *Journal Linnaean London Society Bot.* 46:313–331.

Hubbard, C.E. 1974. *Grasses: A Guide to Their Structure, Identification, Uses and Distribution in the British Isles*. 2nd ed. Penguin Books, Richard Cley Ltd., Suffolk.

Hubbard, J.C.E. 1984. *Grasses: A Guide to Their Structure, Identification, Uses, and Distribution in the British Isles*. 3rd edition. Viking Penguin, Inc., New York.

Huff, D.R. and A.J. Palazzo. 1998. Fine fescue species determination by laser flow cytometry. *Crop Science* 38:445–450.

Hultén, E. 1942. Flora of Alaska and Yukon. Volume 2. *Monocotyledoneae*. *Lunds Univ. Arsskr. Avd.* 2, 38:129–412.

Hultén, E. 1968. *Flora of Alaska and Neighboring Territories*. Stanford University Press, Stanford, CA.

Humphreys, M.O. 1982. The genetic basis of tolerance to salt spray in populations of *Festuca rubra* L. *New Phytologist* 91:287–296.

Huon, A. 1972. La discontinuite genetique dans les populations de *Festuca ovina* de la moitie ouest de la France: etude preliminaire et comparaison avec *Festuca rubra*. *C.R. Hebd. des Seances de l'Academie des Sciences* 274:1648–1651.

Hurley, R.H. 1990. Best turfgrasses for southern winter overseeding? *Grounds Maintenance* 25:108–131.

Hurley, R.H., C.R. Funk, D.C. Saha, and P.M. Halisky. 1984. Endophytes in fine fescues and their role in modifying performance. *Agronomy Abstracts* 76:151.

Hylander, N. 1953. *Nordisk Karlvaxtflora* 1. Almqvist and Wiksell, Stockholm, Sweden.

Jackson, N. 1960. Further notes on the evaluation of some grass varieties. *Journal Sports Turf Research Institute* 10(36):156–160.

Jackson, N. 1962. Further notes on the evaluation of some grass varieties. *Journal Sports Turf Research Institute* 10(38):394–400.

Jackson, N. 1977. Yellow tuft disease of turfgrasses: A review of recent studies conducted in Rhode Island. pp. 265–270. J.B. Beard, Ed. In *Proceedings Third International Turfgrass Conference*. International Turfgrass Society and ASA and CSSA, Madison, WI.

Jauhar, P.P. 1975. Genetic regulation of diploid-like chromosome pairing in the hexaploid species *Festuca arundinacea* Schreb. and *F. rubra* L. (*Gramineae*). *Chromosoma* 52:353–382.

Jenkin, T.J. 1924. The Artificial Hybridization of Grasses. Bulletin Welsh Plant Breeding Station. Series H, Number 2. Aberystwyth, Wales.

Jenkin, T.J. 1933. Interspecific and intergeneric hybrids in herbage grasses. Initial crosses. *Journal Genetics* 28:205–264.

Jenkin, T.J. 1955a. Interspecific and intergeneric hybrids in herbage grasses. IX. *Festuca arundinacea* with some other *Festuca* species. *Journal Genetics* 53:81–93.

Jenkin, T.J. 1955b. Interspecific and intergeneric hybrids in herbage grasses. XI. Some of the breeding interactions of *Festuca pratensis*. *Journal Genetics* 53:100–104.

Jenkin, T.J. 1955c. Interspecific and intergeneric hybrids in herbage grasses. XII. *Festuca capillata* in crosses. *Journal Genetics* 53:105–111.

Jenkin, T.J. 1955d. Interspecific and intergeneric hybrids in herbage grasses. XIII. The breeding affinities of *Festuca heterophylla*. *Journal Genetics* 53:112–117.

Jenkin, T.J. 1955e. Interspecific and intergeneric hybrids in herbage grasses. XIV. The breeding affinities of *Festuca ovina*. *Journal Genetics* 53:118–124.

Jenkin, T.J. 1955f. Interspecific and intergeneric hybrids in herbage grasses. XV. The breeding affinities of *Festuca rubra*. *Journal Genetics* 53:125–130.

Jenkin, T.J. and P.T. Thomas. 1949. Genetic affinities of *Festuca ovina* and *Festuca rubra*. Proceedings 8th International Congress Genetics. Hereditas, Lund. Stockholm, Sweden. pp. 602–603.

Johnson-Cicalese, J., M.E. Secks, C.K. Lam, W.A. Meyer, J.A. Murphy, and F.C. Belanger. 2000. Cross species inoculation of Chewings and strong creeping red fescues with fungal endophytes. *Crop Science* 40:1485–1489.

Johnston, D.T. and J.S. Faulkner. 1986. Countess and Duchess—Aminotriazole-tolerant cultivars of Chewings fescue and browntop bent. *Journal Sports Turf Research Institute* 62:217.

Johnston, D.T., A.J.P. van Wijk, and D. Kilpatrick. 1989. Selection for tolerance to glyphosate in fine-leaved *Festuca* species. pp. 103–105. H. Takatoh, Ed. In *Proceedings of the Sixth International Turfgrass Research Conference*. Japanese Society of Turfgrass Science and the International Turfgrass Society. Tokyo, Japan.

Jones, R.N. 1975. B-chromosome systems in flowering plants and animal species. *Int. Rev. Cytol.* 40:1–100.

Karatanglis, S.S. 1980. Selective adaptation to copper of populations of *Agrostis tenuis* and *Festuca rubra* (Poaceae). *Plant Systematics Evolution* 134:215–228.

Kemp, M.L. and B.B. Clarke. 1991. Pathogenicity of *Gaeumannomyces incrustans* on fine fescues. *Agronomy Abstracts* 83:177.

Kemp, M.L., P.J. Landschoot, B.B. Clarke, and C.R. Funk. 1990. Response of fine fescues to field inoculation with summer patch. *Agronomy Abstracts* 82:176.

Kjellqvist, E. 1961. Studies in *Festuca rubra* L. 1. Influence of environment. *Bot. Notiser* 114:403–408.

Kjellqvist, E. 1964. *Festuca arenaria* Osb.—a misinterpreted species. *Bot. Notiser* 117:389–396.

Klebesadel, L.J. 1971. Nyctoperiod modification during late summer and autumn affects winter survival and heading of grasses. *Crop Science* 11:507–511.

Konarska, B. 1974. Karyological studies on *Festuca rubra* L. s.l. from Poland. *Acta Biologica Cracoviensia Botanica* 17:175–186.

Kreczetovich, V.I. and E.G. Bobrov. 1934. *Festuca* L. s. str. V.L. Komarov, Ed. In *Flora of* the U.S.S.R. Vol. 2. The Botanical Institute of the Academy of Sciences of the U.S.S.R., Leningrad, USSR.

Landschoot, P.J., B.S. Park, A.S. McNitt, D. Livingston. 2000. Performance of fine fescue cultivars and selections (1993–96). 2000 Annual Research Report. Center for Turfgrass Science, The Pennsylvania State University. University Park, PA. pp. 19–26.

Lee, H. and C.E. Wright. 1981. Effective selection for aminotriazole tolerance in *Fesctua* and *Agrostis* turf grasses. pp. 41–46. R.W. Sheard, Ed. In *Proceedings Fourth International Turfgrass Conference*. International Turfgrass Society and Ontario Agricultural College, University of Guelph, Guelph, Ontario.

Levitsky, G.A. and N.E. Kuzmina. 1927. Karyological investigations on the systematics and phylogenetics of the *Festuca* genus. *Bulletin Applied Genetics Plant Breeding* 17:3–36.

Liu, H., J.R. Heckman, and J.A. Murphy. 1995. Screening fine fescues for aluminum tolerance. *HortScience* 30:789.

Liu, H., J.R. Heckman, and J.A. Murphy. 1997. Greenhouse screening of turfgrasses for aluminum tolerance. *International Turfgrass Society Research Journal* 8:719–728.

Löve, A. and D. Löve. 1961. Chromosome numbers of central and northwest European plant species. *Opera Bot.* 5:1–581.

Löve, A. and D. Löve. 1975. *Cytotaxonomic Atlas of the Artic Flora.* J. Cramer, Vaduz.

Luesink, B. and J.J. van Hardeveld. 1975. Chromosomen-aantallen van rassen van roodzwenkgras (2). *Mededelingen van de N.A.K.* 33:8–9.

Lynch, S. (Ed.). 1996. *Seed Specification Manual.* 1st ed. Seed Research of Oregon, Inc. Cascade Printing Company. Corvallis, OR.

Markgraf-Dannenberg, I. 1980. *Festuca.* pp. 125–153. T.G. Tutin et al., Eds. In *Flora Europaea 5.* Cambridge University Press. Cambridge, UK.

McHugh, S., M.S. Johnson, T. McNeilly, M.O. Humphreys, and R. Rowling. 1985. Tolerance to salt in *Festuca rubra* L. pp. 185–193. F. Lemaire, Ed. In *Proceedings Fifth International Turfgrass Research Conference.* International Turfgrass Society and Institute National de la Recherche Agronomy, Paris, France.

McNitt, A.S. and D.V. Waddington. 1992. Evaluation of traction on turfgrass maintained under various cultural practices. *Agronomy Abstracts* 84:173.

Meier, V. 1992. Looking for low maintenance turf? Try fescues. *Lawn and Landscape Maintenance* 13(3):82–83.

Meyer, W.A. 1982. Breeding disease-resistance cool-season turfgrass cultivars for the United States. *Plant Disease* 66:341–344.

Meyer, W.A. 1986. Fine fescues—What is their place in parks and school grounds? *Proceedings of the 40th Northwest Turfgrass Conference* 40:112–114.

Meyer, W.A. 1994. Appreciating the strengths and weaknesses of fine fescues in erosion control. *Proceedings of the 34th Virginia Turf and Landscape Conference and Trade Show.* pp. 63–65.

Meyer, W.A. 2000. Dr. C. Reed Funk—Director of the world's most productive cool-season turfgrass breeding program. *TurfNews.* 24(3):56–59.

Meyer, W.A., C.A. Rose-Fricker, B.L. Rose, and C.R. Funk. 1990. Registration of 'Flyer' strong creeping red fescue. *Crop Sci.* 30:1156–1157.

Meyer, W.A. and F.C. Belanger. 1997. The role of conventional breeding and biotechnical approaches to improve disease resistance in cool-season turfgrasses. *International Turfgrass Society Research Journal* 8:777–790.

Meyer, W.A. and C.R. Funk. 1989. Progress and benefits to humanity from breeding cool-season grasses for turf. pp. 31–48. D.A. Sleper, K.H. Asay, and J.F. Pedersen, Eds. In *Contributions from Breeding Forage and Turf Grasses.* Crop Science Society of America Special Publication Number 15. Madison, WI.

Meyer, W.A., C.A. Rose-Fricker, and C.R. Funk. 1991. Registration of 'Aurora' hard fescue. *Crop Science* 31:1088.

Meyer, W.A., C.A. Rose-Fricker, and S.J. Yoder. 1983. The response of turfgrass cultivars to wear stress. *Agronomy Abstracts* 75:128.

Morgan, A. 1998. The Chewings story. *TurfNews* 22(6):33–35.

Morris, K. and R. Shearman. 1994. National Fine Leaf Fescue Test - 1989. National Turfgrass Evaluation Program NTEP 94-17.

Morris, K. and R. Shearman. 1996. National Fineleaf Fescue Test—1993. National Turfgrass Evaluation Program NTEP 96-8.

Morrish, R.H. and C.M. Harrison. 1948. The establishment and comparative wear resistance of various grasses and grass-legume mixtures to vehicular traffic. *Journal American Society Agronomy* 40:168–179.

Murray, J.J., C.D. Foy, J.P. Knorr. 1976. Differential lime responses of turfgrass cultivars on an acid soil high in exchangeable aluminum. *Agronomy Abstracts* 68:102.

Murray, J.J., A.C. Wilton, and J.B. Powell. 1973. Floral induction and development in *Festuca rubra* L. Differential clonal response to environmental conditions. *Crop Science* 13:645–648.

Nilsson, C. and P. Weibull. 1981. Selection for improved resistance to *Fusarium nivale* (FR) CES. pp. 21–26. R.W. Sheard, Ed. In *Proceedings Fourth International Turfgrass Conference.* Interna-

tional Turfgrass Society and Ontario Agricultural College, University of Guelph, Guelph, Ontario, Canada.
Nilsson, F. 1933. Ein spontaner Bastard zwischen *Festuca rubra* und *Lolium perenne*. *Hereditas* 18:1–15.
Nitzsche, W. and L. Hennig. 1976. Fruchtknotenkultur bei Gräsern. *Zeitschrift für Pflanzenzuchtung* 77:80–82.
OECD. 1990. OECD Schemes for the Varietal Certification of Seed Moving in International Trade. List of Cultivars Eligible for Certification 1989. Paris, France.
Ohba, T., A. Miyawaki, and R. Tuxen. 1973. Plant communities of the Japanese dune littoral. *Vegetatio* 26:3–143.
Ohmura, T., M. Yaneshita, S. Kaneko, Y. Ogihara, and T. Sasakuma. 1993. Turfgrass species and cultivars identification by RFLP analysis of chloroplast and nuclear DNA. *International Turfgrass Society Research Journal* 7:754–760.
Önder, A. and K. Jong. 1977. The occurrence of B chromosomes in *Festuca ovina* L. sensu lato from Scotland. *Watsonia* 11:327–330.
Oregon Agricultural Experiment Station. 1978. Registration of Checker chewings fescue (Reg. No. 14). *Crop Science* 18:912.
Parks, O.C. and P.R. Henderlong. 1967. Germination and seedling growth rate of ten common turfgrasses. *Proceedings of the West Virginia Academy of Science* 39:132–140.
Pavlick, L.E. 1983a. The taxonomy and distribution of *Festuca idahoensis* in British Columbia and northwestern Washington. *Canadian Journal Botany* 61:345–353.
Pavlick, L.E. 1983b. Notes on the taxonomy and nomenclature of *Festuca occidentalis* and *F. idahoensis*. *Canadian Journal Botany* 61:337–344.
Pavlick, L.E. 1984. Studies on the *Festuca ovina* complex in the Canadian Cordillera. *Canadian Journal Botany* 62:2448–2462.
Pavlick, L.E. 1985. A new taxonomic survey of the *Festuca rubra* complex in northwestern North America, with emphasis on British Columbia. *Phytologia* 57:1–17.
Pepin, G.W., W.K. Wiley, D.E. King, B.B. Clarke, and C.R. Funk. 1988a. Registration of 'Victory' Chewings fescue. *Crop Science* 28:1020–1021.
Pepin, G.W., W.K. Wiley, D.E. King, R.W. Duell, and C.R. Funk. 1988b. Registration of 'Spartan' hard fescue. *Crop Science* 28:1020.
Perdomo, P., J.A. Murphy, W.A. Meyer, C.R. Funk, D.A. Smith, R.F. Bara, M.M. Mohr, and E.Watkins. 1999. Performance of fine fescue cultivars and selections in New Jersey turf trials. *1999 Rutgers Turfgrass Proceedings* 31:69–97.
Pfender, W.F. 2001. Is there a difference between the stem rust pathogens from tall fescue and perennial ryegrass? W.C. Young, III, Ed. In 2000 Seed Production Research at Oregon State University, USDA-ARS Cooperating. Department of Crop and Soil Science Extension/CrS 115. Oregon State University, Corvallis, OR. www.css.orst.edu/seed-ext/Pub/2000/22.htm.
Piper, C.V. 1906. North American Species of *Festuca*. *Contribution United States National Herbarium* 10:1–51.
Prentice, H.C., M. Lönn, L.P. Lefkovitch, and H. Runyeon. 1995. Associations between allele frequencies in *Festuca ovina* and habitat variation in the alvar grasslands on the Baltic island of Öland. *Journal Ecology* 83:391–402.
Razmjoo, K., T. Imada, J. Hirano, and S. Kaneko. 1993. Performance of fine-leaved fescues (*Festuca* spp.) in Japan. *Journal Sports Turf Research Institute* 69:90–100.
Rees, H. 1974. B Chromosomes. *Sci. Prog. Oxf.* 61:534–554.
Riordan, T.P. 1997. Low-maintenance turfgrass—The practical choice for golf-course roughs. *Grounds Maintenance* 32(8):G44–G45, G48.
Roberts, E.C. 1987. Fine fescues have a place in your turf program. *Park Maintenance and Grounds Management* 40(2):10–11.
Roberts, E.C. 1990. Fine fescue: A kinder, gentler lawngrass. *Seed World* 128(9):16–19.
Robinson, M.F., B.B. Clarke, D.C. Saha, R.W. Duell, and C.R. Funk. 1989. Registration of 'SR3000' hard fescue. *Crop Science* 29:826.
Roland, A.E. and E.C. Smith. 1969. The Flora of Nova Scotia. The Nova Scotia Museum. The Nova Scotia Department of Education. Halifax, Nova Scotia, Canada.

Rose-Fricker, C.A., M.L. Fraser, and W.A. Meyer. 1999a. Registration of 'Discovery' hard fescue. *Crop Science* 39:1530–1531.

Rose-Fricker, C.A., M.L. Fraser, and W.A. Meyer. 1999b. Registration of 'Tiffany' Chewing's fescue. *Crop Science* 39:1529–1530.

Rose-Fricker, C.A., M.L. Fraser, and W.A. Meyer, and A.J.P. Van Wijk. 1999c. Registration of 'Seabreeze' slender creeping red fescue. *Crop Science* 39:1529.

Rose-Fricker, C.A., M.L. Fraser, and W.A. Meyer. 2000. Registration of 'Shademaster' strong creeping red fescue. *Crop Science* 40:291.

Rose-Fricker, C.A. and W.A. Meyer. 1993. Registration of 'Bighorn' sheep fescue. *Crop Science* 33:206.

Rose-Fricker, C.A. and J.K. Wipff. 2001. Breeding for salt tolerance in cool-season turf grasses. *International Turfgrass Society Research Journal* 9:206–212.

Rossi, F. and A. Sausen. 1995. National Turfgrass Evaluation Program (NTEP): 1993 fine-leaf fescue cultivar evaluations. *Wisconsin Turf Research: Results of 1995 Studies* 13:35–38.

Rozema, J., E. Rozema-Dijst, A.H.J. Freijsen, and J.J.L. Huber. 1978. Population differentiation within *Festuca rubra* L. with regard to soil salinity and soil water. *Oecologia* 34:329–341.

Ruemmele, B.A., L.A. Brilman, and D.R. Huff. 1995. Fine fescue germplasm diversity and vulnerability. *Crop Science* 35:313–316.

Rumball, W. 1982a. 'Grasslands Cook' Chewings fescue (*Festuca rubra* L.)—Bred for amenity areas. *New Zealand Journal of Experimental Agriculture* 10:167–168.

Rumball, W. 1982b. 'Grasslands Tasman' creeping red fescue (*Festuca rubra* L.)—Bred for amenity areas. *New Zealand Journal of Experimental Agriculture* 10:169–173.

Rumball, W. and M.P. Rolston. 1988. New Zealand bred turf grasses—Seed availability. *New Zealand Turf Management Journal* 2:19–20.

Saha, D.C., J.M. Johnson-Cicalese, P.M. Halisky, M.I. Van Heemstra, and C.R. Funk. 1987. Occurrence and significance of endophytic fungi in the fine fescues. *Plant Disease* 71:1021–1024.

Saint-Ives, A. 1925. Contribution a l'étude des *Festuca* (subgen. *Eu-Festuca*) de l'Amerique du Nord et du Mexique. *Candollea* 2:229–316.

Sann, C. 1993. Understanding necrotic ring spot: Undetected chronic infections contribute to a variety of problems. *TurfGrass Trends* 11:1–4, 6.

Sargent, S. 1982. The sod webworm, turfgrass pest. *American Lawn Applicator.* 3(2):4–7.

Schmidt, R.E. and J.F. Shoulders. 1977. Seasonal performance of selected temperate turfgrasses overseeded on bermudagrass turf for winter sports. pp. 75–86. J.B. Beard, Ed. In *Proceedings Third International Turfgrass Conference*. International Turfgrass Society and ASA and CSSA, Madison, WI.

Schmit, R.M., R.W. Duell, and C.R. Funk. 1974. Isolation barriers and self-compatibility in selected fine fescues. pp. 9–17. E.C. Roberts, Ed. In *Proceedings Second International Turfgrass Conference*. International Turfgrass Society and ASA and CSSA. Madison, WI.

Schwendiman, J.L., A.L. Hafenrichter, and A.G. Law. 1964. Registration of Durar hard fescue. *Crop Science* 4:114.

Schwendiman, J.L., A.G. Law, J.R. Carlson, and C.A. Kelley. 1980. Registration of Covar sheep fescue (Reg. No. 16). *Crop Science* 20:669–670.

Sellmann, M.J., A.D. Brede, and A.W. Jacklin. 1997. Registration of 'MX-86' sheep fescue. *Crop Science* 37:1381.

Shildrick, J.P. 1976a. A provisional grouping of cultivars of *Festuca rubra* L. *Journal Sports Turf Research Institute* 52:9–13.

Shildrick, J.P. 1976b. Evaluation of red fescue cultivars, 1973–76: Part III. Preliminary trial with artificial wear. *Journal of the Sports Turf Research Institute* 52:38–51.

Shildrick, J.P. 1977. Turfgrass seed mixtures in the United Kingdom. pp. 57–64. J.B Beard, Ed. In *Proceedings Third International Turfgrass Conference*. International Turfgrass Society and ASA and CSSA, Madison, WI.

Shildrick, J.P. 1980. Species and cultivar selection. pp. 69–97. R. Rorison and R. Hunt, Ed. In *Amenity Grassland—An Ecological Perspective*. John Wiley & Sons. New York.

Shildrick, J.P., R.W. Laycock, and R. Dunn. 1983. Multi-centre trials of turfgrass cultivars in the UK. 3. Fine-leaved fescues, 1978–81. *Journal of the Sports Turf Research Institute* 59:51–72.

Skalinski, M., A. Jankun, and H. Wcsilo. 1971. Studies in chromosome numbers of Polish angiosperms, eighth contribution. *Acta Biol. Cracov.,* Bot. 14:55–102.

Skirde, W. 1970. Ergebnisse zur Salztoleranz von Gräsersorten. *Rasen-Turf-Gazon* 1:12–14.

Skogley, C.R., N. Jackson, B. Ruemmele, J.M. Johnson-Cicalese, and R.H. Hurley. 1993. Registration of Jamestown II Chewings fescue. *Crop Science* 44:875–876.

Smith, J.D. 1971. *Phleospora* stem eyespot of fescues in Oregon and the *Didymella* perfect stage of the pathogen. *Plant Disease* 55:63–67.

Soo, R. 1973. Zeitgemasse taxonomie der *Festuca ovina*-gruppe. *Acta Botanica Academiae Scientiarum Hungaricae* 18:363–377.

Spangenberg, G., Z.Y. Wang, G. Legris, P. Montavon, T. Takamizo, R. Perez-Vicente, M.P. Valles, J. Nagel, and I. Potrykus. 1995a. Intergeneric symmetric and asymmetric somatic hybridization in *Festuca* and *Lolium*. *Euphytica* 85:235–245.

Spangenberg, G., Z.Y. Wang, J. Nagel, and I. Potrykus. 1994. Protoplast culture and generation of transgenic plants in red fescue (*Festuca rubra* L.). *Plant Science* 97:83–94.

Spangenberg, G., Z.Y. Wang, and I. Potrykus. 1998. Biotechnology in fescues and ryegrasses: Methods and perspectives. pp. 223–228. M.B. Striklen and M.P. Kenna, Eds. In *Turfgrass Biotechnology: Cell and Molecular Genetic Approaches to Turfgrass Improvement.* Ann Arbor Press. Chelsea, MI.

Spangenberg, G., Z.Y. Wang, X.L. Wu, J. Nagel, V.A. Iglesias, and I. Potrykus. 1995b. Transgenic tall fescue (*Festuca arundinacea*) and red fescue (*Festuca rubra*) plants from microprojectile bombardment of embryogenic suspension cells. *Journal Plant Physiology* 145:693–701.

Stace, C.A. and M.M. Ainscough. 1984. Continuing addition to the gene-pool of the *Festuca rubra* aggregate (Poaceae: Poeae). *Plant Systematics Evolution* 147:227–236.

Stace, C.A. and A-K.K.A. Al-Bermani. 1989. Earliest records for two x *Festulpia* combinations. *Watsonia* 17:363–364.

Stace, C.A., A-K.K.A. Al-Bermani and M.J. Wilkinson. 1992. The distinction between the *Festuca ovina* L. and *Festuca rubra* L. aggregates in the British Isles. *Watsonia* 19:107–112.

Stace, C.A. and R. Cotton. 1974. Hybrids between *Festuca rubra* L. sensu lato and *Vulpia membranacea* (L.) Dum. *Watsonia* 10:119–138.

Stahlin, A. 1929. Morphologische und zytologische Untersuchungen an Gramineen. *Pflanzenbau* 1330–98.

Stebbins, G.L. 1971. *Chromosomal Evolution in Higher Plants.* Edward Arnold Limited. London.

Stier, J. 1999. Fine fescue fairways. *The Grass Roots* 28(4):27–29, 31.

Taylor, R.L. and B. MacBryde. 1977. Vascular Plants of British Columbia. The Botanical Garden. The University of British Columbia Technical Bulletin Number 4. The University of British Columbia Press, Vancouver.

Taylor, R.L. and G. Mulligan. 1968. Flora of the Queen Charlotte Islands. Part 2. Canada Department Agriculture Monograph No. 4. Ottawa, Ontario, Canada.

Torello, W.A., A.G. Syminton, and R. Rufner. 1984. Callus initiation, plant regeneration, and evidence of somatic embryogenesis in red fescue. *Crop Science* 24:1037–1040.

Tracey, R. 1977. Drei neue arten des *Festuca ovina*-Formenkreises (Poaceae) aus dem Osten Osterreichs. *Systematics and Evolution* 128:287–292.

Turesson, G. 1926. Studien über *Festuc ovina* L. I. Normalgeschlechtliche, halb- und ganzvivipare Typen Nordischer Herkunft. *Hereditas* 8:161–206.

Turesson, G. 1930. Studien über *Festuca ovina* L. II. Chromosomenzahl und viviparie. *Hereditas* 13:177–184.

Tzvelev, N.N. 1976. Grasses of the Soviet Union. Zlaki, SSSR. Nauka, Leningrad, Soviet Union. Parts 1 and 2. Published for the Smithsonian Institution Libraries and the National Science Foundation. Washington, D.C. by Oxonian, New Delhi, India (1983).

Urquhart, C. 1971. Genetics of lead tolerance in *Festuca ovina*. *Hereditas* 26:19–33.

USDA, ARS, National Genetic Resources Program. 2001. *Germplasm Resources Information Network (GRIN).* [Online Database] National Germplasm Resources Laboratory, Beltsville, Maryland. Available: www.ars-grin.gov/cgi-bin/npgs/html/tax_search.pl?red+fescue (August 31, 2001).

Vargas, J.M., Jr., K.T. Payne, A.J. Turgeon, and R. Detweiler. 1980. Turfgrass disease resistance—Selection, development and use. pp. 179–182. B.G. Joyner and P.O. Larsen, Eds. In *Advances in Turfgrass Pathology*. Harcourt Brace Jovanovich, Inc. Duluth, MN.

Villamil, C.B., R.W. Duell, D.E. Fairbrothers, and J. Sadowski. 1982. Isoelectric focusing of esterases for fine fescue identification. *Crop Science* 22:786–793.

Villamil, C.B., D.E. Fairbrothers, and R.W. Duell. 1979. Cultivar identification among Chewings fescues by enzyme isoelectric focusing. *Agronomy Abstracts* 71:124.

Vinall, H.N. and M.A. Hein. 1937. Breeding miscellaneous grasses. p. 1032. In *U.S.D.A. Yearbook of Agriculture*. United States Government Printing Office. Washington, DC.

Walker, C. 1969. Australian cricket grounds through New Zealand eyes. *Newsletter of the New Zealand Institute for Turf Culture* 62.

Wang, Z.Y., G. Legris, M.P. Valles, I. Potrykus, and G. Spangenberg. 1995. Plant regeneration from suspension and protoplast culture in the temperate grasses *Festuca* and *Lolium*. pp. 81–86. M. Terzi et al., Eds. In *Current Issues in Plant Molecular and Cellular Biology*. Kluwer Academic Publications. Hingham, MA.

Watschke, T.L. 1990. Low-maintenance grasses for highway roadsides. *Grounds Maintenance* 25(8):40–41.

Watson, P.J. 1958. The distribution in Britain of diploid and tetraploid races within the *Festuca ovina* group. *New Phytologist* 57:11–18.

Weibull, P., L. Ghatnekar, and B.O. Bengtsson. 1991. Genetic variation in commercial varieties and natural populations of sheep's fescue, *Festuca ovina* s.l. *Plant Breeding* 107:203–209.

Weibull, P., L. Ghatnekar, I. Frykman, and B.O. Bengtsson. 1986. Electrophoretic variation in *Festuca ovina* L. *Agri Hortique Genetica* 44:25–37.

Welsh, S.L. 1974. *Anderson's Flora of Alaska and Adjacent Parts of Canada*. Brigham Young University Press, Provo, UT.

Welty, R.E. and M.D. Azevedo. 1995. Occurrence of *Puccinia graminis* subsp. *graminicola* in Chewings fescue in Oregon. *Plant Disease* 79:1014–1016.

White, D.B. and M.H. Smithberg. 1977. Cold acclimation and deacclimation in cool-season grasses. pp. 149–154. J.B. Beard, Ed. In *Proceedings Third International Turfgrass Conference*. International Turfgrass Society and ASA and CSSA, Madison, WI.

Wild-Duyfjes, B.E.E. de. 1973. On the two subspecies of *Festuca rubra* L. in the Netherlands. *Gorteria* 6:128–131.

Wilkinson, M.J. and C.A. Stace. 1991. A new taxonomic treatment of the *Festuca ovina* L. aggregate (Poaceae) in the British Isles. *Botanical Journal Linnaean Society* 106:347–397.

Willis, A.J. 1975. *Festuca* L. x *Vulpia* C. C. Gmel. = x *Festulpia Melderis ex* Stace & Cotton. pp. 552–554. C.A. Stace, Ed. In *Hybridization and the Flora of the British Isles*. Academic Press. London.

Wilson, B. 1999. Fescue Taxonomy in the Pacific Coast States. Ph.D. Dissertation. Oregon State University. Corvallis, OR.

Wojcik, M.K. and C.R. Skogley. 1991. Fine fescue adaptations to water deficit. *Agronomy Abstracts* 83:165.

Wycherley, P.R. 1954. Vegetative proliferation of floral spikelets in British grasses. *Annals of Botany N.S.* 18(69):119–127.

Young, W.C., III, M.E. Melbye, and G.A. Gingrich. 1997. The Oregon Grass Seed Industry. Presented to the Oregon Seed Council on Jan. 28, 1997. Internet publication: www.css.orst.edu/seed-ext/pub/industry.htm. Crop and Soil Science Department, Oregon State University. Corvallis, OR.

Youngner, V.B. 1957. Fine-leaved fescues as turfgrasses in southern California. *Southern California Turfgrass Culture* 7:7–8.

Zaghmout, O.M.F., and W.A. Torello. 1988. Enhanced regeneration in long-term callus culture of red fescue by pretreatment with activated charcoal. *HortScience* 23:615–616.

Chapter 10

Creeping Bentgrass (*Agrostis stolonifera* L.)

Scott Warnke

Creeping bentgrass is a cool-season grass species best known for its fine texture and adaptation to mowing heights as low as 3 mm, which makes it well suited for use on high quality golf course tees, greens, and fairways. Creeping bentgrass is highly stoloniferous, with rolled vernation and pointed leaf tips. The species has prominent, membranous ligules and prominent veins on the upper side of its leaves, which are useful identifying characteristics.

Creeping bentgrass is primarily adapted to cool, humid regions, but its use in warmer climatic zones is increasing because it forms a higher quality putting surface than bermudagrass (*Cynodon* spp. L.C. Rich). Maintenance of creeping bentgrass requires an intense management regime, specialized mowing equipment, and a high level of turf management skill, making it poorly suited to home lawn situations. Creeping bentgrass has outstanding cold temperature tolerance, and it will generally survive the coldest winters in North America. Creeping bentgrass tolerates a wide range of soil types but is best adapted to fertile, fine-textured soils of moderate acidity and good water holding capacity (Beard, 1973). Creeping bentgrass has good salinity and flooding tolerance but its tolerance of compacted soils is quite poor.

The stoloniferous growth habit of creeping bentgrass can result in excessive thatch formation. The level of thatch accumulation can be minimized through regular topdressing. In addition, vertical mowing can be used to promote juvenile shoot development and rooting at the nodes on stolons. Excess thatch layers may result in poor rooting of the turfgrass, which can cause nutrient deficiencies and reduced water uptake. In addition, thick thatch layers are a good media for turf pathogens and insects. The stress placed on the grass by a thatch layer can leave it susceptible to disease organisms.

Creeping bentgrass is susceptible to a wide range of diseases including dollar spot (*Sclerotinia homeocarpa*), brown patch (*Rhizoctonia solani*), *Fusarium* blight (*Fusarium roseum* and *F. tricinctum*), *Fusarium* patch or pink snow mold (*Microdochium nivale*), *Pythium* blight (*Pythium* spp.), and *Typhula* blight or gray snow mold (*Typhula incarnata* or *T. ishikariensis*). No cultivars show complete resistance to these pathogens; however, differential cultivar susceptibility does exist and can be an effective method of reducing fungicide applications (Vincelli and Doney, 1997).

DISTRIBUTION AND CYTOTAXONOMY

The bentgrasses are native to Western Europe (Harlan, 1992) with the genus *Agrostis* consisting of approximately 200 species (Hitchcock, 1951). Five species, *Agrostis stolonifera*–creeping bentgrass (2n=4x=28), *Agrostis canina* L.–velvet bentgrass (2n=2x=14), *Agrostis capillaris* L.–colonial bentgrass (2n=4x=28), *Agrostis castellana* L.–dryland bentgrass (2n=4x=28), and *Agrostis gigantea* L.–redtop bentgrass (2n=6x=42), have been adapted for turfgrass use. Considerable variation exists in the literature concerning the naming of these species. For example, creeping bentgrass is referred to as *Agrostis palustris* Huds., *Agrostis stolonifera* L., and *Agrostis stolonifera* L. var *palustris* (Huds.) Farw. Hitchcock (1951) considers *A. stolonifera* and *A. palustris* to be two distinct species with the primary difference being more prolific stolon production by *A. palustris*. However, *A. stolonifera* is more universally accepted as the nomenclature for this species.

Cytotaxonomy

The taxonomic classification of creeping bentgrass is:

Family	Gramineae (Poaceae)
Subfamily	Pooideae
Supertribe	Poeae
Tribe	Aveneae
Genus	*Agrostis*
Group	stolonifera
Species	*stolonifera*
Authority	Linneaus

The genus *Agrostis* has been described as one of the most difficult and complicated of the grass genera from the taxonomic point of view. Jones (1956 a,b,c) conducted the most thorough cytogenetic study of relationships among cultivated members of *Agrostis*. *A. stolonifera* is reported to be a strict allotetraploid with a genome designation of $A_2A_2A_3A_3$. The genome designation of *A. capillaris* is reported to be $A_1A_1A_2A_2$ and *A. gigantea* is $A_1A_1A_2A_2A_3A_3$. The proposed genome designations of *A. capillaris, A. stolonifera,* and *A. vinealis* Salisb. and their putative diploid progenitors are presented graphically in Figure 10.1. Wipff and Fricker (2001) have proposed an alternative genomic designation for *Agrostis*, distinguishing the A_1 and A_2 genomes (C_1 and C_2) from the A_3 genome (S), largely based on the homology between the A_1 and A_2 genomes.

Amplified fragment length polymorphic (AFLP) markers verified the genetic similarity of *A. canina* and *A. vinealis* and the similarity of these two species with *A. capillaris* (Vergara and Bughrara, 2001). This study also demonstrated considerable overlap among marker profiles of accessions classified as *A. palustris* or *A. stolonifera*, providing further evidence of genetic homogeneity between these two taxa.

The results of this research provided an excellent initial examination of the species relationships in this complex genus. Supplemental work using DNA-based phylogenetic tools such as chloroplast and mitochondrial markers may add further resolution. A thorough understanding of the genome relationships within this genus may facilitate the use of diploid breeding to incorporate desired characteristics such as turf quality into the polyploid member of the genus, similar to the breeding strategies used to improve tetraploid cultivated potato (*Solanum tuberosum* L.). A number of interspecific hybrids between *A. stolonifera* and *A. capillaris* have been reported (Bradshaw, 1957).

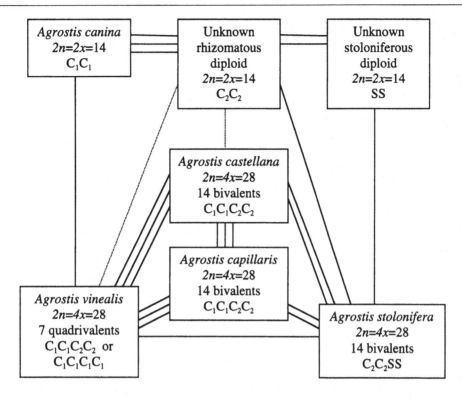

Figure 10.1. The chromosomal and phylogenetic affinities of several important *Agrostis* species. The number of connecting lines indicates relative affinity as judged by chromosome pairing, i.e., the more lines, the closer the affinity. The relationship of the putative diploid ancestors is similarly indicated. The figure is adapted from Jones (1956c) and Wipff and Fricker (2001).

The hybrids have a high degree of sterility, but the perennial nature of these hybrids and the strong vegetative nature of the genus would seem to indicate that gene flow between species is highly likely.

Mode of Inheritance

Jones (1956b) reported that *A. stolonifera* is a strict allotetraploid based on a mean chiasma frequency of 24.86 per cell, which is higher than any individual mean observed for *A. capillaris*. High chiasma frequencies are a good indicator of preferential chromosome pairing, but some species have been inaccurately classified as allopolyploids when the classification is based only on bivalent pairing because random bivalent pairing can lead to tetrasomic inheritance (Krebs and Hancock, 1989). Cultivated *A. stolonifera* does not exhibit random bivalet pairing based on genetic analysis (Warnke et al., 1998). Inheritance of isozyme loci from cultivated creeping bentgrass clones followed only disomic segregation patterns at five different loci. Additionally, several loci exhibited fixed heterozygosity that is a characteristic of disomic polyploid inheritance. The strict disomic inheritance seen in creeping bentgrass may indicate the presence of a pairing control gene similar to the *Ph* gene that has been identified in wheat (*Triticum aestivum* L.).

BREEDING AND CULTIVAR DEVELOPMENT

Major Sources of Germ Plasm

The first bentgrass to be used for turf purposes in the United States was known as South German Mixed Bentgrass (Duich, 1985). South German Mixed Bentgrass was harvested from pastures in present day Austria and Hungary and later from other areas of Europe. South German bentgrass consisted of a mixture of creeping, colonial, velvet and redtop bentgrasses and would segregate into patches of predominately creeping bentgrass and to a lesser extent velvet bentgrass in favorable temperate climates. Many of the better appearing patches were selected and maintained vegetatively by the United States Golf Association Green Section at the Arlington Turf Gardens in Arlington, VA (Duich, 1985). Many hundreds of clones were eventually selected, some of which became known as the C-series creeping bentgrasses and were used for vegetative establishment of putting greens. The most widely used C-series bentgrasses were Toronto (C-15), Cohansey (C-7), Washington (C-50), Arlington (C-1), Congressional (C-19), and Old Orchard (C-52), each of which was well adapted to specific regions of the United States.

The cultivar Seaside was the only available seeded creeping bentgrass from the late 1920s until the 1950s (Duich, 1985). Seaside was discovered growing in tidal flatlands near Coos Bay, Oregon. Seaside was much inferior in turf quality to the vegetatively propagated C-series bentgrasses. Therefore, putting greens in the United States were primarily established vegetatively until the 1950s, when the desire to utilize creeping bentgrass for fairways led breeders to begin developing an acceptable seeded creeping bentgrass. Holt and Payne (1952) studied the growth rate, texture, density, and drought tolerance of seedling progenies from 49 C-series clones of creeping bentgrass to determine the extent of variability within and among strains. The results indicated that seed propagation of most of the available C-series creeping bentgrass clones was not promising, due to high levels of variability in the seeded progeny.

Dr. H. B. Musser released the cultivar Penncross from the Pennsylvania State University in 1955. Penncross established a new level of excellence for seeded creeping bentgrasses and is often referred to as the landmark cultivar of this species (Rogers, 1991). Penncross creeping bentgrass is the first generation seed, produced by the random crossing of three vegetatively propagated clonal strains (Hein, 1958). Penncross is a mixture of three two-clone hybrids and is a good example of the potential for the utilization of heterosis in outcrossing grass species (Brummer, 1999). Since the development of Penncross, breeding efforts have focused on increasing turf quality and widening the adaptability of creeping bentgrass. Breeding efforts in Arizona and Texas have led to the release of cultivars such as SR1020 and Crenshaw that have significantly improved heat tolerance over previous varieties. Interestingly, SR1020 and Crenshaw share three parental clones that were selected for heat tolerance at the University of Arizona (Engelke et al., 1995). However, SR1020 and Crenshaw are not closely related based on isozyme analysis (Warnke et al., 1997), indicating that the additional clones involved in the development of Crenshaw provided enough genetic diversity to significantly differentiate it from SR1020.

The current breeding materials used in the development of new cultivars are phenotypically far removed from wild germ plasm. Wild populations are more prostrate in growth habit and have wider leaves than cultivated material. The crossability of this material is still adequate, based on seed set of controlled crosses (Warnke et al., 1998). However, breeders have made significant advances in quality based on turf color, density, and growth habit. Therefore, most breeders would be reluctant to include wild germ plasm in their current breeding efforts. Wild germ plasm, however, may be a source of important genes for disease and insect resistance that are not present in current breeding populations. The outcrossing nature of this species and the excellent selection efforts that have been practiced by current breeding programs has led to a diverse cultivated germ plasm pool that is not likely to be genetically vulnerable.

Marker Assisted Selection

Marker-assisted selection has the potential to be a valuable tool in the development of creeping bentgrass, particularly in the selection of characters that are difficult and time-consuming to select, such as disease and insect resistance or abiotic stress tolerance. The only genetic markers that have been reported in creeping bentgrass are isozyme markers (Warnke et al., 1998), but Randomly Amplified Polymorphic DNA (RAPD) markers have been utilized for cultivar identification (Golembiewski et al., 1997).

Simple Sequence Repeats (SSRs) have great potential for use in outcrossing species, but they are expensive and time-consuming to develop because sequence information is needed to create these markers. Marker-assisted selection strategies will require the development of mapping populations and accurate phenotyping to establish linkages between markers and genes of interest. With outcrossing species, the most suitable mapping strategies involve the use of double pseudotestcross population structures or three generation pedigrees (Weeden, 1994). These population structures take advantage of the high level of polymorphism in outcrossing species and allow map development without the development of homozygous lines. The ability to vegetatively propagate creeping bentgrass should enhance phenotypic data collection from mapping populations and may even allow replicated plots of individual clones to be established and phenotyped.

Plant Transformation Potentials

Stable transformation was first reported in *A. stolonifera* by Zhong et al. (1993). Direct gene transfer by both protoplast and particle bombardment transformation have been successfully used to produce transgenic *Agrostis* plants. With protoplast transformation, both electroporation (Asano and Ugaki, 1994) and polyethylene glycol (PEG) (Lee et al., 1996) have been used to deliver DNA into protoplasts isolated from embryogenic callus-derived suspension cultures. Microprojectile bombardment has been used to directly deliver DNA into intact embryogenic callus cells and suspension cells (Zhong et al., 1993). The silicon carbide fiber method is an alternative DNA delivery technique that has been used to produce transgenic plants (Dalton et al., 1998).

The genes used for transformation experiments in *Agrostis* include the *gus* reporter and the selectable marker *npt II*, *hph*, and *bar* genes. The *bar* gene is the first useful foreign gene introduced into creeping bentgrass and confers herbicide resistance to transgenic plants. The insecticidal protein *cry IA* gene derived from *Bacillus thuringensis* has been introduced into *A. stolonifera*, but expression has not been observed (Ito et al., 1995). For further information, Asano et al. (2000) provides an excellent review of electroporation-mediated direct gene transfer methods in *Agrostis*.

Gene Escape

Transformation studies in creeping bentgrass have been numerous since the early 1990s and one commercial company is close to the release of a glyphosate-resistant cultivar. However, research concerning the potential escape of transgenes into adjacent nontransgenic fields or naturalized and native populations is limited. The most thorough research to date has been conducted by Wipff and Rose-Fricker (2000), and their findings indicate the following: (1) transgenes of creeping bentgrass can flow to other species of *Agrostis*, (2) gene flow in creeping bentgrass is possible for much longer distances than traditionally theorized, and (3) transgenic bentgrass plants are fertile and stable. This preliminary work indicated that transgenes in creeping bentgrass will be released into wild and cultivated members of this species. Many transgenes have tremendous potential to improve creeping bentgrass cultivars, but they should be released with caution.

GENETIC AND PHYSIOLOGICAL CONTROL OF FLOWERING

Flowering Requirements

Creeping bentgrass is a long-day species that requires vernalization and generally flowers in late spring to early summer in production regions of the United States. The juvenility phase of creeping bentgrass has not been thoroughly studied. Flowering during the first year of production when plants are small can be unpredictable, with some clones producing panicles and others producing no panicles until the second year of production. The juvenility requirement can be a problem for controlled crosses using growth-chamber vernalized plant materials; flowering is usually poor and seed set limited. Controlled crossing should be done using field-grown plant material established in the spring and utilized for crossing the following spring. Reproductive tillers can be cut below a node just prior to flowering and placed in test tubes or beakers with tillers of another clone of interest. Tillers will advance through anthesis and produce seed in about one month. Good seed set with very little self-fertilization has been obtained using this crossing procedure (Warnke et al., 1998). Cooper and Calder (1964) have provided the most thorough study of the inductive requirements for flowering of temperate grasses, and their work indicated that creeping bentgrass has a moderate inductive requirement, mainly requiring short days (<10 hr photoperiod). However, complete induction was only obtained in plants established outdoors. The more recent review by Heide (1994) provided a thorough background of the terminology and theories behind flowering in temperate grasses. However, in many cases even taxonomically closely related grasses may differ considerably in their floral induction requirements.

Photoperiod Sensitivity and Genetic Variation for Maturity

Research has not been conducted on the specific photoperiod sensitivity of individual cultivars or genetic variation for maturity. However, given the importance of these factors to reproductive success, considerable variation likely exists.

No studies have been published on major genes or quantitative trait loci (QTL) influencing flowering in creeping bentgrass. The control of flowering is, however, an important consideration for cultivars because synchronicity of flowering times for parental clones is required for adequate seed set. Utilizing 24 allozymes of creeping bentgrass, Warnke et al. (1997) divided 19 cultivars into two separate clusters (Figure 10.2). The primary allozyme for this separation were the phosphoglucomutase (PGM) allozymes PGM-12 and PGM-13. The PGM enzymes are involved in dark respiration of photosystem II and may play a role in the control of flowering (Rainey et al., 1987). However, further crossing work and genetic map development are required to test this hypothesis.

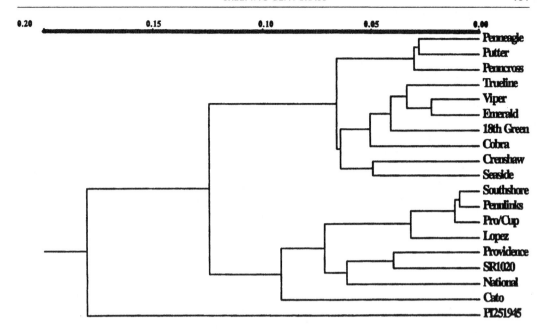

Figure 10.2. Dendrogram of 19 creeping bentgrass cultivars generated by UPGMA cluster analysis. The clustering is based on 24 allozymes with the PGM-1 locus being the primary difference between the two major clusters (Warnke et al., 1998).

MAJOR SELECTION CRITERIA

All cultivars of creeping bentgrass are synthetics with the exception of Penncross, which is a three-parent hybrid. Currently available cultivars vary in the number of parent clones from six to as high as 1,000 in the cultivar Southshore. Cultivars with high numbers of parental clones can exhibit problems with segregation of plant types. The two main objectionable off-type segregants in creeping bentgrass are purple coloring under cool autumn temperatures and poor density. These two types of plants can become more pronounced over time and result in a mottled appearance on older greens. Many of the current cultivars have been extensively studied for a number of years to determine the extent of segregation that is likely to occur. High shoot density at low mowing heights and short internode lengths are two of the primary breeding objectives for new greens-type creeping bentgrasses. However, cultivars with extremely high shoot density will produce high levels of thatch if they are not properly maintained.

Disease Resistance

Several major diseases are capable of significantly lowering the quality of creeping bentgrass greens and fairways. The majority of creeping bentgrass diseases have different levels of importance throughout the United States due to different climatic conditions. For example, dollar spot (*Sclerotinia homoeocarpa* Bennett) is a significant disease problem in areas of the United States that have hot, humid summer conditions such as the upper Midwest, East Coast, and southern states. However, in the Pacific-Northwest and southwestern states that have low humidity, dollar spot is not a significant problem.

Dollar Spot

In temperate and hot humid regions of the United States, more money is spent to manage dollar spot than on any other turfgrass disease (Vargas, 1994). A practical means of reducing dollar spot damage would be to develop host plant resistance to this disease. Research on the host pathogen interaction of *A. stolonifera* and *S. homoeocarpa* is limited. Cole et al. (1969) studied the influence of fungal isolate and bentgrass variety on dollar spot development. Eighteen different creeping bentgrass cultivars were screened with four different dollar spot isolates and a significant fungal isolate x cultivar interaction was found. Reports from regional cultivar trials (Colbaugh and Engelke, 1993; Hsiang and Cook, 1993; Doney and Vincelli, 1993) provide somewhat conflicting information with regard to which cultivars show the highest levels of resistance. However, the varieties SR1020, Crenshaw, Emerald, and 18th Green are consistently highly susceptible, while the colonial bentgrasses have shown the highest levels of resistance.

The improvement of creeping bentgrass resistance to dollar spot will require a more thorough understanding of the methods of pathogenesis utilized by *S. homoeocarpa* as well as the plant defense mechanisms available in creeping bentgrass. The study of plant resistance mechanisms under field conditions can be misleading, because it is difficult to determine if a plant without disease symptoms is resistant or was not exposed to disease pressure. The ability of breeders to produce cultivars with enhanced resistance to this important disease might be significantly enhanced through the use of marker-assisted selection (Bonos and Meyer, 2000).

Brown Patch

The causal agent of brown patch is *Rhizoctonia solani* Kuhn. Brown patch occurs during warm wet weather and is characterized by light brown to tan-colored circular patches up to three feet in diameter. There is no characterized resistance to this disease in creeping bentgrass (Beard, 1973).

Fusarium Patch

Fusarium patch is also called pink snow mold and is caused by *Fusarium nivale* (syn. *Microdochium nivale*). The disease is characterized by pink mycelium on leaves with 2.5 to 5 cm. tan circular patches or a white mycelial mass on the leaves with white to pink circular patches up to half a meter. Fusarium patch is common in cool wet weather or cold moist conditions under a snow cover and can be of particular importance in the Pacific Northwest region of the United States. No genetic resistance has been reported.

Typhula Blight

Typhula blight is also called gray snow mold and is caused by *Typhula itoana* or *T. ishikariensis*. The disease is characterized by light gray mycelium on leaves and whitish gray circular patches up to half a meter in diameter. Typhula blight is common in cold humid conditions, especially under a snow cover, and is of particular importance in northern regions of the United States where it can cause considerable turf loss. Fungicide control of gray snow mold is difficult and the chemicals required are not environmentally friendly. Therefore, new breeding programs in the northern regions of the United States are attempting to identify genetic resistance (Wang et al., 2000).

Abiotic Stress Resistance

The desire to expand the range of adaptability of creeping bentgrass to southern regions of the United States has led many of the breeding programs currently working on creeping bentgrass improvement to focus on abiotic stress resistance. Duncan and Carrow (1999) provide an excellent review of the potential for molecular genetic improvement of turfgrasses. Lehman and Engelke (1993) have provided the most thorough research focusing on drought tolerance in creeping bentgrass. Using parent progeny regression, narrow-sense heritabilities of 0.98 and 1.0 for leaf water content under soil temperature stress were obtained. These high heritabilities indicated that a strong genetic component exists for shoot water content and this trait may be an excellent candidate for improving heat and drought tolerance of creeping bentgrass.

Salinity Tolerance

Response of creeping bentgrass to salt stress is an important selection criteria as seawater encroachment and the use of effluent or recycled water sources increases. Younger et al. (1967) found that substantial creeping bentgrass cultivar differences for salinity tolerance exist and that selection in cultivars such as Seaside might lead to substantial increases in salt tolerance. Smith et al. (1993) have successfully used whole-plant microculture as a method for selection of salt tolerant creeping bentgrass clones.

CONCLUSIONS

The majority of research occurring in creeping bentgrass has focused on management related issues and physiological studies of environmental responses to stress. The study of genetic control of biotic and abiotic resistance is limited. Future genetic improvement of creeping bentgrass will require a much more thorough understanding of the genetic mechanisms behind important selection criteria. Genetic transformation of creeping bentgrass has progressed to a level where it can provide useful genetic variation for cultivar improvement. However, it is dangerous to rely solely on transformation as a cultivar improvement mechanism. At some point it will be necessary to develop genetic-based approaches such as marker-assisted selection. The genetic understanding of this important turfgrass species is in its infancy; future research will continue to result in cultivars with improved utility and performance.

REFERENCES

Asano, Y. and M. Ugaki 1994. Transgenic plants of *Agrostis alba* obtained by electroporation-mediated direct gene transfer to protoplasts. *Plant Cell Rep.* 13:243–246.

Asano, Y., M. Ugaki, Y. Ito, M. Fukami, and A. Fujiie. 2000. Transgenic Bentgrass *(Agorstis spp.) Biotechnology in Agriculture and Forestry, Transgenic Crops* Y.P.S Bajaj, Ed. Springer-Verlag, Berlin, Vol. 46, pp. 127–138.

Beard, J.B. 1973. *Turfgrass: Science and Culture.* Prentice Hall, Upper Saddle River, NJ.

Bonos, S.A. and W.A. Meyer 2000. Evaluation and heritability of dollar spot resistance in creeping bentgrass genotypes. *Agronomy Abstracts.* pp. 161.

Bradshaw, A.D. 1957. Natural hybridization of *Agrostis capillaris* Sibth. and *A. stolonifera* L. *Journal of the Sports Turf Research Institute.* 9(34): 422–429.

Brummer, E.C. 1999. Capturing heterosis in forage crop cultivar development. *Crop Sci.* 39: 943–954.

Colbaugh, P.F. and M.C. Engelke. 1993. Sclerotina dollar spot on bentgrasses. *Biological and Cultural Tests.* 8:111.

Cole, H., J.M. Duich, L.B. Massie, and W.D. Barber. 1969. Influence of fungus isolate and grass variety on Sclerotinia dollar spot development. *Crop Sci.* 9:567–570.

Cooper J.P. and D.M. Calder 1964. The inductive requirements of flowering of some temperate grasses. *Journal of the British Grassland Society.* 19:6–14.

Dalton S.J., A.J.E. Bettany, E. Timms, and P. Morris 1998. Transgenic plants of *Lolium multiflorum. Lolium perenne, Festuca arundinacea* and *Agrostis stolonifera* by silicon carbide fiber-mediated transformation of cell suspension cultures. *Plant Sci.* 132:31–43.

Doney, J.C. Jr. and P.C. Vincelli. 1993. Reactions of bentgrasses to dollar spot and brown patch. *Biological and Cultural Tests.* 8:118.

Duich, J.M. 1985. The bent grasses. *Weeds Trees and Turf.* January. pp. 72–78. Harcourt Brace Jovanovich Publ., New York

Duncan R.R. and R.N. Carrow 1999. Turfgrass molecular genetic improvement for abiotic/edaphic stress resistance. *Advances in Agronomy* 67:233–305.

Engelke M.C., V.G. Lehman, W.R. Kneebone, P.F. Colbaugh, J.A. Reinert, and W.E. Knoop. 1995. Registration of 'Crenshaw' creeping bentgrass. *Crop Sci.* 35:590.

Golembiewski, R.C., T.K. Danneberger, and P.M. Sweeney. 1997. Potential of RAPD markers for use in the identification of creeping bentgrass cultivars. *Crop Sci.* 37:212–214.

Harlan, J.R. 1992. *Crops and Man,* 2nd ed. American Society of Agronomy, Madison, WI.

Heide O.M. 1994. Control of flowering and reproduction in temperate grasses. *New Phytol.* 128:347–362.

Hein, M.A. 1958. Registration of varieties and strains of grasses. *Agron. Jour.* 50:399.

Hitchcock, 1951. Manual of the Grasses of the United States. USDA Misc. Publ. 200. U.S. Gov. Print. Office, Washington, DC.

Holt, E.C. and K.T. Payne. 1952. Variation in spreading rate and growth characteristics of creeping bentgrass seedlings. *Agronomy Journal.* 44:88–90.

Hsiang, T. and S. Cook. 1993. Resistance of bentgrass cultivars to dollar spot disease. *Biological and Cultural Tests.* 8:110.

Ito Y., M. Fukami, Y. Asano, K. Sugiura, M. Ugaki, and A. Fujiie 1995. Transgenic creeping bentgrass plants from electroporated protoplasts. *J. Jpn. Soc. Turfgrass Sci.* 23:128–133.

Jones, K. 1956a. Species differentiation in *Agrostis* I. Cytological relationships in *Agrostis canina* L. *J. Genet.* 54:370–376.

Jones, K. 1956a. Species differentiation in *Agrostis II.* Cytological relationships in *Agrostis canina* L. *J. Genet.* 54:370–376.

Jones, K. 1956b. Species differentiation in *Agrostis* II. The significance of chromosome pairing in the tetraploid hybrids of *Agrostis canina* subsp. Montana Hartm., *A. capillaris* Sibth. and *A. stononifera* L. *J. Genet.* 54:377–393.

Jones, K. 1956c. Species differentiation in *Agrostis* III. *Agrostis gigantea* Roth. and its hybrids with *A. capillaris* Sibth. and *A. stolonifera* L. *J. Genet.* 54:394–399.

Krebs, S. and J. Hancock. 1989. Tetrasomic inheritance of isozyme markers in the highbush blueberry, *Vaccinium corymbosum* L. *Heredity* 63:11–18.

Lee, L., C.L. Laramore, P.R. Day, and N.E. Tumer 1996. Transformation and regeneration of creeping bentgrass (*Agrostis palustris* Huds.) protoplasts. *Crop Sci.* 36:401–406.

Lehman, V.G., M.C. Engelke, and R.H. White. 1993. Leaf water potential and relative water content variation in creeping bentgrass clones. *Crop Sci.* 33:1350–1353.

Lehman, V.G. and M.C. Engelke. 1993. Heritability of creeping bentgrass shoot water content under soil dehydration and elevated temperatures. *Crop Sci.* 33:1061–1066

Rainey, D.Y., J.B. Mitton, and R.K. Monson. 1987. Associations between enzyme genotypes and dark respiration in perennial ryegrass, *Lolium perenne* L. *Oecologia* 74: 335–338.

Rogers, M. 1991. Cool-season turfgrass varieties. *Grounds Maintenance.* p. 22

Smith, M.A.L., J.E. Meyer, S.L. Knight, and S. Chen. 1993. Gauging turfgrass salinity responses in whole-plant microculture and solution culture. *Crop Sci.* 33:566–572.

Vargas, J.M. 1994. *Management of Turfgrass Diseases.* 2nd ed. CRC Press, Inc., Boca Raton FL. pp. 23–26.

Vergara, G.V. and S.S. Bughrara. 2001. Genetic diversity and distances among bentgrass accessions characterized by AFLP markers. Abstract No. 124343P. In *Agronomy Abstracts.* ASA-CSSA-SSSA, Madison, WI.

Vincelli, P. and J.C. Doney Jr. 1997. Variation among creeping bentgrass cultivars in recovery from epidemics of dollar spot. *Plant Disease* 81:99–102.

Wang, Z., M.D. Casler, J.C. Stier, J. Gregos, S.M. Millett, D.P. Maxwell. 2000 Speckled snow mold resistance in creeping bentgrass. *Agronomy Abstracts.* p. 151.

Warnke, S.E., D.S. Douches, and B.E. Branham 1998. Isozyme analysis supports allotetraploid inheritance in tetraploid creeping bentgrass (*Agrostis palustris* Huds.). *Crop Sci.* 38:801–805.

Warnke, S.E., D.S. Douches, and B.E. Branham 1997. Relationships among creeping bentgrass cultivars based on isozyme polymorphisms. *Crop Sci.* 37:203–207.

Weeden, N.F. 1994. Approaches to mapping in horticultural crops: In *Plant Genome Analysis*, P.M. Gresshoff, Ed. CRC Press, Boca Raton, FL, pp. 57–68.

Wipff, J.K. and C. Fricker. 2001. Gene flow from transgenic creeping bentgrass (*Agrostis stolonifera* L.) in the Willamette Valley, Oregon. *Intl. Turf Soc. Res. J.* 9:224–242.

Wipff, J.K. and C. Rose-Fricker. 2000. Determining gene flow of transgenic creeping bentgrass and gene transfer to other bentgrass species. *Diversity.* 16:36–39.

Younger, V.B., O.R. Lunt, and F. Nudge. 1967. Salinity tolerance of seven varieties of creeping bentgrass, *Agrostis palustris* Huds. *Agronomy Journal* 59:335–336.

Zhong, H., M.G. Bolyard, C. Srinivasan, and M.B. Sticklen. 1993. Transgenic plants of turfgrass (*Agrostis palustris* Huds. From microprojectile bombardment of embryogenic callus. *Plant Cell Rep.* 13:1–6.

Chapter 11

Agrostis capillaris (*Agrostis tenuis* Sibth.) Colonial Bentgrass

Bridget A. Ruemmele

Agrostis capillaris L., a bentgrass species known by various common names, including colonial bentgrass and browntop (also written as brown top), has evolved in use as much as nomenclature since its adoption as a mown turfgrass. The colonial bentgrass name was based on its introduction to the United States of America during colonization by European immigrants from England (Madison, 1971). Native to Europe and temperate Asia, this grass is called brown top in Great Britain (Hubbard, 1984).

Confusion exists in actual classification of germ plasm within this species. Within the turfgrass industry, colonial bentgrass and brown top are considered separate species (*A. capillaris* and *A. castellana* Boiss. & Reuter, respectively) by some sources (Brilman, 2001; Hartmann et al., 1988; Hubbard, 1984). *A. castellana* is also commonly referred to as dryland bentgrass. The 'Highland' cultivar is placed in *A. castellana* by Hubbard (1984) and Shildrick (1976), while Brilman (2001) noted that 'Exeter' also belongs in *A. castellana*. Color, ligule size, growth habit, and flower color separate *A. capillaris* from *A. castellana* plants morphologically (Hubbard, 1984; USDA, 1948). Seedheads and seeds are usually distinct between the two species (Brilman, 2001). Ploidy levels also differ, with *A. capillaris* being tetraploid (2n=4x=28), while *A. castellana* is hexaploid (2n=6x=42) (Brilman, 2001). Despite the controversy with classification, inclusion within the United States of America National Plant Germplasm System (NPGS), which placed Highland and Exeter in *A. capillaris*, is used as the basis for germ plasm described herein.

BOTANICAL AND PHYSIOLOGICAL DESCRIPTIONS
Worldwide Use and Management

Originally used for golf greens, bowling greens, tennis courts, and high-grade lawns, this species is more commonly used today in Europe, North America, and New Zealand as a fairway grass (Beard, 1973; Hubbard, 1984), and for erosion control. Robinson (1969) observed the suitability of this species for golf greens and fairways when adequately topdressed. *A. capillaris* is also used on golf course fairways in Japan (Oohara, 1977). Home lawn use has become limited as other species with more choices have replaced *A. capillaris*.

Low-maintenance mixtures, used for roadsides in the United Kingdom, have included *A. capillaris* cultivars (Shildrick, 1977). Studies in France (Henensal et al., 1977) and Germany (Trautmann and Lohmeyer, 1977) concluded *A. capillaris* was not as suitable for roadside use as other grass species.

Festuca spp. (fine fescues) are sometimes combined with *A. capillaris* for both golf course and home lawn purposes (Duyvendak and Vos, 1974; Escritt, 1969; Skogley, 1977; Shildrick, 1977; USDA, 1948; Vos and Scheijgrond, 1969), although some recommend its use alone due to the tendency for *A. capillaris* to dominate many mixtures (Davis, 1958; Elmore et al., 1997; Turgeon, 1987). Increased utilization for low-maintenance conditions is being encouraged as new cultivars are introduced (Brilman, 2001; Ruemmele, 2000).

A. capillaris preferred mowing height ranges between 1.0 to 2.5 cm. Under today's management regimens, this limits the grass's use on golf courses to fairways and tees, rather than greens (Elmore et al., 1997; Ruemmele, 2000; Shildrick, 1977).

Climatic and Edaphic Limitations and Flexibility

While described as ideally requiring high fertility (up to 290 kg ha^{-1}) and supplemental irrigation (Elmore et al., 1997; Skirde, 1969), some *A. capillaris* germ plasm requires less of these inputs compared to *Agrostis stolonifera* L. (creeping bentgrass) (Brilman, 2001; Ruemmele, 2000). *Poa pratensis* L. (Kentucky bluegrass) is not competitive against *A. capillaris* under the low mowing heights tolerated by the latter, even when irrigation and fertility are sufficient to support the former (Brilman, 2001). A Connecticut pasture study reported *A. capillaris* exhibited more dominance under low-fertility conditions compared to *Trifolium repens* L. (white clover) and *P. pratensis* (Brown, 1932). This study documents the possibility of selecting reduced-fertility-tolerant germ plasm from naturalized pastures.

Skirde (1969) determined the greatest differences in plot discoloration between irrigated and nonirrigated plots for *A. capillaris* compared to *Festuca* spp., *P. pratensis*, and *Lolium perenne* L. (perennial ryegrass). Despite this contrast, weed invasion was minimal even in unwatered *A. capillaris* plots.

Another fertility experiment conducted by Adams (1977) demonstrated better cover by *A. capillaris* under medium fertility (88, 34, and 62 kg ha^{-1} N, P, and K, respectively) compared to high fertility (312, 134, and 250 kg ha^{-1} N, P, and K, respectively). These results confirm *A. capillaris* tolerates reduced fertilization practices.

Soil acidity tolerance is good, with tolerance as low as pH 5.0 (Hartwell and Damon, 1917; Musser, 1948; Robinson, 1969). Medium and coarse-textured, well-drained soils are preferred to fine soils (Hubbard, 1984). Vos and Scheijgrond (1969) described widespread *A. capillaris* distribution in the Netherlands on acid, sandy soils. *A. capillaris* showed greater survival than *A. stolonifera* on acid, high-clay soils in Griffin, Georgia (Brilman, 2001).

Temperature tolerances limit *A. capillaris* use in the United States to the Pacific Northwest and New England. Carroll (1943) reported *A. capillaris* ranks among the most cold-tolerant cool-season turfgrasses, comparable to *P. pratensis*. Although winter dormancy may extend for longer periods in *A. capillaris* compared to other cool-season turfgrasses (Skogley, 1977), fall, winter, and spring color of *A. capillaris* may be better than other *Agrostis* species in more temperate climates (Brilman, 2001). Ice cover injury (Beard, 1973) and low temperature survival (Beard, 1969b) varies among *A. capillaris* cultivars. Heat and drought tolerances are considered poor to fair, depending on cultivar (Elmore et al., 1997; Ruemmele, 2000).

Although *Agrostis* species are not known for shade tolerance as a genus, *A. capillaris* tolerates moderate tree shade (Beard, 1969a; Chesnel et al., 1977). Pool (1948) ranked *A. capillaris* shade tolerance even lower.

Wear tolerance, recuperative ability, herbicide sensitivities, and proneness to thatch formation need to be addressed for improvement of this species (Beard, 1973; Elmore et al., 1997; Emmons, 1995; Madison, 1971). Kamps (1969) observed dominance of *A. capillaris* in mixed turfgrass stands during establishment from seed, followed by high sensitivity to sports field action six months after sowing. Moderate wear tolerance was noted for monostands of *A. capillaris* subjected to artificial, football-type wear (Shildrick, 1975). Bourgoin and Mansat (1981) observed less wear tolerance of 'Orbica' and 'Tracenta' compared to other cultivars stressed by a studded wear machine. More recent experiments at Rutgers University demonstrated better wear tolerance among a number of newer *A. capillaris* cultivars compared to several *A. stolonifera* cultivars (Bonos et al., 1999, 2001).

Reduced salinity tolerance and wide-ranging disease susceptibility are additional concerns (Elmore et al., 1997; Emmons, 1995). Emmons (1995) cites problems with submersion tolerance of *A. capillaris*, ranking that characteristic as only fair.

Despite its limitations, *A. capillaris* is less time-consuming and expensive to maintain on golf course fairways and tees (Emmons, 1995). The introduction of *P. pratensis*, *Festuca* spp., and *L. perenne* cultivars after the 1950s have reduced the utilization of *A. capillaris* under close-mowing situations (Skogley, 1977). Poor quality seed problems among early introductions of *A. capillaris* contaminated by *A. stolonifera* seed (Skogley, 1977) have been addressed with improved seed cleaning technology and practices. Interest in *A. capillaris* improvement has been renewed as problems (disease susceptibility and increased cultural input requirements) with other species have made the possible use of *A. capillaris* more attractive.

Morphology

Morphology varies within this perennial species, particularly with respect to presence of stolons and rhizomes. Chase (1971) described *A. capillaris* as having short stolons, but no creeping rhizomes. Others report the presence of short or creeping rhizomes, while still designating the growth habit as bunch-type (Brilman, 2001; Christians, 1998; Elmore et al., 1997). The growth habit results in a fairly dense turf, though slightly less than that of *A. stolonifera* (Elmore et al., 1997). Both *A. canina* (velvet bentgrass) and *A. stolonifera* spread by extensive stolon production. *A. castellana* has more deep and extensive rhizomes compared to *A. capillaris* (Brilman, 2001).

Leaves culminate in a pointed leaf tip, with short ligules, less than 1 to 2 mm long on the culm. The other commonly-used bentgrasses (*A. canina* and *A. stolonifera*) typically have longer ligules. Blades usually range from 5 to 10 mm in length and 1 to 3 mm in width (Chase, 1971). Vernation is rolled (Christians, 1998; Beard, 1973; Elmore et al., 1997). Leaf color spans light to dark green (Elmore et al., 1997) although older *A. capillaris* cultivars were not as dark as *A. stolonifera*.

Shallow rooting is considered the main cause of supplemental irrigation requirements for adequate growth (Elmore et al., 1997). Boeker (1974) reported *A. capillaris* rooting to be comparable to *A. stolonifera*, with *A. canina* rooting to be shallower. Caution is urged regarding this study, which included only one *A. capillaris* sample (cultivar not stated) evaluated on one date, yet assessed eight and 14 cultivars of *A. canina* and *A. stolonifera*, respectively, on three dates. The effects of shallower rooting by *Agrostis* spp. cultivars Tracenta and Highland were demonstrated when turf was easily pulled out of the ground on turf used for horse racing (deChevigny and Dujardin, 1981).

Seeds are borne in panicles on erect, slender culms. These culms are tufted, ranging from 20 to 40 cm in height (Chase, 1971). Panicles are 5 to 10 cm long in an open, 'delicate' formation where spikelets are not crowded (Chase, 1971). Seed size is among the smallest for turfgrasses, ranging from 12.1 to 19.8 million seeds kg^{-1}, depending on cultivar (Lynch, 1996; Elmore et al., 1997). The resulting seeding rate is low, from 24 to 49 kg ha^{-1}.

DISTRIBUTION AND CYTOTAXONOMY

Origin and Natural Distribution

A. capillaris is considered native to both temperate Asia and Europe, with widespread distribution elsewhere in the world, particularly in North and South America, Australia and New Zealand (Hubbard, 1984 and USDA-ARS, 2001). Within Asia, countries of native origin include Afghanistan (eastern), Armenia, Azerbaijan, Georgia, Iran (northern), Kazakhstan (eastern), Russian Federation—Ciscaucasia, Dagestan, Eastern Siberia (western), Western Siberia, and Turkey. European countries of native origin include Albania, Austria, Belarus, Belgium, Bulgaria, Czechoslovakia, Denmark, Estonia, Finland, France (including Corsica), Germany, Greece, Hungary, Ireland, Italy, Latvia, Lithuania, Moldova, Netherlands, Norway, Poland, Portugal; Romania, Russian Federation—European part, Spain, Sweden, Switzerland, Ukraine (including Krym), United Kingdom, and Yugoslavia.

Within a few years after introduction, this species was considered naturalized in Rhode Island and coastal New England (Madison, 1971). It also established rapidly along the coast of Nova Scotia and Prince Edward Island in Canada. Spread of this grass continued when émigrés from Nova Scotia to New Zealand introduced *A. capillaris* to the latter country (Madison, 1971). Rhode Island colonists carried *A. capillaris* alone, or in mixtures of bentgrass known as South German bent, westward in the United States of America across the Oregon Trail to the Astoria, Oregon region. In North America, *A. capillaris* use is limited to coastal regions (Lynch, 1996; Chase, 1971; Elmore et al., 1997; Ruemmele, 2000), although it has been documented in 35 states, including Hawaii (USDA-NRCS, 2001). Cultivar evaluation trials in Maryland and Georgia, as well as germ plasm collections by turfgrass breeders, demonstrate a more widespread usage within North America may be possible (Brilman, 2001). *A. capillaris* specimens have been collected from old turf sites in North America as far south as Georgia, Tennessee, Kentucky, Virginia, and Missouri (Brilman, 2001).

Taxonomy

The taxonomic classification of *A. capillaris* is:

Family	Poaceae (Gramineae)
Subfamily	Pooideae
Tribe	Aveneae
Genus	*Agrostis*
Species	*capillaris*
Authority	Linnaeus

Controversy continues regarding taxonomy within the *Agrostis* species, notably with the turfgrasses commonly referred to as colonial bentgrass and browntop. (Brilman, 2001; Ruemmele, 2000). *A. capillaris* was formerly known as *A. tenuis* Sibth. (Chase, 1971; Watson and Dallwitz, 1992), but is now classified as *A. capillaris* L. (Christians,

1998; USDA-ARS, 2001). As noted in the introduction, some browntop are more properly classified as dryland bentgrass *(A. castellana)* (Brilman, 2001; Hubbard, 1984).

Synonyms associated with this species in GRIN (Germplasm Resources Information Network) include *A. alba* var. *vulgaris* (With.) Coss. & Durieu, *A. tenuis* Sibth. (previously associated with 30 accessions now classified as *A. capillaris*), *A. tarda* Bartl. and *A. vulgaris* With. (previously associated with two accessions now classified as *A. capillaris*) (USDA-ARS 2001). Additional species synonyms have been noted (Chase, 1971; USDA-NRCS, 2001). These include *A.sylvatica* Huds., *A.t.* Sibthorp, *A.t.* Sibthorp var. *aristata*, *A.t.* Sibthorp var. *hispida* (Willd.) Philipson, and *A.t.* Sibthorp var. *pumila* (L.) Druce.

Other common names for *A. capillaris* include fine bentgrass, common bentgrass, New Zealand bentgrass, Prince Edward Island bentgrass, Rhode Island bentgrass, waipu, agrostide commune (French), agrostide ténue (French), gemeines Straußßgras (German), Rotstraußßgras (German), agróóstide comúún (Spanish), and hierba fina (Spanish) (Hubbard, 1984; Madison, 1971; USDA-ARS 2001).

Cytology and Cytogenetics

Cytological analyses are limited for *A. capillaris*. This species is described as a tetraploid ($2n = 4x = 28$) (Brilman, 2001; Clark et al., 1989; Hubbard, 1984; Jones, 1953; Jones, 1956). The genome is listed as $A_1A_1A_2A_2$ or $A_2A_2A_2A_2$ (Clark et al., 1989). In chromosome pairing analyses (Jones, 1956), *A. capillaris* bivalents behaved as a segmental allotetraploid, with bivalents formed at meiosis. Open pollination is the primary means of sexual reproduction.

COLLECTION, SELECTION, AND BREEDING HISTORY

Major Sources of Germ Plasm

Early cultivars of *A. capillaris* were limited to Highland, 'Astoria,' Exeter, 'Holfior,' and 'Boral.' Although some do not consider Highland (Brilman, 2001; Chase, 1971; Hubbard, 1984; USDA, 1948) or even Exeter (Brilman, 2001) to fit within *A. capillaris*, they are designated as such by GRIN and included in the list of accessions for *A. capillaris* (USDA-ARS, 2001).

Released from the Oregon Agricultural Experiment Station in 1934, Highland had improved drought tolerance and a bluish-green color (Ruemmele, 2000). These unique characteristics led to reclassifying Highland as *A. castellana*.

Astoria has been considered essentially the same as common *A. capillaris* (USDA, 1948). This cultivar was also released from the Oregon Agricultural Experiment Station in 1936 by Madison (1971). It was derived from collections during 1926 by Engbretson and Hysop in northwestern Oregon along the coast near the mouth of the Columbia River. Astoria is more yellow-green than other *A. capillaris* cultivars (Brilman, 2001).

Although the seed of Astoria was harvested from uniform stands of naturally pure stands, the seed is considered as wild-type and highly variable. The high amount of genetic variability in this release is likely why it is considered equivalent to a common type of *A. capillaris*. Washington State certifies Astoria only as *A. capillaris*, rather than by the cultivar name (Madison, 1971). This semierect cultivar has short, determinate stolons.

J.A. DeFrance and C.R. Skogley released Exeter in 1963 from the Rhode Island Agricultural Experiment Station (Ruemmele, 2000) after more than 20 years of selection from material collected within New England (Madison, 1971). This cultivar is distinguished by its high density, attractive color even in summer, earlier greenup, good cold tolerance, and some leaf spot resistance (genus and species not stated).

A European cultivar, Holfior, was released by D.J. van der Have of the Netherlands in 1940 from germ plasm collected in southern Holland (Madison, 1971). Unique features of this cultivar include a darker green color than Astoria, improved winter color retention, moderate spread via short stolons, slight *Fusarium* spp. resistance, and upright growth habit.

More recent cultivar introductions include Tracenta, 'Bardot,' 'Egmont,' 'Tendenz,' 'Duchess,' 'Sefton,' 'SR 7100,' and 'Tiger.' Tracenta, for example, was observed to have less winter discoloration than *A. canina* or *A. gigantea* Roth (formerly classified as *A. alba* and commonly named redtop) (Bourgoin, et al., 1974).

Some *A. capillaris* accessions collected and/or developed from sites throughout the world are maintained in the United States at the Western Regional PI Station (W6) (USDA-ARS, 2001), the Non-NPGS (National Plant Germplasm System) Security Backup Collection (NSSB), or the Plant Germplasm Quarantine Office (PGQO). Submitted accessions are listed in Table 11.1. Prior to 1992, most accessions in this system were classified as *A. tenuis*. Since 1992, all accessions listed in Table 11.1 are identified by the NPGS as *A. capillaris*.

Progress in Cultivar Development

Germ Plasm Sources

Early seed of *A. capillaris* for sale came from harvests of practically pure stands in Rhode Island (North and Odland, 1935). It was from these stands that the germ plasm constructing the Exeter cultivar was collected.

While limited improvement has occurred to date with *A. capillaris* germ plasm, current efforts hold promise for advances in stress tolerance or resistance (Brilman, 2001; Ruemmele, 2000).

Naturalized materials collected within New England during the past 10 years exhibit increased brown patch (*Rhizoctonia solani* Kuhn) resistance (Ruemmele, 2000). Many breeding programs continue to rely on new collections of wild germ plasm as sources for cultivar development and improvement (Brilman, 2001). The limited number of accessions in the NPGS makes this a necessity.

Potentially desirable naturalized *A. capillaris* germ plasm is frequently observed on old lawns on village greens, cemeteries, municipal buildings, churches, and homes throughout New England (Skogley, 1977). Many of these locations have been managed under reduced or no supplemental cultural inputs, often only being mown. This provides an excellent opportunity to select materials likely to tolerate reduced management.

Vast regions within the area of reported origin for *A. capillaris* have yet to be explored for possible breeding material. Considering the variability observed within naturalized germ plasm, plants collected from Europe and temperate Asia could significantly increase the possibility for rapid improvement within this species.

Major and/or Unique Selection Criteria

In addition to enhanced brown patch resistance, other diseases of interest for resistance selection include red thread (*Laetisaria fuciformis* McAlp. Burdsall), pink snow mold (*Fusarium nivale* [Fr.] Ces.) (Duyvendak and Vos, 1974; Eschauzier, 1969; Nilsson and Weibull, 1981; Smith, 1977), yellow tuft (*Sclerophthora macrospora* [Sacc.] Thirum., Shaw, Naras) (Jackson, 1977), *Ophiobolus* patch (*Ophiobolus graminis* Sacc.) and dollar spot (*Sclerotinia homeocarpa* F.T. Bennett) (Eschauzier, 1969).

Table 11.1. *Agrostis capillaris* cultivar and germ plasm accessions listed and/or maintained in the United States at the National Plant Germplasm System (NPGS) Western Regional Plant Introduction Station, the Non-NPGS Security Backup Collection (NSSB) or the Plant Germplasm Quarantine Office (PGQO).[a]

Accession	Synonym and/or Site Identifier	Collection (c) and/or Development Sites (d)	Comments;	[Year Submitted]
PS 101101			1	[1932]
PI 111985			1	[1935]
PI 171470	W6 6924	Turkey (c)	2, 3	[1948]
PI 172698	W6 6925	Turkey (c)	2, 3	[1949]
PI 204397	W6 6926	Turkey (c)	2, 3	[1953]
PI 206626	W6 6927	Turkey (c)	2, 3	[1953]
PI 234685	W6 6928	Denmark (c)	2, 3	[1956]
PI 235217	W6 6929		2, 3	[1956]
PI 237717	'ODENWALDER' W6 6930	Germany (d)	2, 3	[1957]
PI 252045	W6 6931	Italy (c)	2, 3	[1958]
PI 283173	W6 6932	Czechoslovakia (c)	2, 3, 4	[1962]
PI 290708	W6 6933		2, 3	[1963]
PI 311011	W6 6934	Romania (c)	2, 3	[1966]
PI 325192	S-190 W6 6935	Stavropol, Russian Federation (c)	2, 3, 5	[1968]
PI 325194	S-157 W6 6935	Stavropol, Russian Federation (c)	2, 3, 5	[1968]
PI 392338	W6 6937	former Soviet Union (c)	2, 3	[1974]
PI 420235	'GOGINAN' W6 6938	England (d)	2, 3	[1977]
PI 420236	'PARYS' W6 6939	England (d)	2, 3	[1977]
PI 440109	D-1327 W6 6940	Russian Federation (c)	2, 3, 5, 6	[1982]
PI 469217	'HIGHLAND' W6 6941	Oregon (c, d)	2, 3, 7	[1982]
PI 491264	HJA 166 W6 6942	Finland (c, d)	2, 3, 8	[1984]
PI 494120	'TENDENZ' W6 6943	Germany (c) Russian Federation (d)	2, 3	[1984]
PI 494121	'BARDOT' V 430 W6 6944	Netherlands (c) Russian Federation (d)	2, 3	[1984]
PI 509437	W6 6945	Romania (c)	2, 3, 5	[1987]
PI 538785	AJC-065 W6 6946	Russian Federation (c)	2, 3, 5, 4	[1990]
PI 578228	'SR 7100'		2, 3, 8	[1994]
PI 578527	'ASTORIA'	Oregon (c, d)	2, 3	[1961]
PI 578528	'EXETER'	Rhode Island (c, d)	2, 3, 7	[1964]
PI 600936	'DUCHESS'	England (c, d)	2, 3, 8	[1982]
PI 601701	'EGMONT'	New Zealand (c, d)	9, 10	[1989]
PI 606339	'TIGER'	United States (d)	9, 8, 10	[1998]
Q 40389	16	Australia (c)	11	[1999]
Q 40390	21	Australia (c)	11	[1999]
Q 40391	27	Australia (c)	11	[1999]
Q 40392	38	Australia (c)	11	[1999]
W6 17688	'BARDOT'		3, 12	[1996]
W6 19421	B96-258	Bulgaria (c)	3, 12	[1996]

(continued)

Table 11.1. (continued)

Notes: 1. Historical record only; no seed is available; 2. 250 grams of seed may be requested from the National Genetic Resources Program; 3. Maintained at the Western Regional PI Station (W6); 4. Originally classified as *A. vulgaris;* 5. Wild material; 6. Originally classified as *A. gigantea;* 7. Considered by some sources to be *A. castellana;* 8. Further descriptive information and ownership of this cultivar is detailed in Table 11.3; 9. Contact the accession owners for available seed; 10. Maintained at the Non-NPGS Security Backup Collection (NSSB); 11. Maintained at the Plant Germplasm Quarantine Office (PGQO). Contact PGQO for availability status; 12. Material not available at this time. Contact the maintenance site for more information.
[a] Information for this table was obtained primarily from USDA-ARS (2001).

A. capillaris improvement must include enhanced stress tolerance to heat, drought, and wear. Reduced thatching tendency and sensitivity to herbicides would also be desirable (Emmons, 1995; Ruemmele, 2000). Differences between *A. capillaris* cultivars regarding herbicide sensitivity were documented (Fisher and Wright; 1977; Johnston and Fisher, 1985). Duchess was found to be less sensitive to a weed herbicide, aminotriazole, compared to Bardot by the end of the third selection cycle (Lee and Wright, 1981).

Good rhizome development with no stolon growth would also be desirable (Meyer and Funk, 1989). Such enhancement will increase usefulness of *A. capillaris* in mixtures, as well as increase recuperative potential.

Current and Potential Breeding Techniques

Breeding efforts for *A. capillaris* have centered on conventional methods of germ plasm collection, via screening for desirable turf traits under greenhouse and field conditions. Synthetic polycrosses of selected phenotypes undergo one or more rounds of selection for refinement of turf qualities and seed yield (Funk, 1998). Assessment under managed conditions occurs for one or more years prior to determination of cultivar release. Although the value of inbreeding was demonstrated (Dudeck and Duich, 1967; Perkins, 1969), it is not utilized widely at the present time.

Due to the spate of conventional or molecular genetic experiments, phenotypic selection is the norm for this species. With increased interest from university and commercial concerns, molecular technology should become a factor in *A. capillaris* cultivar development.

Clark et al. (1989) used polyacrylamide gel electrophoresis of seed proteins to distinguish among and within *Agrostis* species. Highland banding differed significantly from Bardot and Exeter cultivars of *A. capillaris* in two band locations. Unique banding patterns between the latter two cultivars permitted distinction of all three cultivars. The strong difference in the Highland banding pattern lends support to the assertion that Highland does not belong in the same species grouping with other *A. capillaris* cultivars.

Early tissue culture of *A. capillaris* produced callus that was able to be maintained as callus (Atkin and Barton, 1973). Krans (1981) initiated *A. capillaris* callus from the embryo region and along the coleoptile up to the first node. Shoots, but no roots were regenerated from this callus. A complete somatic tissue culture system from seed explants was developed at The Pennsylvania State University (Duich, 1988). Entire *A. capillaris* plants were also regenerated from callus induced from seed in University of Rhode Island Research (2001, unpublished data).

Success with marker-assisted selection and plant transformation of other species in this genus (*A. canina* and *A. stolonifera* [University of Rhode Island, 2001, unpublished data]) provide encouragement that molecular technology could be successfully applied to *A. capillaris* germ plasm.

The varied response of *A. capillaris* germ plasm to environmental conditions related to soil acidity, fertility, and irrigation affects widespread use of cultivars across diverse environments. This has been especially reflected by the restricted adaptability of current cultivars to limited regions within the United States.

Annual flowering and the resultant seed ripening in *A. capillaris* occurs about the same time as *A. gigantea*, earlier than *A. stolonifera*, and later than *A. canina* (North and Odland, 1935). While vernalization of at least three weeks may be necessary for floral induction in *A. capillaris* (C.R. Skogley, 1992, personal communication), some germ plasm has been observed to flower under greenhouse conditions without vernalization. Flower induction was enhanced using up to 84 days of alternating light (8 hours, 10°C) and dark (16 hours, 4.5°C), with one seed cycle generation possible every 20 to 25 weeks (Nelson and Duich, 1988). Cooper and Calder (1962, 1964) observed maximum *A. capillaris* flowering was induced by a short photoperiod with cool temperatures (approximately 5–10°C). They found no floral induction response to cold temperatures.

Wipff and Fricker (2000) report *A. capillaris* freely hybridizes with *A. canina*, *A. castellana*, *A. gigantea*, *A. stolonifera*, and *A. vinealis* Schreber (brown bentgrass), with various degrees of fertility from sterility to the same degree of fertility as the parents. Interspecific hybridization of *A. capillaris* with three other *Agrostis* species produced sterile hybrids (Hubbard, 1984). The species included *A. gigantea*, *A. stolonifera*, and *A. canina*. Bradshaw (1958) found relatively sterile F1 hybrids from natural hybridizations between *A. capillaris* and *A. stolonifera*.

Examples of Breeding Progress

Prior to 1990, limited breeding efforts were described for *A. capillaris*. Germany (Entrup, 1969) and the Netherlands (Eschauzier, 1969) produced one and five cultivars, respectively.

Panella (1977) reported ecotype assessment of Italian *A. capillaris* germ plasm collected near Perugia commenced in 1970 at Italy's Forage Breeding Center of the Italian Research Council operating at the Plant Breeding Institute of the University of Perugia. No known cultivars were introduced from this program.

Several private seed companies in the United States and Europe are actively engaged in *A. capillaris* improvement. Companies with recently-released cultivars or germ plasm include: Cebeco International Seeds, Pure Seed Testing, and Seed Research of Oregon in the United States. A more limited number of university-sponsored programs are active in *A. capillaris* breeding, including Rutgers University, the University of Rhode Island, and the University of Wisconsin. Selected progress is detailed below.

Pure Seed Testing *A. capillaris* breeding work (Anon., 1998) includes fairway trials established in 1994 in Oregon and maintained at 1.6 cm cutting height. Turf performance and take-all patch (*Gaeumannomyces graminis* [Sacc.] Arx & D. Olivier var. *arvenae* [E.M. Turner] Dennis) resistance were evaluated among the germ plasm collected from throughout the United States.

Additional turf trials were initiated at Rutgers University, North Carolina, and Oregon for brown patch disease resistance, close mowing tolerance, and seed yield assessment (Anon., 1998). Table 11.2 describes Pure Seed Testing experimental germ plasm

Table 11.2. Pure Seed Testing, Inc. experimental *A. capillaris* germ plasm undergoing evaluation for cultivar potential.[a]

Synthetic Designation	Comments
9HG	aggressive; good fall and spring density; from Rutgers University collections
9456	bright green color; derived from two collections from England
9SA, 9ET, 98L	polycrosses assembled from more than 20,000 plants screened in the greenhouse for salinity tolerance
9FL	survivors of plot flooding which had the best turf quality
9TR, 9LH	good summer performance

[a] Information for this table was derived from the Turf-Seed, Inc. website: www.turf_seed.com/research/col_bent.asp.

and the strength(s) of each. One synthetic (9TR) showed significantly better general turf quality compared to commercially-available SR 7100.

The Rhode Island Agricultural Experiment Station *A. capillaris* breeding program released Exeter in 1963, though some now classify this as *A. castellana* (Brilman, 2001). Germ plasm collection continues from naturalized sites throughout New England and the Midwest. This material is evaluated for turfgrass stress tolerances as well as seed yield potential.

Current breeding efforts identified germ plasm with enhanced *R. solani* resistance collected from throughout New England over the past 10 years (Zeng, 1995). This material is undergoing evaluation for overall turfgrass quality and seed yield in open-pollinated space plantings.

The National Turfgrass Evaluation Program (Beltsville, MD) began evaluations of cultivar and experimental germ plasm from bentgrass species in 1989. Plots mown at typical fairway and greens heights included one or more *A. capillaris* cultivars or experimental germ plasm. Each cycle of testing spans a maximum of five years. Three *A. capillaris* cultivars were included in the 1989 greens and fairway trials. *A. capillaris* cultivar testing was limited to fairway testing in 1993 and 1998 at the discretion of entry sponsors. Entry numbers of *A. capillaris* germ plasm increased to five in 1993. The most recent evaluation, initiated in 1998, contains eight *A. capillaris* entries submitted for comparison to *A. stolonifera* and *A. idahoensis* Nash. (Idaho redtop) under fairway management.

Although results vary among test sites, the University of Rhode Island location in Kingston, Rhode Island, ranked three of the experimental *A. capillaris* cultivars in the highest group in overall quality in the second year of evaluation, surpassing commercial *A. capillaris* cultivars, Tiger and SR 7100 (Anon., 2001).

During these years of NTEP evaluation, commercial seed companies from both Europe and North America sponsored these *A. capillaris* entries. Some *A. capillaris* cultivars released within the past 20 years are described in Table 11.3. While improvements have been introduced by the release of these cultivars, considerable potential remains for exploitation of *A. capillaries* germ plasm.

Interest is increasing regarding the possibilities for producing improved *A. capillaris* cultivars in both North America and Europe. Recent cultivar releases exhibit performance superior to older cultivars. As cultural inputs continue to remain restricted, the value of cultivar development within this species for reduced-management situations will increase.

Table 11.3. Selected *Agrostis capillaris* cultivar release descriptions, including ownership and/or improvements over older cultivars.[a]

Name	Ownership/PVP Number	Improved Traits
HJA 166	Hankkija Plant Breeding Institute	early spring greenup; good overwintering ability; drought resistant
'SR 7100'	Seed Research of Oregon, Inc.	medium dark green color; very fine texture; upright growth pattern that does not produce a false crown; excellent dollar spot resistance; improved brown patch resistance
'DUCHESS'	Germinal Holdings Ltd.; PVP 8200081	cold resistance; resistant to Fusarium patch, *Corticium fuciforme*, and powdery mildew diseases; resistant to the known lethal dose of the herbicide, aminotriazole
'TIGER'	Cebeco International Seeds, Inc.; PVP 9800388	increased brown patch resistance in 1993 University of Rhode Island NTEP plots.

[a] Some information from: USDA-ARS, National Genetic Resources Program. 2001 *Germplasm Resources Information Network (GRIN)*. [Online Database] National Germplasm Resources Laboratory, Beltsville, MD. Available: www.ars-grin.gov/.

REFERENCES

Adams, W.A. 1977. Effects of nitrogen fertilization and cutting height on the shoot growth, nutrient removal, and turfgrass composition of an initially perennial ryegrass dominant sports turf. pp. 343–350. In *Proceedings Third International Turfgrass Conference*. J.B Beard, Ed. International Turfgrass Society and ASA and CSSA, Madison, WI.

Anon. 1998. Colonial bentgrass breeding. Turf-Seed, Inc. website: www.turf_seed.com/research/col_bent.asp.

Anon. 2001. National Turfgrass Evaluation Program (NTEP) 1998 National Bentgrass (Fairway/Tee) Test 2000 Data. Progress Report NTEP No. 01-2. National Turfgrass Evaluation Program. Beltsville, MD.

Atkin, R.K. and G.C. Barton. 1973. The establishment of tissue culture of temperate grasses. *J. Exp. Bot.* 24:689–699.

Beard, J.B. 1969a. Turfgrass shade adaptation. pp. 273–282. In *Proceedings First International Turfgrass Conference*. International Turfgrass Society and Sports Turf Research Institute, Bingley, Yorkshire, England.

Beard, J.B. 1969b. Winter injury of turfgrasses. pp 226–234. In: *Proceedings First International Turfgrass Conference*. International Turfgrass Society and Sports Turf Research Institute, Bingley, Yorkshire, England.

Beard, J.B. 1973. *Turfgrass: Science and Culture*. Prentice-Hall, Inc.: Englewood Cliffs, NJ.

Boeker, P. 1974. Root development of selected turfgrass species and cultivars. pp. 55–61. E.C. Roberts, Ed. In *Proceedings Second International Turfgrass Conference*. International Turfgrass Society and ASA and CSSA, Madison, WI.

Bonos, S.A., J.A. Murphy, W.A. Meyer, B.B. Clarke, K.A. Plumley, W.K. Dickson, J.B. Clark, J.A. Honig, and D.A. Smith. 1999. Performance of bentgrass cultivars and selections in New Jersey Trials. pp. 33–48. A.B. Gould and B.B. Clarke, Eds. In *Rutgers Turfgrass Proceedings, 1998*, Vol. 30. Rutgers Center for Turfgrass Science and New Jersey Turfgrass Association, New Brunswick, NJ.

Bonos, S.A., E. Watkins, J.A. Honig, M. Sosa, T. Molnar, J.A. Murphy, and W.A. Meyer. 2001. Breeding cool-season turfgrasses for wear tolerance using a wear simulator. *International Turfgrass Society Research Journal* 9:137–145

Bourgoin, B., C. Billot, M. Kerguelen, A. Hentgen, and P. Mansat. 1974. Behavior of turfgrass species in France. pp. 35–40. E.C. Roberts, Ed. In *Proceedings Second International Turfgrass Conference*. International Turfgrass Society and ASA and CSSA, Madison, WI.

Bourgoin, B. and P. Mansat. 1981. Artificial trampling and player traffic on turfgrass cultivars. pp. 55–63. R.W. Sheard, Ed. In *Proceedings Fourth International Turfgrass Conference*. International Turfgrass Society and Ontario Agricultural College, University of Guelph, Guelph, Ontario, Canada.

Bradshaw, A.D. 1958. Natural hybridization of *Agrostis tenuis* Sibth. and *A. stolonifera* L. *New Phytologist* 57:68–84.

Brilman, L.A. 2001. Colonial bentgrass: An option for fairways. *Golf Course Management* 69(1):55–60.

Brown, B.A. 1932. The effects of fertilization on the chemical composition of vegetation in pastures. *J. Am. Soc. Agron.* 24:129–145.

Carroll, J.C. 1943. Effects of drought, temperature, and nitrogen on turfgrasses. *Plant Physiol.* 18:19–36.

Chase, A. 1971. Manual of the Grasses of the United States. USDA Miscellaneous Publication No. 200, Volume 2, 2nd Edition. General Publishing Co., Ltd. Toronto, Ontario, Canada. pp. 341 and 809.

Chesnel, A., R. Croise, and B. Bourgoin. 1977. Tree shade adaptation of turfgrass species and cultivars in France. pp. 431–436. J.B Beard, Ed. In *Proceedings Third International Turfgrass Conference*. International Turfgrass Society and ASA and CSSA, Madison, WI.

Christians, N. 1998. *Fundamentals of Turfgrass Management*. Ann Arbor Press, Chelsea, MI, pp. 46–47.

Clark, K.W., A. Hussain, K. Bamford, and W. Bushuk. 1989. Identification of cultivars of *Agrostis* species by polyacrylamide gel electrophoresis of seed proteins. pp. 121–125. H. Takatoh, Ed. In *Proceedings Sixth International Turfgrass Research Conference*. International Turfgrass Society and the Japanese Society of Turfgrass Science, Tokyo, Japan.

Cooper, J.P. and D.M. Calder. 1962. pp. 20–22. Flowering responses of herbage grasses. Report of the Welsh Plant Breeding Station for 1961.

Cooper, J.P. and D.M. Calder. 1964. The inductive requirements for flowering of some temperate grasses. *J. of the British Grasslands Society* 17:6–14.

Davis, R.R. 1958. The effect of other species and mowing height on persistence of lawn grasses. *Agron. J.* 50:671–673.

deChevigny, Y. and J. Dujardin. 1981. Experimental studies on turf for racing horses. pp. 27–33. R.W. Sheard, Ed. In *Proceedings Fourth International Turfgrass Conference*. International Turfgrass Society and Ontario Agricultural College, University of Guelph, Guelph, Ontario, Canada.

Dudeck, A.E. and J.M. Duich. 1967. Preliminary investigation on the reproductive and morphological behavior of several selections of colonial bentgrass *Agrostis tenuis* Sibth. *Crop Sci.* 7:605–610.

Duich, J.M. 1988. Bentgrass breeding. p. 26. In *1988 Turfgrass Research Summary*. United States Golf Association. Far Hills, NJ.

Duyvendak, R. and H. Vos. 1974. Registration and evaluation of turfgrasses in the Netherlands. pp. 62–73. E.C. Roberts, Ed. In *Proceedings Second International Turfgrass Conference*. International Turfgrass Society and ASA and CSSA, Madison, WI.

Elmore, C.L., C. Wilen, D.W. Cudney, and V. Gibeault. 1997. UC IPM Pest Management Guidelines P: Turfgrass. UC DANR Publication 3365-T.

Emmons, R.D. 1995. *Turfgrass Science and Management*. 2nd Edition. Delmar Publishers. Albany, NY. pp. 75–76, 83.

Entrup, E.L. 1969. Turfgrass breeding in West Germany. pp 65–69. In *Proceedings First International Turfgrass Conference*. International Turfgrass Society and Sports Turf Research Institute, Bingley, Yorkshire, England.

Eschauzier, W.A. 1969. Turfgrass breeding in the Netherlands. pp. 70–79. In *Proceedings First International Turfgrass Conference*. International Turfgrass Society and Sports Turf Research Institute, Bingley, Yorkshire, England.

Escritt, J.R. 1969. Turfgrass conditions in Britain. pp. 14–16. In *Proceedings First International Turfgrass Conference*. International Turfgrass Society and Sports Turf Research Institute, Bingley, Yorkshire, England.

Fisher, R. and C.E. Wright. 1977. The breeding of lines of *Agrostis tenuis* Sibth. and *Festuca rubra* L. tolerant of grass-killing herbicides. pp. 11–18. J.B Beard, Ed. In *Proceedings Third International Turfgrass Conference*. International Turfgrass Society and ASA and CSSA, Madison, WI.

Funk, C.R. 1998. Opportunities for the genetic improvement of underutilized plants for turf. pp. 23–26. *Proceedings of the Seventh Annual Rutgers Turfgrass Symposium*, Center for Turfgrass Science, Cook College, Rutgers, The State University of New Jersey, New Brunswick, NJ.

Hartmann, H.T., A.M. Kofranek, V.E. Rubatzky, and W.J. Flocker. 1988. *Plant Science Growth, Development, and Utilization of Cultivated Plants*. Prentice-Hall, Upper Saddle River, NJ, p. 460.

Hartwell, B.L. and S.C. Damon. 1917. The persistence of lawn and other grasses as influenced especially by the effect of manures on the degree of soil acidity. *Rhode Island Agric. Exp. Stn., Bull.* 1970.

Henensal, P., G. Arnal, and J. Puig. 1977. Research into the establishment of roadside embankments. pp. 391–400. J.B Beard, Ed. In *Proceedings Third International Turfgrass Conference*. International Turfgrass Society and ASA and CSSA, Madison, WI.

Hubbard, J.C.E. 1984. *Grasses: A Guide to Their Structure, Identification, Uses, and Distribution in the British Isles*. 3rd ed. Viking Penguin, Inc. New York, pp. 298–299, 372.

Jackson, N. 1977. Yellow tuft disease of turfgrasses: A review of recent studies conducted in Rhode Island. pp. 265–270. J.B Beard, Ed. In *Proceedings Third International Turfgrass Conference*. International Turfgrass Society and ASA and CSSA, Madison, WI.

Johnston, D.T. and R. Fisher. 1985. Grass weed control in lawns of aminotriazole tolerant vultivars of *Festuca nigrescens* and *Agrostis tenuis*. pp. 727–734. F. Lemaire, Ed. In *Proceedings Fifth International Research Conference*. International Turfgrass Society and Institute National de la Recherche Agronomy, Paris, France.

Jones, K. 1953. The cytology of some British species of *Agrostis* and their hybrids. *British Agricultural Bulletin* 5. p. 316.

Jones, K. 1956. Species differentiation in *Agrostis* II. The significance of chromosome pairing in the tetraploid hybrids of *Agrosts canina* subsp. Montana Hartm., *A. tenuis* Sibth. and *A. stolonifera* L. *J. Genetics*. 54:377–393.

Kamps, M. 1969. Effects of real and simulated play on newly-sown turf. pp. 118–123. In *Proceedings First International Turfgrass Conference*. International Turfgrass Society and Sports Turf Research Institute, Bingley, Yorkshire, England.

Krans, J.V. 1981. Cell culture of turfgrasses. pp. 27–33. R.W. Sheard, Ed. In *Proceedings Fourth International Turfgrass Conference*. International Turfgrass Society and Ontario Agricultural College, University of Guelph, Guelph, Ontario, Canada.

Lee, H. and C.E. Wright. 1981. Effective selection for aminotriazole tolerance in *Fesctua* and *Agrostis* Turf Grasses. pp. 41–46. R.W. Sheard, Ed. In *Proceedings Fourth International Turfgrass Conference*. International Turfgrass Society and Ontario Agricultural College, University of Guelph, Guelph, Ontario, Canada.

Lynch, S., Ed. 1996. Seed Specification Manual. Seed Research of Oregon, Inc. Corvallis, OR.

Madison, J.H. 1971. *Practical Turfgrass Management*. Van Nostrand Reinhold Co., New York.

Meyer, W.A. and C.R. Funk. 1989. Progress and benefits to humanity from breeding cool-season grasses for turf. pp. 31–48. D.A. Sleper, K.H. Asay, and J.F. Pedersen, Eds. In *Contributions from Breeding Forage and Turf Grasses*. Crop Science Society of America Special Publication Number 15. Madison, WI.

Musser, H.B. 1948. Effects of soil acidity and available phosphorus on population changes in mixed Kentucky bluegrass - bent turf. *J. Am. Soc. Agron.* 40:614:620.

Nelson, E.K. and J.M. Duich. 1988. Floral induction and flowering of 'Astoria' and 'Highland' bentgrasses. *Agron. Abst.* 80:154.

Nilsson, C. and P. Weibull. 1981. Selection for improved resistance to *Fusarium nivale* (Fr.) Ces. pp. 21–25. R.W. Sheard, Ed. In *Proceedings Fourth International Turfgrass Conference*. International Turfgrass Society and Ontario Agricultural College, University of Guelph, Guelph, Ontario, Canada.

North, H.F.A. and T.E. Odlund. 1935. The relative seed yields in different species and varieties of bent grass. *J. Am. Soc. Agron.* 27:374–383.

Oohara, Y. 1977. Some ecological observations on turf establishment and culture of turfgrasses in cool regions of Japan. pp. 419–422. J.B. Beard, Ed. In *Proceedings Third International Turfgrass Conference*. International Turfgrass Society and ASA and CSSA, Madison, WI.

Panella, A. 1977. Observations on differently adapted grasses for turf in central Italy. pp. 413–417. J.B Beard, Ed. In *Proceedings Third International Turfgrass Conference*. International Turfgrass Society and ASA and CSSA, Madison, WI.

Perkins, A.T. 1969. Inbreeding and Pathogen Inoculation Response of Colonial Bentgrass, *Agrostis tenuis* Sibth. Ph.D. Dissertation, Pennsylvania State University, College Park, PA.

Pool, R.J. 1948. *Marching with the Grasses*. University of Nebraska Press. Lincoln, NE.

Robinson, G.S. 1969. Greenkeeping problems in New. Zealand. pp. 10–12. In *Proceedings First International Turfgrass Conference*. International Turfgrass Society and Sports Turf Research Institute, Bingley, Yorkshire, England.

Ruemmele, B. 2000. Breeding colonial bentgrass for drought, heat, and wear. *Diversity* 16:34–35.

Shildrick, J.P. 1975. Turfgrass mixtures under wear treatments. *J. Sports Turf Res. Inst.* 51:9–40.

Shildrick, J.P. 1976. Highland bent: A taxonomic problem. *J. Sports Turf Res. Inst.* 52:142–150.

Shildrick, J.P. 1977. Turfgrass seed mixtures in the United Kingdom. pp. 57–64. J.B Beard, Ed. In *Proceedings Third International Turfgrass Conference*. International Turfgrass Society and ASA and CSSA, Madison, WI.

Skirde, W. 1969. Reaction of turfgrasses to watering. pp. 311–322. In *Proceedings First International Turfgrass Conference*. International Turfgrass Society and Sports Turf Research Institute, Bingley, Yorkshire, England.

Skogley, C.R. 1977. Influence of fertilizer rate, mower type, and thatch control on colonial bentgrass lawn turf. pp. 337–342. J.B. Beard, Ed. In *Proceedings Third International Turfgrass Conference*. International Turfgrass Society and ASA and CSSA, Madison, WI.

Smith, J.D. 1977. Snow mold resistance in turfgrasses and the need for regional testing. pp. 275–282. J.B Beard, Ed. In *Proceedings Third International Turfgrass Conference*. International Turfgrass Society and ASA and CSSA, Madison, WI.

Trautmann, W. and W. Lohmeyer. 1977. Studies on the development of turfgrass seedings on roadsides in the Federal Republic of Germany. J.B Beard, Ed. pp. 401–405. In *Proceedings Third International Turfgrass Conference*. International Turfgrass Society and ASA and CSSA, Madison, WI.

Turgeon, A.J. 1987. *Agrostis tenuis* Sibth. Cooperative Extension Service Circular 1105. University of Illinois at Urbana-Champaign.

USDA. 1948. *Grass: The Yearbook of Agriculture*. U.S. Government Printing Office. Washington, DC, p. 648.

USDA-ARS National Genetic Resources Program. *2001. Germplasm Resources Information Network (GRIN)*. [Online Database] National Germplasm Resources Laboratory, Beltsville, Maryland. Available: www.ars_grin.gov/cgi_bin/npgs/html/tax_search.pl?colonial+bentgrass.

USDA-NRCS. 2001. The PLANTS Database, Version 3.1 (plants.usda.gov) National Plant Data Center, Baton Rouge, LA.

Vos, H. and Scheijgrond. 1969. Varieties and mixtures for sports turf and lawns in the Netherlands. pp. 34–44. In *Proceedings First International Turfgrass Conference*. International Turfgrass Society and Sports Turf Research Institute, Bingley, Yorkshire, England.

Watson L. and M.F. Dallwitz. 1992. *The Grass Genera of the World*. CAB Publications, U.K. pp. 223–986.

Wipff, J.K. and Fricker, C.R. 2000. Determining gene flow of transgenic creeping bentgrass and gene transfer to other bentgrass species. *Diversity* 16:36–39.

Zeng, P. 1995. Resistance of Colonial Bentgrass (*Agrostis tenuis*) to Brown Patch (*Rhizoctonia solani*). M.S. thesis, University of Rhode Island, Kingston, RI.

Chapter 12

Velvet Bentgrass (*Agrostis canina* L.)

L. A. Brilman

Velvet bentgrass, *Agrostis canina* (L.), is a cool-season, extremely fine-textured, stoloniferous grass from Europe. New cultivars are currently being developed to take advantage of the high quality turf with reduced inputs demonstrated by this species. Velvet bentgrass has been recognized for many years as forming the most beautiful turf surface due to its fine texture and high shoot density (DeFrance et al., 1952; Sprague and Evaul, 1930). In 1932, 10 professional golfers were invited to putt on the greens at the Arlington Turf Gardens in Virginia. In this trial that included the top vegetative creeping bentgrasses (*A. stolonifera* L. also known as *A. palustris* Huds.), colonial bentgrasses (*A. capillaris* L., previously misclassified as *A. tenuis*), and velvet bentgrasses, all the professionals chose the velvet bentgrass as the best putting surface (Monteith and Welton, 1932).

Sprague and Evaul (1930) described the turf as luxurious green velvet. Common perception, however, has been that this is a high maintenance turf species with its use restricted to temperate-oceanic climates, as in New England and the Pacific Northwest areas of the United States (Turgeon, 1996; Christians, 1998). This common perception has limited its use in much of the United States, although Beard (1973) noted the species has good heat and low temperature tolerance and the drought tolerance is better than other bentgrass species. Other desirable traits were demonstrated in trials by Reid (1932) and North and Odland (1934) showing velvet bentgrass is more tolerant of acidic soils than other bentgrasses. In addition, Reid (1933) noted that velvet bentgrass was the most shade tolerant of the bentgrasses. Sprague and Evaul (1930) observed that velvet bentgrass does not spread as rapidly as some other bents, but is very persistent once established.

A. canina until recently was divided into two subspecies, ssp. *canina* (also previously called var. *fascicularis*) or velvet bentgrass, and ssp. *montana* (also previously called ssp. *arida*) or brown bentgrass. The latter is now recognized as the species *A. vinealis*. This species, however, appears to be an autotetraploid (2n=4x=28) of ssp. *canina*, and is often difficult to separate in field specimens (Jones, 1953). Velvet bentgrass can be distinguished from other species of *Agrostis* by having a longer, more pointed ligule and a shorter palea than either *A. capillaris* or *A. stolonifera* (Hubbard, 1984). It belongs to the sect. *Trichodium* of *Agrostis* distinguished by a lemmatal or trichodium net, in which the epidermal cells of the lemma show a reticulate pattern at high magnification from thickenings in the outer walls of the cells (Bjorkman, 1960; Carlbom, 1967). *A. vinealis* has rhizomes, while *A. canina* does not produce rhizomes (Hubbard, 1984). This difference is correlated with ecological preferences; velvet bentgrass occurs naturally in wet and damp soils, while brown bentgrass is found on heaths and upland grounds (Davies, 1953).

DISTRIBUTION AND CYTOTAXONOMY

Classification of velvet bentgrass is described by Soreng et al. (2001).

Family	Poaceae
Subfamily	Pooideae
Tribe	Poeae
Subtribe	Agrostidinae
Genus	*Agrostis*
Section	Trichodium
Species	*canina*
Authority	Linneaus

Velvet bentgrass shows one of the widest distributions of *Agrostis* species. It occurs in all of western Europe, widely over the European portion of the former Soviet Union and reaches into eastern Asia. It is at considerable less density in eastern Siberia and does not reach the far Orient (Sokolovskaya, 1938). Sokolovskaya (1938) also describes another species, *A. Trinii*, which he states occurs in Siberia and the Orient and has both diploid and tetraploid forms. This is now recognized as part of *A. vinealis*. It may be that the diploid types he called *A. Trinii* should be included in *A. canina* and thus expand the range. He also states that velvet bentgrass does not occur above 65 to 70° N latitude. Davies (1953) states velvet bentgrass occurs throughout Europe and Asia from the Caucasus and Himalayas northward. Other sources, such as The PLANTS database (USDA-NRCS, 2001) and Hitchcock (1950) suggest velvet bentgrass may be a native North American species in at least portions of its American distribution. Many of the other species in sect. *Trichodium* are native to the United States (Carlbom, 1967), where they show the greatest distribution.

Velvet bentgrass is a typical diploid with $x = 7$ and $2n = 2x = 14$ (Bjorkman, 1954) primarily having only bivalents in the first metaphase of meiosis and 1.35 chiasmata per cell (Jones, 1953, 1956a). *A. vinealis* is an autotetraploid ($2n=4x=28$) with quadrivalent formation (Jones 1953, 1956a). Jones (1956a) reported a mean of 2.66 quadrivalents per cell (range of 0 to 7) and 1.67 chiasmata per cell. Triploid hybrids between these two species confirmed that *A. vinealis* is an autotetraploid of *A. canina*. The hybrids showed up to seven trivalents per cell, with usually not more than seven bivalents, although eight bivalents were present in a few cases and some cells had more than seven associations, demonstrating structural heterozygosity of *A. vinealis*.

Jones (1956b) showed chromosome pairing in hybrids between *A. stolonifera* and *A. vinealis* was poor and variable, and the hybrids are sterile. Chromosome pairing in hybrids of *A. capillaris* (*A. tenuis* in Jones's paper) and *A. vinealis* was practically complete and hybrid fertility was high in the F_1, F_2 and backcross generations. It was concluded that *A. capillaris* is a segmental allopolyploid with one ancestor from *A. canina*. Velvet bentgrass has been given the genomic designation of A_1A_1. Brown bent has been given the corresponding genomic designation of $A_1A_1A_1A_1$. Due to the high homology between *A. vinealis* and *A capillaris* it has also been proposed that *A vinealis* originated as an autotetraploid but due to outcrosses with *A. capillaris* it now has a similar genomic structure to the later species (Wipff and Fricker, 2001). Further work on the genomic structure of the *Agrostis* genus should be done using new techniques and a broad range of ecotypes.

Early turf researchers in the United States produced self-pollinated seed from superior vegetative types of *A. canina* they had collected from old turf and evaluated the progeny for uniformity and plant characteristics (DeFrance et al., 1952). Davies (1953) in studying self- and cross-fertility in *Agrostis* found that *A. canina* in 43 isolations produced 0 seedlings on 20 inflorescences, 1 seedling on 16, 2 seedlings on 3, and 5 seedlings on 4 inflorescences. These plants produced significantly more seed when open pollinated, demonstrating that the species is typically self-sterile. *A. vinealis* gave similar results, demonstrating the normal cross-pollinated nature of the species.

COLLECTION, SELECTION, AND BREEDING HISTORY

Velvet bentgrass was introduced to many old golf courses and other turf sites in the United States primarily from seed harvested from mixed stands of bentgrass species in Europe and sold as South German bent. Piper (1918) stated that South German bentgrass usually consisted of about 75% Rhode Island bentgrass (*A. capillaris*, now known as colonial bent), 15% velvet bentgrass, and 1% creeping bentgrass (*A. stolonifera*), besides many impurities. Recent collections of bentgrasses in Europe, plus new European cultivars, suggest most of the bentgrasses from this seed were not adapted to many regions of the United States. Over time the few adapted ecotypes were selected out and spread vegetatively. Certain superior genotypes were selected by early golf course superintendents and turf researchers and increased vegetatively for planting on additional golf courses. These included the cultivars Piper, formerly called B.P.I. 14,276 selected by the U.S. Golf Association (USGA) Greens Section; Merion, a selection from the Merion Cricket Club, Ardmore, PA; and Kernwood, a selection from the Kernwood Country Club, Salem, MA (DeFrance et al., 1952). Other sources of early selections included Mountain Ridge from the USDA; Highland velvet; Yorkshire, from Yorkshire, England; Newport velvet, from Washington Co., OR; Valentine No. 2 from Mr. Joseph Valentine of the Merion Cricket Club, Haverford, PA (may be the same as Merion above); Acme, from the USGA Green Section and USDA; Wykagyl from Wykagyl Country Club in New Rochelle, NY; Nichol Ave. Nos. 1 and 2, from Rutgers University, NJ; Cunningham, from the USGA Green Section; and Elizabeth from Elizabeth, NJ (North and Odland, 1934). In the 1930s Rhode Island farmers also produced some seed from Piper and Kernwood (North et al., 1938), which may have been an additional source of velvet bentgrasses still found on older golf courses. Early reports show a lack of reliable seed supply was the primary reason velvet bentgrass was not used more extensively (Sprague and Evaul, 1930; North et al., 1938). Based on these early reports the velvet bentgrass seed produced in North America primarily came from naturalized stands in Oregon and Prince Edward Island, Canada (North and Odland, 1934). The latter site may have included native ecotypes of velvet bentgrass.

In addition to evaluating vegetative velvet bentgrasses, the early researchers also produced self-pollinated seed from these vegetative types and evaluated the progeny for uniformity and plant characteristics (DeFrance et al., 1952). The utilization of self-pollinated seed is in contrast to methods usually employed today in the development of turfgrass cultivars, since most species are heterozygous, cross-pollinated, and demonstrate self-incompatibility (Davies, 1953) and inbreeding depression. In 1930 the early turf researcher H.B. Sprague planted all available commercial and named cultivars and developed the seeded cultivar Raritan, released by the New Jersey Agricultural Experiment Station in 1940 (Hanson, 1972). Unfortunately this cultivar was lost due to World War II, but Sprague thought this material was the future of turfgrass for all situations from home lawns to golf courses (C.R. Funk, 1998, personal communication). Later the seeded cultivar Kingstown (Kingston) was released by C.R. Skogley and J.A. DeFrance from an inbred selection made by H.F.A. North in 1929 (Hanson, 1972).

Seed production problems in early velvet bentgrass cultivars may have contributed to low utilization of the species. These production problems may have been due to utilization of inbred material or to crossing with other bentgrass species leading to infertile progeny, since velvet bentgrass is a diploid and the other species are tetraploids (L.A. Brilman, 1993, personal observation). Although the vegetative cultivars of velvet bentgrass exhibited a range of color (North and Odland, 1934), Kingstown was lighter green than many other bentgrasses, which may have contributed to excessive fertilization. This excessive fertilization of velvet bentgrasses and excessive watering seem to have contributed to heavy thatch development, disease development, and *Poa annua* encroachment that led to their loss at many turf sites and the misconception of velvets as high maintenance turf (Skogley, 1996, personal communication).

Little germ plasm collection and preservation has been done in this species. Only six accessions in the USDA-NPGS-GRIN system (www.ars-grin.gov/npgs/) trace to natural (noncultivated) stands, these from Iran, Somalia, and England. Until recently the only germ plasm collection and breeding work done in this species was at the University of Rhode Island. New germ plasm collections have been made by both public and private breeders and include material collected in New England, Pennsylvania, New Jersey, New York, Wisconsin, and Michigan. Breeding efforts are underway in Oregon, New Jersey, Wisconsin, Rhode Island, and Michigan. Further collections should be made in Europe and Asia to explore more completely the potential for the species. The species is of interest for breeders and geneticists both due to its low maintenance turf characteristics and since it is a diploid species. Brown bent readily forms hybrids with colonial bentgrass, which may have some level of fertility since they share one genome (Davies, 1953). Both velvet and brown bentgrass form hybrids with creeping bentgrass less readily than with each other.

Studies by Davies (1953) have shown that *Agrostis* is a long-day plant. In the conditions at Aberystwyth, Wales, velvet bentgrass had the earliest heading date of the bentgrasses studied, averaging 23.3 days from heading to anthesis. Anthesis time of the species of bentgrass studies varied from brown bentgrass being the earliest at 4:00 to 5:00 a.m. (with a secondary anthesis at 5:30 to 8:00 p.m.), followed by velvet bentgrass at 4:30 to 5:30 a.m., *A. stolonifera* at 10:00 to 11:30 a.m., *A. gigantea* at 2:30 to 3:30 p.m., and *A. capillaris* at 1:00 to 5:00 p.m. *A. canina* flowers in the coolest time of day when humidity is high, and *A. stolonifera* seems to require higher temperatures. This can provide a partial barrier to prevent interspecific crossing. Vernalization and juvenility requirements have not been determined but velvet bentgrasses that do not reach a proper size before winter, or receive no cold temperatures, will not flower the following summer (L.A. Brilman, 1997, personal observation). Some collected ecotypes show little or no flowering in Oregon and this may be due a longer daylength or cold requirement than is supplied in the Willamette Valley of Oregon. Further studies should be done on actual requirements in different clones.

For many years the only velvet bentgrass material available was vegetative clones passed from golf course to golf course. The current seeded cultivars available have usually been developed utilizing a small number of clones crossed together and evaluated for performance and uniformity in progeny trials. Current cultivar development has concentrated on documenting improvements over creeping bentgrass for various environments. Further improvements need to concentrate on *Pythium* spp. resistance as seedlings and copper spot (*Gloeocercospora sorghi*) resistance. Additional research should be done on management strategies for golf greens, fairways, and possibly lawns.

REFERENCES

Beard, J.B, 1973. *Turfgrass: Science and Culture*, Prentice-Hall, Englewood Cliffs, NJ.
Bjorkman, S.O. 1954. Chromosome studies in *Agrostis*. II. *Hereditas*, Lund. 40:254–258.
Bjorkman, S.O. 1960. Studies in *Agrostis* and related genera. Symbolae botanicae *Upsaliensis* 17:1–112.
Carlbom, C.G. 1967. A Biosystematic Study of Some North American Species of *Agrostis* L. and *Podagrostis* (Griseb.) Scribn. & Merr. Ph.D. Dissertation. Oregon State University.
Christians, N. 1998. *Fundamentals of Turfgrass Management*. Ann Arbor Press, Chelsea, MI.
Davies, W.E. 1953. The breeding affinities of some British species of *Agrostis*. *Brit. Agric. Bull.* 23:313–315.
DeFrance, J.A., T.E. Odland, and R.S. Bell. 1952. Improvement of velvet bentgrass by selection. *Agron. J.* 44:376–378.
Hanson, A.A. 1972. *Grass Varieties of the United States*. Agricultural Handbook No. 170. USDA, Agricultural Research Service, Beltsville, MD.
Hitchcock, A.S. 1950. *Manual of the Grasses of the United States*. pp. 352–353. U.S. Govt. Printing Office, Washington, D.C.
Hubbard. C. E. 1984. Grasses. *A Guide to their Structure, Identification, Uses and Distribution in the British Isles*. 3rd ed. Revised by J.C. E. Hubbard. Penguin Books USA Inc., New York.
Jones, K. 1953. The cytology of some British species of *Agrostis* and their hybrids. *Brit. Agric. Bulletin*, 1953, Vol. V, No. 23.
Jones, K. 1956a. Species differentiation in *Agrostis*. II. The significance of chromosome pairing in the tetraploid hybrids of *Agrostis canina* subspecies *montana* Hatmn. *A. tenuis* Sibth. and *A. stolonifera* L. *J. of Genetics* 56:377–393.
Jones, K. 1956b. Species differentiation in *Agrostis*. III. *Agrostis gigantea* Roth. and its hybrids with *A. tenuis* Sibth. and *A. stolonifera* L. *J. of Genetics* 56:394–399.
Monteith, J. Jr. and Welton, K. 1932. Putting tests upon bentgrasses. *Bulletin of the USGA Green Section* 12: 224–227.
North, H.F.A. and T.E. Odland. 1934. Putting green grasses and their management. *Rhode Island Agric. Exp. Stn. Bulletin* No. 245, pp. 1–44.
North, H.F.A., T.E. Odland, and J.E. DeFrance. 1938. Lawn grasses and their management. *Rhode Island Agric. Exp. Stn. Bulletin* No. 264, pp. 1–36.
Piper, C.V. 1918. The agricultural species of bent grass. Part I. Rhode Island bent and related grasses. *USDA Bulletin* No. 692. pp. 1–14.
Reid, M.E. 1932. The effects of soil reaction upon the growth of several types of bentgrasses. *Bulletin of the USGA Green Section* 12:196–212.
Reid, M.E. 1933. Effects of shade on the growth of velvet bent and Metropolitan creeping bent. *Bulletin of the USGA Green Section* 13:131–135.
Sokolovskaya, A.P. 1938. A caryo-geographical study of the genus *Agrostis*. *Cytologia* 8:452–67.
Soreng, R.J., G. Davidse, P.M. Peterson, F.O. Zuloaga, E.J. Judsiewicz, and T.S. Filgueiras. 2001. Catalogue of New World Grasses (Poaceae). mobot.mobot.org/W3T/Search/nwgc.html.
Sprague, H.B. and E.E. Evaul. 1930. Experiments with Turf Grasses in New Jersey. NJ *Ag.Exp.Stn.Bull.* No. 497.
Turgeon, A.J. 1996. *Turfgrass Management*. 4th ed. Prentice Hall, Upper Saddle River, NJ.
USDA-NRCS. 2001. The PLANTS database. (plants.usda.gov/ plants). National Plant Data Center, Baton Rouge, LA 70874-4490.
Wipff, J.K. and C. Fricker. 2001. Gene flow from transgenic creeping bentgrass (*Agrostis stolonifera* L.) in the Willamette Valley, Oregon. *International Turfgrass Society Res. J.* 9:224–242.

Chapter 13

Three Minor *Agrostis* Species: Redtop, Highland Bentgrass, and Idaho Bentgrass

A. Douglas Brede and Mark J. Sellmann

Three species comprise the principal economic grasses of the *Agrostis* genus: *A. stolonifera* L. (=*A. palustris* Huds.), *A. capillaris* L. (=*A. tenuis* Sibth.), and *A. canina* L. These three grasses are adequately reviewed elsewhere in this book. The present chapter deals with three lesser known and lesser utilized *Agrostis*:

- *A. gigantea* Roth (=*A. alba*) – redtop
- *A. castellana* Boiss. & Reuter – Highland bentgrass
- *A. idahoensis* Nash – Idaho bentgrass

Unlike the *Agrostis* of the earlier chapters, the three *Agrostis* discussed herein have received only minimal cytogenetic study and breeding improvement. Therefore, we will spend as much time discussing their *potentials* for improvement as their current realities. At the conclusion of this chapter, we mention several additional *Agrostis* species that either have potential for breeding improvement, or which intercross with the foregoing grasses and thus may have utility in interspecific hybridization.

THE *AGROSTIS* GENUS

The *Agrostis* genus ranks among the earliest plants classified into species. Linneaus described 12 *Agrostis* in his 1753 text and 5 in his second edition of Flora Suecica (Linneaus, 1755), including *A. canina* and *A. stolonifera*. Linneaus grouped the *Agrostis* based on the presence or absence of awns, panicle shape and color, and orientation of culms and shoots. No mention was made of differences in floret characters, ligules, or the presence of rhizomes or stolons.

Many of today's grass breeders are aware that breeding populations of a single *Agrostis* species can sometimes vary in nearly all of the foregoing characters. Deviations in culm orientation and panicle shape and color are commonplace, even within a cultivar.

Linneaus originally selected no "type" species for the *Agrostis* genus (Linneaus, 1754), leaving that to later authors. Hitchcock (1905) was the first author to describe a type species for *Agrostis*, following the provisions in cannon 15, section f, of the Code of Botanical Nomenclature (*Bul. Torr. Bot. Club* 31:249). The type species he selected was

A. alba L. Unfortunately, the particular plant specimen Hitchcock picked later turned out to be *Poa nemoralis* L., which added to the early confusion. He later found and corrected the mistake in his 1920 paper (Hitchcock, 1920). Years later, Hylander (1953) asserted that the type species should belong to *A. gigantea* Roth., a species not described by Linneaus's earlier texts. *Agrostis gigantea*, or redtop, is now considered to be an *A. alba* synonym.

Clayton and Renvoize (1986) describe the *Agrostis* as "a large genus for which no adequate comprehensive treatment is available." Research has shown that many of the *Agrostis* differ only in ploidy or time of flowering (Widen, 1971). Many freely intercross, making possible an array of intergrades, some fertile and some not (Table 13.1). In the future, it is entirely possible that the species described in this treatise will be revisited and reclassed as scientists learn more about their species boundaries.

Redtop (*A. gigantea*)

Redtop [synonyms: *A. stolonifera* L. subsp. *gigantea* (Roth) Schübl. & G.Martens, *A. nigra* With., *A. graniticola* Klokov, *A. alba* auct., non L. subsp. *gigantea* (Roth) V.Jirásek, *A. korczaginii* Senjan.-Korcz., *A. sabulicola* Klokov, *A. praticola* Klokov, *A. maeotica* Klokov, *A. exarata* var. *mutica* Hicken, *A. semi-nuda* Knapp, *Cinna karataviensis* N. Pavl., *Vilfa alba* Gray, *V. divaricata* (Hoffm.) Gray, *V. gigantea* (Roth) P. Beauv., *V. nigra* (With.) Gray] is a cool-season perennial grass useful for forage, turf, and reclamation purposes, but perhaps best suited to the latter. Up until the 1960s, redtop was a significant economic grass species with thousands of kilograms of seed produced each year in southern Illinois and Missouri. Since that time, redtop production and use has waned. Before 1960, redtop was recommended in most cool-season lawn and forage mixtures, to act as a nurse or companion grass, due to redtop's quick and reliable establishment by seed (Musser, 1950; Wright et al. 1977). Unfortunately, this nurse grass persisted at low levels in the sward long after its nursing duties were needed, creating a undesirable patchwork appearance in the turf. Juska et al. (1955) demonstrated that when redtop was added to mixtures of 'Merion' Kentucky bluegrass (*Poa pratensis* L.) and red fescue (*Festuca rubra* L.), the resulting competition inhibited topgrowth and rhizome production of the desirable species. They concluded that "there was no apparent advantage for using redtop" in mixtures. Levy (1924), a New Zealand scientist, went even further in his assessment: Redtop "as a lawngrass ranks as one of the poorest. Agriculturally redtop is of no significance in New Zealand, and in the writer's opinion not a pound of seed of this grass should be imported into the country."

Botanical Description

Culms are erect, ascending to 40 to 120 cm from a procumbent base, rooting and branching at the lower nodes, with a rather low shoot density (Hubbard, 1984). Leaves are hairless and a characteristic dull green. Ligules are blunt, 1.5 to 1.6 mm in length, toothed, and membranous. Panicles are erect, open, and usually loose, ranging from 8 to 25 cm in length, with a distinctive brilliant red-to-purplish hue at anthesis, thus giving rise to its common name.

Musser (1950) observed that redtop "consists of many individual types, ranging from definitely creeping forms that have well-developed [rhizomes] to plants that are noncreeping and have a decidedly bunchy habit of growth." The majority of plants he observed in seed production were of the latter growth habit.

Table 13.1. Distinguishing characters between Agrostis castellana, A. capillaris, and A. stolonifera (after Batson, 1998).

	Number of Lemma Nerves	Length of Lemma (mm)	Description of Lemma	Callus	Awn	Description of Awn
A. castellana	5	1.84	Outer nerves conspicuously excurrent	Tufts of hair (0.5 mm) either side of callus	+[a]	Arising basally and often twice length of lemma
A. capillaris	3 (5)[b]	1.84	Outer nerves sometimes slightly excurrent	Smooth or few short hairs	−/(+)	If present, arising mid-dorsally and terminal
A. stolonifera	5	1.70	Outer nerves slightly excurrent	Smooth or few short hairs	−	—

	Length of Ligule (mm)	Description of Ligule	Mode of Spread	Shoot Branching	Behavior of Extravaginal Shoots	Ploidy Level
A. castellana	0.7 to 2.0	Longer than wide, minutely ciliate tip	Long, slender rhizomes	Intravaginal	Extravaginal shoots are absent	6
A. capillaris	0.6 to 2.0	Blunt to obtusely rounded	Very short rhizomes	Intravaginal and extravaginal	Greater than 3 scale leaves and growth is subterranean	4
A. stolonifera	2.0 to 6.0	Long, rounded	Long, over-ground stolons	Intravaginal and extravaginal	Up to 3 scale leaves, shoots ascending at once	4, 5, and 6

[a] Only terminal spikes are usually awned.
[b] The lemma is rarely five-nerved with awn arising from median nerve middorsally and terminal.

Physiological Limitations to Management

Prior to 1940, redtop was the second most important pasture grass in the United States, exceeded only by Kentucky bluegrass (Heath et al., 1973). Its use declined rather quickly and today, it is of only minor importance. Redtop exhibits "satisfactory" nutritive quality for livestock though it is less palatable than other species.

Redtop is perhaps best suited to purposes of land reclamation—revegetation of disturbed sites. Redtop can persist on adverse soils, with pH below 5, and under high heavy metal concentrations. Hogan and Rauser (1979) documented that redtop was able to accumulate and immobilize copper in its roots. They speculated that redtop uses a phytochelatin which acts like a metallothionein. They also demonstrated zinc and cobalt tolerance in redtop as well as clonal variations. McLaughlin and Crowder (1988) observed tolerance to iron and nickel as well as plant-to-plant variations, suggesting the future potential for breeding for enhanced tolerance.

Redtop shows moderate resistant to soil salinity. McCarty and Dudeck (1993) classified redtop (cv. 'Streaker') as the "most salt tolerant" of the eight *Agrostis* cultivars they tested. They calculated that 16 g L^{-1} NaCl would be required to reduce redtop's seed germination to 50%—a tolerance equal to that of 'Seaside,' a saline-tolerant cultivar of creeping bentgrass.

In a multistate low maintenance trial, Diesburg et al. (1997) showed that redtop (cv. 'Reton') ranked intermediate in persistence and overall turf quality, topped only by the *Festuca* species. Redtop performed best in cooler locations and with little or no mowing.

Redtop is one of seven *Agrostis* species known to contain a fungal endophyte. White et al. (1992) located a redtop plant growing near the Somerset airport in New Jersey which contained an *Acremonium starrii* White et Morgan-Jones endophyte. *Agrostis hiemalis* (Walt.) B.S.P., *A. perennans* (Walt.) Tuckerm., and *A. scabra* Willd. also may contain an endophyte. None of the current redtop cultivars are known to be endophyte infected, however. Endophytes provide desirable insect-resistance properties but may be toxic to grazing livestock (Funk et al., 1994).

Highland Bentgrass (*A. castellana*)

Highland bentgrass [synonyms: *A. hispanica* Boiss. & Reut., *A. azorica* (Hochst.) Tutin & E.F.Warb., *A. byzantina* Boiss.] is a loose-to-densely tufted perennial, which spreads readily by short, stout rhizomes (Hubbard, 1984). The 'Highland' name represents both the common name of the species and the name of the only cultivar, which is confusing to the newcomer. Perhaps a better species name is dryland bent or browntop, which is preferred by some writers (Batson, 1998).

The Highland cultivar of *A. castellana* is a collection of strains from the hills of western Oregon. But Highland bentgrass is not native to Oregon. It was originally imported in the 1930s and planted throughout the area west of the Oregon Cascades up until the 1970s (Brede, 2000).

Hubbard (1984) was one of the first to recognize Highland as a species distinct from colonial bentgrass, *A. capillaris*. Many but not all scientists have since accepted this classification, including one writer from Oregon State University (www.orst.edu/Dept/hort/turf/common.htm): "I agree with [Hubbard] from a practical field standpoint because Highland bentgrass has distinct morphological and growth characteristics that set it apart from colonial bentgrasses. For example Highland is strongly rhizomatous and requires low mowing heights to avoid severe false crowning. [Figure 13.1] Highland bentgrass forms a dense turf that looks best when mowed at [1.3 to 1.9 cm]. Highland has a dark blue-gray color and generally looks better in winter than other

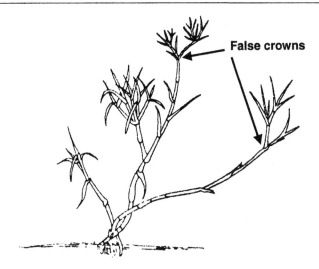

Figure 13.1. Growth habit of Highland bentgrass, exhibiting the typical "false crowns" (after Madison, 1982). Mowing tends to remove these false crowns, resulting in a rather stemmy appearance to the turf.

bentgrasses. It also looks good in early spring but loses color and becomes stemmy during May through mid-June."

Years ago, the production of Highland bentgrass was such a staple to the Oregon economy, that in the late 1950s the Highland Bentgrass Commission was established to promote seed in domestic and foreign markets (www.css.orst.edu/seed-ext/commissions/bentgrass.htm). Over the years, Highland bentgrass has been used for golf course fairways, putting greens, and for unmowed, lower maintenance sites.

Botanical Description

Highland bentgrass has a very slender plant, with culms maturing to 100 cm on fertile soil. Panicles are 4 to 8 cm wide and 17 to 20 cm long. Sheaths are smooth, blades acute, flat, up to 12 cm long, 1.5 to 4.5 mm wide, minutely rough, and closely ribbed. The uppermost ligule can be up to 3 mm long, though the remainder are shorter, prominent, and often jagged (Hubbard, 1984). Table 13.1 shows a comparison between Highland bentgrass and two commercial species.

Physiological Limitations to Management

There are indications that Highland bentgrass originated in a warmer, drier climate than *A. capillaris*, probably along the Mediterranean coast (Hubbard, 1984). Its origin lends itself toward additional management opportunities, notably on droughty soils. Batson (1998) collected *Agrostis* plants from introduced pastures in Tasmania and Victoria, Australia. At the Tasmania sites, which were considerably cooler and more moist, he found an equal number of *A. capillaris* and *A. castellana* plants. At the warmer, drier Victoria sites, however, *A. castellana* plants outnumbered *A. capillaris* plants by 30 to 1.

Madison (1961) compared Highland bentgrass versus Seaside creeping bentgrass under several maintenance treatments. Closer mowing increased the chlorophyll content and shoot density of Highland while reducing root mass. Shoot density of Highland and Seaside, averaged across mowing heights, was 348 and 486 plants dm^{-2}, respectively, in full sun and 236 and 422 plants dm^{-2} at a shade density of 81%.

Newell and Jones (1995) evaluated *A. castellana* and 26 other grass cultivars of six species against "tennis-type" abrasion wear. Before the test began, Highland had a ground coverage equal to the other cultivars. After the 15th day of treatment Highland was reduced to 0% cover. Other bentgrass species fared only slightly better. Cultivars of perennial ryegrass (*Lolium perenne* L.) and Kentucky bluegrass retained up to 70% cover following the wear treatment. Henry et al. (1995) made a similar conclusion following "golf-type" wear. On the positive side, Canaway and Baker (1992) found that Highland ranked second among five turf species for lawn bowling.

Eggens et al. (1993) tested four bentgrass species for resistance to *Poa annua* L. weed invasion. They learned that under both putting green and fairway regimes, *A. castellana* allowed significantly more *Poa* invasion than *A. stolonifera*, *A. capillaris*, or *A. tenuis*.

Highland bentgrass use for turf is generally limited to mowing heights of 1 to 1.9 cm. Below this range, it thins out (i.e., its shoot density becomes unsatisfactorily low). Above this height range, Highland exhibits "false crowns"—secondary crowns and meristems elevated above ground level (Figure 13.1). These false crowns are easily removed by mowing, leading to a stemmy appearance (Madison, 1982).

In disease resistance, Doney et al. (1993) concluded that *A. castellana* appears "to be less susceptible to dollar spot [*Sclerotinia homoeocarpa* F.T. Bennett] and more susceptible to brown patch [*Rhizoctonia solani* Kuhn] than creeping bentgrass. Mowing height undoubtedly interacts."

Agrostis castellana does not tolerate adverse soil conditions as well as *A. gigantea*, but it does show tolerance to arsenate and possibly copper and zinc (De Koe and Jaques, 1993).

Idaho Bentgrass (*A. idahoensis*)

Idaho bentgrass (synonyms: *A. borealis* Hartman var. *recta* [Hartman] Boivin, *A. clavata* auct. non Trin., *A. filicumis* M.E. Jones, *Agrostis tenuis* var. *erecta* Vasey ex Nash, *A. idahoensis* var. *bakeri* [Rydb.] W.A. Weber, *A. tenuiculmis* Nash) is a North American native perennial bunchgrass (lacking lateral runners). One of us (Brede) first discovered the potential of Idaho bentgrass in 1987 when a fieldman brought in a sample of what he thought was dwarf redtop (*A. gigantea*). The plant was growing in a field of Streaker redtop, along a river drainage so polluted with heavy metals that it is within one of the Environmental Protection Agency's (EPA) Superfund sites (Aiken, 1998). Upon examination of the field, I discovered a number of these dwarf plants, which later keyed out to *A. idahoensis*. Subsequent collection trips to that river basin (by fishing boat) netted several hundred specimens which were eventually bred into an improved cultivar, 'GolfStar' (Brede, 1999).

Botanical Description

Idaho bentgrass is a tufted bunchgrass, lacking stolons and rhizomes. Culms are slender, leaves are narrow and basal, with minor evidence of rooting at nodes, and panicles are loose and spreading, 5 to 10 cm long (Hitchcock, 1951).

Distinguishing characters of Idaho bentgrass:

- *The plant*—A tufted, perennial, cool-season bunch grass. Appearance in turf is most similar to colonial bentgrass, although blades are somewhat broader and darker. Unmowed, culms mature to about 30 to 50 cm with a brilliant purple-red panicle at maturity (Hitchcock, 1951).

- *Caryopsis*—Seed characteristically lacks a palea. About one-third of the seed loses its lemma in threshing and appears as naked caryopses (additional characters are listed in Table 13.2).
- *Moisture requirement*—In its native habitat, Idaho bentgrass survives on as little as 380 mm of annual precipitation (Cronquist et al., 1977). It is found most abundantly along streambanks, indicating that it favors moister conditions (DeBenedetti and Parsons, 1984).
- *Applications*—Idaho bentgrass is best adapted to lawns where a softer, fine-textured grass is desired. It also adapts to: low maintenance sites; alkaline, slightly saline, or heavy metal-impacted sites; golf fairways in mixtures with fine fescue; or as an ornamental grass when planted in mass and left unmowed (Brede, 1999). Mintenko and Smith (Mintenko and Smith, 1998; Smith and Mintenko, 2000) found it to be ideally suited to lower maintenance golf course turf when mowed at heights of about 3.8 cm.
- *Management*—The amount of thatch produced is considerably lower than velvet or creeping bentgrass but more than perennial ryegrass. Sod "harvestability" is similar to Kentucky bluegrass, according to one Maryland turf farm (Frank Wilmot, 1999, personal communication).
- *Disease reaction*—Idaho bentgrass is resistant to a number of turf diseases (see Figure 13.2), particularly at midrange cutting heights (5 cm). At closer cuts, however, it becomes increasingly susceptible. Bonos et al. (1998) observed that Idaho bentgrass tended to have better resistance to dollar spot (*Sclerotinia homoeocarpa* Bennett) than other bentgrass species in a four-year turf trial in New Jersey.

DISTRIBUTION AND CYTOTAXONOMY

Distribution

Grasses in the *Agrostis* genus are dispersed across the globe, from the arctic rim to the mountains of the equator (Widen, 1971) (Table 13.2). *Agrostis* can be found on all continents, including some of the Antarctic offshore islands (Edgar and Forde, 1991). Such a wide dispersion suggests that either a prototypic *Agrostis* plant was present in the Cretaceous, prior to the breakup of the continents, or that *Agrostis* is readily wind or water propagated over long distances.

Redtop

Redtop is native to Afghanistan, Albania, Armenia, Austria, Azerbaijan, Belarus, Belgium, Bulgaria, China, Czech Republic, Denmark, Estonia, Finland, France, Georgia, Germany, Hungary, India, Ireland, Iran, Iraq, Italy, Latvia, Lithuania, Moldova, Mongolia, Nepal, Netherlands, Norway, Pakistan, Poland, Romania, Russian Federation, Sweden, Switzerland, Turkey, Ukraine, United Kingdom, and Yugoslavia (USDA ARS, 2001). Agriculture has introduced redtop throughout the remainder of the world. Introduced plants can be found all over North America, except in the deep South, where redtop fails to persist in the humidity and heat. Redtop was introduced into New Zealand as early as 1872 (Edgar and Forde, 1991). It was extensively sown there during the mid 1900s but has not persisted outside of a few naturalized areas in South Waikato and Southland.

In Finland redtop occurs naturally along the shores of lakes and rivers (Widen, 1971). Stands can be found as far north as 70° N latitude, well above the Arctic Circle. Cvelev (1964) observed an arctic form of redtop spread from Russia to Western Siberia. Roseveare

Table 13.2. Botanical and management characteristics of redtop, Highland bentgrass, and Idaho bentgrass, contrasted with those of creeping and colonial bentgrass.

Species	Native to	Growth Habit	Chromosome Number	Typical Mowing Height (cm)	Mixing Ability	Seed per Gram
A. gigantea	Eurasia	Rhizomes	42	5 and above	Yes	11,000
A. castellana	Eurasia, Africa	Rhizomes	42	1.2 to 2.5	Limited	12,600
A. idahoensis	North America	Bunch	28	1.2 to 7.5	Yes	10,500
A. stolonifera	Eurasia	Stolons	28, 56	0.3 to 1.6	No	17,600
A. capillaris	Eurasia	Some rhizomes and stolons	28	1 to 1.9	Limited	19,100

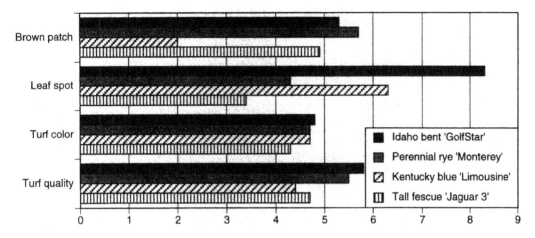

Figure 13.2. Idaho bentgrass compared against three top cultivars of other turf species in a 1995 planting mowed at 5 cm. Higher rating values indicate better disease resistance (leafspot = *Drechslera* spp.; brown patch = *Rhizoctonia* spp.), darker color, and better overall turf quality (after Brede [1999]). Idaho bentgrass cultivar 'GolfStar' was bred in Northern Idaho where diseases like brown patch are uncommon. Thus, it was unexpected that this cultivar would display improved levels of disease resistance when trialed in the so-called "brown patch alley" near Washington, DC. Perhaps it is due to that fact that Idaho bentgrass evolved in North America, in association with local pathogens, while the other species were of exotic origin.

(1948) cataloged plants of redtop in La Picota in the Magellan lands on the extreme southern tip of Chile. He notes that although it flowered abundantly, redtop rarely set seed there and the cattle ate it only as a last resort.

Highland Bentgrass

Highland bentgrass is native to Albania, Azores Islands, Bulgaria, France, Greece, Italy, Portugal, Spain, and Yugoslavia (USDA ARS, 2001). Madison (1982), reporting on studies by the Berlin Herbarium, says that *A. castellana*'s center of origin most likely also includes Turkey and Africa. Madison observed plants of *A. castellana* growing in the Sierra Nevada mountains outside Granada, Spain, above the 2,200-m elevation. He concluded that *A. castellana* is one of the few cool-season grasses with the "ability to compete with bermudagrass in the Transition Zone." One of us (Brede) has found patches of *A. castellana* naturalized in northern Alabama, outcompeting bermudagrass on low-fertility cemetery soils. Some patches were up to 7 m across, originating from a single clone. J.M. Duich (1980, personal communication) scouted Alabama for naturalized *Agrostis* and found similar, highly rhizomatous clones.

Tenacious growth has made *A. castellana* a weed under certain circumstances. *Agrostis castellana* is one of the dominant introduced grasses of southeastern Australia. Hill et al. (1996) commented that it "has become one of the worst pasture weeds of the high rainfall zone of Victoria ... estimated to dominate over 1 million ha." They recommend a glyphosate treatment as the best remedy.

Idaho Bentgrass

Idaho bentgrass is native to the mountains of Arizona, British Columbia, California, Colorado, Idaho, Montana, Nevada, New Mexico, Oregon, Utah, Washington, Wyoming

(Piper and Beattie, 1914; USDA ARS, 2001) and north to Fairbanks, AK (Hitchcock, 1951). It has become naturalized in South America (USDA ARS, 2001).

Cytotaxonomy

The cytotaxonomy of the minor *Agrostis* species has never been comprehensively investigated. Information exists only where these grasses were used as "comparison" species in conjunction with the major *Agrostis*.

The base chromosome number for *Agrostis* is n=7. Differences in ploidy level often determine species boundaries. *Agrostis alpina, A. elegans, A. flaccida, A. canina,* and *A. nebulosa* are examples of the diploid form with 2n=2x=14 chromosomes. Most diploids originate from Europe; many have arctic adaptations. The largest majority of *Agrostis* are tetraploids (2n=4x=28 chromosomes), including many temperate species. These include the three economic species covered in other chapters, along with *A. idahoensis, A. semiverticillata, A. mongolica,* and *A. rossae. Agrostis castellana* is usually found in the tetraploid form, but hexaploids are known to occur in nature. Redtop (*A. gigantea*) is a hexaploid (42 chromosomes), along with *A. exarata, A. clavata, A. diegoensis,* and *A. hallii*. Octaploids (2n = 56) and even decaploids (2n = 70) are known to occur (Widen, 1971). *Agrostis borealis, A. lepida,* and *A. retrofracta* are octaploids, with origins ranging from the Arctic to Polynesia (Darlington and Wylie, 1956).

Bonos et al. (1999) validated the chromosome numbers of several *Agrostis*, including Highland bent and redtop, using laser flow cytometry. She found an r=0.94 correlation between published chromosome counts and observed DNA content. Bonos concluded that "laser flow cytometry is a quick, reliable tool to distinguish certain species of *Agrostis*."

As a rule, only euploid numbers are known to occur in somatic cells of *Agrostis*; aneuploids are rare (Björkman, 1954; Stuckey and Banfield, 1946). Aneuploids have been found in tetraploid *A. castellana*, with up to three additional chromosomes. Aneuploidy also occurs in the hexaploid *A. gigantea* and *A. nevadensis*, with up to four added chromosomes in the former and 10 in the latter (Björkman, 1954; Darlington and Wylie, 1956). Most aneuploids are sterile with respect to both pollen and seed set, while some show degrees of fertility (Widen, 1971).

Most *Agrostis* set abundant seed. But vegetative reproduction is also an important means of reproduction in the genus, via rhizomes or stolons. This vegetative nature allows the possibility for aneuploidy to persist and propagate.

Self-incompatability may play a role in the minor bentgrass species, just as it does in major species (Widen, 1971). J.M. Duich (1980, personal communication) found that most of his Alabama clones (presumably *A. castellana*) exhibited self-incompatibility, while a very few were able to self-fertilize.

Flowering time (also known as maturity date) varies considerably among the *Agrostis*. Flowering time is valuable in species differentiation. It is a mechanism used in nature to keep species and populations genetically isolated. Many *Agrostis* are rather late maturing, ripening seed in later summer to early autumn. This late maturation results in problems in short-season alpine areas, where many plants exhibit barren panicles (Widen, 1971).

Widen (1971), in southern Finland, observed the following sequence in flowering time from earliest to latest: *A. canina, A. stricta* Muhl., *A. capillaris, A. stolonifera,* and *A. gigantea*. He noted that northern populations tended to flower earlier than southern populations. And all species tended to overlap in flowering when planted in extreme northern locations. Flowering in the north also took place over a shorter period.

Jones (1955c) used interspecific hybrids of *A. tenuis*, *A. stolonifera*, and *A. gigantea* and their resulting offspring to help decode the genome constitution in *Agrostis*. He concluded that the genome of *A. tenuis* was $A_1A_1A_2A_2$, *A. stolonifera* was $A_2A_2A_3A_3$, and *A. gigantea* was $A_1A_1A_2A_2A_3A_3$. He adds the caveat, "assigning this constitution to *A. gigantea* enables the pairing in the hybrids to be explained, but at the same time it is not implied that the A_2 genomes of the three species are absolutely identical."

Björkman (1951) recorded the presence of B-chromosomes in *A. gigantea*. Jones (1955b) observed these B-chromosomes in his Aberystwyth [Wales, U.K.] selections. He reported that they are "about one-third to half the size of the normal chromosomes, and the centromere was invariably median or submedian." He found one or two of these chromosomes in all somatic and generative tissue.

GENETIC IMPROVEMENT

The breeding of turf and forage grasses has taken tremendous strides forward in the past 30 years. Yet even today, the breeding sophistication of top grass crops like perennial ryegrass or Kentucky bluegrass probably lags an order of magnitude behind that of wheat (*Triticum*) and maize (*Zea*). Utilization of gene maps, identification of major yield or disease-resistance genes, calculations of quantitative inheritance, and even the use of breeding techniques beyond mass selection are still the exception rather than the rule.

By comparison, the minor bentgrass species discussed in this chapter are probably an order of magnitude lower yet in breeding sophistication than even ryegrass or bluegrass. Most cultivars among these grasses are mere landraces.

Therefore, in the remainder of the chapter, we will attempt to emphasize the breeding potential for these grasses, as much as past breeding accomplishments. We will also mention several untapped *Agrostis* species that may hold potential for commercialization.

Breeding of Redtop

Many of the named cultivars on the Germplasm Resources Information Network (GRIN) catalog (USDA ARS, 2001) originate from Poland, where cultivation of redtop landraces is popular. Current redtop cultivars worldwide include 'Bela,' 'Barracuda,' 'Fireball' (PI 590428), 'Freja,' 'Gosta,' 'Karmos,' 'Kita' (PI 305494), 'Listra,' 'Reton' (PI 406637), 'Streaker' (PI 527690), 'Venture,' and 'Zygma' (Anonymous, 1993). In the United States, only Streaker is consistently in certified seed production (AOSCA, 1996). Streaker and Fireball are the only cultivars with a documented breeding history.

Streaker originated as seedlot P-501.2 chosen from 21 redtop fields in southern Illinois. Jacklin et al. (1989) screened the 21 lots for uniformity and agronomic characters before choosing the best lot, which later became Streaker. Streaker averages 1.1 m in height, and has dull green, hairless leaves. Its lemmas are silvery translucent as opposed to deep gold, and the palea has no observable notch, which differentiates it from common redtop.

Redtop is used occasionally in mixtures for winter overseeding of dormant bermudagrass (*Cynodon* spp.) putting greens, where it is sown in mixtures with perennial ryegrass at 20% by weight (Brede, 1989). Streaker was tested in the 1984 National Turfgrass Evaluation Program fairway overseeding study (Morris, 1986) and showed the lowest turf quality of all the cultivars in the trial. However, it had one of the finest leaf textures and earliest spring transitions. Redtop, like creeping bentgrass, is better suited to overseeding on putting greens than on fairways.

P. Salon and J. Dickerson (Alderson and Sharp, 1994) developed Fireball (experimental 9051629) at the Big Flats Plant Material Center, Corning, NY. Fireball was developed from 9046772, a seedlot of common redtop, and from accession PI 443037, collected in August 1975 from Orleans County, NY, by W. Oaks (USDA ARS, 2001). Fireball is winter hardy and strongly rhizomatous, with an erect base. Seedling growth is rapid with excellent establishment from spring and fall seedlings. Fireball is adapted to temperate regions of United States with adequate rainfall. It tolerates acid soils down to pH 4.0 and adapts to soils that are excessively well drained to poorly drained.

Redtop has the potential for development into either an improved turf or forage cultivar. Present-day redtop germ plasm is probably at a similar point in development to where perennial ryegrass was 35 years ago, when only the cultivar 'Linn' was available. Today's top perennial ryegrass cultivars bear little resemblance to Linn. They are a testimony to what can be accomplished from years of breeding.

Forage potential of redtop may be more limited than its turf potential, due to its stemmy sward and relatively low palatability. However, its soil adaptations and wide geographical range should make it a target for forage improvement.

Breeding of Highland Bentgrass

Aside from Highland, there are no other recognized cultivars of *A. castellana*. The Highland cultivar is a naturalized landrace of western Oregon (Brede, 2000).

Nelson (1987) studied the floral induction requirements of *A. castellana* as a precursor to a future breeding program. He found a significant interaction between plant age and inductive (light/temperature) treatment. Longer induction periods coupled with older plants produced the greatest number of panicles. An alternating 8-hr, 10°C photoperiod and 16-hr, 4.5°C dark period for six weeks provided the optimum floral initiation.

The Instituto Forestal de Investigaciones y Experiencias in Madrid, Spain, has contributed many of the 33 *A. castellana* accessions to the United States plant introduction collection (USDA ARS, 2001). In order to mount a serious breeding program for this species, it might be necessary to collect additional accessions from the Mediterranean rim, where the species is native. *Agrostis castellana* holds the potential as a bonafide reclamation or low maintenance grass, due to its superior drought and heat tolerance and strong rhizomatous habit.

Breeding of Idaho Bentgrass

GolfStar, presently the only cultivar of Idaho bentgrass, was bred from plants collected in the Coeur d'Alene river basin in North Idaho over the years of 1988 through 1990 (Brede, 1999). Seeds from hundreds of accessions were grown in a 15,000-plant selection nursery. Only a small fraction showed turf potential. In general, wild-type plants of *A. idahoensis* performed similar in turf to that of redtop, averaging 2 to 3 on a 1 to 9 quality scale, with 9 equal to ideal turf. However, the species responded favorably to population improvement, increasing in quality with each generation.

The germ plasm was improved through a series of two to three cycles of phenotypic selection. Breeder seed of GolfStar was created from the best six clones from a 1992 turf plot planting: 92-1549, 92-1543, 92-1231, 92-1163, 92-1097, and 92-1165. GolfStar was optimized for a 3-cm mowing height and is therefore best suited where the mowing height is above 2 cm.

Plants of *A. idahoensis* resemble *A. castellana*, but lack rhizomes and stolons and the false crowns of the exotic bentgrasses. GolfStar was entered into the 1998 National Turfgrass Evaluation Program fairway/tee trial (Morris, 2000) where it performed similar to colonial bentgrass. GolfStar has been tested for winter overseeding of bermudagrass putting greens, where it exhibited rapid germination and early spring transition (Anderson and Dudeck, 1995; Kopec and Gilbert, 1995). GolfStar and all germ plasm derived from it are protected by a U.S. utility patent 5,981,853.

The native North American *Agrostis*, exemplified by GolfStar, seem to possess natural resistance to disease that many of the European-derived species lack. Figure 13.2 shows ratings from a high temperature, high humidity field trial near Washington, DC. Results indicate that when managed appropriately, native bentgrasses hold potential for improving the pest-resistance profile of the current *Agrostis* germ plasm.

Application of Biotechnology

Yamamoto and Duich (1994) tested an isoenzyme electrophoresis system to differentiate Highland bentgrass, redtop, and 27 colonial, creeping, and velvet bentgrass cultivars. They used peroxidase, triosephosphate isomerase, glutamate oxaloacetate transaminase, and phosphoglucose isomerase enzyme systems and concluded that the latter system alone could differentiate redtop from all other bentgrasses in their trial. Highland could be differentiated from all bentgrasses except 'Exeter' (*A. capillaris*), which had identical banding of all four enzymes, suggesting that perhaps Exeter was misclassed as to species.

Nelson (1987) tested five *A. tenuis* cultivars and one *A. castellana* for the ability to generate callus for tissue culture. He found that *A. castellana* was one of the easiest to culture, exceeded only by *A. tenuis* cv. 'Astoria.' He then refined the protocols of the induction media and concluded that 4 to 8 mg L^{-1} concentrations of 2,4-D were optimal, resulting in 100% and 72% regeneration, respectively.

Nelson (1987) also investigated the generation of haploids of *A. castellana* through pollen culture. He determined that the best time to sample pollen grains was at the beginning of meiosis, which was "triggered by floret emergence from the flag leaf sheath since premeiotic pollen mother cells were only found among ensheathed florets." The best pollen recovery occurred between maturity Stages 3 and 4 in a system similar to that used in barley (*Hordeum vulgare* L.).

Yoshito Asano (Asano and Unaki, 1994; Asano et at., 1994) of Chiba University in Japan claims to have been the first to successfully transform a plant of *Agrostis*. Asano and his colleagues developed a method of weekly subculturing redtop tissue colonies at 25°C in the dark for two weeks prior to protoplast isolation and transformation. Cells were transformed using a selectable marker for hygromycin resistance, which was later verified in mature plants by PCR and Southern hybridization.

Wipff and Rose-Fricker (2000) tested *A. stolonifera* clones, transformed with herbicide resistance, to study pollen flow into populations of *A. capillaris*, *A. gigantea*, and other bents. The 286 source plants contained a gene for phosphinothricin acetyl transferace. Nontransgenic receptor plants were scattered across the field at distances of up to 300 m. A sample of the resulting seed was screened using Southern Blots to document gene flow. They observed pollen introgression from transgenic *A. stolonifera* into *A. capillaris* and *A. gigantea*. In one year of the two-year study, they recorded transgene levels (presumably in *A. stolonifera* receptor plants) at the maximum distance of 300 m, at levels exceeding the 0.02% maximum recommended by USDA/APHIS.

Unexploited *Agrostis* Species and Hybrids

Among the *Agrostis*, tremendous potential exists for future improvement of previously untapped species. These candidates can be developed through a startup breeding

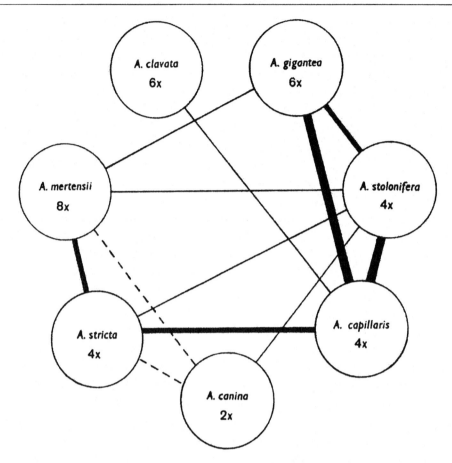

Figure 13.3. Diagram summarizing the current knowledge of hybridization pathways between the *Agrostis* species. Thick lines indicate frequent hybridization paths. Thin lines indicate where hybridization is rare. A broken line indicates that only artificial hybrids are known. After Widen (1971).

program, as exemplified by Idaho bentgrass cv. GolfStar. Or wild-type germ plasm can be introduced into existing cultivars via hybridization. Nearly all of the *Agrostis* share a base chromosome number of 14, differing only in ploidy. Today, ploidy level can be manipulated by the breeder using either anther culture to halve the chromosome number or colchicine to double it (Fehr, 1987).

The earliest example of interspecific hybridization in *Agrostis* dates back to the mid 1800s (Widen, 1971). Since that time, a number of *Agrostis* species have been shown to intercross, some resulting in viable offspring. In a survey of *Agrostis* in Finland, Widen (1971) concluded that some interspecific "hybrids are rather common [and] in some areas they even form an important part of the vegetation" (see Figure 13.3).

Hybrids of *A. gigantea* x *stolonifera* are among the easiest to produce, according to hybridization experiments by Davies (1953) using germ plasm from the United Kingdom. The resulting offspring even yielded seed with good germination and up to 30% viable pollen. Using Swedish germ plasm, Björkman (1954) made successful crosses between *A. gigantea* and tetraploid *A. stolonifera*. Jones (1955a,b,c) studied meiosis in 2n=35-chromosome crosses in natural habitats of these two species and regularly found 14 bivalents and 7 univalents. According to Jones, the constitution of this hybrid is $A_1A_2A_2A_3A_3$.

Widen (1971) discovered a previously unknown hybrid of *A. gigantea* x *A. mertensii* in the Murmansk region of Finland. The low-growing (50 cm culm), highly rhizomatous plant had a somatic chromosome number of 2n = 49 and was infertile.

Edgar and Forde (1991) created hybrids of *A. castellana* x *capillaris*. They found a mean pollen fertility of 92% of plants that flowered, with a heading time intermediate between the two species.

In future breeding improvement programs, other native *Agrostis* should be explored and used in hybridization. In particular, the large *A. exarata* complex should hold a number of traits valuable in withstanding harsh conditions and endemic diseases (Simpson, 1967).

REFERENCES

Aiken, K. 1998. The Bunker Hill Mine Superfund site at Kellogg, Idaho. www.environwest.uidaho.edu/Issues/bunkerhill/bunker.htm (Verified 3-13-01).

Alderson, J. and W.C. Sharp. 1994. Grass Varieties in the United States. USDA SCS Ag. Handbook No. 170.

Anderson, S.F. and A.E. Dudeck. 1995. 1994–1995 Overseeding Trials in North Florida. Univ. of Florida, Gainesville.

Anonymous. 1993. OECD Schemes for the Varietal Certification of Seed Moving in International Trade. OECD, Paris, France.

AOSCA. 1996. Acres Applied for Certification in 1996 by Seed Certification Agencies. Assoc. of Official Seed Certifying Agencies, Mississippi State, MS.

Asano, Y. and M. Unaki. 1994. Transgenic plants of *Agrostis alba* obtained by electroporation-mediated direct gene transfer into protoplasts. *Plant Cell Reports* 13:243–246.

Asano, Y., Y. Ito, K. Sugiura, and A. Fujiie. 1994. Improved protoplast culture of bentgrass (*Agrostis* L.) using a medium with increased agarose concentration. *J. Plant Physiol.* 143:122–124.

Batson, M.G. 1998. *Agrostis castellana* (Poaceae), dominant *Agrostis* species, found in bent grass pastures in south-eastern Australia. *Aust. J. Bot.* 46:697–705.

Björkman, S.O. 1951. Chromosome studies in *Agrostis* (preliminary report). *Hereditas* 37:465–468.

Björkman, S.O. 1954. Chromosome studies in *Agrostis* II. *Hereditas* 40:254–258.

Bonos, S.A., J. Murphy, W.A. Meyer, B.B. Clarke, K.A. Plumley, W.K. Dickson, J.B. Clark, J.A. Honig, and D.A. Smith. 1998. Performance of bentgrass cultivars and selections in New Jersey turf trials. *1998 Rutgers Turfgrass Proc.* 30:33–38.

Bonos, S.A., K.A. Plumley and W.A. Meyer. 1999. The use of laser flow cytometry for species determination in *Agrostis*. *Agronomy Abstr.* [ASA/CSSA/SSSA] 91:140.

Brede, A.D. 1989. If overseeding gets you bent out of shape...Redtop. *The Florida Green*. Summer, pp. 56, 60.

Brede, A.D. 1999. Idaho bentgrass: A new species for turf. *1999 Agronomy Abstr.* [ASA/CSSA/SSSA]. Vol. 91, p. 134.

Brede, A.D. 2000. Four more unconventional grasses to know and love. *Turfgrass Trends*. Vol. 9, No. 12, December, pp. 8–13.

Canaway, P.M. and S.W. Baker. 1992. Ball roll characteristics of five turfgrasses used for golf and bowling greens. *J. Sports Turf Res. Inst.* 68:88–94.

Clayton, W.D. and S.A. Renvoize. 1986. Genera graminum; grasses of the world. *Kew Bulletin Additional Series* 13:1–389. London, HMSO.

Cronquist, A., A.H. Holmgren, N.H. Holmgren, J.L. Reveal, and P.K. Holmgren. 1977. *Intermountain Flora*. Columbia University Press, New York.

Cvelev, N.N. 1964. *Agrostis* L. In A.I. Tolmacev, Ed. *Flora Arctica URSS II*, pp. 43–51, Moscow, Russia.

Darlington, C.D. and A.P. Wylie. 1956. *Chromosome Atlas of Flowering Plants*. pp. 436–437. Macmillan Publishing Company, New York.

Davies, W.E. 1953. The breeding affinities of some British species of *Agrostis*. *Brit. Agric. Bull.* 5:313–316.

De Koe, T., and N.M.M. Jaques. 1993. Arsenate tolerance in *Agrostis castellana* and *Agrostis delicatula*. *Plant and Soil*. 151:185–191.

DeBenedetti, S.H. and D.J. Parsons. 1984. Postfire succession in a Sierran subalpine meadow. *Amer. Midland Naturalist* 111:118–125.

Diesburg, K.L., N.E. Christians, R. Moore, B. Branham, T.K. Danneberger, Z.J. Reicher, T. Voigt, D.D. Minner, and R. Newman. 1997. Species for low-input sustainable turf in the U.S. upper Midwest. *Agron. J.* 89:690–694.

Doney, J.C. Jr., P.C. Vincelli, and A.J. Powell Jr. 1993. Reactions of bentgrasses to dollar spot and brown patch, 1992. Biological and cultural tests for control of Plant. 8:118 (Available on-line with updates at sun3.lib.msu.edu/cgi-bin/starfinder/11985/tgif.txt) (Verified 15 Mar 2001).

Edgar, E. and M.B. Forde. 1991. *Agrostis* L. in New Zealand. *New Zealand J. of Bot.* 29:139–161.

Eggens, J.L., K. Carey, and N. McCollum. 1993. Evaluation of Bentgrass Cultivars Managed as Fairway and Putting Green Turf. Univ. of Guelph, Dep. of Hort. Sci., Guelph, Ontario, Canada.

Fehr, W.R. 1987. *Principles of Cultivar Development*. Vol. 1. Macmillan Publishing Co., New York.

Funk, C.R., F.C. Belanger, and J.A. Murphy. 1994. Role of endophytes in grasses used for turf and soil conservation. pp. 201–209. In C.W. Bacon et al., Eds. *Biotechnology of Endophytic Fungi of Grasses*. CRC Press, Boca Raton, FL.

Heath, M.E., D.S. Metcalfe, and R.F. Barnes. 1973. *Forages, the Science of Grassland Agriculture*. 3rd ed. pp. 252–253. Iowa State University Press, Ames, IA.

Henry, J.M., A.J. Newell, and A.C. Jones. 1995. Effects of abrasive wear on close mown amenity grass species and cultivars. *J. Sports Turf Res. Inst.* 71:52–60.

Hill, R.D., D.J. Missen, and R.J. Taylor. 1996. Use of glyphosate to prevent development of reproductive tillers and extend vegetative growth of bent grass (*Agrostis castellana*). *Aust. J. Exp. Agric.* 36:661–664.

Hitchcock, A.S. 1905. North American species of Agrostis. *USDA Bull.* 68:1–64.

Hitchcock, A.S. 1920. The genera of grasses of the United States. *USDA Bull.* 772.

Hitchcock, A.S. 1951. Manual of the Grasses of the United States. USDA Misc. Publ. 200.

Hogan, G.D. and W.E. Rauser. 1979. Tolerance and toxicity of cobalt, copper, nickel, and zinc in clones of *Agrostis gigantea*. *New Phytol.* 83:665–670.

Hubbard, C.E. 1984. *Grasses*. Penguin Books, Middlesex, England.

Hylander, N. 1953. Taxa et nomina nova in opere meo: Nordisk kärlväxtflora I. (1953) *inclusa*. *Bot. Not.* 1953:353–359.

Jacklin, A.W., A.D. Brede, and R.H. Hurley. 1989. Registration of 'Streaker' redtop. *Crop Sci.* 29:1089

Jones, K. 1955a. Species differentiation in *Agrostis*. I. Cytological relationships in *Agrostis canina* L. *J. of Genetics* 54:370–376.

Jones, K. 1955b. Species differentiation in *Agrostis*. II. The significance of chromosome pairing in the tetraploid hybrids of *Agrostis canina* subsp *montana* Hartm., *A. tenuis* Sibth. and *A. stolonifera* L. *J. of Genetics*. 54:377–393.

Jones, K. 1955c. Species differentiation in *Agrostis*. III. *Agrostis gigantea* Roth. and its hybrids with *A. tenuis* Sibth. and *A. stolonifera* L. *J. of Genetics*. 54:394–99.

Juska, F.V., J. Tyson, and C.M. Harrison. 1955. The competitive relationship of Merion bluegrass as influenced by various mixtures, cutting heights and levels of nitrogen. *Agron. J.* 47:513–518.

Kopec, D.M. and J.J. Gilbert. 1995. Overseed Greens Performance Trial 1994–1995. University of Arizona, Tucson.

Levy, E.B. 1924. The *Agrostis* species—Red-top, brown-top, and creeping-bent. *New Zealand J. Agric.* 28:73–91.

Linneaus, C. 1754. *Genera Plantarum*. 5th ed. XXXII, Holmiae.

Linneaus, C. 1755. *Flora Suecica*. 2nd ed., XXXX, Stockholmiae.

Madison, J.H. 1961. Turfgrass ecology. effects of mowing, irrigation, and nitrogen treatments of *Agrostis palustris* Huds., 'Seaside' and *Agrostis tenuis* Sibth., 'Highland' on population, yield, rooting and cover. *Agron. J.* 54:407–412.

Madison, J.H. 1982. *Principles of Turfgrass Culture*. pp. 60–63. Van Nostrand Reinhold Co., New York.

McCarty, L.B. and A.E. Dudeck. 1993. Salinity effects on bentgrass germination. *HortScience* 28:15–17.

McLaughlin, B.E and A.A. Crowder. 1988. The distribution of *Agrostis gigantea* and *Poa pratensis* in relation to some environmental factors on a mine-tailings area at Copper Cliff, Ontario. *Can. J. Bot.* 66:2317–2322.

Mintenko, A.S. and S.R. Smith. 1998. Turfgrass evaluation of native grasses. *Agron. Abstr.* p. 129.

Morris, K.N. 1986. National Dormant Bermudagrass Overseeding Test, 1984. National Turfgrass Evaluation Program, USDA ARS, Beltsville, MD.

Morris, K.N. 2000. Fairway/Tee Bentgrass Test 1998. NTEP No. 00-2 Progress Report, National Turfgrass Evaluation Program, USDA ARS, Beltsville, MD.

Musser, H.B. 1950. *Turfgrass Management*. McGraw-Hill, New York.

Nelson, E.K. 1987. Effect of Controlled Environment Floral Induction and Plant Age on Subsequent Flowering Responses of Two Species of Colonial Bentgrass, *Agrostis tenuis* Sibth. and *A. castellana* Bois. et Reut., Ph.D. dissertation, Penn State University, University Park, PA.

Newell, A.J. and A.C. Jones. 1995. Comparison of grass species and cultivars for use in lawn tennis courts. *J. Sports Turf Res. Inst.* 71:99–106.

Piper, C.V. and R.K. Beattie. 1914. *Flora of SE Washington and Adjacent Idaho*. New Era Printing, Lancaster, PA.

Roseveare, G.M. 1948. The Grasslands of Latin America. In *Bulletin 36, Imperial Bureau of Pastures and Field Crops*, Aberystwyth, England.

Simpson, D.R. 1967. A Study of Species Complexes in *Agrostis* and *Bromus*. Ph.D. dissertation, Dept. of Botany. University of Washington, Seattle, WA.

Smith, S.R. and A. Mintenko. 2000. Developing and evaluating North American native grasses for turf use. *Diversity* 16:43–45.

Stuckey, I.H. and W.G. Banfield. 1946. The morphological variations and the occurrence of aneuploids in some species of *Agrostis* in Rhode Island. *Amer. J. Bot.* 33:185–190.

USDA ARS, National Genetic Resources Program. Germplasm Resources Information Network (GRIN). [Online Database] National Germplasm Resources Laboratory, Beltsville, MD. Available at www.ars-grin.gov/cgi-bin/npgs/html/taxon.pl?2018 (Verified 11 April 2001).

White Jr., J.F., P.M. Haliskey, S. Sun, G. Morgan-Jones, and C.R. Funk. 1992. Endophyte-host associations in grasses. XVI. Patterns of endophyte distribution in species of the tribe *Agrostideae*. *Amer. J. Botany* 79:472–477.

Widen, K.G. 1971. The Genus *Agrostis* L. in Eastern Fennoscandia, Taxonomy and Distribution. Helsinki - Helsingfors, Finland.

Wipff, J.K. and C. Rose-Fricker. 2000. Determining gene flow of transgenic creeping bentgrass and gene transfer to other bentgrass species. *Diversity* 16:36–39.

Wright, D.L., R.E. Blaser, and J.M. Woodruff. 1977. Seedling emergence as related to temperature and moisture tension. *Agron. J.* 70:709–712.

Yamamoto, I and J.M Duich. 1994. Electrophoretic identification of cross-pollinated bentgrass species and cultivars. *Crop Sci.* 34:792–798.

Chapter 14

Hairgrasses
(*Deschampsia* spp.)

Leah A. Brilman and Eric Watkins

Grasses in the genus *Deschampsia* P. Beauv., hairgrass, have not commonly been considered turfgrasses, with most literature referencing their importance as native wetland grasses in much of the world and their importance as a forage in northern and alpine environments. Species of *Deschampsia* are found in temperate, arctic, and antarctic regions of both the northern and southern hemispheres and also occur at high altitudes in the tropics (Gould and Shaw, 1983). They have adapted to many diverse habitats including coastal marshes, stream banks, and mountain meadows. They are often considered key species for wetland restoration, riparian plantings, and reclamation (Alderson and Sharp, 1994).

Hairgrasses, in particular tufted hairgrass, *D. caespitosa* (L.) P. Beauv., have many attributes that suggest potential value as an important turfgrass species, following some breeding and selection. They are one of the first colonizers of disturbed habitats (Brown et al., 1988), especially under acid soil conditions (pH 3.00 to 5.00), or where high heavy-metal concentrations occur (Brown and Johnston, 1978; Cox and Hutchinson, 1981a; Coulaud and McNeilly, 1992; Bush and Barrett, 1993; Von Frenckell-Insam and Hutchinson, 1993). Tufted hairgrass has also been shown to tolerate very low light intensities (photosynthetic photon flux of 25 µmol m^{-2} s^{-1}), low nitrogen sites, and waterlogged sites with high concentration of ammonium ions (Gloser and Holub, 1996). Evolutionarily, this genus appears to be the ancestor of the *Agrostis* or bentgrass genus (Clayton and Renvoize, 1986).

A broad array of species are found in the genus with the species *D. caespitosa* (or *D. cespitosa*), *D. flexuosa* (L.) Trin., and *D. elongata* (Hook.) Munro ex Benth. showing the greatest potential for turf usage. The species *D. caespitosa*, tufted hairgrass, is the only member of the genus that has had cultivars developed for turf and reclamation. These programs were started when turf breeders found natural ecotypes of the species performing well in mown turf environments where more traditional species had failed. Ecotypes were collected to form the basis for the first cultivars in the species. These cultivars have shown usefulness in high wear situations in northern Europe and the United States, particularly under low light situations. These cultivars have problems in high summer stress areas of the United States due to billbug (*Spenophorus* spp.) damage and possibly heat stress (William Meyer, personnel communication). Recently, however, a number of turf breeders have started collecting this species and the range of variation suggests extensive potential for turfgrass usage.

DISTRIBUTION AND CYTOTAXONOMY

Classification

Family	Poaceae
Subfamily	Pooideae
Tribe	Poeae
Subtribe	Agrostidinae
Genus	*Deschampsia* (Soreng et al., 2001)

Deschampsia is a small genus of perhaps two dozen species, the exact number of which depends on how many species or subspecies are considered part of the *D. caespitosa* complex. Table 14.1 shows the currently accepted species in the Catalogue of New World Grasses (Poaceae) (Soreng et al., 2001) representing *Deschampsia* in the Americas. Three species occur in the British Isles: *D. caespitosa*, *D. setacea* (Hudson) Hackel, and *D. flexuosa* (Stace, 1995). There exist few genera which have such a wide distribution from north to south. *Deschampsia antarctica* E.Desv. occurs very close to the South Pole and other species occur in North Greenland (Hagerup, 1939).

The species *D. caespitosa* has the widest range of any species in this genus and has many distinct populations and subspecies. In the Catalogue of New World Grasses (Soreng et al., 2001) 84 homonyms and infraspecific taxa are currently listed for this species, with probably at least 30 distinct subspecies. Many of these have been considered separate species at different points of time. This species has a holoarctic distribution, extending around the globe in the northern hemisphere but rarely north of the Arctic Circle. At middle latitudes it is primarily a montane species. In the tropics it is almost absent except on the highest mountains in southern Mexico and in parts of Africa. Ecotypes or subspecies thought to be part of this species also occur in Australia, New Zealand, and Argentina (Lawrence, 1945).

Deschampsia setacea is a slender species, which grows in small isolated clumps in very narrow ecological conditions. It is found growing in very acid conditions, in heath soils, in areas usually submerged in winter and with some summer moisture. Its geographic range is comparatively larger, occurring along the Atlantic coast from Spain to southern Norway and northern England, but never common in any area. It decreases in frequency rapidly as the distance from the ocean decreases. This species is a diploid with $2n=2x=14$ ($x = 7$), which must be considered the base number of the genus. It appears closely allied to *D. flexuosa*, wavy hairgrass (Hagerup, 1939).

Deschampsia flexuosa, wavy hairgrass, is a tetraploid with $2n= 4x = 28$ chromosomes in the few reports involving this species (Hagerup, 1939; Hubbard, 1984). *D. flexuosa* has a large ecological and geographical range, occurring almost circumpolar in the temperate zone of the northern hemisphere and in South America. It occurs throughout the eastern United States and in Europe from the Arctic to the Mediterranean. It does best on acid soils but does not require low pH condition as does *D. setacea*. In open spaces in woodlands it may form a dense, monotypic sod, but is also found under the shade of trees. It thrives in dry soil but may also be found in moist places in acid bogs (Hagerup, 1939).

Deschampsia caespitosa is usually reported to have a chromosome number of $2n = 26$, with varying numbers of B-chromosomes, and is often referred to as a diploid at this level (Lawrence, 1945; Kawano, 1963, 1966; Rothera and Davy, 1986). The karyo-

Table 14.1. Species of *Deschampsia* accepted by Catalog of New World Grasses (Soreng et al., 2001).

Species of *Deschampsia* P. Beauv. 1812	New World Distribution
D. airiformis (Steud) Benth., 1883	Argentina, Chile
D. antarctica E. Desv., 1854	Argentina, Chile
D. atropurpurea (Wahlenb.) Scheele, 1844	Argentina, Canada, Chile, United States
D. berteroana (Kunth) Trin., 1836	Chile
D. brasiliensis (Louis-Marie) Valencia, 1941	Brazil
D. caespitosa (L.) P. Beauv., 1812	Argentina, Bolivia, Brazil, Canada, Chile, Mexico, United States
D. conferta (Pilg.) Valencia, 1941	Ecuador
D. congestiformis W.E. Booth, 1943	United States
D. cordilleranum Hauman, 1918	Argentina
D. danthonioides (Trin.) Munro ex Benth., 1857	Canada, Mexico, United States
D. domingensis Hitchc. & Ekman, 1936	Caribbean
D. elongata (Hook.) Munro ex Benth., 1857	Argentina, Canada, Chile, Mexico, United States
D. flexuosa (L.) Trin., 1836	Argentina, Canada, Chile, Costa Rica, Mexico, United States
D. juergensii (Hack.) Valencia, 1941	Argentina
D. kingii (Hook f.) Desv., 1854	Argentina, Chile
D. liebmanniana (E. Fourn.) Hitchc., 1913	Mexico
D. looseriana Parodi, 1949	Chile
D. mendocina Parodi, 1949	Argentina
D. micrantha Phil., 1896	Chile
D. monandra Parodi, 1953	Chile
D. parvula (Hook. f.) Desv., 1854	Argentina, Chile
D. patula (Phil.) Pilg. ex Skottsb., 1916	Argentina, Chile
D. setacea (Huds.) Hack., 1880	
D. venustula Parodi, 1949	Argentina, Chile

type consists of two large pairs of metacentric chromosomes (5 to 7 um), five pairs of medium sized submetacentrics (3 to 4 um) and six pairs of medium to small-sized strongly acrocentric chromosomes (2 to 5 um) (Kawano, 1963; Albers, 1972; Rothera and Davy, 1986). Occasional reports of $2n = 28$ exist (Nielson and Humphrey, 1937; Hagerup, 1939; Bowden, 1960) and, due to the broad range of this species, different cytotypes probably exist. The subspecies *bottnica, borealis, arctica, paramushirensis* and *brevifolia* are all reported to be $2n = 28$ or to have both 26- and 28-chromosome cytotypes (Kawano, 1963; Hagerup, 1939; Aiken et al., 1995). The existence of two cytotypes, and not a miscount due to the presence of B chromosomes, is also probable since the base number of the genus is $x = 7$. It appears the predominant chromosome number of $2n = 26$ resulted from a polyploid origin followed by reduction of a chromosome pair (more probable than the origin from ancestors with 6 and 7 chromosome pairs as hypothesized by Lawrence (1945). Darlington (1956) cited many examples of the doubling of the chromosome followed by loss of a pair of chromosomes, which he has termed the "polyploid drop." No reports of a diploid with $2n = 2x = 12$ have been found.

Isozyme studies on the genetics of *D. caespitosa* that had invaded contaminated mine sites, compared with originating populations, supports a diploidized tetraploid origin with considerable gene duplication (Bush and Barrett, 1993). However, isozymes studies on alpine subpopulations from Colorado all showed a disomic inheritance pattern, but this may be due to more advanced diploidization or to the isozymes selected for study (Gehring and Linhart, 1992). Species and/or plants at the diploid level of 2n = 2x =14 are rare in the genus since most species with reports are at higher ploidy levels. Species and subspecies of the *Deschampsia* genus or the *D. caespitosa* complex have not been thoroughly examined cytologically so origin of the ancient tetraploid is difficult to determine. Hedberg (1986) reported sporophytic chromosome numbers of 14, 21, and 28 for *D. caespitosa* as reported in the Index of Plant Chromosome Numbers published by the Missouri Botanical Gardens.

In *D. caespitosa* subsp. *caespitosa* (or *genuina*) and other subspecies, hexaploids (2n = 39) and octoploids (2n = 52) (forms of the 2n = 26 biotypes) are reported (Rothera and Davy, 1986; Kawano, 1966). Variations of all the chromosome numbers occur frequently (2n = 27, 48, 41, 49, and 56, the latter likely being an octoploid), especially in the viviparous forms and subspecies (Kawano, 1963). In a study of 114 populations of seminiferous *D. caespitosa* in the British Isles, Rothera and Davy (1986) found that the 2n = 26 cytotype (called a diploid in their study) and the 2n = 52 cytotype (called a tetraploid in their study) occurred in different ecological areas. The 26-chromosome form was mainly restricted to woodlands of mostly ancient, seminatural origin while the 52-chromosome form was found where colonization had occurred in meadows, pastures, verges, waste grounds, and woodlands. They concluded that the proliferous *D. alpina*, 2n = 52, which reproduces by small plantlets which replace the seeds, should be considered a subspecies of *D. caespitosa* since diagnostic characters used to separate it from *D. caespitosa* are inconsistent and some floral differences are the result of prolifery itself.

Tufted hairgrass, *D. caespitosa*, is a common grass in many diverse habitats. In the northern hemisphere it is abundant in all areas that are not dry (Hagerup, 1939). It grows well in acid sites but is found in soils from pH 3.7 to 7.9 (Rothera and Davy, 1986). It is a natural colonizer and the ability to tolerate heavy metals makes it valuable for reclamation of mine spoil sites (Brown and Johnston, 1978; Cox and Hutchinson, 1981a; Coulaud and McNeilly, 1992; Bush and Barrett, 1993; Von Frenckell-Insam and Hutchinson, 1993). It is considered a native wetlands grass in the United States (Hitchcock, 1950) and dominates moist areas of the Rocky Mountain alpine tundra (Gehring and Linhart, 1992). Lawrence (1945) recognized five distinct ecotypes of the species in the western United States from coastal types to alpine types.

Deschampsia caespitosa is a long-lived, tussock-forming, self-incompatible, wind-pollinated species (Davy, 1980; Bush and Barrett, 1993; Gehring and Linhart, 1992). Studies by Bush and Barrett (1993) showed the self-incompatibility may not be complete as in most grasses. Both seminiferous, seed-bearing, and proliferous—which have plantlets in place of the seeds at the tips of the spikelets—types occur. Prolifery is a form of vegetative multiplication where the plantlets are genetically the same as the female parent. Proliferous types have been known to produce seminiferous spikelets with viable pollen (Rothera and Davy, 1986) and under shortened photoperiods seminiferous plants can become proliferous (Davy, 1980; Lawrence, 1945; Kawano, 1966). This can result in changes in chromosome numbers and allows proliferous types to rapidly colonize a site.

Deschampsia caespitosa has shown the ability to rapidly colonize sites with heavy metal contamination, including sites with copper, nickel, aluminum, cobalt, arsenic, and zinc contamination (Cox and Hutchinson, 1980, 1981a, 1981b; Coulaud and McNeilly, 1992; Bush and Barrett, 1993). Populations that had developed on nickel/copper contaminated soil near Sudbury, Ontario, Canada, also showed increased levels of resistance to aluminum, zinc, and lead, even though they were not exposed to high levels of these heavy metals. This suggests a common mechanism of resistance to these metals in these grasses (Cox and Hutchinson, 1980). In addition, populations from two different contaminated sites in Ontario, Canada apparently evolved resistance independently from each other (Bush and Barrett, 1993). *D. caespitosa* plants growing under zinc pylons had levels of resistance equal to those from zinc mines and some plants from the control group had high zinc tolerance, suggesting preexisting genes or gene combinations (Coulaud and McNeilly, 1992). The genetics of resistance in these plants deserves further study and they may be useful in other soils with heavy metal contamination.

Deschampsia elongata, slender hairgrass, a native of the western United States, occupies many of the same sites that *D. flexuosa* occupies in the eastern United States. It is found in open ground from Alaska to Wyoming south to Arizona and California, Mexico and Chile (Hitchcock, 1950). It is a facultative wetlands grass that can tolerate low fertility and survive in the 20 to 60 cm rainfall zone. It has not been studied as extensively as other members of the genus.

COLLECTION, SELECTION, AND BREEDING HISTORY

The few cultivars developed in this genus are all from *D. caespitosa* and are closely related to plants collected from natural or disturbed habitats. Nortran, released in 1986 by the Alaska Agricultural and Forestry Experiment Station, is a composite of four lines: two selected from plants collected in Iceland, one from a selection from the Talkeentna Mountains in Alaska and one from Galena, along the Yukon River in Alaska. Peru Creek was selected from a native collection at Peru Creek near Dillon, CO, on an alpine meadow (Alderson and Sharp, 1994). SR 6000 and Barcampsia were both developed for turf usage from plants collected in England from mown turf. Norcoast is a cultivar of *D. caespitosa* subsp. *beringensis* developed from bulk seed collections from native stands near the Cook Inlet of Palmer, AK by the Alaska Agricultural and Forestry Experiment Station (Alderson and Sharp, 1994). All current breeding populations are closely related to wild germ plasm. For many potential uses significant selection is not desired, because the resultant cultivars will be used for revegetation.

The genus has just recently been collected by turfgrass breeders, with significant collections created by Rutgers University from Scandinavian germ plasm and other locations. More extensive collections should be made in native sites, as well as turf areas and colonized areas of North America. Genotype x environment interaction has been demonstrated in studies by Lawrence (1945), utilizing the climactic gardens at Stanford, Mather, and Timberline, California maintained by the Division of Plant Biology of the Carnegie Institute of Washington. Plants performed best in an environment most similar to where they were collected, with often only the Timberline populations flowering sufficiently early to set seed at the high altitude of the Timberline garden. Pearcy and Ward (1972) planted *Deschampsia* collections at three elevationally spaced transplant gardens in Colorado. Again, the elevation of the original collection was negatively correlated to the dates of anthesis. Ward (1969) studied *Deschampsia* collections from 12 locations in Colorado and showed plants from high elevations had earlier anthesis dates than those from low elevations. As germ plasm collections are made and utilized in breeding programs, the influence of altitude and latitude of each collection must be considered in crossing plants as well as for seed production characteristics. Many of the plants from recent collections in the United Kingdom have heading dates much later than desirable for economic seed production in the Willamette Valley of Oregon. It may be possible to produce cultivars developed from this material in other areas such as Denmark.

Vernalization requirement of the species is another important consideration in breeding. Davy (1982) showed that 14 to 19 weeks of cold temperatures in the field were required for *Deschampsia* plants to produce at least one panicle. This study also showed that extended exposure to winter conditions increased the number of panicles on each plant. There may also be a daylength requirement. Devernalization does not occur under mild temperatures. Temperature optima, length of the vernalization requirement, and length of the juvenility period have not been determined. Different populations will probably vary in their vernalization requirement, depending on the area of origin.

Breeding in this genus has only recently been initiated. Initial efforts should concentrate on identifying sources of resistance to known problems, such as increased heat tolerance and billbug resistance, as well as improving turf quality. Identification of useful germ plasm and utilization in a recurrent breeding program should receive emphasis. Breeding is currently underway in New Jersey, Oregon, France, and Alaska, in both public and private breeding programs. Due to the inability to reliably distinguish the different ploidy numbers of *D. caespitosa* based on morphology, flow cytometry, supplemented by chromosome counts, should be utilized to ensure that plants of the same chromosome number are used in development of any cultivar. Crosses between the various subspecies of *D. caespitosa* to expand the gene base should probably be utilized before interspecific crosses are utilized. The genus, and in particular the species *D. caespitosa*, shows great potential for use as both a turfgrass with lower nitrogen and light requirements, especially for heavy wear sites, and for erosion control and reclamation.

REFERENCES

Aiken, S.G., L.L. Consaul, and M.J. Dallwitz. 1995 onward. Poaceae of the Canadian Arctic Archipelago: Descriptions, Illustrations, Identification, and Information Retrieval. Version: 27th February 2001. www.mun.ca/biology/delta/arcticf/'.

Albers, F. 1972. Cytotaxonomie und B-chromosomes bei Deschampsia caespitosa (L.) *Beitrage zur Biol. der Pflanzen.* 48: 1–62.

Alderson, J. and W.C. Sharp. 1994. Grass Varieties of the United States. Agricultural Handbook No. 170. USDA, Soil Conservation Service, Washington, DC.

Bowden, W.M. 1960. Chromosome number and taxonomic notes on northern grasses. III. Twenty-five genera. *Can. J. Botany* 38 : 541–557.

Brown, R.W., J.C. Chambers, R.M. Wheeler, E.E. Neely, and M.I. Kelrick. 1988. Adaptations of *Deschampsia cespitosa* (tufted hairgrass) for revegetation of high elevation disturbances: Some selection criteria. pp. 147–172. In W. R. Kaemmerer and L.F. Brown, Eds. *Proc.: High Altitude Revegetation Workshop No. 8*. March 3–4, Colorado State University, Fort Collins, CO. Infor. Series No. 59.

Brown, R.W. and R.S. Johnston. 1978. Rehabilitation of high elevation mine disturbance. pp. 116–129. In S.T. Kenny, Ed. *Proc.: High-Altitude Revegetation Workshop No. 3*. Environ. Resources Center. Infor. Series No. 28, Colorado State University, Fort Collins, CO.

Bush, E.J. and S.C.H. Barrett. 1993. Genetics of mine invasion by *Deschampsia cespitosa* (Poaceae). *Can. J. Bot.* 71: 1336–1348.

Clayton, W.D. and S.A. Renvoize. 1986. Genera Graminum, Grasses of the World. KEW Bulletin Additional Series XIII. Her Majesty's Stationary Office, London.

Coulaud, J. and T. McNeilly. 1992. Zinc tolerance in populations of *Deschampsia cespitosa* (*Gramineae*) beneath electricity pylons. *Pl. Syst. Evol.* 179: 175–185.

Cox, R.M. and T.C. Hutchinson. 1980. Multiple and co-tolerance in the grass *Deschampsia cespitosa* (L.) Beauv. from the Sudbury smelting area. *New Phytol.* 84 : 631–647.

Cox, R.M. and T.C. Hutchinson. 1981a. Environmental factors influencing the rate of spread of the grass *Deschampsia cespitosa* invading areas around the Sudbury nickel-copper smelters. *Water Air Soil Pollution* 16:83–106.

Cox, R.M. and T.C. Hutchinson. 1981b. Multiple and co-tolerance to metals in the grass *Deschampsia cespitosa*: Adaptation, preadaptation and 'cost.' *J. Plant Nutr.* 3:731–741.
Darlington, C.D. 1956. *Chromosome Botany*. Allen and Unwin, London.
Davy, A.J. 1980. Biological flora of the British Isles - *Deschampsia caespitosa*. *J. Ecol.* 68:1075–1096.
Davy, A.J. 1982. Flowering competence after exposure to naturally fluctuating temperatures in a perennial grass, *Deschampsia caespitosa* (L.) Beauv. *Ann. Bot.* 50:705–715.
Gehring, J.L. and Y.G. Linhart. 1992. Population structure and genetic differentiation in native and introduced populations of *Deschampsia caespitosa* (Poaceae) in the Colorado alpine. *Am. J. Bot.* 79:1337–1343.
Gloser, J. and P. Holub. 1996. Interactive effect of radiation and nitrogen availability on growth and photosynthesis of *Deschampsia caespitosa*. Acta scientiarum naturalium *Academiae scientiarum Bohemicae*. 30(1):1–34.
Gould, F.W. and R.B. Shaw. 1983. *Grass Systematics*, 2nd ed. Texas A&M University Press. College Station, TX.
Hagerup, O. 1939. Studies on the significance of polyploidy III. *Deschampsia* and *Aira*. *Hereditas*. 25:430–440.
Hedberg, O. 1986. On the manifestation of vivipary in *Deschampsia caespitosa* s. lat. *Symb. Bot., Upsalla.* 27:183–192.
Hitchcock, A.S. 1950. Manual of the Grasses of the United States. 2nd ed. revised by A. Chase. U.S. Government Printing Office. Washington, D.C.
Hubbard, C. E. 1984. Grasses. *A Guide to Their Structure, Identification, Uses and Distribution in the British Isles*. 3rd ed. revised by J.C.E. Hubbard. Penguin Books. London.
Kawano, S. 1963. Cytogeography and evolution of the *Deschampsia caespitosa* complex. *Can. J. Bot.* 41:719–742.
Kawano, S. 1966. Biosytematic studies of the *Deschampsia caespitosa* complex with special reference to the karyology of Icelandic populations. *Bot. Mag. Tokyo* 79:293–307
Lawrence, W.E. 1945. Some ecotypic relations of *Deschampsia caespitosa*. *Am. J. Bot.* 32:298–318.
Nielsen, E. and L.M. Humphrey. 1937. Grass studies. I. Chromosome numbers in certain members of the tribes *Festuceae, Hordeae, Aveneae, Agrostideae, Chorideaea, Phalarideae* and *Tripsaceae*. *Am. J. Bot.* 24:276–279.
Pearcy, R.W. and R.T. Ward. 1972. Phenology and growth of Rocky Mountain populations of *Deschampsia caespitosa* at three elevations in Colorado. *Ecology* 53:1171–1178.
Rothera, S.L. and A.J. Davy. 1986. Polyploidy and habitat differentiation in *Deschampsia caespitosa*. *New Phytol.* 102:449–467.
Soreng, R.J., G. Davidse, P.M. Peterson, F.O. Zuloaga, E.J. Judsiewicz, and T.S. Filgueiras. 2001. Catalogue of New World Grasses (Poaceae). mobot.mobot.org/W3T/Search/nwgc.html.
Stace, C. 1995. *New Flora of the British Isles*. Cambridge University Press.
Von Franckell-Insam, B.A.K. and T.C. Hutchinson. 1993. Occurrence of heavy metal tolerance and co-tolerance in *Deschampsia cespitosa* (L.) Beauv. from European and Canadian populations. *New Phytol.* 125:555–564.
Ward, R.T. 1969. Ecotypic variation in *Deschampsia caespitosa* (L.) Beauv. from Colorado. *Ecology* 50:519–522.

Part 3
Warm-Season Grasses

Chapter 15

Bermudagrass (*Cynodon* (L.) Rich)

Charles M. Taliaferro

Bermudagrass is the most common of various names applied to several taxa of the genus *Cynodon* (L.) Rich. Other commonly used names include "couchgrass" or "green couchgrass" in Australia, "doobgrass" in Bangladesh and India, and "quickgrass" and "kweekgrass" in South Africa (Chippindall, 1947; Kneebone, 1966; Skerman and Riveros, 1990). These long-lived perennial, warm-season, sod-forming grasses are widely used for turf throughout milder climatic regions of the world. Their use extends to a lesser extent to warmer temperate regions, often described as regions of transition between warm- and cool-season grass species. Selected bermudagrass cultivars provide attractive durable turf for many purposes including home and institutional lawns, parks, athletic fields, and golf course fairways, tees, putting greens and roughs (Beard, 1973). Bermudagrass is also widely used in cemeteries, on roadside right-of-ways, and other comparable areas.

Bermudagrass remains in a nondormant state and maintains active growth throughout the year in climatic regions where temperatures never, or seldom, reach 0°C. Under such conditions, growth is slowed during the cool season and low temperatures may temporarily cause discoloration of foliage by inducing anthocynanin pigmentation or tissue damage from light frost. In colder regions where temperatures frequently reach 0°C or below during winter months, the aboveground foliage is killed and regenerative tissues are dormant. New growth is initiated in the spring from crown buds, rhizomes, and any stolons that survive the winter. Bermudagrass plants used as turf are typically relatively low-growing and have refined texture relative to taller growing and more robust forms of the genus, but otherwise are morphologically similar. Plants have flat, linear leaf blades, sometimes rolled, with a membranous ligule often with hairs on the upper edge. Leaves are rolled or once folded in the bud (Watson and Dallwitz, 1992). Culm nodes are glabrous and culm internodes hollow. The inflorescence is comprised of digitate one-sided spikes borne in one or more closely spaced whorls. Spikelets are narrowly ovate, have one flower, no awns, and are subtended by two glumes of unequal length and shorter than the floret. The caryopsis is ellipsoid and laterally compressed. Plants spread by prostrate stolons that may establish roots and shoots at each node. The major *Cynodon* turfgrass taxa bear underground rhizomes, though some *Cynodon* taxa are void of these organs.

Bermudagrass has the C4 photosynthetic pathway (Watson and Dallwitz, 1992). It is generally regarded as having excellent tolerance to heat and drought, but low tolerance to freezing temperatures (Beard, 1973). However, some bermudagrass forms are capable of surviving very low temperatures and persist in cold climatic regions (Harlan and de Wet, 1969). It is best adapted to moderately well-drained fertile loamy soils but grows well over a wide range of edaphic conditions and tolerates periodic inundation from flooding. It is very intolerant of shading (McBee and Holt, 1966). The cultural requirements for maintenance of good turf quality are relatively high for bermudagrass. Though very drought tolerant, adequate soil moisture is necessary for sustained growth, color, and sod density critical to high turf quality. It is highly responsive to fertilizer, with nitrogen being the nutrient of primary importance. The maintenance of good turf quality generally requires application of nitrogen in the range of 39 to 88 kg ha^{-1} (0.8 to 1.8 lb/1,000 ft^2) per growing month (Beard, 1973). The decumbent growth habit of bermudagrass makes it tolerant of close mowing. Bermudagrass cultivars have been selected for use on golf course putting greens that tolerate mowing as close as 3.2 mm (1/8 in.) above the soil surface. For other turf uses, mowing heights typically vary from 1.3 to 2.5 cm (1/2 to 2 in.). Some general purpose bermudagrass turf areas are mowed at heights up to 7.6 cm (3 in.). Bermudagrass is subject to thatch accumulation that can be minimized or avoided by periodic topdressing or vertical mowing or both. Frequent mowing is required to avoid scalping.

TAXONOMIC CLASSIFICATION AND SPECIES DISTRIBUTION

The genus *Cynodon* Rich. is a member of the family *Gramineae* (*Poaceae*), subfamily *Chloridoideae*, tribe *Cynodonteae*, and subtribe *Chloridinae* (Clayton and Renvoize, 1986). Harlan et al. (1970a) revised the taxonomic classification of *Cynodon* to include nine species and ten varieties (Table 15.1). The Royal Botanic Gardens, Kew (1999) lists eight *Cynodon* species, omitting *C. X magennisii* Hurcombe. Harlan (1970) and de Wet and Harlan (1970) omitted *C. X magennisii* in listings published shortly before their final published revision (Harlan et al., 1970a). The taxa of predominant turf importance are *C. dactylon* (L.) Pers. var. *dactylon* and *C. transvaalensis* Burtt-Davy. Most of the turf bermudagrasses of wide-scale economic importance emanate from these two taxa. Forms of the narrowly endemic *C. X magennisii* and *C. incompletus* var. *hirsutus* (Stent) de Wet et Harlan taxa have had fairly wide use as turfgrasses. Taxa of minor turf importance are *C. arcuatus* J. S. Presl. Ex C. B. Presl., *C. barberi* Rang. Et Tad. and *C. dactylon* var. *polevansii*. Other taxa have little, or no, value as turf, but in some cases will hybridize with the turf species, thereby representing potentially important contributors of genes for the breeding improvement of turf bermudagrass. Discussions in this treatment will be limited largely to the aforementioned bermudagrass taxa having importance as turf. Brief descriptions of the turf taxa and discussion of their origin and distributions follows.

C. dactylon var. *dactylon* contains enormously variable plant types ranging from the small, fine-textured, plants used as turf (Figure 15.1) to the large robust plants with high biomass production capability that are used for cultivated pasture. Plants of the taxon have rhizomes varying from short and slender to stout and fleshy. Inflorescences are borne in one, rarely two, whorls with two to several slender, stiff, racemes. The stolons range from very fine to coarse. Harlan and de Wet (1969) characterized the taxon as a ubiquitous, cosmopolitan weed of the world and gave a detailed description of its distributional patterns and variation. It is found over all continents and ocean islands having climatic conditions conducive to its survival. Ecotypes of the taxon occur as far north as 53° N latitude and at elevations from sea level to 3,000 m. Though

Table 15.1. A revised classification of the genus *Cynodon*.[a]

Epithet	Chromosome Number	Distribution
C. aethiopicus Clayton et Harlan	18, 36	East Africa: Ethiopia to Transvaal
C. arcuatus J. S. Presl ex C. B. Presl	36	Malagasy, India, S.E. Asia, S. Pacific to Australia
C. barberi Rang. et Tad.	18	India
C. dactylon (L.) Pers.		
var. *dactylon*	36	Cosmopolitan
var. *afghanicus* Harlan et de Wet	18, 36	Afghanistan
var. *aridus* Harlan et de Wet	18	South Africa northward to Palestine, East to South India
var. *coursii* (A. Camus) Harlan et de Wet	36	
var. *elegans* Rendle	36	Madagascar
var. *polevansii* (Stent) Harlan et de Wet	36	Southern Africa, south of lat. 13°S. near Baberspan, S. Africa
C. incompletus Nees		
var. *incompletus*	18	Transvaal to Cape
var. *hirsutus* (Stent) de Wet et Harlan	18, 36	Transvaal to Cape
C. nlemfuensis Vanderyst		
var. *nlemfuensis*	18, 36	Tropical Africa
var. *robustus* Clayton et Harlan	18, 36	East Tropical Africa
C. plectostachyus (K. Schum.) Pilger	18	East Tropical Africa
C. transvaalensis Burtt-Davy	18	South Africa
C. x magennisii Hurcombe	27	South Africa

[a] After Harlan et al., 1970a.

cosmopolitan, Harlan and de Wet (1969) noted that it does not occur in pure stands over vast areas except under the most artificial conditions. Rather, it thrives only in disturbed areas and does not invade natural grasslands or forest vegetation. They suggested that the variety was a Eurasian grass until recent times and that an area from West Pakistan to Turkey was a center of evolutionary activity from which current aggressive weedy races likely emerged. They described three major races of var. *dactylon*. The tropical race has pan tropical distribution with plants that are short in stature and producing a loose turf. Plants of this race are adapted to the leached, acid soils and seasonal fluctuations in rainfall common to the tropics. Plants of the temperate race are similar in appearance to those of the tropical race, differing principally in adaptation characteristics. Plants of the temperate race are much more winter hardy, more disease susceptible, and are less tolerant to seasonally waterlogged soils and soils of low pH and fertility. They typically form a more dense turf than plants of the tropical race. The seleucidus race has its center of diversity from Pakistan to Turkey, corresponding closely to the geographic area of the original Seleucid Empire. Plants of this race are described as remarkably coarse and robust, usually glaucous, and frequently hairy. Plants of the seleucidus race are very freeze tolerant, tall, and generally produce large amounts of biomass on fertile soils. They have heavy, thick stolons and rhizomes with short internodes. The rhizomes often emerge to form stolons and then grow into the soil again to reform rhizomes. The three races are sometimes sympatric.

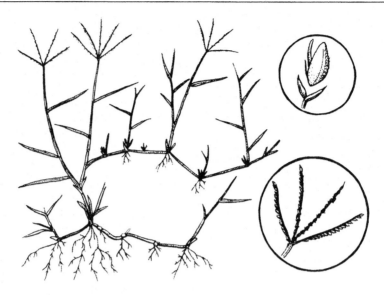

Figure 15.1. *Cynodon dactylon* var. *dactylon*.

C. transvaalensis, African bermudagrass (Figure 15.2), contains plants easily distinguished on the basis of their very fine texture, narrow yellowish-green erect leaves, and small inflorescences with spikelets loosely arranged on three, rarely two or four, short racemes (Harlan et al., 1970a; de Wet and Harlan, 1971). Stolons are slender with short internodes, and typically develop red pigmentation in response to cool temperatures. Plants bear many shallow, short but fleshy rhizomes. The taxon is narrowly endemic to the Transvaal and Orange Free State of South Africa, where its natural habitat is in moist areas along streams and near ponds. African bermudagrass has been widely distributed as a turfgrass and has more winter hardiness than is required in its endemic range. 'Uganda' and some other clonal African bermudagrass plants have demonstrated good winter hardiness to 39° N latitude in the United States.

C. X magennisii is based on a single clonal plant presumed to have originated in South Africa from natural crossing of *C. dactylon* and *C. transvaalensis* parent plants (Harlan et al., 1970a). Morphological features are similar to those of *C. transvaalensis*. It is a sterile triploid (2n=3x=27 chromosomes) plant with very fine texture, producing a low-growing, dense turf. The Alabama, Arkansas, Oklahoma, and South Carolina Agricultural Experiment Stations released this clonal plant (PI 184339) as 'Sunturf' in 1956 (Huffine, 1957).

C. incompletus var. *hirsutus* plants are variable in size, but generally low-growing and producing a relatively dense fine turf when mowed or grazed (Harlan et al., 1970a). Leaves are short and moderately to densely hairy on both sides. Inflorescences have two to several slender, stiff, sometimes reflexed racemes in one whorl. Stolons are slender with red color under cool conditions. Plants have no rhizomes. The taxon is endemic to southern Africa, below 23° S latitude. The taxon includes the lawn grass classified by Stent (1927) as *C. bradleyi* Stent (Harlan, 1970). Plants have been relatively winter hardy in the United States to about 36° N latitude.

C. arcuatus is distinct and easily recognized from the other *Cynodon* taxa because of conspicuous features including large ovate-lanceolate leaf blades and long slender racemes (3–6) in one whorl (Harlan et al., 1970a; de Wet and Harlan, 1970). It has no rhizomes and little freeze tolerance. It forms a loose open mat of inferior turf quality. It is distributed from Madagascar and other islands off east Africa to Sri Lanka, across

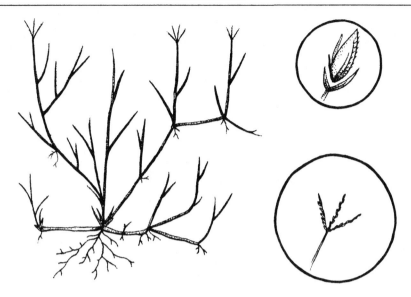

Figure 15.2. *Cynodon transvaalensis.*

south India to lowland Nepal and eastward across southeast Asia to northern Australia (Harlan et al., 1970a).

C. barberi is described as similar in appearance to *C. arcuatus* except for very different inflorescence characteristics (Harlan, 1970). The racemes (3–6 per whorl) of *C. barberi* are shorter, more delicate, and the spikelets are less densely spaced compared to *C. arcuatus*. It is endemic to south India. The small, nonrhizomatous plants occur sparsely mainly in moist sites along ditches and watercourses. It has little winter hardiness.

C. dactylon var. *polevansii* contains small, dark green plants with fine culms and stolons bearing harsh stiffly ascending leaves. Stolons have short internodes. Inflorescences bear two to several small, slender racemes. Plants have small rhizomes. Harlan et al. (1970a) noted that plants had good winter hardiness in Oklahoma and had potential as turfgrasses. The taxon is a narrow endemic to a localized area near Barberspan, South Africa. It was originally described by Stent (1927) as *C. polevansii*.

CYTOGENETIC AND REPRODUCTIVE CHARACTERISTICS

Chromosome Numbers

Bermudagrass has a base chromosome number of $x=9$, with reported 2n chromosome numbers ranging from 18 to 54 in multiples of 9 (Forbes and Burton, 1963; Malik and Tripathi, 1968; Harlan et al., 1970b; de Silva and Snaydon, 1995). Several early investigators reported the genomic number as $x=10$ based on their determinations of plants with $2n=20$ (Hurcombe, 1946, 1947, 1948), $2n=30$ (Hunter, 1934; Hurcombe, 1946, 1947; Rochecouste, 1962), or $2n=40$ chromosomes (Hurcombe 1946, 1947, 1948; Moffett and Hurcombe, 1949). Forbes and Burton (1963) demonstrated the presence of long satellites on some bermudagrass chromosomes, which often appeared as fragments in somatic cell squashes. They noted that the fragments could easily be misinterpreted as whole chromosomes resulting in erroneous counts of $2n=20$, 30, or 40 chromosomes. Accessory chromosomes have been reported in some bermudagrass plants (Gould, 1966; Powell et al., 1968a; Malik and Tripathi, 1968), and, if present, are subject to misinterpretation.

Chromosome numbers for *Cynodon* taxa were listed by Harlan et al. (1970a) in their revised classification (Table 15.1). For the turfgrass taxa, *C. barberi* and *C. transvaalensis* were indicated as strictly diploid. *C. arcuatus* and *C. dactylon* vars. *dactylon* and *polevansii* were listed as strictly tetraploid. *C. incompletus* var. *hirsutus* was indicated as mainly diploid with rare tetraploid forms. Malik and Tripathi (1968) reported 2n=18, 36, and 54 chromosome races for *C. dactylon* from Udaipur, India, but did not specify taxonomic variety or the numbers of plants studied within the respective races. de Silva and Snaydon (1995) determined chromosome numbers in *C. dactylon* populations collected from three habitat types and five climatic regions in Sri Lanka. They reported that fully diploid populations occurred only in very acid (pH<5.0) soils, and fully tetraploid populations only in nonacid (pH>6.5) soils. Diploid and tetraploid plants were sympatric in areas with soil pH between 6.0 and 6.5. They did not classify plants as to taxonomic variety.

Triploid (2n=3x=27) bermudagrass plants occur naturally and have been artificially produced in abundance through both intra- and interspecific hybridization (Burton, 1951; Harlan and de Wet; 1969; Harlan et al., 1970b). Only a few reports of pentaploid (2n=5x=45) and hexaploid (2n=6x=54) bermudagrass plants were found. Powell et al. (1968b) reported a hexaploid plant found among the putative progeny of a cross between *C. dactylon* var. *dactylon* cv. Midland (4x) by *C. transvaalensis*. Johnston (1975) produced pentaploid and aneuploid (2n=45 ± 3) plants by crossing a hexaploid plant found among putative progeny of a *C. dactylon* (4x) by *C. barberi* (2x) cross with a 4x *C. dactylon*. *C. dactylon* cv. Tifton 10 (PI 539857) turf bermudagrass, is a naturally occurring hexaploid introduced from Shanghai, China (Hanna et al., 1990). Burton et al. (1993) reported that 'Tifton 85,' a bermudagrass by stargrass (*C. nlemfuensis* Vanderyst) interspecific F_1 hybrid, is a pentaploid. Hexaploid plants may originate through the functioning of unreduced gametes or through chromosome doubling in early embryonic stages of triploid hybrids, or both. Harlan and de Wet (1969) attributed a high incidence of 36 chromosome hybrids from crosses of 4x *C. dactylon* var. *dactylon* by 2x *C. dactylon* var. *afghanicus* to frequent functioning of unreduced male and female gametes.

Flow cytometry has proven a fast reliable means of estimating ploidy in bermudagrass. Taliaferro et al. (1997) reported mean 2C nuclear DNA contents for five diploid (1.10 pg), five triploid (1.60 pg), five tetraploid (2.25 pg) and three hexaploid (2.80 pg) bermudagrasses. Arumuganathan et al. (1999) reported 2C nuclear DNA contents for one diploid (1.03 pg), two triploid (1.37 and 1.61 pg), and one tetraploid (1.95 pg) bermudagrass cytotypes.

Chromosome Pairing and Karyotype

Forbes and Burton (1963) reported that chromosomes paired mainly as bivalents (II) in diploid *C. dactylon*, *C. transvaalensis*, and *C. incompletus* plants. Univalent (I) chromosomes were observed in some of the diploid plants at prophase I or as laggards at anaphase I. They studied chromosome behavior in one 3x hybrid from 2x by 4x *C. dactylon* parents, and five 3x hybrids from crosses of *C. transvaalensis* by 4x *C. dactylon*. Meiosis was irregular in all six hybrids, with multivalent associations or univalents lagging at anaphase I, or sticky chromosomes, or combinations of these factors. They reported up to eight and nine trivalent (III) associations in the one interspecific and five intraspecific hybrids, respectively. Hanna and Burton (1977) reported mean II, I and quadrivalent (IV) chromosome associations at diakinensis for *C. dactylon* forage cvs. Suwanee (17.45, 0.10, and 0.34 cell^{-1}), Midland (16.13, 1.72, and 0.39 cell^{-1}) and Coastal (16.58, 0.89, and 0.26 cell^{-1}). A few trivalent associations (0.15 cell^{-1}) were found in Midland. Harlan et al. (1970b) reported results from extensive analyses of chromosome pairing in intra- and inter-specific bermudagrass hybrids. Diploid by diploid hybrids had mainly II pairing ranging from 8.4 to 9.0 cell^{-1}. The II pairing in 4x by 4x hybrids ranged from about 14 to 17 cell^{-1}. Conversely to the results of Forbes and Burton (1963), they reported that chromosomes tended to associate as nine II and nine I in hybrids from 2x by 4x crosses. They found II pairing in hybrids from parents of *C. dactylon* var. *dactylon* to frequently be less than in hybrids involving different varieties. Hybrids involving parents from the different races of *C. dactylon* var. *dactylon* tended toward meiotic irregularity and high sterility, though seldom were completely sterile. They noted that meiosis was most regular in hybrids from parents of similar geographic origin, but cautioned that exceptions occurred. They hypothesized that meiotic irregularity in hybrids was mainly due to chromosomal structural differences between populations and that there is only one basic genome in the *Cynodon* taxa that can be crossed.

Pachytene chromosome karyotypes were constructed for 2x *C. dactylon* by Ourecky (1963) and Brilman et al. (1982). Ourecky's karyotype was based on a single clone (PI 220385) from Afghanistan. The nine chromosomes of PI 220385 ranged in length from 16.8 to 36.2µ and had arm ratios (long/short) from 1.12 to 1.88. The chromosomes were further distinguished by differences in number and distribution of chromomeres or terminal knobs, or both. Brilman et al. studied three clones of *C. dactylon* var. *aridus* and two F$_1$ hybrid clones from *C. dactylon* var. *afghanicus* by *C. dactylon* var. *aridus*. The results were in general agreement with those of Ourecky, differing mainly in arm length measurements of chromosomes three and four.

Hybridization Potential

The turf bermudagrass taxa *C. arcuatus* and *C. barberi* are reported to be genetically isolated from each other and from all other taxa of the genus (de Wet and Harlan, 1970, 1971; Harlan et al., 1970a, b). *C.* X *magennisii* is a sterile triploid. Extensive crossing by Harlan et al. (1970b) between plants of *C. dactylon* vars. *dactylon* and *polevansii*, *C. transvaalensis*, and *C. incompletus* var. *hirsutus* produced hybrids except for *C. transvaalensis* crossed to *C. dactylon* var. *polevansii* or *C. incompletus* var. *hirsutus*. Many crosses involving *C. dactylon* var. *polevansii* produced a substantial number of seeds that failed to germinate. Among the crosses producing hybrids, substantial differences in hybridization potential were indicated. *C. dactylon* var. *dactylon* crossed rather easily with *C. transvaalensis*, but to a much lesser degree with *C. dactylon* var. *polevansii* and *C. incompletus* var. *hirsutus*. The taxa *C. dactylon* var. *polevansii* and *C. incompletus* var. *hirsutus* crossed sparingly. Interestingly, *C. transvaalensis* is apparently isolated from turfgrass taxa other than *C. dactylon* var. *dactylon*, but according to the Harlan et al.(1970b) results, crosses readily with the robust East African stargrass taxon *C. nlemfuensis*.

Reproductive Traits

Bermudagrass reproduces sexually and is easily clonally propagated. Asexual propagation of plants is achieved by planting stolons, rhizomes, crown buds, or combinations thereof. Intact sod, cut and rolled, is a principal means of commercial distribution of most widely used turf bermudagrass clonal cultivars. Sexual reproduction in bermudagrass occurs through the development of male and female gametophytes. The embryo-sacs contain an egg, two polar, two synergid, and three antipodal nuclei at maturity and thus are of the normal polygonium type (Burson and Tischler, 1980). Mature pollen grains contain a vegetative (tube) nucleus and a generative nucleus. The generative nucleus enters the growing pollen tube and divides to produce two nuclei that respectively combine with the egg and two fused polar nuclei to produce the 2n embryo and 3n endosperm. The fertility (seed set) of bermudagrass plants is highly contingent on the level of meiotic regularity. Plants with good seed set have mainly bivalent chromosome pairing and normal disjunction of paired homologous chromosomes at anaphase I of meiosis (Forbes and Burton, 1963; Hanna and Burton, 1977).

Bermudagrass is highly outcrossed as a result of cross-pollination and self-incompatibility (Burton, 1947; Burton and Hart, 1967). Flowers are perfect, bearing one pistil and three anthers that dehisce following extrusion from the flower. Flowering usually occurs during early morning hours (H"0500 – 0800 hr), but may vary some with local or regional climatic conditions. According to A. A. Baltensperger (personal communication, 2001), in the United States desert Southwest, flowering may be initiated during the middle of the night from about 2300 to 0200 hr.

The generally strong self-sterility in *C. dactylon* is thought to result from genetic self-incompatibility. Studies by Burton and Hart (1967), Richardson et al. (1978) and Kenna et al. (1983) indicated plants to typically vary in selfed seed set from about 0.5 to 3%. Richardson et al. (1978) reported one clone with 2-yr mean selfed seed set of 8%. Taliaferro and Lamle (1997) demonstrated that self-pollination results in pollen tubes that grow slower and seldom reach the micropylar region of ovaries in comparison to tubes emanating from cross-pollination. Burton and Hart's (1967) research did not define the nature of self-incompatibility in bermudagrass, but led them to suggest a compatibility relationship of the diploid personate type of multiple oppositional factors.

Little information could be found regarding controlled studies on the self-fertility of other bermudagrass taxa. Burton (1965) reported self-incompatibility in four *C. transvaalensis* introductions. Baltensperger (1962) indicated that giant bermudagrass, *C. dactylon* var. *aridus*, plants under greenhouse isolation averaged about 10% selfed seed set in comparison to 2% selfed seed set of *C. dactylon* var. *dactylon* plants grown under the same conditions.

Controlled crosses between bermudagrass plants can be achieved by hand emasculation and pollination but the technique is tedious and time-consuming (Burton, 1965; Richardson, 1958). Burton (1965) used a fog chamber and manipulated light to facilitate controlled crossing. The usually strong self-incompatibility of bermudagrass plants enables the use of mutual pollination as a means of crossing. Growing such plants in isolated field plots or placing detached flowering culms of the respective plants together in a container of water isolated from other bermudagrasses results in mostly hybrid seed (Burton, 1965).

GERM PLASM COLLECTION, SELECTION, AND BREEDING HISTORY

Early Germ Plasm Dispersal

The widespread geographic distribution of bermudagrass, particularly the cosmopolitan *C. dactylon* var. *dactylon*, probably occurred through the combined efforts of man and nature over hundreds or perhaps thousands of years. From its apparent centers of diversity in Africa and Eurasia, bermudagrass was exported, by one means or another, to most landmasses around the world having climatic conditions compatible with its survival. Beehag (1992) stated that early records indicated common couch (*C. dactylon* var. *dactylon*) to be widely distributed in Australia at the time of European settlement and was found in remote areas not apparently disturbed by man. He indicated that it generally is considered to be indigenous to Australia.

The introduction of bermudagrass can be more confidently traced to the new world than can its introduction to regions of the old world beyond its putative regions of origin. Its introduction to America is of historical interest and of importance in providing the varied germ plasm leading to turf cultivar development and breeding improvement. Georgia Governor Henry Ellis is credited in the diary of Thomas Spalding, a prominent antebellum agriculturist and owner of Sapeloe Island, GA, as first introducing bermudagrass to the contiguous United States (Burton and Hanna, 1995; Kneebone, 1966). Kneebone (1966) indicated that bermudagrass may have been introduced to the new world, probably the West Indies, soon after its discovery by Columbus, and possibly on ships commanded by Columbus. He also stated that it may have been introduced to the Savannah, GA, area prior to 1751 by Robert Miller, a botanist employed by the Lords Proprietor to collect plants from the Caribbean Islands and Central America from 1733 to 1738. Early records imply that it spread rapidly in southern Colonial America, probably coinciding with the migrations of settlers and westward expansion of agriculture. Mease (1807) described it as one of the most important grasses of the South. Its early use as a turfgrass is exemplified by the story of a boatload of sprigs being shipped around 1830 from Baton Rouge, Louisiana, to Ft. Smith, Arkansas, to plant on a parade ground (Staten, 1952).

Affleck (1844), McCaughan (1843), and Spalding (1844) were enthusiastic supporters of bermudagrass for livestock herbage and soil stabilization. Howard (1881), as cited by Burton and Hanna (1995), indicated it to be the most important pasture grass of the South. Tracy (1917) noted its distribution in southern states from Maryland to California and described it as the most important perennial grass in southern states. He cited 'St. Lucie' as a common nonrhizomatous form used on Florida lawns because of its dwarf growth and refined texture. Kneebone (1966) provided evidence of bermudagrass in California in the mid-1800s and of its introduction into Arizona from sources including importation from California as a lawn grass, as contaminant seed in Mexican sesbania (*Sesbania* sp.) seed, and as viable seed passed through the digestive tracts of cattle imported to the region. By the early 1900s, bermudagrass had become a serious pest of irrigated alfalfa (*Medicago sativa* L.) in Yuma Co. Arizona and the Imperial Valley of California (Kneebone, 1966; Baltensperger et al., 1993). The more salinity tolerant invasive bermudagrass in alfalfa seed production fields produced seed that was sold in modest amounts as a by-product by 1917 (Baltensperger et al., 1993). Tracy (1917) stated that part of the commercial supply of bermudagrass seed in the United States at that time was produced in southern California, Arizona, and New Mexico. He noted that the price of bermudagrass seed had steadily declined due to the competition between domestic and imported sources and could be bought for $0.50 lb^{-1} ($1.10 kg^{-1}). Commercial bermudagrass seed was imported into the United States from Australia in the late 1800s and early 1900s (Tracy, 1917; Kneebone, 1966). By the 1930s, bermudagrass seed production had become a major agricultural enterprise along the Colorado river in Arizona and California. Most of the world's supply of bermudagrass seed continues to be produced in this area.

Bermudagrass was planted from sprigs, seed, or both over much of the United States during the last and first quarters of the nineteenth and twentieth centuries, respectively. Some of these plantings were in central and northern states beyond its area of best adaptation. From such plantings occasional plants survived and gained a foothold (Staten, 1952; Tracy, 1917). Today, bermudagrass is ubiquitous in the southern United States and is sparsely distributed in most central and northern states.

Early Turf Bermudagrass Development

Though bermudagrass probably has been incidentally used as turf for hundreds of years, actions leading to its widespread, intensive use for this purpose began around the start of the twentieth century. Records from this period indicated that turfgrass aficionados in different parts of the world actively sought bermudagrass germ plasm with superior turf attributes. The literature indicated that turfgrass enthusiasts in South Africa actively collected bermudagrasses with superior characteristics from around 1900 to 1930 (Roux, 1969). Examples include 'Florida' and 'Bradley,' respectively discovered about 1907 and 1910 near Johannesburg, 'Magennis,' discovered in 1920, and 'Royal Cape' discovered about 1930 on the Royal Cape Golf Course at Wynberg, Cape Town. These early discoveries were included in formal turfgrass research initiated in 1933 at the Frankenwald Botanical Research Station, University of Witwatersrand, Johannesburg, which subsequently added additional accessions to the bermudagrass collection (Hall et al., 1948). Other African Agricultural Research Stations, principally the Kitale Research Station in Kenya and the Henderson Research Station in Rhodesia (now Zimbabwe), collected bermudagrasses that ultimately were introduced into the United States. Many of these plant introductions are listed by Juska and Hanson (1964).

The search for superior turf bermudagrasses in the United States also gained momentum in the early 1900s. Bermudagrasses used as turf to that time were relatively coarse-textured, and on average probably little different from the "common" bermudagrasses presently ubiquitous to the region. Plantings of turf areas with bermudagrasses emanating from initial introductions and from imported and domestic seed identified the need for types better suited to specific turf uses. The early search for better turf types was led by practitioners such as U.S. Golf Association Green Section agronomists and golf course superintendents. For example, an early planting of experimental bermudagrasses by USGA agronomist C. V. Piper on the East Lake Country Club in Florida resulted in the 'Atlanta' cultivar (Latham, 1966).

Burton (1977) noted that the first bermudagrass putting greens in the South were planted with seed. The resulting heterogeneous populations produced putting surfaces of inferior quality and the stands quickly thinned due to various biotic and abiotic stresses to which they were subjected. These conditions, however, acted as a very effective screen for plants in the populations that could survive the stresses. Tolerant plants were selected, as in the case of the 'U-3' cultivar, one of many fine strains selected by Superintendent D. Lester Hall from putting greens on a golf course in Savannah, GA, (Juska and Hanson, 1964; Hanson, 1972; Burton, 1977). Plants collected under such conditions, along with many foreign introductions, principally from Africa, contributed to the germ plasm pool that evolved during the first half of the twentieth century. A worldwide *Cynodon* collection was amassed at the Oklahoma State University in the 1960s by J. R. Harlan and colleagues for use in biosystematic investigations of the genus (Harlan and de Wet, 1969). The available germ plasm pool has been broadened over the past 40 years with accessions of domestic and foreign origin. Still, the available genetic variation in *Cynodon*, particularly *C. dactylon* var. *dactylon*, is inadequately sampled and much potential exists to add valuable new germ plasm to the collection.

The vulnerability of bermudagrass taxa to erosion of natural genetic variation differs in magnitude, but is generally considered low because of their ability to thrive on disturbed sites as typical of secondary succession plants. The very narrow taxa endemic to southern Africa would logically seem to be at greatest risk for genetic erosion, but there is no evidence of such occurrence. High vulnerability is associated with the widespread use of very few clonally propagated turf bermudagrass cultivars (Taliaferro, 1995). Examples include the industry standard 'Tifgreen' and 'Tifway' cultivars released in 1956 and 1960, respectively (Hanson, 1972). These two cultivars continue to be the predominant ones used in the southern United States, and have been distributed worldwide where turf bermudagrass is grown. Such widespread use provides the potential for pest agents to develop new virulent strains as in the case of *Helminthosporium maydis* Nisik. & Miy. that caused the southern corn (*Zea mays* L.) leaf blight disease epidemic in 1970. The new physiologic strain of the pathogen was exceptionally virulent on corn hybrids with Texas male-sterile cytoplasm, which constituted about 90% of United States production at that time (Ullstrup, 1977)

Organized Breeding of Turf Bermudagrass

Scientific breeding of turf bermudagrass is of relatively recent origin and of small magnitude in terms of numbers of programs and scientists involved. The first documented turf bermudagrass breeding program was started in 1946 by Glenn Burton, USDA-ARS Geneticist at the Coastal Plains Experiment Station, Tifton, Georgia (Burton, 1991). This immensely successful breeding program has produced several cultivars, including the previously mentioned industry standards Tifway and Tifgreen. The program presently continues under the direction of W. W. Hanna. Breeding for improved seeded turf bermudagrass cultivars was conducted by W. R. Kneebone at the University of Arizona in the 1960s and 1970s (Kneebone, 1973). The Kansas Agricultural Experiment Station conducted turf bermudagrass breeding under the direction of R. A. Keen and H. A. Haekrott during the 1950s and 1960s. Turf bermudagrass breeding was initiated at the Oklahoma State University in the mid 1980s. Additional breeding programs were developed beginning in the mid 1980s, mainly by private industry and directed toward improvement of seed-propagated turf bermudagrass. Eighteen private seed companies were listed as sponsors of bermudagrass cultivars in the National Turfgrass Evaluation Program bermudagrass tests of 1986, 1992, and 1997, though not all companies were developers of sponsored cultivars (Anonymous, 1993, 1997, 1999).

Frey (1996) reported numbers of projects and scientific years (SYs) devoted to turf bermudagrass genetic improvement by State Agricultural Experiment Stations (SAES), USDA-ARS, and private industry. At the time of his report, the number of SYs and (projects) were: SAES—1.65 (3); USDA—ARS – 0.2 (1); private—2.7 (6).

Cultivar Composition and Breeding Techniques

Until recently, virtually all of the turf bermudagrass cultivars were clonally propagated plants, the major exception being the common bermudagrass grown for seed in Arizona and California. The seed-propagated cultivars developed over the past 20 years are predominantly synthetic cultivars produced by intercrossing two to several parent plants. Baltensperger and Klingenberg (1994) classified two seed-propagated cultivars, each having two parents, as F_1 hybrids.

The F_1 hybrid clonal cultivars have been developed by inter- and intraspecific hybridization and by mutation breeding. Burton (1973) used interspecific hybridization involving tetraploid *C. dactylon* parent plants to produce 'Tiflawn.' Intraspecific hybridization of tetraploid *C. dactylon* and diploid *C. transvaalensis* plants has been more extensively used to produce sterile triploid clonal cultivars. 'Tiffine' and Tifgreen were produced by controlled crosses of *C. dactylon* and *C. transvaalensis* parents (Burton, 1977, 1991; Hein, 1953). Tiflawn was the *C. dactylon* parent of Tiffine. Tifgreen resulted from crossing a fine-textured *C. dactylon* plant selected from a putting green on the Charlotte Country Club, Charlotte, NC, with a *C. transvaalensis* plant (Burton, 1991; Hein, 1961). Tifway is a putative chance hybrid of *C. transvaalensis* and *C. dactylon* found among plants started from a packet of *C. transvaalensis* seed from South Africa (Burton, 1966). 'Midway,' 'Midiron,' 'Midlawn,' and 'Midfield' are triploid intraspecific hybrids of the two taxa bred by the Kansas State University (Alderson and Sharp, 1995). Breeders have generally relied on mutual pollination procedures to produce inter- and intraspecific hybrids. Crosses controlling both parents have been effected mainly by collecting culms with inflorescences just before flowering and placing them together in a container of water in isolation from other bermudagrass. If placed away from air currents, agitation of the inflorescences during the flowering period will facilitate pollen dispersal. Such crosses have also been made by growing plants as mixtures or in close proximity within isolated field crossing blocks. Crosses controlling only one parent have been achieved principally by field polycrossing. For all of these methods, some selfing occurs, but the outcrossing nature of both *C. dactylon* and *C. transvaalensis* assures that many of the resulting plants are hybrids.

Ionizing radiation has been used to induce variation among clonal propagules of bermudagrass plants as a means of cultivar improvement. Powell et al. (1974) found 71 mutant plants of Tifgreen and Tifdwarf resulting from treatment of rhizomes with 90 or 113 Gy (9000 or 11300 rads) of Cobalt-60 g-radiation. Several mutational phenotypes were recovered from each cultivar and at treatment frequencies from 3.1 to 6.3%. The predominant changes were in color hue and dwarfed growth. Similar results were reported from subsequent experiments in which rhizomes and/or stolons of Tifgreen, Tifway, and 'Tufcote' were treated with Cobalt-60 g-radiation at dosages from 57 to 115 Gy (5700 to 11500 rads) (Powell, 1973). This work produced the mutant clones ultimately released as 'Tifgreen II' and 'Tifway II.' Tifgreen II was described as having lighter green color, greater cold tolerance, lower maintenance requirements, but being coarser in texture than Tifgreen (Burton, 1991). Tifway II was described as being superior to Tifway in resistance to root knot, ring and sting nematodes, frost tolerance, and spring greenup (Burton, 1985, 1991). Hanna (1986) induced fine-textured mutations in cold hardy, but coarse textured, Midiron, and dwarf mutations in Tifway and Tifway II. 'Tift 94' (now named 'TifSport') is one of 66 finer-textured mutants induced with 80 Gy (8000 rads) of Cobalt-60 g-radiation from Midiron (Hanna et al., 1997). Several year's field evaluation of the 66 mutant plants identified TifSport on the basis of tolerance to close mowing, turf quality, resistance to southern mole cricket (*Scapteriscus borellii* Giglio-Tos; syn *S. acletus* Rehn & Hebard), and greenup characteristics. TifSport has significantly higher turf quality than Midiron and sufficient cold tolerance for use in the northern part of the bermudagrass belt. 'TifEagle' was identified as an off-type genotype within a plot of an induced mutation (Mutant No. 2) of Tifway II. Mutant No. 2 was one of 48 putative mutants resulting from 79 Gy (7000 rads) of Cobalt-60 g-radiation of dormant Tifway II rhizomes (Hanna and Elsner, 1999; Hanna, 1986). Early screening of the 48 mutant plants included mowing at 6 mm height three times week^{-1} over two growing seasons. Subsequent evaluations of TifEagle showed it to be superior to Tifdwarf in putting green quality under close mowing (3 to 4 mm). It was described as producing more stolons and having shorter and narrower leaves than Tifdwarf.

Turf bermudagrass cultivars propagated by seed require breeding procedures effective in improving plant populations. Phenotypic, genotypic, and phenotypic-genotypic recurrent selection procedures have been used in breeding seeded turf bermudagrass (Baltensperger et al., 1993). These procedures have been used on the basis of studies indicating the presence of ample heritable variation in bermudagrass for many of the major characters needed in seeded turf bermudagrass. Wofford and Baltensperger (1985) reported moderate to high narrow sense heritability (H_n^2) estimates based both on progeny analyses and parent-offspring covariance for several characteristics. Visually assessed characters included color, density, appearance, growth, and seedhead production, and measured traits included leaf length and width, stem and stolon internode length, clipping weight, and regrowth. They concluded that for many of the characters, H_n^2 was sufficiently high for suitable genetic gains to be made without progeny testing. In a similar study, Coffey and Baltensperger (1989) concluded, on the basis of genetic variance and H_n^2 estimates, that chlorophyll concentration and associated visual color could be altered by either phenotypic recurrent selection or mass selection techniques. They further concluded that progeny testing would be required to alter plot (sod) density and clipping (biomass) weight. Cluff and Baltensperger (1991) reported heritabilities for seed yield and seed yield components in bermudagrass. On the basis of realized heritability estimates that they considered most accurate, they concluded that seed yield was partially determined by additive gene action and could be increased by phenotypic selection methods. Percentage seed set was the only component trait correlated with seed yield. Because the realized heritability for seed yield (0.42) was higher than that for percent seed set (0.31), they recommended direct selection for the former as the preferred means of increasing seed production.

Recurrent selection procedures have been used in the development of several synthetic turf bermudagrass cultivars (Baltensperger et al., 1993). Phenotypic recurrent selection was used to produce 'NuMex Sahara,' which compared to the seeded common bermudagrass produced in Arizona and California has shorter internodes, greater density, and better summer color. 'Sonesta' was developed by initially selecting six clonal plants on the basis of progeny testing. The six clones were polycrossed and populations then subjected to four cycles of phenotypic recurrent selection for turf quality and two cycles of phenotypic recurrent selection for seed yield. Phenotypic recurrent selection for turf quality and seed yield was used in the development of two synthetic cultivars, 'Yukon' (OKS 91-11) and 'OKS 95-1,' recently released by the Oklahoma State University.

Flowering Characteristics

No published information was found on the genetic and physiological mechanisms controlling flowering in bermudagrass. Bermudagrass typically flowers under field conditions in the spring in response to longer daylengths and perhaps increased temperature. Depending on latitude and climatic conditions, flowering generally begins from about late April to early June. In general, plants of *C. dactylon* var. *dactylon* are indeterminate and will produce inflorescences throughout the summer and fall if growth is maintained. Much variation exists within this taxon for flowering characteristics such as earliness and prolificacy of seedhead production. This variation extends to plants that set few or no seed. Seed production from the common bermudagrass grown in Arizona and California is enhanced by careful regulation of nitrogen fertilizer and soil moisture. Following the initial seedhead crop, mild water deficit stress is induced by withholding irrigation to promote a new flush of seedheads. Such stress may be repeated one to three times to produce two to four seedhead flushes that are harvested in midsummer, usually July. Nitrogen fertilizer is applied in quantities to maintain nonluxuriant growth. Summer and fall seed crops are typically harvested.

C. transvaalensis has somewhat different flowering habit than *C. dactylon*. Typically, plants will flower prolifically over a 4 to 5 week period in the spring, then produce few or no seedheads until fall. Seed set is generally good and, unlike *C. dactylon* var. *dactylon*, the seed shatter at maturity. During peak flowering, seedheads may be produced even under close mowing, detracting from turf quality.

Bermudagrass is generally recalcitrant to induction of flowering by manipulating lighting in growth chambers or greenhouses. A few photoperiod insensitive plants have been identified. One such plant, maintained at the Oklahoma State University for many years, flowers profusely in the greenhouse without artificial light during winter months as well as under greenhouse or field conditions during the summer.

Selection Criteria

Turf quality and adaptation to environmental conditions are the major criteria on which turf bermudagrass cultivar performance is based. These represent global interrelated traits conditioned by many component traits. Turf quality generally is visually assessed and represents a composite of factors such as plant color, texture, density, and uniformity. The major criteria associated with the environmental adaptation of cultivars are their response to abiotic and biotic stresses. The major abiotic factors are temperature extremes (particularly low temperatures), edaphic conditions (soil type and fertility), precipitation (amount and distribution), and suboptimal light. Major biotic influences include disease agents, nematodes, and insect pests. Genetic variation within bermudagrass has been documented for most of the traits conditioning turf quality and adaptation.

Growth Morphology and Color

Bermudagrass varies greatly in plant size as related to textural characteristics such as stem diameter, internode length, leaf length and leaf width. Variation for these traits exists in both major turfgrass taxa, *C. dactylon* var. *dactylon* and *C. transvaalensis*, though it is more extensive in the former. In general, plants of *C. dactylon* var. *dactylon* are relatively coarser in texture than those of the very fine textured *C. transvaalensis*. While *C. transvaalensis* contains plants of smallest size and most refined texture, undesirable characteristics include typically yellow green color, a propensity to thatch and scalp, and a general decline in quality in hot summer months. *C. transvaalensis* plants typically have good cold tolerance. Accordingly, hybridization between plants of the two taxa has been the principal means of developing cultivars that combine the desirable characteristics of the two species. Variation for plant type and other performance characteristics generally varies widely within F_1 plant populations.

Plant size and morphology determine performance traits that may differ in cultivars bred for different uses. The ability to quickly generate new growth for repair of physical injury is important in cultivars used on playgrounds, sports fields, and golf course fairways. The ability to tolerate frequent very close mowing is of paramount importance in cultivars for golf putting greens, bowling greens, croquet courts, and similar uses.

The merchandizing of clonally propagated turf bermudagrass cultivars as cut sod has grown dramatically over the past few decades. Accordingly, sod tensile strength is an important performance trait that must be considered in a breeding program. Wide differences exist among plants for this trait as a function of rhizome and shoot density and crown morphology.

Systematic selection to alter frequencies of genes conditioning turf morphological traits within populations has been restricted to *C. dactylon* var. *dactylon* in association with the breeding of seeded turf bermudagrass cultivars. The previously discussed studies of heritable variation for morphological traits in *C. dactylon* point to the potential for continued progress through recurrent selection procedures.

Abiotic Influenced Traits

Diversity exists within bermudagrass for freeze tolerance and chilling tolerance. Response of plants to freezing temperatures is perhaps the most important trait influencing adaptation. The wide diversity in bermudagrass for freeze tolerance has been well documented (Beard et al., 1980; Anderson et al., 1988, 1993). Freezing tolerance is a function of the ability of plant reproductive organs (rhizomes, crown, and sometimes stolon buds) to withstand temperatures less than or equal to 0°C. Though the inheritance and genetic and physiological mechanisms of freeze tolerance have not been elucidated, crosses between plants differing in freeze tolerance produce populations exhibiting a wide range for the trait. Significant variation also exists in plant response to chilling (0 to 10°C) temperatures that cause cessation of growth and discoloration (Dudeck and Peacock, 1985; White and Schmidt, 1989). The ability of cultivars grown in subtropical environments to respond to chilling temperatures with minimal loss of color and growth is a desired quality.

Though generally considered highly drought tolerant, substantial variation has been demonstrated in bermudagrass for dehydration avoidance and drought tolerance (Beard and Sifers, 1997) and water use (Kneebone and Pepper, 1982; Beard et al., 1992). Beard and Sifers (1997) reported substantial differences among 26 bermudagrass genotypes for dehydration avoidance and drought tolerance. They found that 12 *C. dactylon* x *C. transvaalensis* hybrids generally ranked in the middle for drought avoidance. In contrast, the 14 *C. dactylon* genotypes split about evenly into groups ranked highest and lowest in both drought avoidance and drought resistance. *C. dactylon* genotypes selected under subtropical conditions were generally highest for both traits, while those selected under temperate conditions were lowest. Key characteristics associated with superior drought avoidance and drought resistance were much greater root depth, root density, and root biomass. Carrow (1996) presented evidence emphasizing the importance of selecting for rooting characteristics as a means of increasing drought avoidance in turfgrasses.

Variation in bermudagrass of a magnitude permitting selection has been reported for salinity tolerance (Dudeck et al., 1983; Francois, 1988), iron chlorosis (McCaslin et al., 1981), and response to nitrogen and potassium fertilization influences on vegetation establishment (Trenholm et al., 1997) and growth, nonstructural carbohydrate concentration, and quality (Trenholm et al., 1998). Gaussoin et al. (1988) reported differences among 32 bermudagrass clones for response to reduced light intensity and concluded that enough variability exists to select for shade tolerance.

Biotic Influenced Traits

Bermudagrass is host to many diseases (Smiley, 1987), but few are considered serious. Leaf, stem, and crown rots caused by *Bipolaris cynodontis* (Marig.) Shoem. are among the more serious diseases of bermudagrass in humid regions. Spring dead spot is a major disease of bermudagrass with greatest incidence in the northern half of the use zone in the United States. Spring dead spot disease is caused by at least three ectotrophic, root-rotting fungi, including *Ophiosphaerella herpotricha* (Fr.) J. Walker, *O. korrae* Walker and Smith, and *O. narmari* n. comb. Bermudagrasses vary in response from highly tolerant to highly susceptible to both diseases (Wells, 1963; Baird et al., 1997). *Bipolaris cynodontis* has been shown to produce a selective phytotoxin called bipolaroxin that causes injury to tissues of susceptible cultivars (Sugawara, 1985). Other diseases of substantial importance in bermudagrass are brown patch caused by *Rhizoctonia* sp. and dollar spot caused by *Sclerotinia homoeocarpa* F. T. Bennett (Martin, 1994; Wells, 1963). Bermudagrasses are known to differ in susceptibility to these diseases. Bermudagrass decline caused by *Gaeumannomyces graminis* (Sacc.) Arx & Oliver var. *graminis* has become a serious disease of bermudagrass cultivars used on golf course putting greens in Florida (Elliott, 1993). The extent of variation among bermudagrasses in susceptibility to this disease has not been elucidated.

Nematodes capable of causing serious damage to bermudagrass include the lance (*Hoplolaimus* sp.), root-knot (*Meloidogyne* sp.), spiral (*Helicotylenchus* sp.), and sting

(*Belonolaimus* sp.). Tarjan and Busey (1985) documented variation among bermudagrass cultivars grown in soil containing at least eight different nematode species.

Important insect pests capable of causing serious injury to bermudagrass include the southern mole cricket (*Scaptericus acletus* Rehn and Hubard), tawny mole cricket (*S. vicinus* Scudder,) and tropical sod webworm (*Herpetogramma phaeopteralis* Guenee) in the southeastern United States and the fall armyworm (*Spodoptera frugiperda* J. E. Smith) over much of the bermudagrass belt (Reinert, 1982; Quisenberry, 1990). Moderate levels of resistance have been reported for mole crickets (Reinert and Busey, 1984; Busey, 1986; Braman et al., 2000), fall armyworm (Leuck et al., 1968; Quisenberry and Wilson, 1985; Croughan and Quisenberry, 1989) and tropical sod webworm (Reinert and Busey, 1983). A wide range of responses among cultivars to the bermudagrass mite (*Eriophyes cynodoniensis* Sayed.), including high levels of resistance, have been reported (Reinert, 1982; Tashiro, 1987).

Field screening under conditions imposing multiple stresses (abiotic and biotic) has identified bermudagrass genotypes incorporating strong tolerance or resistance to a range of abiotic and biotic induced stresses (Busey, 1986; Dudeck et al., 1994).

Breeding Progress

The development of the high quality turf bermudagrass cultivars, principally Tifgreen, Tifway, and Tifdwarf, in the mid 1900s represented landmark achievements that continue to serve as industry standards with mean turf quality not substantially different from the best of newer cultivars in NTEP trials (Table 15.2). However, increased breeding research with bermudagrass over the past quarter century has produced additional high quality clonal cultivars adapted to a broader range of environments. The development of Midlawn (Pair et al., 1994) and TifSport (Hanna et al., 1997) has helped satisfy the need for more freeze tolerant and higher quality cultivars adapted to zones of transition between cool- and warm-season species. FloraTex™ (Dudeck et al., 1994) represents a cultivar developed for its ability to perform well under low maintenance conditions. Over the past decade, new "super dwarf" bermudagrass cultivars have been developed with enhanced tolerance to lower mowing heights (White, 1999).

Dramatic improvement of seeded turf bermudagrass is indicated by NTEP data as a result of breeding effort over the past 15 years (Table 15.2). The best of the seeded cultivars in the most recent (1997) NTEP test has overall quality not significantly different from the best of the clonal cultivars. Additionally, some of the seeded cultivars have much greater freeze tolerance than Arizona Common, enabling their use in colder climates.

REFERENCES

Affleck, T. 1844. Comment to the editor. *Amer. Agric.* 3:335–336.

Alderson, J. and W.C. Sharp. 1995. *Grass Varieties in the United States*. CRC Lewis Publishers, Boca Raton, FL.

Anderson, J.A., M.P. Kenna, and C.M. Taliaferro. 1988. Cold hardiness of 'Midiron' and 'Tifgreen' bermudagrass. *HortSci.* 23:748–750.

Anderson, J.A., C.M. Taliaferro, and D.L. Martin. 1993. Evaluating freeze tolerance of bermudagrass in a controlled environment. *HortSci.* 28:955.

Anonymous. 1993. National Bermudagrass Test—1986. NTEP No. 93-1. National Turf Evaluation Program. U.S. Dep. Agric., Agric. Res. Service., Beltsville, MD.

Anonymous. 1997. National Bermudagrass Test—1992. NTEP No. 97-9. National Turf Evaluation Program. U.S. Dep. Agric., Agric. Res. Service., Beltsville, MD.

Table 15.2. Mean turfgrass quality ratings of bermudagrass cultivars in National Turfgrass Evaluation Program tests. Ratings were on a scale of 1 to 9 where 9 represents ideal turfgrass quality.

Comparison	1986 Test[a]			1992 Test[b]			1997 Test[c]		
	N	Mean	Range	N	Mean	Range	N	Mean	Range
Clonal entries									
All	21	6.32	5.5–6.7	10	5.56	4.5–6.3	10	6.29	5.6–6.8
Standards	5	6.32	5.7–6.3	4	5.85	5.4–6.1	3	6.50	6.4–6.6
Others	16	6.01	5.5–6.7	6	5.68	4.5–6.3	7	6.20	5.6–6.8
Seeded entries									
All	7	4.84	4.4–5.4	16	4.58	4.2–5.4	18	5.39	4.7–6.7
Standards	2	4.40	4.4–4.4	3	4.60	4.2–5.0	2	4.85	4.7–5.0
Others	5	4.04	4.7–5.4	13	4.89	4.4–5.4	16	5.46	5.0–6.7

[a] Data from NTEP No. 93-1, National Bermudagrass Test-1986, Final Report 1986–1991. Test conducted at 21 locations. Clonal standard cultivars: Tufcote, Tifgreen, Tifway, Tifway II. Seeded standard cultivars: Arizona Common, Guymon.

[b] Data from NTEP No. 97-9, National Bermudagrass Test-1992, Final Report 1992–1996. Test conducted at 25 locations. Clonal standard cultivars: Midiron, Tifgreen, Tifway, Texturf 10. Seeded standard cultivars: Arizona Common, NuMex Sahara, Guymon.

[c] Data from NTEP No. 00-4, National Bermudagrass Test-1997, Progress Report 1999. Test conducted at 19 locations. Clonal standard cultivars: Midlawn, Tifway, Tifgreen. Seeded standard cultivars: NuMex Sahara, Arizona Common.

Anonymous. 1999. National Bermudagrass Test—1997. NTEP No. 99-1. National Turf Evaluation Program. U.S. Dep. Agric., Agric. Res. Service., Beltsville, MD.

Arumuganathan, K., S.P. Tallury, M.L. Fraser, A.H. Bruneau, and R. Qu. 1999. Nuclear DNA content of thirteen turfgrass species by flow cytometry. *Crop Sci.* 39:1518–1521.

Baird, J.H., D.L. Martin, C.M. Taliaferro, M.E. Payton, and N.A. Tisserat. 1997. Bermudagrass resistance to spring dead spot caused by *Ophiosphaerella herpotricha*. *Plant Dis.* 82:771–774.

Baltensperger, A.A. 1962. Breeding bermudagrass. pp. 7–9 In *Report of Joint Meeting of The Western Grass Breeders Work Planning Conf. and 19th Southern Pasture and Forage Crop Imp. Conf.* 27–28 June 1962, Texas A&M University, College Station, TX. U.S. Dept. Agric., Agric. Res. Ser. Rpt. CR-25-63.

Baltensperger, A.A. and J.P. Klingenberg. 1994. Introducing new seed-propagated F_1 hybrid (2-clone synthetic) bermudagrass. *USGA Green Section Record* 32:14–19.

Baltensperger, A.A., B. Dossey, L. Taylor, and J. Klingenberg. 1993. Bermudagrass, *Cynodon dactylon* (L.) Pers., seed production and variety development. pp. 829–838 In R.N. Carrow, N.E. Christians, and R.C. Sherman, Eds. *International Turfgrass Soc. Res. J.* Vol. 7. Intertec Publ. Corp., Overland Park, KS.

Beard, J.B. 1973. *Turfgrass Science and Culture*. Prentice-Hall, Inc., Englewood Cliffs, NJ.

Beard J.B and S.I. Sifers. 1997. Genetic diversity in dehydration avoidance and drought resistance within the *Cynodon* and *Zoysia* species. *Intl. Turfgrass Soc. Res. J.* 8:603–610.

Beard, J.B, R.L. Green, and S.I. Sifers. 1992. Evapotranspiration and leaf extension rates of 24 well-watered turf-type *Cynodon* genotypes. *HortSci.* 27:986–988.

Beard, J.B, S.M. Batten, and G.M. Pittman. 1980. The Comparative Low Temperature Hardiness of 19 Bermudagrasses. Texas Agr. Expt. Stn. Progr. Rpt. 3835.

Beehag, G.W. 1992. Couchgrass culture in Australia. *Turf Notes* 11:10–11.

Braman, S.K., R.R. Duncan, W.W. Hanna, and W.G. Hudson. 2000. Evaluation of turfgrasses for resistance to mole crickets (Orthoptera: Gryllotalpidae). *HortSci.* 35:665–668.

Brilman, L.A., W.R. Kneebone, and J.E. Endrizzi. 1982. Pachytene chromosome morphology of diploid *Cynodon dactylon* (L.) Pers. *Cytologia* 47:171–181.

Burson, B.L. and C.R. Tischler. 1980. Cytological and electrophoretic investigations of the origin of 'Callie' bermudagrass. *Crop Sci.* 20:409–410.

Burton, G.W. 1947. Breeding bermuda grass for the southeastern United States. *J. Amer. Soc. Agron.* 39:551–569.

Burton, G.W. 1951. Intra- and inter-specific hybrids in bermudagrass. *J. Heredity* 42:153–156.

Burton, G.W. 1965. Breeding better bermudagrasses. pp. 93–96 In *Proc. IX Intl. Grassl. Cong.*, Jan. 7–20, 1965, Sao Paulo, Brazil.

Burton, G.W. 1966. Tifway (Tifton 419) bermudagrass. *Crop Sci.* 6:93–94.

Burton, G.W. 1991. A history of turf research at Tifton. *Green Section Record* 29(3):12–14. United States Golf Association, Far Hills, NJ.

Burton, G.W. 1973. Breeding bermudagrass for turf. pp. 18–22, In E. C. Roberts, Ed. *Proceedings of the Second International Turfgrass Research Conference*, Blacksburg, VA. Amer. Soc. Agron. and Crop Sci. Soc. Amer.

Burton, G.W. 1977. Better turf means better golf; the bermudagrasses, past, present, and future. pp. 5–7. In *Green Section Record*, March 1977. United States Golf Association, Far Hills, NJ.

Burton, G.W. 1985. Registration of Tifway II bermudagrass. *Crop Sci.* 25:364.

Burton, G.W. and W.W. Hanna. 1995. Bermudagrass. pp. 421–429. In Barnes, R.F., D.A. Miller, and C.J. Nelson. *Forages—An Introduction to Grassland Agriculture*, Vol. 1. Iowa State University Press, Ames, IA.

Burton, G.W. and R.H. Hart. 1967. Use of self-incompatibility to produce commercial seed-propagated F_1 bermudagrass hybrids. *Crop Sci.* 7:524–527.

Burton, G.W., R.N. Gates, and G.M. Hill. 1993. Registration of 'Tifton 85' bermudagrass. *Crop Sci.* 33:644–645.

Busey, P. 1986. Bermudagrass germplasm adaptation to natural pest infestation and suboptimal nitrogen fertilization. *J. Am. Soc. Hort. Sci.* 111:630–634.

Carrow, R.N. 1996. Drought resistance aspects of turfgrasses in the southeast: root-shoot responses. *Crop Sci.* 36:687–694.

Chippindall, L.K.A. 1947. The Common Names of Grasses in South Africa. Union of South Afr. Dep. Agric. Bull. 265. Union S. Afr. Government Printer, Pretoria.

Clayton, W.D. and S.A. Renvoize. 1986. Genera Graminum—Grasses of the World. Kew Bull. Additional Series XIII. Royal Botanic Gardens, Kew, UK.

Cluff, G.J. and A.A. Baltensperger. 1991. Heritability Estimates for Seed Yield and Seed Yield Components in Bermudagrass. Bull. 759. College of Agric. and Home Economics, New Mexico State University.

Coffey, B.N. and A.A. Baltensperger. 1989. Heritability estimates for selected turfgrass characteristics of bermudagrass evaluated under shade. pp. 117–119. In H. Takatoh, Ed. *Proc. 6th Int. Turfgrass Res. Conf.*, 31 July–5 Aug., 1989. Tokyo, Japan. Intl. Turfgrass Res. Conf. and Jpn. Soc. Turfgrass Soc.

Croughan, S.S. and S.S. Quisenberry. 1989. Enhancement of fall armyworm (Lepidoptera: Noctuidae) resistance in bermudagrass through cell culture. *J. Econ. Entomol.* 82:236–238.

de Silva, P.H.A.U. and R.W. Snaydon. 1995. Chromosome number in *Cynodon dactylon* in relation to ecological conditions. *Annals of Bot.* 76:535–537.

de Wet, J.M.J. and J.R. Harlan. 1970. Biosystematics of *Cynodon* L. C. Rich. (Gramineae). *Taxon* 19:565–569.

de Wet, J.M.J. and J.R. Harlan. 1971. South African species of *Cynodon* (Gramineae). *J. So. Afr. Bot.* 37:53–56.

Dudeck, A.E. and C.H. Peacock. 1985. Tifdwarf bermudagrass growth response to carboxin and GA_3 during suboptimum temperatures. *HortSci.* 20:936–938.

Dudeck, A.E., J.B Beard, J.A. Reinert, and S.I. Sifers. 1994. FLoraTeX™ Bermudagrass. Univ. Fla. Bull. 891.

Dudeck, A.E., S. Singh, C.E. Giordano, T.A. Nell, and D.B. McConnel. 1983. Effects of sodium chloride on *Cynodon* turfgrasses. *Agron. J.* 75:927–930.

Elliott, M.L. 1993. Bermudagrass decline: Transmission of the causal agent *gaeumannomyces graminis* var. *graminis* by vegetative planting material. *Intl. Turfgrass Res. J.* 7:329–334.

Forbes, I. and G.W. Burton. 1963. Chromosome numbers and meiosis in some *Cynodon* species and hybrids. *Crop Sci.* 3:75–79.

Francois, L.E. 1988. Salinity effects on three turf bermudagrasses. *HortSci.* 23:706–708.

Frey, K.J. 1996. National Plant Breeding Study—1. Human and financial resources devoted to plant breeding research and development in the United States. Special Report 98. Iowa State University. Ames, IA.

Gaussoin, R.E., A.A. Baltensperger, and B.N. Coffey. 1988. Response of 32 bermudagrass clones to reduced light intensity. *HortSci.* 23:178–179.

Gould, F.W. 1966. Chromosome numbers of some Mexican grasses. *Can. J. Bot.* 44:1683–1696.

Hall T.D., D. Meredith, S.M. Murray, and R.E. Altona. 1948. The effect of fertilizers on turf grasses at the Frankenwald Botanical Research Station. pp. 1–16. In T.D. Hall, D. Meredith, H. Weinmann, H.B. Gilliland, S.M. Murray, R.E. Altona, E. Goldsmith, G. Weinbrenn, R.E. Hurcombe, and M. le Roux, Eds. Experiments with *Cynodon dactylon* and Other Species at the South African Turf Research Station. African Explosives and Chemical Industries Unnumbered Publication.

Hanna, W.W. 1986. Induced mutations in Midiron and Tifway bermudagrasses. p. 175. *Agronomy Abstracts*. Amer. Soc. Agron., Madison, WI.

Hanna, W.W. and G.W. Burton. 1977. Cytological and fertility characteristics of some hybrid bermudagrass cultivars. *Crop Sci.* 17:243–245.

Hanna, W.W., G.W. Burton, and A.W. Johnson. 1990. Registration of 'Tifton 10' turf bermudagrass. *Crop Sci.* 30:1355–1356.

Hanna, W.W., R.N. Carrow, and A.J. Powell. 1997. Registration of 'Tift 94' bermudagrass. *Crop Sci.* 37:1012.

Hanna, W.W. and J.E. Elsner. 1999. Registration of 'TifEagle' bermudagrass. *Crop Sci.* 39:1258.

Hanson, A.A. 1972. Grass Varieties in the United States. USDA-ARS Agricultural Handbook No. 170. U.S. Gov. Printing Office, Washington, D.C.

Harlan, J.R. 1970. *Cynodon* species and their value for grazing and hay. *Herbage Abs.* 40:233–238.

Harlan J.R. and J.M.J. de Wet. 1969. Sources of variation in *Cynodon dactylon* (L.) Pers. *Crop Science* 9:774–778.

Harlan, J.R., J.M.J. de Wet, W.W. Huffine, and J.R. Deakin. 1970a. A Guide to the Species of *Cynodon* (Gramineae). Okla. Agric. Exp. Sta. Bull. B-673.

Harlan, J.R., J.M.J. de Wet, K.M. Rawal, M.R. Felder, and W.L. Richardson. 1970b. Cytogenetic studies in *Cynodon* L. C. Rich. (*Gramineae*). *Crop Sci.* 10:288–291.

Hein, M.A. 1953. Registration of varieties and strains of bermudagrass, II. [*Cynodon dactylon* (L.) Pers.]. *Agron. J.* 45:572–573.

Hein, M.A. 1961. Registration of varieties and strains of bermudagrass, III. [*Cynodon dactylon* (L.) Pers.]. *Agron. J.* 53:276.

Howard, C.W. 1881. *Manual of Cultivated Grasses and Forage Plants of the South.* James Harrison, Atlanta, GA.

Huffine, W.W. 1957. Sunturf Bermuda—A New Grass for Oklahoma Lawns. Okla. Agric. Exp. Stn. Bull. B-494.

Hunter, A.W.S. 1934. A karyosystematic investigation in the *Gramineae*. *Canadian J. Res.* 11:213–241.

Hurcombe, R. 1946. Chromosome studies in *Cynodon*. *S. Afr. J. Sci.* 42:144–146.

Hurcombe, R. 1947. A cytological and morphological study of cultivated *Cynodon* species. *J. So. Afr. Bot.* 13:107–116.

Hurcombe, R. 1948. A cytological and morphological study of cultivated *Cynodon* species. pp. 36–47. In *Better Turf through Research—Experiments with* Cynodon dactylon *and Other Species at the South African Turf Research Station.* African Explosives and Chemical Industries.

Johnston, R.A. 1975. Cytogenetics of Some Hexaploid x Tetraploid Hybrids in *Cynodon*. M.S. thesis. Oklahoma State University, Stillwater.

Juska, F.V. and A.A. Hanson. 1964. Evaluation of Bermudagrass Varieties for General-Purpose Turf. U.S. Dep. Agric., Agric. Res. Ser., Agric. Handbook No. 270.

Kenna, M.P., C.M. Taliaferro, and W.L. Richardson. 1983. Comparative fertility and seed yields of parental bermudagrass clones and their singlecross F_1 and F_2 populations. *Crop Sci.* 23:1133–1135.

Kneebone, W.R. 1966. Bermudagrass—Worldly, wily, wonderful weed. *Econ. Bot.* 20:94–97.

Kneebone, W.R. 1973. Breeding seeded varieties of bermudagrass for turfgrass use. pp. 149–153. In *Proc. Scotts Turfgrass Res. Conf. Vol. 4, Turfgrass Breeding.*

Kneebone, W.R. and I.L. Pepper. 1982. Consumptive water use by sub-irrigated turfgrasses under desert conditions. *Agron. J.* 74:419–423.

Latham, J.M. Jr. 1966. Better Bermudagrass Greens and Tees. Milwaukee Sewerage Commission. Bull. No. 6.

Leuck, D.B., C.M. Taliaferro, G.W. Burton, R.L. Burton, and M.C. Bowman. 1968. Resistance in bermudagrass to the fall armyworm. *J. Econ. Entomol.* 61:1321–1322.

Malik, C.P. and R.C. Tripathi. 1968. Cytological evolution within the *Cynodon dactylon* complex. *Biologisches Zentrablatt* 87:625–627.

Martin, B. 1994. Diseases of Turfgrasses in the Southeast. Clemson Univ. Coop. Ext. Ser. Pub. EB 146.

McBee, G.G. and E.C. Holt. 1966. Shade tolerance studies on bermudagrass and other turfgrasses. *Agron. J.* 14–17.

McCaslin, B.D., R.F. Samson, and A.A. Baltensperger. 1981. Selection of turf-type bermudagrass genotypes with reduced iron chlorosis. Commun. In *Soil Sci. & Plt. Anal.* 12:189–204.

McCaughan, J.J. 1843. Letter to the editor. *Amer. Agric.* 2:201.

Mease, J. 1807. A Geological Account of the United States, Comprehending a Short Description of Their Animal, Vegetable, and Mineral Productions. Birch and Small. Philadelphia, PA.

Moffett, A.A. and R. Hurcombe. 1949. Chromosome numbers of South African grasses. *Heredity* 3:369–373.

Ourecky, D.K. 1963. Pachytene chromosome morphology in *Cynodon dactylon* (L.) Pers. *The Nucleus* 6:63–82.

Pair, J.C., R.A. Keen, C.M. Taliaferro, D.L. Martin, J.F. Barber, and R.N. Carrow. 1994. Registration of 'Midlawn' bermudagrass. *Crop Sci.* 34:306–307.

Powell, J.B. 1973. Induced mutations in turfgrasses as a source of variation for improved cultivars. pp. 3–8. In E.C. Roberts, Ed. *Proceedings of the Second International Turfgrass Research Conference*, Blacksburg, VA. Amer. Soc. Agron. and Crop Sci. Soc. Amer.

Powell, J.B., G.W. Burton, and C.M. Taliaferro. 1968a. A hexaploid clone from a tetraploid x diploid cross in *Cynodon*. *Crop Sci.* 8:184–185.

Powell, J.B., G.W. Burton, and C.M. Taliaferro. 1968b. A hexaploid clone from a tetraploid x diploid cross in *Cynodon*. *Crop Sci.* 8:184–185.

Powell, J.B., G.W. Burton, and J.R. Young. 1974. Mutations induced in vegetatively propagated turf bermudagrasses by gamma radiation. *Crop Sci.* 14:327–330.

Powell, J.B., I. Forbes, and G.W. Burton. 1968b. A *Cynodon dactylon* (L.) Pers. (bermudagrass) clone with five accessory chromosomes. *Crop Sci.* 8:267–268.

Quisenberry, S.S. 1990. Plant resistance to insects and mites in forage and turf grasses. *Fla. Entomol.* 73:411–421.

Quisenberry, S.S. and H.K. Wilson, 1985. Consumption and utilization of bermudagrass by fall armyworm (Lepidoptera:Noctuidae) larvae. *J. Econ. Entomol.* 78:820–824.

Reinert, J.A. 1982. A review of host resistance in turfgrasses to insects and Acarines with emphasis on the southern chinch bug, pp. 3–12. In H.D. Niemczyk and B.G. Joyner, Eds. *Advances in Turfgrass Entomology*. Hammer Graphics, Inc., Piqua, OH.

Reinert, J.A. and P. Busey. 1983. Resistance of bermudagrass selections to the tropical sod webworm (Lepidoptera:Pyralidae). *Environ. Entomol.* 12:1844–1845.

Reinert, J.A. and P. Busey. 1984. Resistant varieties. pp. 35–40. In T.J. Walker, Ed. Mole Crickets in Florida. Univ. Fla. Agr. Expt. Stn. Bull. 846.

Richardson, W.L. 1958. A technique of emasculating small grass florets. *Indian J. Genet. and Plant Breeding* 18:69–73.

Richardson, W.L., C.M. Taliaferro, and R.M. Ahring. 1978. Fertility of eight bermudagrass clones and open-pollinated progeny from them. *Crop Sci.* 18:332–334.

Rochecouste, E. 1962. Studies on the biotypes of *Cynodon dactylon* (L.) Pers. *Weed Res.* 2:1–23.

Roux, W.M. 1969. *Grass, a Story of Frankenwald*. Oxford University Press, Cape Town, South Africa.

Royal Botanic Gardens, Kew. 1999. World grasses database. Published on the Internet; www.rbgkew.org.uk/herbarium/gramineae/wrldgr.htm [accessed 14 June 2001; 14:25 GMT]

Skerman, P.J. and F. Riveros. 1990. Tropical grasses. FAO Plant Production and Protection Series, No 23. Food and Agricultural Organization of the United Nations. Rome, Italy.

Smiley, R.W. 1987. Compendium of turfgrass diseases. *Amer. Phytopath. Soc. Press*, St. Paul, MN.

Spalding, T. 1844. Letter to the editor. *Amer. Agric.* 3:335.

Staten, H.W. 1952. *Grasses and Grassland Farming*. The Devin-Adair Co., New York.

Stent, S.M. 1927. South African species of *Cynodon*. *Bothalia* 11:274–288.

Sugawara, F. 1985. Bipolaroxin, a selective phytotoxin produced by *Bipolaris cynodontis*. *Proc. Nat. Acad. Sci.* 82:8291–8294.

Taliaferro, C.M. 1995. Diversity and vulnerability of Bermuda turfgrass species. *Crop Sci.* 35:327–332.

Taliaferro, C.M., A.A. Hopkins, J.C. Henthorn, C.D. Murphy, and R.M. Edwards. 1997. Use of flow cytometry to estimate ploidy level in *Cynodon* species. *Intl. Turfgrass Soc.* 8:385–392.

Taliaferro, C.M. and J.T. Lamle. 1997. Cytological analysis of self-incompatibility in *Cynodon dactylon* (L.) Pers. *Intl. Turfgrass Res. J.* (Part 1) 8:393–400.

Tarjan, A.C. and P. Busey. 1985. Genotypic variability in bermudagrass damage by ectoparasitic nematodes. *HortSci.* 20:675–676.

Tashiro, H. 1987. *Turfgrass Insects of the United States and Canada*. Cornell Univ. Press. Ithaca, NY.

Tracy, S.M. 1917. Bermuda grass. United States Department of Agriculture. Farmers Bulletin 814. U.S. Government Printing Office, Washington, DC.

Trenholm, L.E., A.E. Dudeck, J.B. Sartain, and J.L. Cisar. 1998. Bermudagrass growth, total nonstructural carbohydrate concentration, and quality as influenced by nitrogen and potassium. *Crop Sci.* 38:168–174.

Trenholm, L.E., A.E. Dudeck, J.B. Sartain, and J.L. Cisar. 1997. *Cynodon* responses to nitrogen, potassium, and day-length during vegetative establishment. *Intl. Turfgrass Soc. Res. J.* 8:541–552.

Ullstrup, A.J. 1977. Disease of corn. pp. 391–500. In G.F. Sprague, Ed. Corn and Corn Improvement. Monograph No. 18, Amer. Soc. Agron., Madison, WI.

Watson, L. and M.J. Dallwitz. 1992. *The Grass Genera of the World*. C.A.B. International, University Press, Cambridge, UK.

Wells, H.D. 1963. Georgia Turfgrass Diseases and Their Control. Ga. Agric. Expt. Stn. Cir. N. S. 39.

White, R. 1999. Unleash the full potential of new bermudagrass cultivars. *USGA Green Section Record* 37:16–18.

White, R.H. and R.E. Schmidt. 1989. Bermudagrass response to chilling temperatures as influenced by iron and benzyladenine. *Crop Sci.* 29:768–773.

Wofford, D.S. and A.A. Baltensperger. 1985. Heritability estimates for turfgrass characteristics in bermudagrass. *Crop Sci.* 25:133–136.

Chapter 16

Buffalograss, *Buchloe dactyloides* (Nutt.) Engelm

T.P. Riordan and S.J. Browning

Buchloe dactyloides (Nutt.) Engelm is the only native grass that is being used extensively as a turfgrass in the United States. Buffalograss is a perennial, stoloniferous, drought resistant, warm-season species. This grass has long been an important component of the central and southern Great Plains, thriving and spreading under semiarid conditions with heavy to moderate grazing (Quinn, 1987). Spreading by profusely branching stolons, buffalograss formed a dense sod that sustained vast herds of buffalo (American bison) before the turn of the century and is still highly regarded among stockmen for good palatability and high nutritional quality (Beetle, 1950; Huff and Wu, 1987; Wenger, 1943). Buffalograss sod provided the raw materials used by settlers to build sod houses and service buildings (Beetle, 1950; Pozarnsky, 1983; Wenger, 1943).

DISTRIBUTION AND TAXONOMY

Adaptation and Distribution

Reeder and Stebbins suggested that buffalograss originated in central Mexico (Shaw et al., 1987). However, buffalograss is native to an area that extends from southern Canada to central Mexico and includes approximately 492,100 km^2 (Beetle, 1950; Wenger, 1943). Buffalograss is largely co-dominant with blue grama, *Bouteloua gracilis* (H.B.K.) Laq. Ex Steud in an area characterized by annual rainfall of 381–635 mm (Savage and Jacobson, 1935; Shaw et al., 1987; Wenger, 1943). These two grasses comprise 90% of the native vegetation on nonsandy soils (Wenger, 1943). An area of secondary importance for buffalograss, which includes 606,060 km^2, exists around the perimeter of the primary zone of adaptation (Beetle, 1950; Wenger, 1943). Buffalograss occurs in these areas where favorable soils and extensive grazing allow it to thrive and replace less adapted grasses. Both the primary and secondary zones of adaptation extend to an altitude up to 1,824 m (Beetle, 1950).

Records for various states have shown a stabilized range of adaptation for buffalograss, with little change occurring since the time of settlement except where directly disturbed by plowing or building (Beetle, 1950). However, seasonal shifts in the eastern boundary have been found to occur due to fluctuations in annual rainfall. Buffalograss moved eastward during dry years in overgrazed areas and replaced the taller bluestem grasses (Beetle, 1950; Wenger, 1941; Wenger, 1943). The bluestems (*Andropogoneae* tribe, various species) are bunch grasses and are neither as drought nor wear resistant as buffalograss (Beetle, 1950). Under permanent grazing conditions, the bluestems are eliminated and replaced by dense stands of buffalograss (Beetle, 1950). With the return of favorable moisture, the bluestems reclaimed the area left to buffalograss (Pozarnsky, 1983; Wenger, 1943). Buffalograss cannot compete with the taller species under favorable conditions and is of little importance in the true tall grass prairie (Pozarnsky, 1983).

Buffalograss usually occurs in mixtures with other native grasses such as blue grama and side-oats grama, *Bouteloua curtipendula* (Michx.) Torr (Savage and Jacobson, 1935). Buffalograss prefers heavy soils and is dominant in those areas, while side-oats grama prefers and will increase its percentage of the population on sandy soils. Buffalograss does not thrive on sandy soils, and for this reason is not common in the sand hills areas. Blue grama prefers a medium weight soil and will fluctuate within the species mix, decreasing on heavy soils and increasing with lighter soils (Beetle, 1950). Buffalograss occurs only in pure stands in areas that are continuously overgrazed, areas that were once cultivated and have been revegetated, or in pastures that were so severely damaged during the drought of 1932–1934 that buffalograss recovered the fastest with the return of favorable moisture, reclaiming a majority of the land (Wenger, 1943). During prolonged drought, buffalograss is stimulated both during the initial dry years and during the first years of recovery. This is because western wheatgrass (*Agropyron smithii* Rydb.), blue grama and other less drought resistant species are less competitive and are slower to recover. In both cases competition for buffalograss is lessened (Beetle, 1950).

Erosion Control

In the late 1800s buffalograss was generally believed to no longer be a useful species and was expected to be extinct or occur only in isolated patches before the end of the century. C.E. Bessey from the University of Nebraska may have started this theory in 1893. He wrote, "It is the most valuable native pasture grass, but is rapidly passing toward extinction .... while it may endure in small isolated patches here and there for perhaps many years, it will ere long cease to have any agricultural interest" (Beetle, 1950). Overgrazing, control of yearly fires, and plowing of the prairie for farmland were the primary factors credited for the decrease of buffalograss (Beetle, 1950). Although these practices influenced the growth and spread of buffalograss, the simple misunderstanding of vegetative species fluctuation of vegetation due to seasonal changes in rainfall was probably at the base of this theory. Before 1932 many acres of buffalograss were plowed under for use as farmland before its true value to agriculture and soil conservation was known (Wenger, 1943).

Following the drought of the 1930s, buffalograss reestablished itself as a grass of great agricultural and conservation importance, having survived the combined effects of dust, drought, plowing, and overgrazing (Beetle, 1950; Pozarnsky, 1983). Its aggressive stoloniferous habit and dense sod-forming capabilities proved very effective at binding the soil to prevent wind and water erosion (Pozarnsky, 1983; Webb, 1941; Wenger, 1943). An erosion control study done by the Soil Conservation Service at Hays, Kansas from 1930 to 1938 determined that buffalograss sod lost an average of 1.0 Mg ha^{-1} of top soil annually (Ahring et. al., 1964). This was a substantial reduction compared to common grain crop rotations of the region. Buffalograss sod is ranked first among the native grasses for wind erosion control and is particularly valuable on heavy soils or where bermudagrass, *Cynodon dactylon* (L.) Pers., is not winter hardy (Beetle, 1950; Wenger, 1943). Buffalograss sod also exhibits high water-holding capacities, usually 57–60% of field capacity (Beetle, 1950).

Drought Resistance

Buffalograss also emerged as an excellent drought resistant native grass. Both buffalograss and blue grama are considered highly drought resistant and fully adapted to the driest sections of the central Great Plains (Briggs, 1914; Savage and Jacobson, 1935; Wenger, 1943). Beard and Kim (1989) estimated that buffalograss grown under optimum conditions within its area of adaptation has an evapotransporation rate of less than 6 mm d^{-1}, less than any other commonly used warm- or cool-season turfgrass. Some of the characteristics responsible for the drought resistance of buffalograss are extensive fine-branched root systems, aggressive low-growing aerial parts and the ability of the leaf blades to limit transpiration by rolling tightly during periods of stress (Engelke and Hickey, 1983; Savage and Jacobson, 1935). These characteristics enable buffalograss to go dormant sooner than other native grasses under stress conditions and to revive quickly with the first good rain (Beetle, 1950; Savage and Jacobson, 1935).

Survival of buffalograss to heat and drought has also been shown to increase in direct proportion to the closeness of cut (Savage and Jacobson, 1935). Clipping also reduces weed competition and favors spread (Savage and Jacobson, 1935). A study done by Savage and Jacobson (1935) at the Fort Hays Branch Experiment Station in the fall of 1934 showed the effects of heat and drought on buffalograss and blue grama. The climatic conditions that prevailed during the drought of 1933–1934 were the hottest and driest on record since 1894–1895. The period was characterized by high temperatures, hot winds, excessive evaporation, and widely interspersed ineffective light or torrential showers. Annual precipitation averaged more than a 153 mm deficit during this 2-year period compared to the average annual precipitation from the last 67-year period. The average mean temperatures for May, June, and July were the highest ever recorded at Fort Hays in 1934, and those for August were second highest. During that year, the daily air temperatures exceeded 38°C on 4 days in May, 10 days in June, 24 days in July, and 15 days in August. The highest evaporation over a growing season, April to September, was also recorded in 1934 at 1676 mm, compared to 1435 mm in 1933 and 1212 mm as the average for the last 28 years (Savage and Jacobson, 1935). As Shear observed, "Buffalograss will survive the greatest hardships and is about the last species to succumb under excessive use" (Beetle, 1950).

Taxomomy

Buffalograss is the only member of its genus and it is a member of the *Chlorideae* tribe (Hitchcock and Chase, 1935). *Bouteloua, Chloris,* and *Trichloris* are other genera in this tribe; the relationships among these four genera have not been described. Buffalograss and blue grama grass (*Bouteloua gracilis* [H.B.K.] Lag. ex Steud.) are commonly found together in the short-grass plains of the United States, and they are sometimes misidentified (Beetle, 1950). Although found together, they have substantially different characteristics. Buffalograss spreads by stolons and blue grama spreads by tillers. Also, blue grama has perfect flowers and buffalograss is dioecious. Buffalograss, with a base chromosome number of $n=10$ is found as diploid, tetraploid, and hexaploid clones (Johnson et al., 2000).

BUFFALOGRASS USE

Turf-Type Buffalograss

The resurgence of buffalograss is evident through its increased use in minimum care areas in the Central Great Plains. It has been used successfully on highway shoulders or right-of-ways, airfield runways, cemeteries, parks, golf courses and other athletic fields (Beetle, 1950; Pozarnsky, 1983; Wenger, 1943). 'Hays,' a cultivar produced by the Fort Hays, Kansas, Agricultural Experiment Station, has been used on airfields and landing strips in Africa and the South Pacific as well as in the United States for its superior wear resistance. Other uses for buffalograss include lining terrace channels and gullies, protecting earth fills around ponds, reservoirs, or garbage dumps (Beetle, 1950).

Aside from these uses and its perennial use as a forage grass, buffalograss has great untapped potential as a home lawn turf (Huff and Wu, 1987; Pozarnsky, 1983; Wenger, 1943). A turf-type buffalograss differs from a forage-type in several ways. The turf-type should have shorter, denser leaves and more stolons, with short internodes, which branch profusely, providing a dense, highly resilient turf (Wenger, 1941, 1943). It should also green up early in the spring and remain green later in the fall than a common buffalograss. As a dioecious species, the buffalograss pistillate plant, rather than the staminate plant, has been identified as most desirable for turf use due to a better correlation with the above-mentioned characteristics (Wenger, 1943). Staminate plants produce their inflorescence above the canopy of the grass plant which detracts from the appearance of the lawn. A male-biased turf would increase the frequency of mowing required to maintain a uniform turf (Wenger, 1943).

Over most of the Great Plains region, where it is naturally adapted and distributed, buffalograss produces an attractive lawn with a minimum of care. It reduces the need for frequent mowing due to a relatively slow vertical elongation rate and survives with a minimum of moisture. With evenly spaced summer rains, buffalograss will maintain its gray-green color throughout the growing season without supplemental irrigation (Wenger, 1943). In fact, forcing buffalograss into a faster growth rate through weekly irrigations can seriously harm the turf by encouraging a shallow root system, which reduces drought resistance, and by speeding up the natural tendency of buffalograss to become sod bound (Beetle, 1950; Wenger, 1943). Three to five years of irrigation will reduce the vigor of the turf and increase weed competition (Wenger, 1943). Buffalograss controls weeds by the very denseness of its stand; thus, weeds generally are not a serious problem on a properly managed buffalograss turf (Wenger, 1943).

Buffalograss is the only turf that can withstand all combinations of cold, heat, and drought and yet maintain an attractive turf under maximum usage with minimum care (Wenger, 1943). Its growth slows down with short days and low temperatures but buffalograss is not as affected by low temperatures as other warm-season grasses and can survive even the coldest winters except in the extreme north (Wenger, 1943). One disadvantage is the slow greenup response of buffalograss that imparts a brownish-tan color to the turf fairly late into spring (Beard, 1973; Beetle, 1950; McGinnies, 1979; Turgeon, 1980; Wenger, 1943). Buffalograss is a warm-season species and does not green up as early or retain its color as late in fall as cool-season turfgrasses (Pozarnsky, 1983; Wenger, 1943). Buffalograss does not thrive in dense shade or sandy soils and may be limited in its landscape uses for these reasons (Wenger, 1943). Buffalograss is adapted to a wide range of soil types, exhibiting a high alkali tolerance that makes it especially well suited to fine textured, alkaline soils (Beetle, 1950; Pozarnsky, 1983; Turgeon, 1980; Wenger, 1943). It can survive on badly eroded soils and has been found to adapt to even poorer soil conditions in moist climates, but drought resistance of buffalograss is especially high on heavy, fertile soils (Beetle, 1950; Pozarnsky, 1983; Wenger, 1943).

Buffalograss has also been found on rocky calcareous slopes in mixtures with blue grama, sand dropseed (*Sporobolus cryptandrus* [Torr.] Gray), side-oats grama, hairy grama (*Bouteloua hirsuta* Lag.), and little bluestem (*Andropoqon scoparius* Michx.) (Wenger, 1943).

It has been said that "no grass is more characteristic of or more nearly confined to the shortgrass plains area than buffalograss" and although it would be unwise to extend the use of buffalograss into an area where excessive care was required to maintain it, surprisingly good results have come from this turf outside its natural region of adaptation (Beetle, 1950; Wenger, 1941). Buffalograss was introduced to Virginia in 1856 according to the United States Patent Office Book of Agriculture (Pozarnsky, 1983). It was also grown successfully in New York where it showed greater resistance to chinch bug, *Blissus occiduus* Barber (*Hemiptera: Lygaeidae*), attacks than other lawn grasses in use at that time (Crocker, 1945). Buffalograss grown in central South Dakota, an area of secondary adaptation, proved to be more competitive than zoysia grass (*Zoysia japonica* Steud.) tested at the same time (Pozarnsky, 1983). These and other results give a favorable indication that a superior turf-type buffalograss could be used more widely than its specific zone of adaptation. Buffalograss regional evaluation trials have been conducted using six experimental lines of buffalograss. The regional trials evaluated the performance of each line for specific turfgrass characteristics (Huff et al., 1987).

Morphology

Buffalograss is a fine-textured perennial that forms a dense gray-green sod and turns a light straw to purplish red color when dormant (Beetle, 1950; Wenger, 1943). The leaf blades are curled and sparsely ciliate on the upper and lower surfaces (Beetle, 1950; Milby, 1971; Pozarnsky, 1983; Turgeon, 1980; Wenger, 1943). The leaves are usually 10–15 cm tall with a broad collar. Auricles are absent and the ligule is a fringe of hairs with short hairs in the center and longer hairs at the edges. The vernation is rolled.

Buffalograss is rapidly spread by many stolons, aboveground stems, which root freely at the nodes when in contact with moist soil (Pozarnsky, 1983; Wenger, 1941). Each stolon, in the common cultivars, is usually 60 to 75 cm long and can grow as much as 1 to 5 cm d^{-1} (Pozarnsky, 1983). The root system of buffalograss is extensive and fibrous. The majority of roots are in the upper 60–120 cm of the soil but roots have been excavated at depths up to 300 cm (Beetle, 1950; Pozarnsky, 1983; Webb, 1941).

As a dioecious species, buffalograss has separate male and female flowers that most commonly occur on separate plants (Beetle, 1950; Bur, 1951; Wenger, 1943; Wu et al., 1984). The staminate inflorescence is borne above the foliage on a culm usually 10–20 cm tall (Beetle, 1950; Quinn, 1987; Wu et al., 1984). It consists of one to three short spikes that hold the flowers in a flag-like shape (Wenger, 1940). The pistillate inflorescence is produced inside the canopy of leaves, so is normally hidden from view (Beetle, 1950; Quinn, 1987; Webb, 1941; Wenger, 1940; Wu et al., 1984). The female inflorescence produces caryopses enclosed within a hard bur that serves as a dispersal unit (Quinn, 1987; Wenger, 1940). One to five caryopses are usually contained in each bur (Webb, 1941). When moisture is abundant, burs are produced from midsummer until frost (Pozarnsky, 1983). The bur enclosure anchors and protects the caryopses from fire damage, a regular feature of the original prairie, and premature germination when sufficient moisture is not present (Quinn, 1987). The thick hull enclosing the caryopses requires considerable moisture for saturation and therefore germination, ensuring adequate moisture for the seedling (Quinn, 1987). Monoecious plants are occasionally found in which separate male and female flowers are produced on the same plant. Perfect flowers, or hermaphrodites, and other inflorescence variations have also been reported; however, monoecious and hermaphrodite inflorescences are self-incompatible (Beetle, 1950; Bur, 1951; Huff and Wu, 1987; Quinn and Engle, 1986; Wenger, 1940; Wu et al., 1984). Males and females do not differ significantly in either total biomass or in relative biomass allocation to crown, stolon, or sexual reproduction tissues (Quinn and Engle, 1986).

Sex Expression

The true sexual nature of buffalograss is rather obscure throughout its history, largely due to the difference in appearance of male and female inflorescences and the great amount of variation in sex expression. Originally, the male and female plants were assigned to separate genera. The more conspicuous, upright males were first to be discovered (Bur, 1951). In 1818 Nuttall examined a male plant in Missouri and assigned it to the genus and species *Sesleria dactyloides*. The female plant was not discovered until 1854. Found by Drummond in Texas, the female was described by Steudel who assigned it to the genus and specie *Antephora axilliflora* (Beetle, 1950; Bur, 1951; Huff and Wu, 1987). Not until 1959, when Engelmann examined a monoecious plant containing both male and female inflorescence, was it understood that the two were actually one species. He proposed a new name, *Buchloe dactyloides*, to include both forms (Beetle, 1950; Bur, 1951).

However, a great deal of controversy still exists in the scientific world over the true nature of buffalograss. Schaffner in 1920, Webb in 1941, and Ahring in 1964 accepted buffalograss as dioecious (Bur, 1951; Quinn and Engle, 1986). Savage in 1934, and Anderson et al. in 1937, and Wenger in 1943 considered it to be unisexual and only occasionally dioecious (Bur, 1951; Quinn and Engle, 1986). Savage considered the perfect condition to be the true ancestral form of buffalograss (Bur, 1951). Others believed that unisexual stolons that spread and intermingled accounted for the seemingly dioecious nature of the plant (Plank, 1892). Plank in 1892, Hitchcock in 1885, Gernert in 1937, and Durham in 1942 vehemently insisted on a monoecious nature (Bur, 1951; Plank, 1892; Quinn and Engle, 1986). Buffalograss is now excepted to be dioecious with occasional monoecious plants occurring (Quinn and Engle, 1986). Studies done by Wu et al. (1984) have determined the ratio among male, female, and monoecious plants to be 1:1:1.

Shaw et al. (1987) suggested that dioecy may have developed in buffalograss as a colonizing mechanism in response to increasingly arid environment following the rise of the Rocky Mountains in the early tertiary period. Synaptospermy, the keeping together of caryopses with a bur until germination, counteracts a potential problem of dioecy through the concurrent dispersal of both a male and a female to a new site simultaneously (Quinn, 1987). In observations from an Oklahoma native and a Kansas cultivar, Quinn found at least 50% of burs contained both male and female caryopses (Quinn, 1987; Quinn and Engle, 1986). The presence of both male and female progeny from the caryopses of a single bur make possible the colonization of an area by one bur (Quinn, 1987; Quinn and Engle, 1986; Webb, 1941). However, observations by Quinn and Engle (1986) of a native clone and a cultivar showed marked differences in the sex ratios of offspring from a single bur. Oklahoma native buffalograss exhibited a roughly 1:1 ratio, while the Kansas cultivar showed a female bias of progeny from a single bur. The cultivar progeny also showed 20% monoecious expression. Quinn and Engle (1986) estimated that sexual reproduction could occur following dispersal of 63% of the cultivar burs studied compared to only 47% of the Oklahoma native burs studied. In general, the differences exhibited by the two clones conformed to those shown by the overall population samples (Quinn and Engle, 1986). In addition, multiple caryopses within a bur provide several chances at establishment through differing germinability of each caryopsis. This also reduces competition between seedlings (Quinn, 1987; Quinn and Engle, 1986).

Establishment Methods

Establishment of a buffalograss turf in areas suited to its growth and spread can be accomplished by three methods: natural revegetation, vegetative propagation, and seeding (Pozarnsky, 1983; Wenger, 1943). Vegetative propagation and seeding are the most practical methods since natural revegetation of an area by buffalograss can take from 25 to 40 years (Beetle, 1950; Clark, 1945; Wenger, 1943). Of these two methods, vegetative propagation through plugs or sod has been used most often due to the difficulties associated with the production of seed and its prohibitive cost. However, the cost of revegetation of an area with buffalograss plugs or sod can also be prohibitive due to the large amounts of labor involved even with a small lawn (Clark, 1945; Wenger, 1943).

One recommendation made by Tom Pozarnsky, a soil conservationist with the United States Department of Agriculture Natural Resource Conservation Service, indicated that sod pieces or plugs about 8–10 cm in diameter should be spaced 30 cm apart and will usually result in complete sod cover in one growing season (Pozarnsky, 1983). However, even wider spacings can result in a complete sod cover in one season as indicated in a study done by Webb (1941). This study showed that plants placed 60 cm apart formed a solid cover in 4 months, and individual plants were indiscernible. Plants spaced at 90 cm also formed a solid, although less dense, cover. Spacings of 120 and 150 cm did not result in a solid turf cover in one growing season. This indicated that a plug spacing from 30–60 cm would produce a satisfactory turf in one season, while the 60 cm spacing maximized plant and labor costs. Preliminary results of research done at the University of Nebraska-Lincoln on buffalograss vegetative establishment by Schwarze et al. (1987) indicated that prerooting plugs in a greenhouse or under clear plastic in the field may increase establishment rates even further. Soil type should be considered before revegetating an area, and some control of weeds would be highly desirable while the turf is becoming established (Pozarnsky, 1983; Wenger, 1943). Recent commercial efforts by sod producers indicated that 2.5-cm diameter plugs can be greenhouse grown in a soilless mix and be marketed by mail (Thorson, 2001). These small plugs planted at 30 cm spacing can establish in a single growing season.

Several obstacles prevent the widespread use of buffalograss caryopses for the establishment of turf. These include high cost of burs and limited availability of deburred caryopses. Buffalograss burs are limited in availability primarily due to the short female inflorescence which makes harvesting difficult (Clark, 1945; Huff and Wu, 1987; Wenger, 1941). Attempts at developing harvesting equipment for buffalograss burs has been difficult and largely unsuccessful (Wenger, 1941). Low bur yield per plant is another limiting factor in buffalograss availability (Wenger, 1943). Natural stands have yielded as much as 112 kg ha^{-1} of clean burs but normal yields will vary from 22–112 kg ha^{-1} (Beetle, 1950). Native stands, however, are not a reliable source of seed (Wu et al., 1984).

At the Fort Hays, Kansas, Agricultural Experiment Station buffalograss selections were chosen for seed borne at heights at least 102 mm to facilitate harvesting. In this study, selections that produced a yield in burs of 336 kg ha^{-1} were considered favorable and worthy of further testing. Unfortunately, higher seed heights were usually accompanied by taller foliage (Wenger, 1941). Similarly, seed produced at good heights was often of low quality. Plants with good seed culm height were usually forage types. However, exceptions were found. Although buffalograss caryopses are not adversely affected by mechanical processing to remove them from the bur, bur shattering during harvesting results in the loss of many valuable caryopses (Ahring et al., 1964).

Poor germination of buffalograss burs presents another drawback to widespread use. Initially buffalograss seed was thought to be of such low quality due to its poor germination that satisfactory stands could never be obtained (Wenger, 1941; 1943). In reality, mature buffalograss burs enter a very pronounced dormancy stage that allows only a very small percentage of seed to germinate the first year (Wenger, 1943). The hard outer bracts of the bur contain oils that prevent the absorption of water and other gases required for germination (Ahring and Todd, 1977). This results in very poor germination rates from areas seeded with freshly harvested buffalograss burs (Wenger, 1943). Quinn suggested that differences in the intensity of dormancy may exist among cultivars of buffalograss (Quinn and Engle, 1986). Various methods of increasing germination have been tried including natural weathering, mechanical treatments, acid and chemical dust treatments, soaking and chilling treatments, and embryo culture techniques that bypass bur dormancy (Ahring and Todd, 1977; Ahring et al., 1964; Brown, 1945; Hickey et al., 1983; Parkey, 1952; Pladeck, 1940; Wenger, 1943).

Mechanical removal of the caryopses from the bur eliminates the germination problem but increases the cost of seed (Ahring and Todd, 1977). Approximately 40,000 to 50,000 pure caryopses can be obtained from 454 g of buffalograss burs. Chemical treatments are also available that dramatically improve the germination rates of caryopses or burs (Beetle, 1950). Good quality seed with less than 40% germination should be treated before planting (Beetle, 1950; Wenger, 1943). The most effective and commonly used treatment is a soaking and chilling method called the 'Hays' treatment (Wenger, 1943). The burs are soaked for 24 hours in a 0.5% solution of saltpeter (KNO_3) followed by six weeks of moist chilling between 32° and 40°F. This treatment increases germination to 95% or 100%. Gentian violet can also be used along with the KNO_3 treatment at concentrations of 1 part dye to 40,000 parts liquid for marking the treated burs (Wenger, 1943). Other recommendations for chemical treatment include salt (NaCl) solutions of 0.02% following the same procedure as the 'Hays' treatment method but using variable chilling temperatures, and 5.25% sodium hypochlorite solution (full strength commercial bleach) soaking burs for 12 hours, followed by a five-minute cool tap-water rinse (Hickey et al., 1983; Wenger, 1943). Both of these methods should give 50% or better germination.

Clark (1945) indicated that germination of buffalograss burs and caryopses improves for at least the first three years in storage, and that processed seed cannot be stored as long as unprocessed seed. Partial viability of unprocessed seed has been documented over long periods of time. Specifically, A.E. Lowe at the Garden City Branch of the Kansas Experiment Station found 28% viability from buffalograss burs recovered from a sod house 25 years after harvest (Wenger, 1943). However, results from research done by J.L. Svoboda (1987) at the University of Nebraska on buffalograss storage show good germination, 96–110% (more than one caryopsis per bur), of buffalograss burs and caryopses after 3 months under cold storage. Longer storage periods, 9 and 15 months, revealed a decrease in germination that was most dramatic in buffalograss burs.

Buffalograss should be seeded in the late spring, and seedling emergence usually coincides with the last date of killing frost (Beetle, 1950; Pozarnsky, 1983; Wenger, 1943). Early plantings may be troubled by weeds, but will provide a solid lawn sooner than late plantings (Wenger, 1943). Optimum moisture availability is the most important factor in the timing of buffalograss plantings when considering spring or fall planting dates (Beetle, 1950). Seeding rates range from 2.5 to 10 g m^{-2}, and require high quality seed with a minimum of 85% purity (Beetle, 1950; Wenger, 1943). Seeds should be planted at 1.25 to 2.50 cm depths. Seeding depth is very important since most complete failures in seeding have been directly traceable to seeding too deeply (Wenger, 1943). When moisture and temperature are favorable, 1.25-cm planting depths are optimum, while 2.50-cm planting depths should be used in soils with high sand content (Beetle, 1950; Wenger, 1943).

Buffalograss germination is affected by some herbicides. Specifically, germination is totally suppressed by applications of picloram (*4-amino-3,5,6-trichloropicolinic acid*) and triclopyr (*[(3,5,6-trichloro-2-pyridinyl)oxy]acetic acid*) at rates greater than 1.1 kg ha^{-1} and by *2,4,5-trichloroacetic acid* at 9.0 kg ha^{-1} (Huffman and Jacoby, 1984). The use of nurse grasses with buffalograss plantings is not recommended due to the competition for moisture and shading. In addition, the texture and appearance of a buffalograss turf is unique and would not be compatible with any remnant nurse grass that might persist. Tillering usually begins 10–14 days after seedling emergence and stolons appear 30 days after emergence (Beetle, 1950).

Management

Buffalograss has been considered an ideal low maintenance species, requiring reduced amounts of water, fertilizer, and pesticides. It also requires less mowing and it produces fewer clippings for disposal (Frank et al., 1997). Research in Nebraska, Kansas, and Utah by Frank et al. (1997), using vegetative and seeded turf-type buffalograss cultivars, and mowing (2.5, 5.0, and unmowed) and fertilizer treatments (0, 98, and 195 kg N ha^{-1}) have shown that although buffalograss is a lower maintenance requiring species, quality can be improved by a moderate management program. Although the results were different in the three locations, there were some results that were similar. The nitrogen rate by year interaction showed that the 98 kg N ha^{-1} rate sustained quality, color, and density over a three-year period. Also, depending upon the use and mowing requirements, the optimal mowing height ranged from 2.5 cm for NE 91-118, a low mowing tolerant buffalograss, to 5.0 to 7.5 cm for seeded cultivars (Frank et al., 1997)

Potential Pest Problems

Buffalograss is susceptible to several diseases although the true economic importance of each is not known. None of the presently known diseases appear to cause excessive damage where the grass is used strictly for grazing, but their effect on a buffalograss turf is not known (Wenger, 1943). Variation among buffalograss selections in disease resistance is not extensive (Wenger, 1943). False smut, caused by the organism *Cercospora seminalis* Ellis & Everh. affects the burs and destroys infected caryopses (Wenger, 1943). However, not all caryopses within a bur are infected. Incidence of false smut is higher in areas of high rainfall and is aggravated by irrigation (Wenger, 1941, 1943). Leaf blotch is a foliar disease that causes premature drying of the leaves (Wenger, 1941). The disease is caused by the fungus *Helminthosporium inconseicuum* var. *Buchloe*, and it destroys the ability of the plant to cure into a palatable nutritious winter forage, causes buffalograss turf to look dull and unthrifty, and detracts from the uniform appearance of the lawn (Wenger, 1943). Leaf rust caused by *Puccinia kansensis* Ellis and Barth is generally considered less serious than leaf blotch. Under the maintenance conditions of a lawn or golf course, the infections are seldom severe even during the wettest seasons.

Relatively few insects are known to cause significant damage to buffalograss. Arthropods previously reported as pests of buffalograss include white grubs, *Phyllophaga crinita* (Burmesiter); grasshoppers; leafhoppers; mound-building prairie ants; buffalograss webworm, *Surattha indentella* (Kearfott); the rhodesgrass mealybug, *Antonina graminis* (Maskell); an eriophyid mite, *Eriophyte slykhuisi* (Hall); and two grass-feeding mealybugs, *Tridiscus sporoboli* (Cockerell) and *Trionymus* sp. (Baxendale et al., 1994; Chada and Wood, 1960; Crocker et al., 1984; Pfadt, 1984; Reinhard, 1940; Sorenson and Thompson, 1979; Wenger, 1943). Several beneficial arthropods have also been collected from buffalograss, including ants, big-eyed bugs, ground beetles, rove beetles, spiders, and numerous hymenopterous parasitoids (Heng-Moss et al., 1998).

Recently the buffalograss mealybug has been identified as a new problem on buffalograss: *Tridiscus sporoboli* and *Trionymus* spp. Two mealybug species have been found in large quantities on buffalograss. No comparable sample of the *Trionymus* species has previously been identified. Both species of the buffalograss mealybug are associated with a gradual decline in turf vigor accompanied by thinning and browning of the turf, especially during mid- to late summer. A species difference has been indicated for the level of damage found on selected buffalograss lines and the number of mealybugs present. Some buffalograss lines have shown severe damage with relatively few mealybugs present while other lines are largely undamaged at high infestation levels. Management practices for the buffalograss mealybug have not yet been developed but may include the use of resistant buffalograss lines, systemic insecticides, and cultural practices such as mowing (Baxendale et al., 1994).

Also, the chinch bug, *Blissus occiduus* Barber (Hemiptera: Lygaeidae), has emerged as an important insect pest of buffalograss in Nebraska (Baxendale et al., 1999). *B. occiduus* was first described by Barber in 1918 from specimens collected in Ft. Collins, CO, and Geronimo, NM (Barber, 1918). Currently, the reported distribution of *B. occiduus* includes California, Colorado, Montana, Nebraska, and New Mexico in the United States, and Alberta, British Columbia, Manitoba, and Saskatchewan in Canada (Bird and Mitchener, 1950; Slater, 1964; Baxendale et al., 1999). Chinch bugs injure buffalograss by withdrawing sap from plant tissues in the crown area and stolons. Feeding initially results in reddish discoloration of plant tissues, followed by irregular patches of yellowing and/or browning turf. At higher infestation levels, chinch bug feeding has the potential to cause severe thinning or even death of buffalograss stands (Baxendale et al., 1999).

BUFFALOGRASS IMPROVEMENT

Breeding and Selection

Early improvement of buffalograss, begun in 1936, emphasized selection of superior types (Huff and Wu, 1987; Wenger, 1943). Large collections of buffalograss from the Great Plains were brought into observation research sites where great variability was found within these collections (Wenger, 1943). Selections from northern sites were found to be more desirable as turf-types due to their short, dense canopy and slow spreading habit (Beetle, 1950; Wenger, 1941, 1943). Northern selections were also found to require fewer mowings; however, they went dormant in the fall earlier than the southern selections and produced seed on very short pistillate inflorescences (Beetle, 1950; Wenger, 1943). Southern selections exhibited increased vigor, longer growing seasons, taller seed stalks, and increased disease resistance, coupled with greater susceptibility to winter injury and a tall, thin, rapidly spreading growth habit (Beetle, 1950; Wenger, 1943). Desirable characteristics included a short, dense, slowly spreading growth habit; heavy seed production; taller seed stalks; increased disease resistance and female-biased population. Turfs with a female-biased population were found to be more desirable due to the absence of staminate inflorescence that detracts from the appearance of the turf (Wenger, 1941, 1943).

The two methods used for the early improvement of buffalograss included selection and hybridization, with selection done under dry land conditions so that no loss of drought resistance resulted (Wenger, 1941, 1943). One difficulty faced by breeders was finding many desirable characteristics combined in one plant, and it was even more difficult to retain desirable characteristics in succeeding generations after hybridization (Wenger, 1943). However, many superior qualities resulting from selection could be transmitted to a high degree to the offspring after relatively few generations of selections (Wenger, 1941). Early improvement of buffalograss during the 1930s focused mainly on the development of improved forage types. Although the improved types developed did not breed true, this was not a serious drawback since their main use was forage. These efforts at Fort Hays, Kansas, resulted in the release of 'Texoka,' Hays 'One-Eye,' and 'Sharps Improved,' three improved forage types of buffalograss (Table 16.1; Ahring et al., 1964).

Another difficulty addressed by breeders more recently is the apparent instability of buffalograss sex expression (Quinn, 1987; Quinn and Engle, 1986; Wenger, 1943). Sex ratios between the three commonly occurring sex forms—male, female, and monoecious—has been reported to be 1:1:1, or slightly female biased (Quinn and Engle, 1986; Shaw et al., 1987). A female biased sex ratio has been reported in the cultivar Sharps Improved (Quinn and Engle, 1986). Reports of male biases within buffalograss populations were all made from field observations and may reflect either differing flower phenology or greater staminate inflorescence visibility (Plank, 1892; Quinn and Engle, 1986). One theory presented by Wilson to account for biased adult sex ratios stated that they may result from differing mortality rates, length of reproductive life, and/or differences in vegetative growth (Shaw et al., 1987). Female biased sex ratios in buffalograss populations are supported by the evidence of varying sex ratios found in other grasses (Shaw et al., 1987).

Initially, environmental conditions were thought to influence sexual instability. A study done by Huff and Wu (1987) examined the response of sex ratios within two buffalograss lines to varying environmental conditions. The cultivars used were a Texas native population that showed 17% monoecious expression and a Colorado cultivar with 38% monoecious expression. The most common form of sex expression was dioecious in both lines. If sex expression could be altered by environment, the treatments used would reveal the nature of the instability. In true male or female plants, sex expression was not affected by any environmental condition; plants showing monoecious expression were the only plants affected by environment. No significant difference was found in sex expression between years. Among monoecious plants, warm temperatures, high light intensity, and low nitrogen levels promoted greater female expression. Male expression was promoted by cool temperatures, low light intensity, and high nitrogen levels. In a related study, Wu et al. (1984) found that mowing increased the number of female inflorescences while encouraging males to reproduce vegetatively.

Another study conducted by Quinn and Engle (1986) used the Kansas cultivar Sharps Improved and an Oklahoma native population of buffalograss. The plants were grown both in the greenhouse and under field conditions for several growing seasons. Sex ratios of the offspring were markedly different for the two populations, even though sex expression was consistent between replication and between greenhouse and field grown plants within each line. The Oklahoma native showed a 1:1 ratio between male and female offspring, with complete stability of expression throughout the study. No

Table 16.1. Commercially available buffalograss cultivars.

Cultivar	Year Released	Sex	Adaptation[a]	Propagation Method	Performance[b]
Legacy	1999	F	N/T	Sod/Plug	Excellent
Scout	1999	F	S/T	Sod	Excellent
609	1993	F	S/T	Sod	Excellent
315	1995	F	N/T	Sod/Plug	Good
378	1995	F	N/T	Sod/Plug	Excellent
Prairie	1991	F	S/T	Sod	Good
Stampede	1997	F	S/T	Sod	Good
Cody	1997	M/F	N/S/T	Seed	Excellent
Bison	1996	M/F	N/T	Seed	Good
Texoka	1965	M/F	N/T	Seed	Fair
Sharps Improved	1965	M/F	N/T	Seed	Fair
Topgun	1997	M/F	S/T	Seed	Good
Plains	1997	M/F	S/T	Seed	Fair

[a]N = Northern U.S.; S=Southern U.S.; T=Transition Zone.
[b]This is performance observed in the National Turfgrass Evaluation Program results (NTEP, 1999).

monoecious plants were observed within the Oklahoma native population. However, the Kansas cultivar varied in sex expression of progeny although no detectable correlation was found between environmental conditions. Thirteen percent of the cultivar plants were monoecious and data supported the evidence that Sharps Improved exhibits a female bias among unisexual plants.

Increases in sex expression instability have been shown to correlate positively with inbreeding. This is supported by observations from Quinn and Engle (1986), which showed some wild populations with 1:1 male to female sex ratios and no monoecious plants. The increase in monoecious plants in buffalograss cultivars is a result of selection during the process of seed improvement, as evidenced by two cultivars, Sharps Improved and the Colorado cultivar used by Huff and Wu (1987), which showed an increase in monoecy. However, differences in the frequency distribution of monoecious plants naturally occurs in wild populations. Huff and Wu (1987) suggested that sex expression is basically a genetically determined character. Again, this theory is supported by evidence from Quinn and Engel (1986) that showed a high heritability of sex expression in stolon transplants from their parent clumps, even when subjected to differing environmental conditions.

In the mid 1980s environmental concern and ongoing support from the United States Golf Association led to increased emphasis on the breeding and development of buffalograss at Texas A & M University, Dallas; University of California, Davis; and the University of Nebraska. The primary objective of these programs was to develop cultivars of buffalograss that would reduce water and energy inputs by 50% in golf course and other turf situations (DeShazer et al., 1992). A number of new vegetative and seed propagated cultivars were developed and tested because of this renewed interest.

Buffalograss Cultivars

'Prairie,' released by the Texas A & M Agricultural Experiment Station in 1989, was the first buffalograss developed specially for turfgrass use (Table 16.1; Engelke et al., 1990). This release had improved turfgrass quality, a light-green color, and uniformity. The University of Nebraska Agricultural Experiment Station released '609' buffalograss in 1991. This cultivar has a darker green color and slightly improved turfgrass performance compared to Prairie (Riordan et al., 1992). Both Prairie and '609' are vegetatively propagated cultivars and they were the first buffalograss cultivars marketed using conventional sod techniques. 'Legacy' buffalograss was released by Nebraska in 2000 and it has improved quality, color, and low mowing tolerance (Johnson et al., 2000).

REFERENCES

Ahring, R.N., G.L. Duncan, and R.D. Morrison. 1964. Effect of processing native and introduced grass seed on quality and standard establishment. Technical Bulletin. T-113.

Ahring, R.M. and G.W. Todd. 1977. The bur enclosure of the caryopses of buffalograss as a factor affecting germination. *Agronomy Journal* 69:15–17.

Barber, H.G. 1918. A new species of *Leptoglossus*: a new *Blissus* and varieties. *Bull. Brook. Ent. Soc.* 13:36.

Baxendale, F.P., J.M. Johnson-Cicalese, and T.P. Riordan. 1994. *Tridiscus sporoboli* and *Trionymus* sp. (Homoptera: Pseudococcidae): Potential new mealybug pests of buffalograss turf. *J. Kans. Entomol. Soc.* 67:169–172.

Baxendale, F.P., T.M. Heng-Moss, and T.P. Riordan. 1999. *Blissus occiduus* Barber (Hemiptera: Lygaeidae): A new chinch bug pest of buffalograss turf. *J. Econ. Entomol.* 92:1172–1176.

Beard, J.B. 1973. *Turfgrass: Science and Culture*. Prentice Hall, Englewood Cliffs, NJ.

Beard, J.B and K.S. Kim. 1989. Low-water-use turfgrasses. *USGA Green Section Record*. 27:12–13.

Beetle, A.A. 1950. Buffalograss-Native of the Shortgrass Plains. Agricultural Experiment Station, University of Wyoming, Laramie. Bulletin 293, pp. 1–31.

Bird, R.D. and A.V. Mitchener. 1950. Insects of the season 1949 in Manitoba. *Can. Insect Pest Rev.* 28:41.

Briggs, L. 1914. Relative water requirements of plants. *Agric. Res.* 3:1–63.

Brown, E.O. 1945. Problems of testing seed stocks of buffalograss. *Proceedings of the Association of Official Seed Analysts*. 35:155–158.

Bur, R.D. 1951. Uncommon occurrence in buffalograss. *Journal of Range Management*. 4:267–269.

Chada, H.L. and E.A. Wood. 1960. Biology and Control of the Rhodesgrass Scale. U.S. Dept. Agr. Tech. Bull. No. 1221. p. 21.

Clark, M. 1945. Lo, a new buffalo. *Southern Seedsman*. p. 15.

Crocker, R.L., V.G. Hickey, M.C. Engelke, H.L. Cromroy, and R.L. Toler. 1984. An eriophyid mite, *Eriophytes slykhuisi* infests buffalograss in Texas. *Tex. Agr. Exp. Stn. Prog. Rep.* 4287, pp. 128–131.

Crocker, W. 1945. Will buffalograss be useful in eastern United States? *Boyce Thompson Institute for Plant Research*. 13:411–414.

DeShazer, S.A., T.P. Riordan, F.P. Baxendale, and R.E. Gaussoin. 1992. Buffalograss: A Warm-Season Native Grass for Turf. Coop. Ext., University of Nebraska-Lincoln, Lincoln, NE. EC92-1245-C.

Engelke, M.C. and V.G. Hickey. 1983. Buffalograss germplasm diversity and development for semi-arid turf. *Texas Turfgrass Proceedings*. PR 4150, pp. 21–22.

Engelke, M.C. and V.H. Lehman. 1990. Registration of 'Prairie' buffalograss. *Crop. Sci.* 30:1360.

Frank, K.W., R.E. Gaussoin, and T.P. Riordan. 1997. Nitrogen fertilization and mowing height effects on buffalograss. *1998 Turfgrass Research Report* [Nebraska]. pp. 29–30.

Heng-Moss, T.M., F.P. Baxendale, and T.P. Riordan. 1998. Beneficial arthropods associated with buffalograss. *J. Econ. Entomol.* 91:1167–1172.

Hickey, V.G., F.A. Miller, and M.C. Engelke. 1983. Buffalograss seed pretreatment germination. *Texas Turfgrass Proceedings*. PR 4169, pp. 126–128.

Hitchcock, A.S. 1935. Manual of the Grasses of the United States. USDA Misc. Publ. No. 200. U.S. Govt. Printing Office, Washington, DC.

Hitchcock, A.S. and A. Chase. 1971. *Manual of Grasses of the United States*, 2nd ed. Dover Publishers, New York.

Huff, D.R. and L. Wu. 1987. Sex expression in buffalograss under different environments. *Crop Science* 27:623–626.

Huff, D.R., T.P. Riordan, and M.C. Engelke. 1987. Buffalograss Regional Evaluation Trial.

Huffman, A.H. and P.W. Jacoby. 1984. Effects of herbicide on germination and seedling development of three native grasses. *Journal of Range Management.* 37(1):40–43.

Johnson, P.G., T.P. Riordan, and K. Arumuganathan. 1998. Ploidy level determinations in buffalograss clones and populations. *Crop Science* 38:478–482.

Johnson, P.G., T.P. Riordan, J.M. Johnson-Cicalese, F.P. Baxendale, R.E. Gaussoin, R.C. Shearman, and R. V. Klucas. 2000. Registration of '61' Buffalograss. *Crop Sci.* 40(2): 569–570.

McGinnies, W.J. 1979. Dryland grasses for turf. *Proceedings of the 25th Annual Rocky Mountain Regional Turfgrass Conference.* pp. 5–13.

Milby, T.H. 1971. The leaf anatomy of buffalograss, *Buchloe dactyloides* (Nutt.) Engelm. *Botanical Gazette.* 132:308–313.

NTEP. 1999. National Turfgrass Evaluation Program Buffalograss Trial. National Turfgrass Evaluation Program. Beltsville, MD.

Parkey, W. 1952. Certain Aspects of Embryo Culture and Progeny Evaluation in Buffalograss. Ph.D. dissertation. University of Nebraska-Lincoln.

Pfadt, R.E. 1984. Species richness, density, and diversity of grasshoppers (Orthoptera: Acrididae) in a habitat of the mixed grass prairie. *Can. Entomol.* 116:703–709.

Pladeck, M.M. 1940. The testing of buffalograss "seed," *Buchloe dactyloides* Englem. *Journal of the American Society of Agronomy* 32:486–494.

Plank, E.N. 1892. *Buchloe dactyloides*, Englem., not a dioecious grass. *Bulletin of the Torrey Botanical Club* 19:303–306.

Pozarnsky, T. 1983. Buffalograss: Home on the range, but also a turfgrass. *Rangelands* 5:214–216.

Quinn, J.A. 1987. Relationship between synaptosPermy and dioecy in the life history strategies of *Buchloe dactyloides* (gramineae). *American Journal of Botany* 74:1167–1172.

Quinn, J.A. and J.L. Engle. 1986. Life-history strategies and sex ratios for a cultivar and a wild population of *Buchloe dactyloides* (gramineae). *Amer. J. of Bot.* 73:874–881.

Reinhard, H.J. 1940. The life history of *Phyllophaga lanceolata* (Say) and *Phyllophaga crinita* Burmeister. *J. Econ. Entomol.* 33:572–578.

Riordan, T.P., S.A. DeShazer, F.P. Baxendale, and M.C. Engelke. 1992. Registration of '609' buffalograss. *Crop Sci.* 32:1511.

Savage, D.A. and L.A. Jacobson. 1935. The killing effect of heat and drought on buffalograss and blue grama grass at Hay's, Kansas. *Journal of the American Society of Agronomy.* 22:566582.

Schwarze, D.S., J.L. Svoboda, and T.P. Riordan. 1987. Buffalograss Vegetative Establishment. *USGA Annual Report.*

Shaw, R.B., C.M. Bern, and G.L. Winkler. 1987. Sex ratios of *Buchloe dactyloides* (Nutt.) Engelm. along catenas on the shortgrass steppe. *Botanical Gazette* 148:85–89.

Slater, J.A. 1964. A Catalogue of the Lygaeidae of the World. University of Connecticut, Storrs, CT.

Sorenson, K.A., and H.E. Thompson. 1979. The life history of the buffalograss webworm, *Surattha indentella* Kearfott, in Kansas. *J. Kans. Entomol. Soc.* 52: 282–296.

Svoboda, J.L. 1987. *UGSA Annual Report.* Buffalograss seed treatment evaluation.

Thorson, W. 2000. Personal communication

Turgeon, A.J. 1980. *Turfgrass Management.* Reston Publishing Company, Reston, VA.

Webb, J.J. 1941. The life history of buffalograss. *Transactions Kansas Academy of Science.* 44:58–71.

Wenger, L.E 1940. Inflorescence variation in buffalograss, *Buchloe dactyloides*. *J. of the Amer. Soc. of Agron.* 32:274–277.

Wenger, L.E. 1941. Improvement of buffalograss in Kansas, Kansas State Board of Agriculture. Thirty-Second Biennial Report. pp. 211–224.

Wenger, L.E. 1943. Buffalograss. Kansas Agricultural Experiment Station, Manhattan, Kansas. Bulletin 321, pp. 1–78.

Wu, L., A.H. Harvindi, and V.A. Gibeault. 1984. Observations on buffalograss sexual characteristics and potential for seed production improvement. *HortScience.* 19:505–506.

Chapter 17

Zoysiagrasses (*Zoysia* spp.)

M.C. Engelke and S. Anderson

Zoysia Willdenow is indigenous to the countries of the western Pacific Rim and westward into the Indian Ocean. The native distribution of the 11 recognized species in the genus extends from New Zealand to the island of Hokkaido in Japan, and from French Polynesia through Malaysia west to Mauritius (Figure 17.1; Anderson, 2000). The recognized species include: *Z. japonica* Steudel, *Z. macrantha* Desvaux, *Z. macrostachya* Franchet & Savatier, *Z. matrella* (L.) Merrill, *Z. minima* (Colenso) Zotov, *Z. pacifica* (Goudswaard) Hotta & Kuroki, *Z. pauciflora* Mez, *Z. planifolia* Zotov, *Z. seslerioides* (Balansa) Claton & Richardson, *Z. sinica* Hance, and *Z. tenuifolia* Thiele. The geographical center of distribution is Southeast Asia and Indonesia. Native habitats are primarily the coastal foredunes in its distribution, but *Zoysia* also may be found in the mangroves and estuaries of these countries, and in the mountains of Korea and Japan (Schmid, 1974; Anderson, 2000).

Among these 11 species, three species have been used as turfgrass. These are *Z. japonica*, *Z. matrella*, and *Z. pacifica*. In the turfgrass literature, *Z. pacifica* has been previously referred to as *Z. tenuifolia* (Anderson, 2000). Collections of *Z. japonica*, *Z. matrella*, and natural and man-made hybrids of *Z. japonica* with *Z. matrella*, of *Z. japonica* with *Z. pacifica*, and of *Z. matrella* with *Z. pacifica* have been used as turf throughout the world. Introductions have been made into the United States, Europe, South America, and India (Hitchcock, 1971; Hoover et al., 1948; Bor, 1960; Anderson, 2000). Indigenous distribution of *Zoysia* germ plasm accessions that have been used for turfgrass extends from midlatitudes of Honshu (Japan) south into the Philippine Islands. Thus, most accessions are tropical to southern temperate (Northern hemisphere) in origin, and are adapted to warm climates.

Zoysia pacifica and *Z. matrella* are indigenous to the Malaysian islands. During work on a taxonomic revision of the genus, a photograph accompanying a herbarium specimen of *Z. pacifica* indicated the vernacular 'buru-buru' (lumpy grass). The photo shows *Z. pacifica* as a 'lawn' around the hut of a New Caledonia islander. Because *Z. pacifica* and *Z. matrella* are especially slow to establish, it is probable that islanders build their huts in an already established sward of these mat-forming grasses. The natural appearance of ungrazed and unmown *Zoysia* is a lumpy sward, much like the tussocks of the alpine sedges.

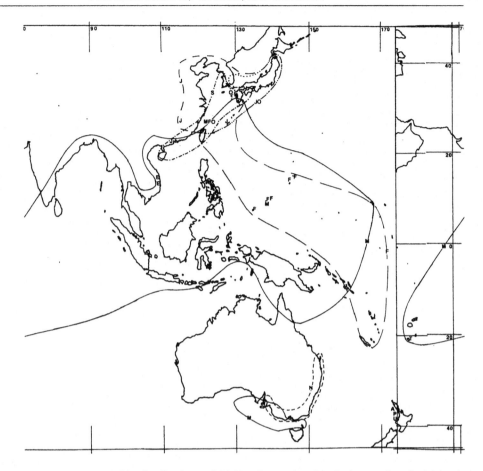

Figure 17.1. Indigenous geographic distributions of 11 Zoysia species. Symbols associated with each taxon are *Z. japonica* (J), *Z. macrocantha* (N), *Z. macrostachya* (O), *Z. matrella* (M), *Z. minima* (A), *Z. pacifica* (F), *Z. pauciflora* (U), *Z. planifolia* (P), *Z. seslerioides* (E), *Z. sinica* (S), and *Z. tenufolia* (T). Line codes for species boundaries are *Z. japonica* (— — — —), *Z. macrantha* (- - - -), *Z. macrostachya* (. . . .), *Z. matrella* (————), *Z. pacifica* (—— ——), and *Z. sinica* (- . - . - . -).

Distribution of *Z. matrella* and *Z. pacifica* in the Pacific Islands is probably a result of inadvertent transfer of the spikelets or sprigs by migrating peoples. Because it is a shore, foredune, and coastal marsh grass (Yang and Chen, 1995), the spikelet was probably picked up in hairs, clothing, and belongings of people as they left shore.

In Australia, two species occur as part of the vegetation of the coastal foredunes. *Zoysia* is indigenous to the coasts of Queensland, Victoria, and New South Wales, Australia. In addition, collections of likely indigenous *Z. matrella* have been made from South Australia and Tasmania. Collections of introduced genotypes of *Z. matrella, Z. pacifica,* and *Z. japonica* were made in the last 40 years.

Zoysia japonica and *Z. matrella* have been cultivated in China, Japan, and Korea for centuries. A survey of herbarium collections revealed *Z. japonica, Z. sinica,* and *Z. matrella* as native to the coasts of China, and *Z. japonica* and *Z. matrella* as common throughout Korea and Japan. Use of *Z. japonica* in Korea has additional significance as Christian burial tombs, being essentially aboveground, are sprigged with selections of *Z. japonica* that impart a 'golden color' when dormant. Hence, the Korean vernacular name 'Golden Grass' is used in honor of the deceased. *Zoysia japonica* is native in the coastal grasslands of these countries, and is used for horse pasture in Japan (Okamoto et al., 1993).

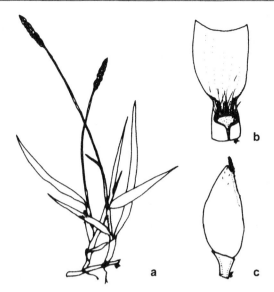

Figure 17.2. Generalized morphology of *Zoysia* turfgrass: a. growth habit of *Z. japonica*; b. collar and ligule; c. spikelet and distal end of pedicel, abaxial view.

It is probable these species originally were coastal and were carried inland as sprigs or spikelets.

Zoysia japonica was introduced to the United States from Japan in 1902 (Lim, 1982; Halsey, 1956). The earliest herbarium collection is of Wilcox, collected 1907, from Washington, D.C. *Zoysia japonica, Z. matrella*, and *Z. pacifica* have been used in the development of zoysiagrasses for turf in the United States.

Anderson (2000) reported that among the over 1,000 herbarium collections of *Zoysia* examined, 25 collections were from Latin American countries. Among the 16 countries represented in the collections were Colombia, Ecuador (the Galapagos Islands), Nicaragua, and Mexico. The grasses were collected from lawns of gardens or private residences, and thus were probably introduced. Among the South American specimens, the earliest collection date was 1949, a *Z. matrella* collected in Costa Rica.

MORPHOLOGY, TAXONOMY, AND GENETICS

Morphology

Zoysia species used for turfgrass are rhizomatous, mat-forming perennials. Culms range from 5 to 40 cm tall (Anderson, 2000). Culms, rhizomes, and stolons are characterized by alternating two short/one long internode lengths in compound node organization for both rhizome and culm (Shoji, 1976, 1983). Roots are adventitious. Figure 17.2 illustrates the collar and ligule, generalized habit for *Z. japonica*, and basic *Zoysia* collar-and-ligule, raceme, and spikelet features.

Blades are usually glabrous abaxially except for a ciliate callus at the base of the blade; glabrous, scabrous, or sparsely pilose adaxially; and the apices often sharply pointed. Blade ptyxis (rolling of mature blades) is supervolute (rolled with overlapping margins) in *Z. pacifica*, canaliculate (canoe-shaped) in *Z. japonica* x *pacifica* hybrids and in *Z. matrella* x *Z. pacifica* hybrids, and flat in *Z. matrella* and *Z. japonica*. Blade diameters of stressed and nonstressed plants of *Z. pacifica* range from 0.3 to 0.5 mm; diameters of canaliculate blades of *Z. pacifica* hybrids range from 0.5 mm to 2 mm; and diameters and widths range from 1 to 5 mm in *Z. matrella* and *Z. japonica*. When plants are stressed, blades will roll, with canaliculate and flat blades becoming canaliculate to supervolute. Stressed plants are commonly misidentified. Vernation in newly emerging blades is terete (rolled) (Anderson, 2000).

In *Zoysia*, the ligule is up to 0.3 mm long and truncate, and is composed of a line of hairs. A line or cluster of longer hairs is often present at the base of the blade immediately behind the ligule (Anderson, 2000). The inflorescence is a terminal, solitary spike-like raceme. Racemes have 5 to 12 spikelets in *Z. pacifica,* have 20 to 45 spikelets in *Z. matrella,* and have 30 to 50 spikelets in *Z. japonica*. Interspecific hybrids may have intermediate numbers of spikelets in a raceme. The spikelets are short pedicellate, with pedicels 0.5 to 2.5 mm long. The pedicel and one side of the spikelet are appressed to the rachis. Disarticulation is beneath the glumes, but is lacking in the three New Zealand endemic *Zoysia* species. Spikelets are laterally compressed, with a single floret. Lower glumes are rare. Upper glumes enclose the floret, and are chartaceous to coriaceous and 1-veined, with the vein evident only in the upper third to half of the glume fold. The upper glume is usually awned, the awns are 2.5 mm long. Lemmas are thin, lanceolate or linear, acute to emarginate, and are 1-veined. The vein of the lemma is usually coriaceous. Paleas are rare, and are thin when present (Anderson, 2000).

Taxonomy and Nomenclature

The taxonomic classification of *Zoysia* is described by Campbell (1985) and Gould and Shaw (1983).

Family	Gramineae (Poaceae)
Subfamily	Chloridoideae
Tribe	Zoysieae
Genus	*Zoysia*
Authority	Willdenow

In a monographic treatment of the genus *Zoysia* (Zoysieae tribe), 11 species were recognized (Anderson, 2000). Below is a key to the species commonly used for turfgrass, which was developed from the information in the monograph. Anderson (2000) addressed improper application of the name *Z. tenuifolia*. Publications of Koyama (1988) and others have used the name *Z. tenuifolia* for specimens collected in the Pacific Rim countries. *Zoysia tenuifolia* is endemic to Mauritius, is not represented in *Zoysia* breeding collections, and has not been used in the development of zoysiagrasses. The specimens collected in Pacific Rim countries have been examined and belong to the species *Z. pacifica* (Goudswaard) Hotta & Kuroki. The capillary-bladed zoysiagrasses used as turfgrass are of the species *Z. pacifica*.

Key to *Zoysia* species used for turfgrass

1. Blades to 0.5 mm diameter; racemes with 12 or fewer spikelets; peduncles exserted less than 1 cm beyond flag leaf sheath .... *Z. pacifica*
1'. Blades greater than 0.5 mm width or diameter, stressed or fully hydrated; racemes with more than 12 spikelets; peduncles exserted at least 1 cm beyond flag leaf sheath
   2. Pedicels longer than 1.75 mm; spikelet ovate, 1.0 to 1.35 mm wide; culm internodes not over 14 mm long; blades ascending .... *Z. japonica*

2'. Pedicels shorter than 1.75 mm long; spikelet lanceolate, 0.65 to 1.0 mm wide; culm internodes, at least a few, over 14 mm long; blades patent .... *Z. matrella*

Cytology

Most literature reports 2n=40 for *Z. japonica, Z. matrella,* and *Z. pacifica* (reported in publications as *Z. tenuifolia,* but collection locations are indicative of *Z. pacifica* distribution). The base number for the tribe has been reported as x=10 (Gould and Shaw, 1983). A diploid specimen of *Z. matrella* was collected from a sandy beach west of Galle in Sri Lanka (Ceylon) by Gould in 1970 (collection #13428). Gould and Soderstrom (1974) recorded 2n=20 for this collection, for which herbarium voucher specimens are housed at S. M. Tracy Herbarium, Department of Rangeland Ecology and Management, Texas A&M University, College Station, Texas (TAES); United States National Herbarium, Smithsonian Institution, Washington, DC (US); Royal Botanic Gardens, Kew, London (K); Australian National Herbarium, Canberra, Australia (CANB). The diploid location is west of the geographical center of distribution for the genus. Insufficient evidence has been reported to estimate the center of origin for the genus. Table 17.1 presents chromosome counts for all *Zoysia* species that have been studied. Aneuploid series have not been reported for *Zoysia*. Allotetraploidy was proposed based on a restriction fragment length polymorphism (RFLP) genetic map derived from a mapping population resulting from a self-pollinated individual and suggesting an interspecific hybrid of *Z. matrella* and *Z. japonica* (Yaneshita et al., 1999).

Yaneshita et al. (1999) published an RFLP linkage map for 115 nuclear loci for 22 linkage groups. A chloroplast restriction map for *Z. japonica* was published by Yaneshita et al. (1993). Molecular markers for agronomic traits have not been mapped in *Zoysia*.

INTERSPECIFIC HYBRIDIZATION

Forbes (1952) examined anthocyanin pigmentation of the stolons and glumes in self, outcrosses, and reciprocal crosses of *Z. japonica* x *Z. matrella, Z. japonica* x *Z. pacifica* (cited in Forbes as *Z. tenuifolia*), and *Z. matrella* x *Z. pacifica* (cited as *Z. tenuifolia*). Forbes concluded anthocyanin production in stolons and glumes was a dominant trait, and not useful for detection of interspecific hybrids. Forbes interpreted an intermediate number of spikelets as evidence for production of interspecific hybrids for crosses of *Z. matrella* x *Z. japonica* and *Z. matrella* x *Z. pacifica* (cited as *Z. tenuifolia*). In addition, Forbes inferred that *Z. japonica, Z. matrella,* and *Z. pacifica* were completely interfertile. Hong (1985) determined Korean collections of *Z. japonica, Z. sinica,* and *Z. matrella* were intercompatible, based on caryopsis fill and percentage germination of filled caryopses. Varying levels of interpopulation compatibilities were observed for Japanese populations of *Z. japonica* (Fukuoka, 1981). Cross-compatibility was measured as the percentage of caryopses that filled per raceme.

Based on personal experience, low germination percentages were observed for interspecific crosses between type specimen forms of the seven *Zoysia* species in the Texas A&M University at Dallas (TAES-Dallas) zoysiagrass collection. *Zoysia macrostachya* as the female parent gave the highest germination percentages of all interspecific crosses.

Esterase (EST) and acid phosphatase (APT) isozyme banding patterns supported interpretation of interspecific hybrids for observed intermediate morphologies (Yang et al., 1995). Continuum in form was observed for blade angle, blade width, presence of blade hairs, raceme length, and spikelet length among accessions of *Z. japonica* and *Z. sinica*.

Table 17.1. Literature reports of chromosome counts for *Zoysia*.

Species	Plant Origin	Tissue[a]	Division Type	Count	Reference
Z. japonica	Manchuria, Japan, Korea	PMC	Meiosis I (anaphase I, diakinesis)	n=20	Forbes, 1952
Z. matrella	Philippine Islands	PMC	Meiosis I (metaphase I)	n=20	Forbes, 1952
Z. tenuifolia[b]	Guam	PMC	Meiosis (metaphase I)	n=20	Forbes, 1952
Z. matrella	Veli, India	PMC		n=20	Christopher and Abraham, 1974
Z. matrella	Sri Lanka	PMC	Meiosis (diakinesis)	2n=20	Gould and Soderstrom, 1974
Z. tenuifolia[b]	Taiwan	PMC		n=20	Chen and Hsu, 1962
Z. matrella	Thailand	[c]	[c]	2n=40	Larsen, 1963
Z. japonica	[c]	[c]	[c]	2n=40	Tateoka, 1954
Z. macrostachya	[c]	[c]	[c]	2n=40	Tateoka, 1954
Z. japonica	[c]	PMC	[c]	2n=40	Lim, 1982

[a] PMC = Pollen mother cell.
[b] Misapplication of name *Z. tenuifolia*. Specimen has not been examined. However, location of collection is indicative of *Z. pacifica* or *Z. matrella*. See text for discussion of *Z. tenuifolia* nomenclatural problem.
[c] No information provided in publication.

Variation within populations was low in *Z. japonica* populations sampled from disparate locations in Japan (Yamada and Fukuoka, 1984). Peroxidase isozyme banding patterns of populations within Honshu and Shikoku islands were relatively similar compared to populations from Hokkaido and in the southern tip of Kyushu. Interspecific hybridization was not characterized. Levels of variation among populations is important in decisions on intensity of collections needed in a region, and to sample the best available genetic variation of wild populations.

In a series of papers on flavonoid patterns in Japanese *Zoysia*, Nakamura and Nakamae (1970, 1972, 1973, 1974) inferred lineage relationships among Japanese populations of *Z. japonica*, *Z. matrella*, *Z. pacifica* (cited as *Z. tenuifolia*), and 'Emerald.' Flavonoid and isozyme banding patterns must be evaluated with caution in estimating lineages and species relationships because of high intraspecific variation and effects of environment on expression (Richardson et al., 1986; Steussy, 1990). Current efforts to examine interspecific hybridization rely on DNA markers, which lack the environmentally influenced expression of the marker.

Nuclear RFLP analyses of five *Zoysia* species provided supporting evidence for interspecific hybrids between *Z. japonica* and *Z. pacifica*, between *Z. japonica* and *Z. macrostachya*, and between *Z. matrella* and *Z. sinica* (Anderson, 2000). Data supported a cline in morphological variation between *Z. pacifica* and *Z. japonica*, with some intermediates sharing several morphological traits with the nomenclatural type specimen form of *Z. matrella*. Nuclear DNA cluster analysis supported introgression among *Z. pacifica*, *Z. matrella*, and *Z. japonica* (Anderson, 2000). Chloroplast phylogeny was estimated for *Z. matrella* and *Z. pacifica* (cited as *Z. tenuifolia*) of the Ryukyu Islands (Yaneshita et al., 1997). Six chloroplast haplotypes were identified among the 16 accessions. Nuclear and chloroplast RFLP analyses provided evidence that natural hybridization between *Z. matrella* and *Z. pacifica* has occurred in these islands. Intermediate morphologies were observed for individuals from Ishigaki Island, and based on chloroplast type and morphology was hypothesized as interspecific hybrids between *Z. matrella* and *Z. pacifica*. Interspecific hybrids among *Z. matrella*, *Z. japonica*, and *Z. macrostachya* were not supported by the nuclear and chloroplast RFLP data (Yaneshita et al., 1997).

GENETIC AND PHYSIOLOGICAL CONTROL OF FLOWERING

Flowering requirements for *Zoysia* species relate to daylength and juvenility. Forbes (1952) reported short-day for *Z. japonica* collected from Manchuria, Korea, and Japan, and for *Z. matrella* from the Philippine Islands and a *Z. pacifica* (cited in Forbes as *Z. tenuifolia*) from Guam. Youngner (1961) reported *Z. matrella* (source not reported) and *Z. japonica* ('Meyer') required short daylength and high temperatures for flowering. A minimum number of nodes in the upright culm was required for flowering in *Z. japonica* (greater than or equal to 12), *Z. matrella* (greater than or equal to 12), *Z. pacifica* (cited as *Z. tenuifolia*) (greater than or equal to 18) (Yeam et al., 1984). In a survey of photoperiod requirements for flowering in five *Zoysia* species, Morton et al. (1997) observed latitude-associated daylength requirements for flowering. A cline from day-neutral was reported in accessions at 10 to 15°N latitude to short- or long-day in latitudes above 30°. In addition, a change in daylength, rather than a given daylength, may be required for flower induction in *Z. macrostachya*. Bud initiation and flower production are controlled by different cues (Nakamura and Nakamae, 1984; Morton et al., 1997).

Pollen release has been evaluated for three members of the Zoysieae, *Perotis indica* (L.) O. Kuntze, *Tragus roxburghii* Panigrahi, and *Zoysia matrella*. *Perotis* pollen release occurred during August with peak shedding for 4 hours at 0600 hr. *Tragus* pollen release peak occurred from 0100 to 0300 hr during August and September. *Zoysia matrella* had a 2-hr release with peak shedding from 0600 to 0700 hr during September. All work was performed at 17° 42' N latitude on the East Coast of India (Reddi et al., 1988).

The presence of apomixis is unstudied in *Zoysia*. Protogyny occurs in at least six *Zoysia* species. In *Zoysia*, it is characterized by the oldest spikelets at the terminal end of the raceme. Protandry has been induced in *Z. matrella* (Genovesi and Engelke, 2000, unpublished). *Zoysia minima*, *Z. pauciflora*, and *Z. planifolia* are likely cleistogamous since the anthers were rarely exserted from spikelets observed on herbarium specimens. Anther and stigma exsertion have been observed for six of the seven species represented in the TAES-Dallas Breeding Collection, and for the 11 species examined as herbarium specimens for the taxonomic revision (Anderson, 2000).

COLLECTION, SELECTION, AND BREEDING HISTORY

The introduction of zoysiagrass germ plasm to North America dates back to the late 1800s—early 1900s with additional seed sources brought in through the efforts of Frank N. Meyer. Only modest collections were accumulated in the 1950s with reports of Mel Anderson, Golf Course Superintendent, Alvamar CC, Lawrence, Kansas seeding fairways to Chinese Common zoysiagrass. The collection of Vic Youngner, University of California provided the germ plasm base from which El Toro (PP5,851), DeAnza (PP9,127), and Victoria (PP9,135) arose. Vic Gibeault (personal communication, 2000) suggested much of this material was from either the USDA-NPGS (National Plant Germplasm System) collection or directly from Japan and/or Korea in the mid-late 1960s. The GRIN (Germplasm Resources Information Network) system (www.ars-grin.gov/cgi-bin/npgs) identifies only 35 accessions held by NPGS. The United States Golf Association and the U.S. Department of Agriculture supported a collection trip in 1982 into the Pacific Rim countries of Japan, Korea, Taiwan, and the Philippines. Murray and Engelke (1982) returned approximately 700 accessions of five species including *Z. japonica*, *Z. matrella*, *Z. sinica*, *Z. macrostaycha*, and *Z. pacifica* (originally thought to be *Z. tenuifolia*). The original collection is maintained as vegetative propagules in greenhouses at Texas A&M University—Dallas (TAES-Dallas) with field plantings of the collection having been made in Maryland, Florida, Oregon, Michigan, and Texas.

International cooperative studies with the scientific communities beginning in the early 1980s plus additional collection trips have expanded the germ plasm resources in the United States in support of zoysiagrass development. The *Zoysia* species remain one of the target species upon which to continually expand the collection. Much of the early collection by Engelke and Murray represented "wild" germ plasm sources, but a significant portion of the group was collected from 'turfed' areas of cemeteries, roadsides, parks, gardens, and lawns. In 1993, a modest collection was assembled from portions of the Peoples Republic of China. The economic development and urbanization of China has notably eliminated many of the natural grasslands. These areas should be considered prime sites of origin that are rapidly being lost. The urgency in germ plasm collection was to preserve as much of the 'wild' resources since these still represent significant contributions to plant genetics. Through the work of Anderson (2000), species were also identified from herbarium collections that are not available in any of the working collections. Species such as *Z. seslerioides* are void, yet hold considerable potential to aid in improving seed production potential of the genus. *Zoysia macrostachya* is a more northern type with potential for improved cold hardiness and demonstrated excellent salinity tolerance (collections made directly for salt beds of S. Korea) is only modestly represented in the collection. Internationally, contacts have been made with scientific groups in Korea, China, and Japan, where active breeding programs are progressing for both forage and turf use.

Most breeding efforts in the United States in the mid 1980s concentrated on developing vegetatively propagated cultivars (Table 17.2). Much effort continues in this area at TAES-Dallas, Southern Illinois University, and the University of Florida. Screening programs in other regions include Michigan and Illinois plus a limited private sector selection in Texas. Many of the private sector programs in Idaho, North Carolina, New Jersey, Oregon, and Georgia are concentrating on developing seeded cultivars. Simplot appears to be among the most active in expanding genetic resources through the development and introduction of Z. sinica from China. Zoysia japonica and Z. sinica are more suited to seed production because each has a high frequency of large and multiple spikelets. Most other species of turf interest have a high frequency of seedheads, however the small size and reduced number of spikelets reduces their seed yield potential.

Characterization of flowering habit, climatic conditions for floral induction, management criteria to optimize seed production and shattering resistance have been difficult to define and hence have resulted in economic restrictions to the development of a viable Zoysia seed industry.

Many of the commercial cultivars presently in use in the United States today are not type specimens of a single species, but rather intermediate forms as hybrids across species. Previously thought to be Z. japonica, Anderson (2000) suggests that cultivars such as Crowne and Palisades are Z. japonica-types but most likely a result of hybridization between Z. japonica and Z. pacifica. Based on morphological and DNA characterization, this appears to be true of many of the other cultivars listed in Table 17.2. The Z. matrella cultivars Cavalier, Diamond, Royal, and Zorro are most likely intermediate hybrids between Z. matrella and Z. pacifica, where Cavalier, Royal, and Zorro are Z. matrella-types and Diamond may have tendencies toward the finer leafed Z. pacifica-type.

The Texas A&M Collection has been subjected to extensive greenhouse and field evaluations. Major emphasis over the past two decades has targeted water-use, winter dormancy and survival, and reproductive growth habit. Excellent success has been made through limited breeding and extensive selection for drought tolerance, disease and insect resistance. For instance, Cavalier (PP10,788), a distinct Z. matrella type, has been identified for its excellent tolerance to a host of chewing insects— most notable being the fall armyworm Spodoptera frugiperda (J.E. Smith). Cavalier has significant tolerance to as many as five different chewing insects including the tawny mole cricket (Scapteriscus vicinus Scudder), tropical sod webworm (Herpetogramma phaeopteralis Guenee), differential grasshopper (Melanoplus differentialis Thomas), and hunting billbug (Spenophorus venatus vestitus Chittenden). Several hundreds of single cross hybrids have been generated through controlled crosses with Cavalier as the male and female parent. These populations will ultimately lead to the characterization and identity of gene complex(es) responsible for these traits. Many of the advanced generation hybrids demonstrate multiple resistance expression for biotic stresses of insect and disease as well as abiotic stresses of low light, high salinity, and prolonged periods of drought. Reduction in pesticide use is desirable for all crops. Use of cultivars with pest and disease resistance is one method of reducing pesticide use. For selected accessions of Z. japonica and Z. matrella from the TAES-Dallas breeding germ plasm collection, Reinert et al. (1993, 1997, 1998, 2001) has identified a wide range of resistance to fall armyworm, zoysiagrass eriophyid mite (Eriophyes zoysiae), and tropical sod webworm.

Table 17.2. Commercially available *Zoysia* varieties as of May 2001.

Variety Name	Experimental Designation	Breeder-Developer	Institution	Species	Year	Status	PP No.ª	Date Issued	Propagation	Reference
Meyer	Z-52	Forbes, Ferguson, Grau	USDA, USGA, Beltsville, MD	*Z. japonica*	1951	Clone	NA		Vegetative	Hanson, 1972
Sunburst	Z-73	Grau	USDA Plant Industry St. Beltsville, MD	*Z. japonica*	1952	Hybrid			Vegetative	Hanson, 1972
Emerald	34-35	Forbes	USDA Plant Industry St. Beltsville, MD	*Z. japonica* x *Z. tenuifolia*	1955	Hybrid	NA		Vegetative	Hanson, 1972
F.C. 13521			Alabama Agric. Exp. Station	*Z. matrella*	?				Vegetative	Hanson, 1972
Midwest		Daniel, LeCroy	Indiana Agric Exp. Station	*Z. japonica*	1963		NA		Vegetative	Hanson, 1972
El Toro	UCR#1	Younger	UC System	*Z. japonica*	1986	Hybrid	PP5,845	Dec 1986	Vegetative	USPTO
ZT-26	ZT-26	Whiting	Proprietary	*Z.*	1988		PP6,345	1988	Vegetative	USPTO
ZT-4		Whiting	Proprietary	*Z. japonica*	1989		PP6,516	1989	Vegetative	USPTO
Cashmere	P-1	Pursley	Proprietary	*Z. matrella*	1989		PP6,529	1989	Vegetative	USPTO
Belair		Murray	USDA	*Z. japonica*	1987				Vegetative	Murray and O'Neil, 1987
DeAnza	Z88-8	Gibeault et al.	UC System	*Z. japonica*	1995	Hybrid	PP9,127	1995	Vegetative	USPTO
Victoria	Z88-14	Gibeault et al.	UC System	*Z. japonica*	1995	Hybrid	PP9,135	1995	Vegetative	USPTO
Miyako	CM-15	Yaneshita et al.	Japan	*Z. japonica*	1998		PP10,187	1998	Vegetative	USPTO

GNZ	ZT-11	Whiting	Proprietary	*Z. japonica*	1989		PP7,074	1989	Vegetative	USPTO
Zoyboy	Z-3	Staton	Hawaii – Proprietary	*Z. japonica*	1994	Clone	PP8,553	1994	Vegetative	USPTO
GS90-18	GS90-18	Yamagishi et al.	Japan	*Z. matrella*	1995		PP9,089	1995	Vegetative	USPTO
Diamond	DALZ8502	Engelke et al.	TAMU	*Z. matrella*	1996	Hybrid	PP10,636	1996	Vegetative	USPTO
Cavalier	DALZ8507	Engelke et al.	TAMU	*Z. matrella*	1996	Clone	PP10,788	1996	Vegetative	USPTO
Crowne	DALZ8512	Engelke et al.	TAMU	*Z. japonica*	1996	Hybrid	PP11,570	2000	Vegetative	USPTO
Palisades	DALZ8514	Engelke et al.	TAMU	*Z. japonica*	1996	Hybrid	PP11,515	2000	Vegetative	USPTO
JaMur		Murray	Bladerunner Farms	*Z. japonica*	1996	Clone	Pending		Vegetative	Personal communication
Zeon		Murray	Bladerunner Farms	*Z. matrella*	1996	Clone	Pending		Vegetative	Personal communication
Empire	SS-500	Ito, Gurgel	Sod Solutions	*Z. japonica*	2000		PP11,466	2000	Vegetative	USPTO
Empress	SS-300	Ito, Gurgel	Sod Solutions	*Z. japonica*	2000		PP11,495	2000	Vegetative	USPTO
Royal	DALZ9006	Engelke et al.	TAMU	*Z. matrella*	2001	Hybrid	Pending		Vegetative	
Zorro	DALZ9601	Reinert et al.	TAMU	*Z. matrella*	2001	Hybrid	Pending		Vegetative	
Seeded	Cultivars									
Compatibility	ZMB-2		Patten Seed	*Z. japonica*	2000	Synthetic	PVP pending		Seeded	PVPO
Zenith	ZNW-1		Patten Seed	*Z. japonica*	2000	Synthetic	PVP pending		Seeded	PVPO
J-37			J. R. Simplot	*Z. japonica*		Synthetic	PVP pending		Seeded	PVPO
J-36			J. R. Simplot	*Z. japonica*		Synthetic	PVP Pending		Seeded	PVPO

[a] PP no. refers to Plant Patent Number assigned by USPTO.

[b] Abbreviations for institutions and references are: PVPO = United States Plant Variety Protection Office, TAMU = Texas A&M University System, UC System = University of California System, USDA = United States Department of Agriculture, and USPTO = United States Patent and Trademark Office.

Braman et al. (1994) has identified moderate resistance to tawny mole cricket among the advanced breeding lines of the TAES—Dallas collection.

Zoysia matrella and *Z. japonica* accessions have been screened for relative susceptibility to brown patch (*Rhizoctonia solani*). Brown patch is widespread among most grass species and *Zoysia* is generally no exception. Brown patch is exacerbated by overfertilization and watering during initial establishment. Recent evidence of strong resistance to brown patch has been reported by P. F. Colbaugh (personal communications, 2001) during his assessment of genetic resources of the NTEP trials and the TAES—Dallas zoysiagrass collection.

Shade tolerance screening of elite lines has identified accessions with adaptability for shaded conditions. Diamond has been documented to have excellent tolerance to low light conditions with persistence at continuous 87% artificial shade for periods exceeding 24 months. Persistence and performance under heavily simulated traffic studies presently underway appear quite promising.

Selection Criteria

Major parameters evaluated during selection of zoysiagrasses include turf quality and characters contributing to low maintenance, such as drought tolerance. Turf quality parameters evaluated include uniformity, density, smoothness, texture, and color. Uniformity (of height) is a measurement of the response of the sward to mowing. *Zoysia* is known for its 'lumpy' habit. Thus, selection is for accessions that are without this lumpy habit, or that can be culturally manipulated to greater uniformity. Uniformity may also be interpreted as the diversity of forms present in the sward. Clonal cultivars and lines should be uniform, as all plants should look the same. In seeded cultivars, depending on the degree of genetic variability, the diversity of forms may be high.

Density expresses the relative amount of airspace between culms. Smoothness (evenness) describes the pattern of culm density in the sward. A regular distribution of shoots in the sward is desirable. Zoysiagrasses may have a clumpy distribution, with regions of higher and lower densities throughout the sward. Clumpiness may be a product of variability in establishment of the sward from plugs, sprigs, or seed. Clumpiness may also be a product of localized decline, resulting from disease or other stresses. The blade texture of zoysiagrasses ranges from fine capillary blades of *Z. pacifica* to coarse (wide) blades of *Z. macrostachya*. Blade color ranges from yellow green to dark green.

Drought resistance is a major selection criterion in many turfgrass breeding programs, to address water conservation concerns in municipalities where turfgrasses are employed as the primary ground cover. Studies of water-relation parameters indicate different mechanisms for *Z. matrella* and *Z. japonica* (White et al., 2001). Bulk modulus of elasticity calculations provided evidence that *Z. matrella* cell walls are more elastic than those of *Z. japonica*. Osmotic potential measurements showed *Z. japonica* was more likely than *Z. matrella* to accumulate solutes to maintain turgor. Root length was correlated with ability to retain green cover for accessions of *Z. matrella* and *Z. japonica*. Deeper rooting was observed for *Z. japonica* than for *Z. matrella* (Marcum et al., 1995).

Unique selection criteria for zoysiagrass include growth rates, salinity tolerance, and cold hardiness. Zoysiagrasses are traditionally classified as having relatively slow growth and recovery rates. This perception alone has limited its use relative to other warm-season grasses. The slower establishment may initially increase the need for herbicides to reduce competition with weeds, but the high density of established zoysia turf provides considerable natural competition to weed invasion. Wide latitudinal distribution in the genus suggests genetic control of cold-hardiness can be found and would improve the adaptation and geographical range of the species. Commercially, Meyer and Midwest are reported to be the most cold-hardy within the *Zoysia*. Greater genetic resources are needed to improve cold-hardiness, as well as more attention to using Meyer and Midwest in hybridization programs to advance the adaptation of *Zoysia* further north.

Zoysia macrostachya is indigenous to coastal salt marshes of China (Yang and Yu, 1995). Salinity trials of selected accessions from the TAES-Dallas zoysiagrass collection have demonstrated a range of salinity tolerances. Marcum et al. (1998) showed that *Z. matrella* generally tolerates higher salinity levels than does *Z. japonica*. *Zoysia matrella* secretes salt through salt glands on the leaf surface at a greater rate than *Z. japonica*, and maintains lower solute levels in shoot cells. Drought and salinity tolerance has also included the ability to retain green vigorous shoots and leaves. Leaf firing, blade rolling, and blade death were each noted as macroscopic responses to various stresses.

General Development Criteria

Zoysiagrass management research is performed as part of elite line development. Some zoysiagrasses appear to support a limited number of green blades, with senescence of blades below the sixth leaf position from the apex. Increased growth rates may increase the number of leaves produced, but will also increase the amount of thatch accumulated in these zoysiagrass genotypes. Cultural practices are being developed to optimize the performance of the high-thatching cultivars such as Emerald, Cavalier, Zeon Royal, and Zorro and other fine-textured turfs when used for athletic fields. Preliminary research suggests that more frequent vertical mowing will be required to optimize turf performance for golf course fairways and athletic fields. Studies of frequency and height of mowing are being conducted on the newer grasses. Simultaneous vertical and reel mowing using "conditioning reels" has been demonstrated to maintain turf at its highest quality standard.

CONCLUSION

The diversity of the species is quite promising across the biological traits under study. The greatest limitations appear to be in: (1) economically viable seed production, (2) cold hardiness, (3) winter color retention, and (4) disease resistance. Other factors will limit species use. However, the considerable genetic diversity across the 11 species as well as within the species is quite promising. Cultivars are commercially available with low water use, excellent salinity and shade tolerance, adequate cold hardiness and responsiveness to low cultural inputs, multiple resistances to chewing insects, and promises of improved disease resistance.

REFERENCES

Anderson, S. 2000. Taxonomy of *Zoysia* (Poaceae): Morphological and Molecular Variation. Ph.D. dissertation. Texas A&M University, College Station, TX.

Bor, N.L. 1960. *The Grasses of Burma, Ceylon, India, and Pakistan*. Vol. 1. Pergamon Press, New York.

Braman. S.K., A.F. Pendley, R.N. Carrow, and M.C. Engelke. 1994. Potential resistance in Zoysiagrasses to tawny mole crickets (Orthoptera: Gryllotalpidae). *Florida Entomologist* 77:301–305.

Campbell, C.S. 1985. The subfamilies and tribes of Gramineae (Poaceae) in the southeastern United States. *J. Arnold Arboretum* 66:123–129.

Chen, C.C. and C.C. Hsu. 1962. Cytological studies of Taiwan grasses. 2. Chromosome numbers of some miscellaneous tribes. *J. of Japanese Bot.* 37:300–313.

Christopher, J. and A. Abraham. 1974. Studies on the cytology and phylogeny of South Indian grasses. II. Sub-family Eragrostoideae. *Cytologia* 39:561–571.
Engelke, M.C. and J. Murray. 1984. Zoysia the golden species. *Grounds Maintenance,* January 1984.
Forbes, I. 1952. Chromosome numbers and hybrids in *Zoysia. Agronomy Journal* 44:194–199.
Fukuoka, H. 1981. Compatibility, germination and fertility in *Zoysia japonica. Japanese Agriculture Department Progress Report* 1981–1983.
Gould, F.W. and R.B. Shaw. 1983. *Grass Systematics.* 2nd ed. Texas A&M University Press, College Station, TX.
Gould, F.W. and T.R. Soderstrom. 1974. Chromosome numbers of some Ceylon grasses. *Canadian Journal of Botany* 52:1075–1090.
Halsey, H.R. 1956. The *Zoysia* lawn grasses. *National Horticultural Magazine* 35:152–161.
Hanson A.A. 1972. Grass Varieties in the United States. Agricultural Handbook 170, U.S. Department of Agriculture, Washington, DC.
Hitchcock, A.S. 1971. *Manual of the Grasses of the United States.* 2nd ed., revised by A. Chase. Dover Publications, New York.
Hong, K.H. 1985. Studies on Interspecific Hybridization in Korean Lawngrasses (*Zoysia* spp.). M.S. thesis. Seoul National University, Seoul, South Korea.
Hoover, M.M., M.A. Hein, W.A. Dayton, and C.O. Erlanson. 1948. The *Zoysia* grasses. *Yearbook of Agriculture* 1948: 699–700.
Koyama, T. 1988. Grasses of Japan and its neighboring regions: An identification manual. Kodansha, Tokyo.
Larsen, K. 1963. Studies in the Flora of Tailand. *Dansk Botanik Arkiv* 20:211–275.
Lim, Y.P. 1982. Studies on the Colchicine-Induced Mutant in Zoysiagrass (*Zoysia japonica* Steud.). M.S. thesis, Seoul National University, Seoul, South Korea.
Marcum, K.B., M.C. Engelke, S.J. Morton, and R.H. White. 1995. Rooting characteristics and associated drought resistance of zoysiagrasses. *Agronomy Journal* 87:534–538.
Marcum, K.B., S.J. Anderson, and M.C. Engelke. 1998. Salt gland ion secretion: A salinity tolerance mechanism among five Zoysiagrass species. *Crop Sci.* 38:806–810.
Morton, S.J., M.C. Engelke, and J.A. Arnau. 1997. Photoperiod effects on flowering in *Zoysia. Texas Turfgrass Research Reports* 97-25.
Murray, J.J. and O'Neil. 1987 Registration of 'Belair' zoysiagrass. *Crop Sci.* 27:151.
Nakamura, N. and H. Nakamae. 1970. The studies on the taxonomy of grasses by the paper chromatography: 1. *Zoysia. Turf Res. Bull. K.G.U. Green Experiment Inst. Res. Report* 19:35–40.
Nakamura, N. and H. Nakamae. 1972. The studies on the taxonomy of grasses by the paper chromatography. 2. *Zoysia. Turf Res. Bull. K.G.U. Green Experiment Inst. Res. Report* 22:25–32.
Nakamura, N. and H. Nakamae. 1973. The studies on the taxonomy of grasses by the paper chromatography. 3. *Zoysia japonica. Turf Res. Bull. K.G.U. Green Experiment Inst. Res. Report* 24:29–37.
Nakamura, N. and H. Nakamae. 1974. The studies on the taxonomy of grasses by the paper chromatography. 4. *Zoysia japonica. Turf Res. Bull. K.G.U. Green Experiment Inst. Res. Report* 26:35–41.
Nakamura, N. and H. Nakamae. 1984. Time of flower buds initiation of *Zoysia japonica* and *Z. matrella. Journal Japanese Society of Turfgrass Science* 13(2):117–122.
Okamoto, C., N. Hattori, K. Kabata, and M. Kikuchi. 1993. Relationship between vegetation and productivity of grazing native pasture in Aso area. *Proceedings of Faculty of Agriculture Kyushu Tokai University* 12: 39–48.
Reddi, C., N.S. Reddi, and B. Atluri Janaki. 1988. Circadian patterns of pollen release in some species of Poaceae. *Review of Palaeobotany and Palynology* 54:11–42.
Reinert, J.A., M.C. Engelke, and S.J. Morton. 1993. Zoysiagrass resistance to the zoysiagrass mite, *Eriophyes zoysiae* (Acari: Eriophyidae) (Ch. 45). *Int. Turfgrass Soc. Res. J.* 7:349–352.
Reinert, J.A., M.C. Engelke, J.C. Read, S.J. Maranz, and B.R. Wiseman. 1997. Susceptibility of cool and warm season turfgrasses to fall armyworm, *Spodoptera frugiperda. Int. Turfgrass Soc, Res. J.* 8:1003–1011.
Reinert, J.A., J.C. Read, M.C. Engelke, P.F. Colbaugh, S.J. Maranz, and B.R. Wiseman. 1998. Fall armyworm, *Spodoptera frugiperda,* resistance in turfgrass. Mededelingen, Faculteit

Landbouwkundige en Toegepaste Biologische Wetenschappen. *Proc. 50th Inter. Sym. Crop Protection, Gent, Belgium* 63(2b):467-471.

Reinert, J.A. and M.C. Engelke. 2001. Resistance in zoysiagrass, *Zoysia* spp., to the tropical sod webworm, *Herpetogramma phaeopteralis* Guenee. *Int. Turfgrass Soc. Res. J.* 9:798-801.

Richardson, B.J., P.R. Baverstock, and M. Adams. 1986. *Allozyme Electrophoresis: A Handbook for Animal Systematics and Population Studies.* Academic Press, Inc., San Diego, CA.

Schmid, M. 1974. *Vegetation du Viet-Nam.* Memoires Orstom no. 74. Orstom, Paris.

Shoji, S. 1976. Ecological studies on the *Zoysia* type grassland: 4. Development and growth of the stolon of *Z. japonica* Steud. *Rep. Inst. Agric. Res. Tohoku University* 27:49-59.

Shoji, S. 1983. Species ecology on *Zoysia* grass. *J. Japan Turfgrass Research Association.* 12(2):105-110.

Steussy, T.F. 1990. *Plant Taxonomy: The Systematic Evaluation of Comparative Data.* Columbia University Press, New York.

Tateoka, T. 1954. Karytoxonomic studies in Poaceae. II. *Annual Report of the National Institute of Genetics (Japan)* 5:68-69.

White, R.H., M.C. Engelke, S.J. Anderson, B.A. Ruemmele, K.B. Marcum, and G.R. Taylor II. 2001. Zoysiagrass water relations. *Crop Science* 41:133-138.

Yamada, T. and H. Fukuoka. 1984. Variations in peroxidase isozyme of Japanese lawn grass (*Zoysia japonica* Steud.) populations in Japan. *Japanese J. of Breeding* 34(4):431-438.

Yaneshita, M., T. Ohmura, T. Sasakuma, and Y. Ogihara. 1993. Phylogenetic relationships of turfgrasses as revealed by restriction fragment analysis of chloroplast DNA. *Theor. Appl. Genet.* 87:129-135.

Yaneshita, M., R. Nagasawa, M.C. Engelke, and T. Sasakuma. 1997. Genetic variation and interspecific hybridization among natural populations of zoysiagrasses detected by RFLP analyses of chloroplast and nuclear DNA. *Genes Genet. Syst.* 72:173-179.

Yaneshita, M., S. Kaneko, and T. Sasakuma. 1999. Allotetraploidy of *Zoysia* species with 2n=40 based on a RFLP genetic map. *Theor. Appl. Genet.* 98(5):751-756.

Yang, G.M., B.J. Ahn, and J.S. Choi. 1995. Identification of native zoysiagrasses (*Zoysia* spp.) using morphological characteristics and esterase isozymes. *J. Kor. Soc. Hor. Sci.* 36(2):240-247.

Yang, S.L. and C.J. Yu. 1995. Coastal salt marshes and mangrove swamps in China. *Chinese Journal of Oceanology and Limnology* 13(4):318-324.

Yeam, D.Y., K.H. Hong, and I.S. Han. 1984. The relation between leaf-node stage and flower initiation in *Zoysia* species. *J. Kor. Soc. Hort. Sci.* 25(2):182-185.

Youngner, V.B. 1961. Growth and flowering of *Zoysia* species in response to temperatures, photoperiods, and light intensities. *Crop Science* 1:91-93.

Chapter 18

Centipedegrass (*Eremochloa ophiuroides*)

W.W. Hanna and J. Liu

Centipedegrass is frequently referred to as *poor man's grass* or *lazy man's grass* because it requires less fertility and management to produce an acceptable turf than most warm-season grasses. At the same time, it could be referred to as a *sleeping giant* for turfgrass, since it is probably very underused. The concerns for water use and variable water quality plus the need for lower maintenance turfgrasses make this turf species an attractive choice to use in lawns (Figure 18.1), on roadsides, and in landscaping.

This species is mainly used as a turfgrass in the southern and southeastern United States, even though it is not native to this area. It is well-adapted to sandy acid soils with low fertility. However, it grows well on heavy soils. This species usually shows iron chlorosis on soils above pH 7.5. In the early part of the twentieth century this species was planted as pastures for cattle to graze. Centipedegrass grew better than other forage grasses without added fertility on the poor droughty soils of the coastal plain in the southeastern United States. Although centipedegrass grew well, cattle did not perform well because of the poor quality of the forage. Centipedegrass can still be found growing in pastures today. The earliest centipedegrass seed for commercial sales was harvested from pastures in northern Florida and southern Georgia. However, today thousands of hectares of centipedegrass are cultivated in the United States for seed and sod production for the turfgrass industry.

Centipedegrass has a naturally lighter green color. It becomes dark green with added nitrogen. However, too much nitrogen can cause decline in centipedegrass stands. It is generally recommended that no more than 100 kg ha^{-1} N per year be used. Johnson and Carrow (1988) found that 100 kg ha^{-1} applied as a single application in April caused stand decline after 3 years. However, no stand decline was observed if the same amount of N was split over 3 or 4 equal applications. Centipedegrass does not perform well with high P levels.

Centipedegrass begins growth slowly in the spring, but grows rapidly in the summer. Unruh et al. (1996) showed that this grass has a higher basal temperature requirement and a higher growth rate constant than most warm-season turfgrasses.

This turfgrass is mainly cultivated in the southeastern United States in the sandy coastal plain area. However, it can be found growing as far north as northern Tennessee and around Stillwater and Oklahoma City, OK (unpublished). The most limiting factor for turf use farther north in the United States is cold resistance. Johnson and

Figure 18.1. Lawn of centipedegrass in Tifton, Georgia.

Dickens (1977) showed that hardening centipedegrass plants before subjecting to cold temperatures reduced top-kill low temperature tolerance by 2.3°C, but no survival differences were observed among genotypes. Hardening beyond 10 days did not increase survival and only slightly reduced top-kill (Johnson and Dickens, 1976). Plants receiving two days of favorable growing conditions after being hardened were no more winter hardy than unhardened plants. Cold resistance can be improved through proper turf management such as higher mowing heights, use of less N, less frequent mowing, and deep and less frequent irrigation, etc. Early research indicates that it may be possible to improve cold resistance through breeding.

DISTRIBUTION, TAXONOMY, AND GENETICS

The taxonomic classification of centipedegrass:

Family	Gramineae
Tribe	Andropogoneae
Subtribe	Rottboelliinae
Genus	*Eremochloa*
Species	*ophiuroides*
Authority	(Munro) Hack.

The Andropogoneae are distributed throughout the tropics, in savanna zones, and extending into warm temperature climates. *Eremochloa* grows in short grasslands from India to China and Australia (Clayton and Renvoize, 1986). The *Eremochloa* genus is composed of at least eight species (Bor, 1960). Origins of herbarium specimen observed by the senior author at Kew Herbarium (Kew, England) in 1983 are summarized in Table 18.1. *E. ophiuroides* or centipedegrass (Figure 18.2) is the only species used for turf and/or cultivated with commercial value. Before 1889, this genus was known as *Ischaenum* (Bouton et al., 1983). Hackel (1889) placed this grass in the genus *Eremochloa* and described *E. ophiuroides* as spreading by rhizomes. Silveus (1933) described centipedegrass as spreading by stolons. Hitchcock (1951) described centipedegrass as a

Table 18.1. Origins of *Eremochloa* herbarium specimen observed by the senior author at Kew Herbarium (Kew, England) in 1983.

Eremochloa Species	Country in Which Collected
E. ophiuroides (Munro) Hack.	Peoples Republic of China, Taiwan
E. ciliaris (L.) Merr.	Peoples Republic of China
E. muricata Hack.	India, Sri Lanka, Burma
E. bimaculata Hack.	India, Thailand, Australia, Papua New Guinea
E. zeylanica Hack.	Sri Lanka
E. ciliatifolia Hack.	Sri Lanka
E. eriopoda C.E. Hubb.	Sri Lanka
E. petelotii (Men.)	Thailand

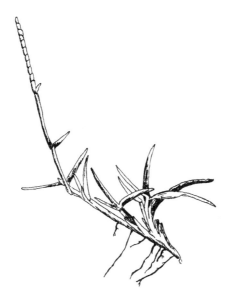

Figure 18.2. Drawing of centipedegrass plant at anthesis.

"low perennial, creeping by thick short-noded stolons...." Bouton et al. (1983) combined the work of others and their studies and published the following description:

"*Perennial* grass; creeping by thick, leafy stolons, branching and rooting at nodes; *Leaf blade* shiny, 1.5 to 4 cm long, 1.5 to 4 mm wide, numerous, bunched on stolons or base of flowering culm, ciliated at base, weakly so toward apex, margins papillose, tip acute to obtuse, sometimes boat-shaped; *Sheaths* glaborous, except for dense pubescence at junction of collars, margins overlapping with edges wide and charteous; *Ligule* truncate, membranous with fringe of hairs; *Collar* somewhat constricted by fused keel, broad, pubescent: *Auricles* absent; *Racemes* solitary, erect and slender, spike-like, 12–24 cm tall, spikelets in pairs at each joint of an articulate rachis with each spikelet overlapping the next joint at 1/3 distance, one sessiled and two flowered, the other pedicellated; *Sessiled spikelet* has lower floret which is sterile or can be staminate, upper florets perfect, stamens 3–6; *Glumes* nearly equal, second shorter, indurate, lower margin enrolled, outer glume dorsally flattened, broadened at apex by broad wing, with nerve endings making margins rough or toothed, (margins inflexed); lemmas (2) and Paleas (2) of nearly equal length, hyaline; *Pedicellated spikelet* rudimentary, linear to oblong."

Southern China is believed to be the center of origin for centipedegrass. In a 1999 plant exploration trip, the authors observed centipedegrass across southern China to 103° 41' East and 34° 44' North. The senior author collected centipedegrass seed on the entire island of Taiwan in 1985. Studies are not complete, but it appears, based on observed distribution and frequency of this species, that the center of origin may be central-southern China.

Centipedegrass is a sexual (Brown and Emery, 1958; Hanna and Burton, 1978) diploid with $2n=2x=18$ chromosomes (Brown, 1950; Hanna and Burton, 1978). Hanna and Burton (1978) reported regular chromosome pairing with mainly nine bivalents at metaphase I. The regular chromosome behavior was reflected in pollen stainabilities ranging from 93 to 97%.

E. ciliaris (L.) Merr. is the only other species related to centipedegrass with some reported genetic information (Hanna and Shu-Liang, 1986). It is a sexual diploid ($2n=18$) perennial bunchgrass. However, it behaved as an annual at Tifton, GA due to winterkill. This species was observed by the authors in 1999 growing north of Kunming, China.

Most centipedegrass has red anthers and red stigmas and varing amounts of red color in the stems and leaves (mainly noticeable in the fall when temperatures become cooler) due to anthocyanin(s) in the plant. A yellow stem characteristic is the most frequently mentioned color variant in centipedegrass (Bouton et al., 1983; Hanna and Burton, 1978). Yellow-stemmed plants have yellow anthers and white stigmas, no red color in stems and leaves, and do not become reddish in the fall when temperatures become cooler. Genetics studies (Bouton et al., 1983; Hanna and Burton, 1978) have shown this yellow-stemmed characteristic to be a recessive, simply inherited genetic characteristic. Although referred to as 'yellow-stem' because plants with this recessive gene tend to be yellowish, a more descriptive term probably should be 'green-stem' since the mutant genotype lacks anthocyanin and appears yellowish due to the naturally lighter green color of centipedegrass. The mutant is not easily recognized in sod except at seedhead formation and in the fall when temperatures become cooler and the plants remain green while red-stem plants become reddish. In either case, it is not known to be objectionable.

Hanna and Burton (1978) showed that a self-incompatibility mechanism(s) is present in centipedegrass. Open-pollinated seed-set in their study ranged from 59 to 90% with 80% or higher seed set on all but one accession. Selfed seed-set ranged from 0 to 58%. Self-incompatibility should be an important consideration in breeding centipedegrass, especially seed-propagated cultivars.

Little is known about the genetics and physiology of flowering. Centipedegrass will usually flower in August. This is probably due to shortening day lengths. It will also flower in June in some years.

COLLECTION, SELECTION, AND BREEDING HISTORY

Source of Germ Plasm

The first record of centipedegrass seed being introduced into the United States was noted in a plant introduction record. The record showed "From China. Collected by Mr. Frank N. Meyer, agricultural explorer of the Department of Agriculture. Received August 12, 1918. This is the last collection of plant material to be made by the late Frank N. Meyer, our agricultural explorer, who was drowned in the Yangtze River on

June 1, 1918." Centipedegrass was included in this collection. Hanna (1995) cited the 1916 introduction date recorded by Hanson et al. (1969), which is incorrect. Hanna (1995) collected and/or introduced centipedegrass from Taiwan and China. Other nonrecorded introductions of small amounts of germ plasm has probably occurred over the years.

The first commercial centipedegrass seed was harvested from a 20 ha pasture by Ray Jensen and B.P. Robinson in 1950 (Jensen, 1996, personal communication). The pasture, owned by Riley Renfoe in Brooks County, GA, produced 114 kg seed. The seed was sold to L.L. Patten for commercial seed sales in 1951. Over the years, centipedegrass has been planted over a broad area in diverse environments. Ecotypes with resistance to such factors as cold and drought have evolved and been selected.

Breeding Techniques

Although centipedegrass is a low maintenance grass with tremendous potential for expanded use, only limited breeding and/or selection efforts by anyone around the world to improve this species has occurred. The reasons for this include: this species appears to be relatively problem-free and little morphological variation has been observed in collected germ plasm.

The first commercial centipedegrass, usually referred to as 'Common,' came from surviving plants in a pasture as discussed above. However, commercial seed producers have indicated to the senior author that beginning in the 1970s, selections for cold resistance made from more northern locations have been mixed with the 'old' Common in production fields, but is still sold as Common.

Oklahoma released 'Oklawn' in 1965 by selecting plants from the turf testing program that showed persistence under adverse environmental conditions (Alderson and Sharp, 1994). Oklawn did not become commercially successful, probably because it did not produce enough seed.

In general, centipedegrass does not appear to have a large amount of morphological variation. However, if individual plants are space-planted, one can observe some variations for internode length as well as leaf and heading characteristics. Radiation has been used in an attempt to broaden the genetic and morphological variation in centipedegrass (Johnston et al., 1978; Hanna, 1995). 'AU Centennial' is a vegetatively propagated clone induced with mutagens in Common, that has shorter internodes, higher leaf density, shorter seedheads, and darker green color than Common (Pedersen and Dickens, 1985).

'TifBlair' is a seed-propagated cultivar with improved resistance to cold and low soil pH (Hanna et al., 1997). It consists of the plants produced from five cycles of recurrent radiation of Common that survived $-28°C$. This cultivar is an excellent seed producer and produces significantly more growth on a soil at pH 4.3 than Common.

'TennTurf' is a more recent vegetatively propagated cultivar selected for cold resistance (Callahan, 1999). It was selected as a single clone that survived from a lawn planted in 1955 (presumably from Common).

Weaver et al. (1995) developed a DNA amplification fingerprinting technique to identify centipedegrass cultivars. They were able to distinguish five cultivars using two primers. The research indicated that the cultivars studied were closely related and were probably all derived from Common.

Major and/or Unique Selection Criteria

Seed production is the most important characteristic for this species. Although it is widely used for sod, the sod fields are reseeded each time the sod is cut and removed. Regeneration from stolons is too slow for economic production of sod. It is usually more expensive to sod an area than to seed it. Therefore, a cultivar must be a good seed producer to be successful, since centipedegrass is considered a low input and low maintenance grass.

Cold resistance is important in the transition and more northern zones of use. Reliability of stands is important since it is considered a low input grass. Cold resistance would also improve the reliability on roadsides where it is usually not practical to reseed frequently.

Low soil pH and high Al resistance again adds reliability to a grass that is expected to grow under usually poor soil conditions. Although it may not produce the best turf under the poor growing conditions, it will provide ground cover and stability to the soil. Many times centipedegrass is planted on new home sites where most, if not all, of the top soil is removed.

Centipedegrass is considered a drought resistant turf (Carrow, 1996; Hook et al., 1992; Huang et al., 1997a). Research by Hook and Hanna (1994) showed that selection within Common centipedegrass may improve drought resistance. Huang et al. (1997b) showed that TifBlair centipedegrass and a *Paspalum vaginatum* accession showed superior drought resistance due to enhanced root growth at deeper soil layers and maintenance of root activity when drought occurred in the upper soil profile. Improved drought resistance is important as water supplies for human consumption become more limited, especially if desirable turf quality is an important factor.

Centipedegrass is generally pest-free. Holcomb (1993) reported the centipedegrass mosaic virus on this species in lawns and sod fields in Louisiana and Georgia. Although this virus is found in centipedegrass, it apparently does not cause major damage to the plant.

Prolific stolon producers tend to produce quicker stands. Initial studies indicate that the germ plasm accessions collected across southern China in 1999 have a large amount of variation for number and length of stolons.

Genetic Markers

The yellow-stem (or green-stem) characteristic (which also imparts yellow anthers and white stigmas) is the most prominant marker in centipedegrass (Hanna and Burton, 1978; Bouton et al., 1983). Bouton et al. (1983) also reported variation for esterases, peroxidases, inflorescence density, and floret density. Various degrees of dwarfness can be observed in this species. Except for the stem color, nothing is known about the genetics of these characteristics.

REFERENCES

Alderson, J. and W.C. Sharp, Eds. 1994. Grass Varieties of the United States. Agric. Handbook No. 170. Soil Conservation Service, USDA, Washington, DC.

Bor, N.L. 1960. *Grasses of Burma, Ceylon, India, and Pakistan*. Pergamon Press, New York.

Bouton, J.H., A.E. Dudeck, R.L. Smith, and R.L. Green. 1983. Plant breeding characteristics relating to improvement of centipedegrass. *Proceed. Soil and Crop Sci. Soc. of Florida* 42:53–58.

Brown, W.V. 1950. A cytological study of some Texas Gramineae. *Bull. Torrey Bot. Club* 77:63–76.

Brown, W.V. and W.H.P. Emery. 1958. Apomixis in the Graminease: Panicoideae. *Amer. J. Bot.* 45:253–263.

Callahan, L.M. 1999. Registration of 'TennTurf' centipedegrass. *Crop Sci.* 39:873.

Carrow, R.N. 1996. Turfgrass science: drought resistance aspects of turfgrasses in the Southeast: Root-shoot responses. *Crop Sci.* 36:687–694.

Clayton, W.D. and S.A. Renvoize. 1986. *Genera Graminum—Grasses of the World.* Royal Botanic Gardens, Kew; Her Majesty's Stationery Office, London.

Hackel, E. 1889. *Eremochloa ophiuroides* (Munro) Hack. p. 261. In A.P. de Candolle and C. de Candolle, Eds. *Monographic Planerogamarum.* Paris, France.

Hanna, W.W. 1995. Centipedegrass—Diversity and vulnerability. *Crop Sci.* 35:332–334.

Hanna, W.W. and G.W. Burton. 1978. Cytology, reproductive behavior, and fertility characteristics of centipedegrass. *Crop Sci.* 18:835–837.

Hanna, W.W., J. Dobson, R.R. Duncan, and D. Thompson. 1997. Registration of TifBlair centipedegrass. *Crop Sci.* 37:1017.

Hanna, W.W. and Hu Shu-Liang. 1986. Chromosomal, reproductive, and breeding behavior of Eremochloa ciliaris (L.) Merr. *J. Hered.* 77:52–53.

Hanson, A.A., F.V. Juska, and G.W. Burton. 1969. Species and varieties. In A.A. Hanson and F.V. Juska Eds. pp. 370–409. *Turfgrass Science, Agron. Mongr. 14,* ASA, Madison, WI.

Hitchcock, A.A. 1951. Manual of the Grasses of the United States. U.S.D.A. Misc. Pub. 200.

Holcomb, G.E. 1993. Occurrence and identification of the virus causing centipedegrass mosiac in Louisiana. *Proc. Louisiana Acad. Sci.* 56:13–17.

Hook, J.E. and W.W. Hanna. 1994. Drought resistance in centipedegrass cultivars. *HortScience* 29:1528–1531.

Hook, J.E., W.W. Hanna, and B.W. Maw. 1992. Quality and growth response of centipedegrass to extended drought. *Agron. J.* 84:606–612.

Huang, B., Duncan, R.R., and R.N. Carrow. 1997a. Drought-resistance mechanisms of seven warm-season turfgrasses under surface soil drying. II. Root aspects. *Crop Sci.* 37:1863–1869.

Huang, B., Duncan, R.R., and R.N. Carrow. 1997b. Root spatial distribution and activity of four turfgrass species in response to localized drought stress. *International Turfgrass Soc. Res. J.* 8:681–690.

Johnson, B.J. and R.N. Carrow. 1988. Frequency of fertilizer application and centipedegrass performance. *Agron. J.* 80:925–929.

Johnson, W.J. and R. Dickens. 1976. Centipedegrass cold tolerance as affected by environmental factors. *Agron. J.* 68: 83–85.

Johnson, W.J. and R. Dickens. 1977. Cold tolerance of several centipedegrass selections. *Agron. J.* 69:100–103.

Johnston, W.J., R. Dickens, and R. Haaland. 1978. Using nuclear science to improve centipedegrass, *Eremochloa ophiuroides,* turf. *Ala. Agric. Exp. Stn. Highlights. Agric. Res.* 25:10.

Pedersen, J.F. and R. Dickens. 1985. AU Centennial centipedegrass. *Crop Sci.* 25:364.

Silveus, W.A. 1933. *Texas Grasses.* The Clegg Co., San Antonio, TX.

Unruh, J.B., R.E. Gaussoin, and S.C. West. 1996. Basal growth temperature and growth rate constants of warm season turfgrass species. *Crop Sci.* 36:997–999.

Weaver, K.R., L.M. Callahan, G. Caetano-Anolles, and P.M. Gresshoff. 1995. DNA amplification fingerprinting and hybridization analysis of centipedegrass. *Crop Sci.* 35:881–885.

Chapter 19

Seashore Paspalum (*Paspalum vaginatum* Swartz)

R.R. Duncan

One grass that will play a dominant role in turf ecosystems in the twenty-first century will be seashore paspalum or *Paspalum vaginatum* (Duncan, 1999a,b; Duncan and Carrow, 2000; Morton, 1973). This grass has multiple stress resistance: salinity tolerance (up to ocean water salt levels for some ecotypes: ECe = 54 dSm^{-1}, TDS = 34,400 ppm, SAR = 57); pH range of 3.6 to 10.2; ecotypes with drought resistance equal to the best drought tolerant centipedegrass (*Eremochloa ophiuroides* Munro cv. 'TifBlair'), and better than the best bermudagrasses (*Cynodon* spp.) with proper management; waterlogging/low-oxygen tolerances; and low light intensity tolerance when exposed to monsoonal or prolonged cloudy/foggy/smoggy conditions, or reduced light in domed stadiums (limited tree shade tolerance = no better than bermudagrasses). *P. vaginatum* requires minimal pesticides and only judicious applications of fertilizers for long-term maintenance. It is very efficient in the uptake of critical fertilizer nutrients and can tolerate a wide range of recycled water resources. Consequently, this grass can also be used for bioremediation (to clean up the environment via phytoaccumulation/ hyperaccumulation of metals/organic chemicals or via rhizofiltration). Turf quality traits and performance are equal to or better than most bermudagrasses, especially in environments subjected to multiple abiotic/biotic stresses, low water quality, or saline-affected conditions.

ADAPTATION AND MORPHOLOGY

Worldwide Dispersion and Use

P. vaginatum is a littoral, warm-season perennial grass found normally between 30–35° N–S latitudes near sea level in tropical and subtropical to warm temperate regions (Morton, 1973; Skerman and Riveros, 1990). The species is considered helophytic and mesophytic (Webster, 1987). This hydrophilous grass is suited to aquatic, semiaquatic, and moist environments (Skerman and Riveros, 1990). The grass will tolerate waterlogged conditions and periodic mesohaline flooding in salt swamps, coastal flats, and tidal marshes (Colman and Wilson, 1960).

P. vaginatum has been vegetatively transported around the world for two different reasons. First, the grass was used as bedding and ballast in the bottom of slave boats as they moved between Africa, North America, South America, Central America, and the Caribbean Islands. This would account for the discovery of diverse paspalum ecotypes at key staging areas for off-loading slaves along the Georgia (Sea Island, Ft. Pulaski) and South Carolina (Sullivan Island) coasts (Gray, 1933), but not universally throughout eastern U.S. coastal venues and only on specific Caribbean islands where the ships docked.

Secondly, the grass was introduced into salt-affected areas as the need for forages, land reclamation, and turf increased. Australia received their first *P. vaginatum* from South Africa during 1935 (Beehag, 1986; McTaggart, 1940; Trumble, 1940), with an introduction into South Australia for use in soil stabilization and pasturage on saline soils.

Some botanists consider *P. vaginatum* indigenous to Asia, Africa, and Europe (Judd, 1979), while others consider it native to the New World but introduced and naturalized in the Old World (Bovo et al., 1988; Chase, 1929; Echarte et al., 1992; Morton, 1973). Based on ecotype collection of paspalums along the eastern coastal United States, coarse-, intermediate-, and fine-textured types are found primarily in Georgia and South Carolina at off-loading sites for slave boats from Africa that arrived during the 1700s and 1800s (Gray, 1933). Sites such as Sea Island and Ft. Pulaski, GA, and Charleston, SC, have provided diverse fine-textured vegetation. Collection activity along the U.S. gulf coast has produced only scattered coarse or ornamental (extremely coarse) types that are supposedly indigenous to the region. Only introduced intermediate types have been found along the west coast in Southern California, reflecting the introduction of 'Adalayd' and 'Futurf' from Australia in the late 1960s (see Table 19.1). Most Caribbean islands have paspalums of varying textures, depending on boat movement via the trade winds from various African coastal countries. Some Central American countries have *P. vaginatum* on their Atlantic coasts. Brazil and Argentina could be a possible secondary center of origin, since slave boat traffic between Africa and those countries was significant during the 1800s and the grass found a compatible environment for optimum growth in that region.

Paspalum vaginatum has been used as a utility turf for erosion control and coastal environmental stabilization in South Africa, Australia, New Zealand, and the southeastern United States. The grass has been used for sand dune stabilization and revegetation in Florida; sports turf in Australia, Israel, Argentina, Brazil, and California; residential/commercial/and park landscapes in the Middle East, Australia, California, Hawaii, and Argentina. Paspalum has been used on golf courses in California, Texas, Florida, Georgia, South Carolina, the Middle East, South Africa, Argentina, China, Thailand, Indonesia, and the Philippines. It has also been used as the primary grass on a polo field in Rancho Santa Fe, CA, and on lawn bowling greens in Australia.

Its use as a forage has been documented in Africa, Australia, South America, and the United States (Alderson and Sharp, 1995; Colman and Wilson, 1960; Dirven, 1963; Loxton, 1977; Maddaloni, 1986; Malcolm, 1962; Mejia, 1984). Dry matter production can range from 1.3 to 7.0 T ha^{-1} yr^{-1} for the coarse types on saline fields (Malcolm, 1986). With fertilizer, dry matter yields have reached almost 23 T ha^{-1} (Hill, 1978a,b), and some nutritional data are available (Dirven, 1963).

Coarse leaf-textured paspalums occur wild on seacoasts of both hemispheres. In the Americas, it is found almost exclusively along the Atlantic coast in marshy, brackish ecosystems.

Vegetation regions in Australia where *P. vaginatum* grows include tropical heaths, tropical and subtropical rainforests, tropical and subtropical wet sclerophyll forests, dry sclerophyll forests, tropical and subtropical subhumid wetlands, semiarid shrub woodlands, and acacia shrub lands (Webster, 1987). It can also be found in mangrove

Table 19.1. Paspalum cultivars and their origin.

Year	Name	Origin	Distributor/Company	Leaf Texture
1951	Saltene	Western Australia	The Turf Farm, Wanneroo Turf Farm	intermediate
	Salpas	Adelaide	—	intermediate
1972	Futurf[a]	Toguay, Victoria, Australia	—	intermediate
1975	Adalayd[a] (Excalibur)	Adelaide, Australia	Pacific Sod (CA) Coastal Turf (TX)	intermediate
—	Fidalayel	California	—	intermediate
1991	Tropic Shore	Oahu, Hawaii	Plant Materials Center-Molokai, Hoolehua, HI	coarse
—	Mauna Kea	Sea Island, GA via Hawaii	—	intermediate
1998	Salam	Sea Island, GA via Hawaii	Southern Turf Hawaii, Alabama, Florida, Egypt	fine
1999	Sea Isle 2000[b]	Florida	[c]	fine
1999	Sea Isle 1[b]	Argentina	[c]	fine
—	Durban Country Club	South Africa	Superlawn, Natal, S. Africa	fine
1999/2000	Seaway,[b] Seadwarf, Seagreen, Seagreen II, Millennium Seafine	Alden Pines CC, Bookeelia, FL	Environmental Turf Solutions, Emerald Isle Turf, Punta Gorda, FL	fine
2001	223	Cuba	J. W. Turf, Loxahatchee, FL	fine

[a] Patented.
[b] Patent or PVP Pending.
[c] Release through the University of Georgia. Licensees can be found on 'seashore paspalum,' 'growers.'

swamps. It is useful for erosion control on salinity-sensitive lands (soil stabilization) and areas subjected to tidal influences (sand-binder, beach protection) (Skerman and Riveros, 1990).

In southern Africa, *P. vaginatum* is found widely distributed in South Africa, Namibia, and Swaziland. In South Africa, it occurs mainly, but not exclusively, along the coast from the northwestern extremity of KwaZulu-Natal (abutting Mozambique) to the furthermost southwestern tip of the country (Cape Town vicinity). In the southwest, the ecotypes are very fine-leaf textured and have been referred to in *Flora capensis* as *P. distichum* L. var. *nanum* J. Doll Cape types (Chippendall, 1955). A different fine-leaved ecotype grows adjacent to a salt marsh behind the dunes near East London in the Eastern Cape. This ecotype does not appear along the western coastline of the country, but can be found in the vicinity of Swakopmund, Mariental Free State, and southern Gauteng regions in Namibia. The distribution extends over much of the Mozambique coast and northward toward East Africa.

P. vaginatum occurs on sandy beaches, on the banks of estuaries frequently inundated by salt water, and along the banks of coastal rivers. It can be found inland near the edge of saline water pools or lakes on sandy soils. It is considered the most salt tolerant warm-season turfgrass (Carrow and Duncan, 1998), with ecotypes that range from ocean water salinity levels to approximately one-fifth that level (similar to the best bermudagrasses) (Lee, 2000).

Nomenclature

Paspalum vaginatum O. Swartz is primarily known as seashore paspalum, or simply paspalum. Other less commonly used names include siltgrass (Morton, 1973) or sand knotgrass. In Australia, it is traditionally referred to as saltwater couch. In Latin American countries, the species name characteristically involves the words grama, gramon, or gramilla, with additional localized phrases or words (Duncan and Carrow, 2000).

From the early 1900s to 1976, *P. vaginatum* was the predominately used scientific name (Chase, 1929; Hitchcock, 1971). In 1976 taxonomists tried to switch the name to *Paspalum distichum* C. Linnaeus (Loxton, 1974; Guedes, 1976). The ensuing debate in the literature (Fosberg, 1977; Renvoize and Clayton, 1980; Guedes, 1981) lasted until 1983 (Report, 1983, Taxon 32:281) when the nomenclature committee for Spermatophyta officially designated *Paspalum vaginatum* O. Swartz as the correct scientific name.

Paspalum distichum

The same confusion occurred for *P. distichum*. The dispute among taxonomists regarding *P. distichum* or *P. paspaloides* was finally resolved in 1983 (Report, 1983, Taxon 32: 281) when the nomenclature committee for Spermatophyta officially designated *Paspalum distichum* L. as the correct scientific name.

Common names for *P. distichum* include knotgrass, joint grass, eternity grass (United States), water couch, couch paspalum (Australia); mercer grass (New Zealand); chepica, grama colorada, grama de aqua, pasto dulce (Spanish); gramilla blanca, pata de gallina, salaillo (Peru); groffe doeba (Suriname); gharib (Iraq); and sacasebo (Cuba).

P. distichum is used for nonsaline stabilization purposes in the Sydney, Australia, area. This species can be found in a similar role on earthen dams in moist nonsaline, inland sites in the Americas.

Morphological Description and Differentiation

Both *P. vaginatum* and *P. distichum* are members of the Disticha group, but can be differentiated by several key attributes. Botanically, both species have rhizomes and stolons, membranous ligules, and rolled leaf buds (Beehag, 1986; Hitchcock, 1971; Hitchcock and Chase, 1917). *Paspalum distichum* has pubescent (hairy) glumes while *P. vaginatum* has glabrous (hairless) glumes. Both species inhabit aquatic biotopes, waste places, and follow rotation crops or perennial crops (Hafliger and Scholtz, 1980). However, *P. distichum* occurs primarily in freshwater habitats, i.e., swamps, irrigation canals, earthen dams, and drainage outlets, while *P. vaginatum* dominates saline soils and brackish swamps in esturine sands and marshy muds near the seashore (Barnard, 1969; Bor, 1960; Skerman and Riveros, 1990).

At least 20 morphological characteristics can be used to differentiate *P. vaginatum* and *P. distichum* (Echarte and Clausen, 1993; Ellis, 1974; Loxton, 1974) involving the inflorescence and leaves, with hairy (pubescent) glumes (*P. distichum*) and hairless (glabrous) glumes (*P. vaginatum*) being a primary distinction. Both *distichum* and *vaginatum* have two conjugate racemes in the inflorescence and are considered psammophilous species (Barreto, 1957).

Chorology

The geographical distribution of the *Paspalum* genus is tropical and subtropical venues and harsh, stressful environments. This large genus is ecologically aggressive and many species possess apomicts that may account for their evolutionary survival (Chapman, 1992). *Paspalum* has a C_4 NADP-ME biochemical-type photosynthetic pathway, which is characteristic of malate-forming grasses that occur in moist ecosystems (Chapman and Peat, 1992). The NAD-ME malate-formers are normally found in arid environments, while PEP-CK aspartate-formers are intermediate (Hattersley, 1992).

P. vaginatum can be grouped into a specific phytochoria (ecological habitat) based on salt-affected and moist biotopes. It is the most salt-tolerant warm-season turfgrass that is known (Carrow and Duncan, 1998), with several ecotypes tolerating ocean water levels (Duncan and Carrow, 2000).

Propagation

P. vaginatum is an ecologically aggressive, littoral warm-season perennial grass. The species is both rhizomatous and stoloniferous (Morton, 1973; Webster, 1987). Flowering culms are erect or basally decumbent, vary in height from 8 to 60 cm with 5–13 glabrous (hairless) nodes. Stolon nodes are distinctly pubescent. Mid-culm leaves do not have sheath or blade auricles, are distinctly distichous, 50–220 mm long, 1–4 mm wide, are linear and glabrous, gradually tapering to a narrow apex. The prophyllum is 20–40 mm long. The 1 mm ligule is membranous and truncate with a pubescent collar. The inflorescence is fully exserted at maturity, is composed mainly of two primary branches (racemes) 20–60 mm in length, with 16–32 twin-rowed spikelets on the primary branch. Each spikelet is solitary, plano-convex, subsessile, elliptic, 2.5 to 4.5 mm long, and 0.9 to 1.5 mm wide. Anthers are 1.2 to 1.6 mm long. Caryopses (seeds) are 2.5 to 3.0 mm long and 1.5 mm wide, narrowly obovate, subacute, and slightly concavo-convex (Hafliger and Scholtz, 1980; Silveus, 1933; Vegetti, 1987; Webster, 1987). The range in leaf textures among *P. vaginatum* ecotypes includes: very coarse, ornamental types; coarse types (resembling St. Augustinegrass); intermediate types (resembling common bermudagrass); and fine-leaf types (resembling dwarf bermudagrass).

P. vaginatum must be propagated from sprigs or sod, since seed production has not been reliable. Sprigging rates can vary from a minimum of 5 bushels per 1,000 ft^2 (19 m^3/ha) to normal warm-season grass rates of 10–20 bushels per 1,000 ft^2 (38–76 m^3/ha). Generally the finer-textured paspalum ecotypes establish better than the coarse-textured types because of a greater node volume (node ratio roughly 13:5) (Webster, 1987).

In order to maintain genetic purity and prevent or minimize cross-contamination in the paspalum breeding program at the University of Georgia-Griffin, very strict protocols in handling vegetatively propagated material have been implemented. Following initial field evaluation, promising ecotypes are taken to the greenhouse, where a single stolon is selected and planted in soil-less media. Each experimental ecotype is continually increased in the greenhouse and planted in a breeder block. A foundation field is subsequently planted in anticipation of possible release for commercial use, with the source vegetative material always tracing back to the single stolon breeder material. All evaluations on golf courses, sports fields, and home lawns involve vegetative material that can be traced back to the single-stolon increase in the greenhouse.

CYTOTAXONOMY

Classification

Family	Gramineae (Poaceae)
Subfamily	Panicoideae
Supertribe	Panicodae
Tribe	Paniceae
Subtribe	Setariinae
Genus	*Paspalum*
Group	Disticha
Species	*vaginatum*
Authority	Swartz

The Panicoideae includes a diverse group of grasses found primarily in tropical and subtropical latitudes (Chapman and Peat, 1992), but extending into temperate climates (Watson and Dallwitz, 1992). The supertribe Panicodae can be found in diverse pantropical to temperate habitats with widely variable rainfall requirements. Besides *Paspalum*, the tribe Paniceae also includes *Axonopus* (carpetgrass), *Digitaria* (crabgrass), *Panicum* (torpedograss), *Pennisetum* (kikuyugrass, fountaingrass, buffelgrass), *Setaria* (bristlegrass), and *Stenotaphrum* (St. Augustinegrass) (Watson and Dallwitz, 1992). The subtribe Setariinae also includes *Paspalum*, *Panicum*, *Brachiaria* (signalgrass), *Echinochloa* (barnyardgrass), *Setaria*, and *Axonopus* (Chapman and Peat, 1992; Hafliger and Scholtz, 1980) as well as 60 additional genera. The group Disticha includes *P. vaginatum* that inherently colonizes saline ecosystems such as along sea coasts and on brackish sand, and *P. distichum* can be found dispersed over a wider geographical area away from coastal venues, but growing in freshwater, moist habitats in more temperate and colder venues (Silveus, 1933). *P. vaginatum* is thought to have evolved directly from *Panicum* (Chapman, 1992). *Paspalum* and *Digitaria* have closely related spikelet inflorescence morphology (Hafliger and Scholtz, 1980).

Ploidy

The genus *Paspalum* contains either 320 species (Watson and Dallwitz, 1992), 330 (Clayton and Renvoize, 1986), or 400 species (Chase, 1929, 1951), depending on the source. Approximately 220 species can be found in practically all herbaceous communities within various ecosystems in Brazil (Valls, 1987). A substantial number of *Paspalum* species are characterized by an apomictic, autotetraploid race and a sexually reproducing self-incompatible diploid race (Quarin, 1992). Phenotypically, cospecific tetraploid and diploid biotypes do not normally differentiate and usually form sympatric populations in South America (Norrmann et al., 1989). Diploids provide the genetic variability during the evolution of apomictic tetraploid *Paspalum* species (Espinoza and Quarin, 1997). The diploid *Paspalums* are generally self-incompatible and non-apomictic (Bovo et al., 1988). The basic chromosome number for most *Paspalums* is $x = 10$ (Bovo et al., 1988; Pitman et al., 1987).

P. vaginatum is a sexually reproducing decumbent diploid species with some self-sterility and a propensity for cross-pollination between clones of diverse origin (Carpenter, 1958). The chromosome number for *P. vaginatum* is predominately diploid ($2n = 2x = 20$) (Bashaw et al., 1970; Burson et al., 1973; Echarte and Clausen, 1993; Fedorov, 1974; Llaurado, 1984; Okoli, 1982; Quarin and Burson, 1983). One report of a tetraploid ($2n = 4x = 40$; Taxon 25: 155–164) and a hexaploid ($2n = 6x = 60$; Taxon 24: 367–372) has been documented for *P. vaginatum*. This species has the D genome (Bashaw et al., 1970; Burson, 1981a, 1983), which is unique among the diploid *Paspalums*, with the exception of *P. indecorum* Mez. whose chromosomes exhibited partial homology with members of the **D** genome (Quarin and Burson, 1983). Interspecific hybridization using *P. vaginatum* has been difficult, but limited success in producing hybrids has been accomplished using *P. vaginatum* as the pollen parent and *not the female* parent (B.L. Burson, personal communication). All hybrids between *P. dilatatum* J. Poiret (2n = 4x = 40) and *P. vaginatum* were completely sterile (Burson et al., 1973). Five triploid hybrids (2n = 3x = 30) were produced between *P. urvillei* Steud. (2n = 4x = 40) and *P. vaginatum* (1.3% cross ability) (Burson and Bennett, 1972). Twenty-eight diploid *P. jurgensii* x *P. vaginatum* hybrids have been produced (Burson, 1981b). Eight diploid hybrids were made between *Paspalum notatum* J. Flugge (2n = 2x =20) and *P. vaginatum*, but no viable seed were produced (Burson, 1981b). Seven diploid *P. pumilum* Nees x *P. vaginatum* hybrids and two reciprocal hybrids have been produced (Burson, 1981b). Four *P. indecorum* x *P. vaginatum* diploid hybrids (two had 2n = 2x = 20 and two had 2n = 2x = 21 chromosomes) revealed partial homology between these species (Quarin and Burson, 1983), but no hybrid produced viable seed. The unique **D** genome in *P. vaginatum* would account for this difficulty. Attempts to consistently produce viable seed of this species have not yet been successful, although research efforts are continuing.

Investigations into the self-incompatibility problem in the sexual diploids *P. simplex*, *P. chaseanum*, and *P. plicatulum* Michx. indicated that these species were highly self-sterile but cross-fertile (Espinoza and Quarin, 1997). The low self-pollinated seed set was caused by failure of the pollen tube to reach the ovule. The pollen germinated and actually penetrated the stigma, but the pollen tube failed to grow in the style. With cross-pollination, the pollen tubes penetrated both the stigma and style, grew through the ovary and reached the micropyle. No research has been conducted on *P. vaginatum* to determine the exact nature of the self-incompatibility.

For *P. distichum*, two cytotypical groups have been documented: tetraploids (2n = 40) in one group and pentaploids (2n = 50), hyperpentaploids (2n = 52, 54, 57, 58), and hexaploids (2n = 60) in the other group (Echarte et al., 1992). Tetraploid and hexaploid cytotypes reproduce by aposporous apomixis (Quarin and Burson, 1991). The hexaploids are the dominant cytotype in Buenos Aires province, Argentina (Echarte et al., 1992). Phenograms indicate a distinct, nonintegrated clustering pattern between the two species and probable unique genomes (Echarte and Clausen, 1993). The genome designation for *P. distichum* is currently unknown. No crosses between *P. vaginatum* and *P. distichum* have been documented.

COLLECTION, SELECTION, AND BREEDING HISTORY

A small collection of 28 *Paspalum vaginatum* accessions from the province of Buenos Aires, Argentina, was assembled from 1986 to 1988 (Clausen et al., 1989) and is maintained at the germ plasm bank of the Estacion Experimental Agropecuaria, INTA, Balcarce, Argentina. Ten accessions from primarily the southeastern United States were assembled at the Ft. Lauderdale, FL agricultural center by 1977 (Busey, 1977). An unknown number of *P. vaginatum* accessions are located in the germ plasm collection of *Paspalums* in Brazil. A larger collection has been assembled at Griffin, GA, involving coarse, intermediate (Adalayd types), and fine-textured accessions from nine countries and seven U.S. states (Duncan, 2000).

Breeding Approach

P. vaginatum is a self-incompatible, sexual diploid. Cross-fertilization is difficult because its genome (**D**) is different from most other paspalum species. Cross-pollination is accomplished at 18–21°C (mid 60s°F) during early morning hours (0500–0730) just prior to sunrise (Burson, 1985).

Several hybrids with other *Paspalum* species have been made (Burson, 1981a,b; Quarin and Burson, 1983), but most hybrids did not produce viable seed. Viable seed production is extremely low (Carpenter, 1958; Malcolm, 1983) due to sterility, ergot (Raynal, 1996), and the self-incompatibility problem.

Seed germination occurs at temperatures >20°C or >68°F (optimum 25–30°C) preceded by a cold vernalization treatment (Carpenter, 1958; Skerman and Riveros, 1990). Even though commercial seed is not available, seed production is possible in swards of mixed genetic backgrounds when conditions are favorable for cross-pollination. Dual spikes in the raceme can range from 10 to 65 mm in length, with the coarse types having the longest spike length (Duncan, 1999a). Consequently, the potential for economical seed production is available, but problems associated with (1) locating the best environment for maximum seed production, (2) finding the best parental combinations for effective cross-pollination, (3) mechanical harvesting challenges in field situations, and (4) seed dormancy problems will govern the eventual success of producing and marketing seed for this species.

Approximately 1,000 viable hybrid seed have been germinated; the seedlings were transplanted to small pots in the greenhouse for increase, and subsequently were transplanted to turf field plots. These new hybrids were immediately subjected to close mowing stress (<13 mm cutting height) with judicious applications of fertilizer and water. Only one hybrid (HYB5) has emerged from the hybrid evaluation program for advanced testing thus far in Georgia.

Additional genetic variability was created by subjecting paspalum to *in vitro* propagation (Cardona, 1996; Cardona and Duncan, 1997, 1998; Cardona et al., 1997) using nine different ecotypes - PI 509021 (Argentina); HI-1, Mauna Kea, K-3, K-7 (Hawaii); Adalayd (Australia); PI 299042 (Zimbabwe); SIPV-1 (Sea Island, GA); and AP-6 (Florida). Over 5,500 tissue-culture-regenerated plants were planted in turf field plots and subjected to mowing stress (3–16 mm). Approximately 100 selections were made of "improved" somaclonal variant plants with turf traits that appeared to be superior to the donor parents. Subsequent evaluations produced four selections for additional evaluation on greens, tees, or fairways (TCR 1, 3, 4, 6). None have been released from the breeding program.

The philosophy of the evaluation program is to subject prospective ecotypes to real world (end-use) conditions as fast as possible: (1) mowing at 3 mm in 3 x 3 m plots on a USGA - specification green, (2) mowing at 15–16 mm height on a fairway, (3) mowing at 4 mm on a push-up tee, or (4) mowing at 13 mm on a native, heavy clay soil in a sports field setting. The final exam is always conducted on golf courses, sports fields, or home landscape situations.

Genetic Analysis

An RFLP analysis of 51 accessions from 29 *Paspalum* species has revealed six cluster groups (Jarret et al., 1998). Cluster group 5 includes *P. distichum, P. unispicatum, P. conjugatum* Berg., and *P. vaginatum*. RAPD profiles for selected *P. vaginatum* ecotypes revealed seven groups within *P. vaginatum* (Liu et al., 1994).

Simple sequence repeat (SSR) analysis of (GA)n and (CA)n repeats using five primers produced 47 loci with repeats of n greater than or equal to 3. The number of alleles resolved per locus ranged from 6 to 16, with an average of 14 (Liu et al., 1995). The SSR dendogram for the 46 ecotypes was similar to the RAPD profiles (Liu et al., 1994). No (AT)n, (ATT)n, (CTT)n, or (GATA)n repeats were detected.

Specific SSR analysis using five markers (Brown et al., 1998) and 10 accessions (50 electropherograms) detected an average of 8.4 fragments and had a diversity index of 0.79 for each primer pair. Overall fragment similarity for the SSRs varied from 0.16 to 0.26. The dendogram depicting the genetic relationship among the 10 ecotypes revealed Excalibur, Fidalayel, and FR-1 (Fairbanks Ranch) had identical DNA profiles, and all three had an 85% similarity to Adalayd.

Sea Isle 1 and AP-14 had a 55% similarity to each other, and Sea Isle 1 and Sea Isle 2000 had a 45% genetic similarity to each other. Sea Isle 2000 and AP14 were supposedly Adalayd derivatives, but their genetic similarities were 35% and 30%, respectively, probably revealing significant genetic mutations toward finer-textured ecotypes and away from the intermediate-textured parental Adalayd. Sea Isle 2000 and AP14 were only 30% genetically similar and morphological traits plus turf performance had revealed their differences in research plots. Directional mutation in *P. vaginatum* has supposedly been from coarse to intermediate to fine leaf texture. Field observations within fine-textured ecotypes and regenerated plants from tissue culture of fine-textured types has not produced any additional mutations. The diploid genome apparently is quite stable at the point (fine-textured status) of evolution. Genetic analysis of *Paspalum* species is listed chronologically to date in Table 19.2.

RAPD and SSR analyses were similar for the 46 *P. vaginatum* ecotypes. Using the grouping format for RAPDs, the following conclusions can be drawn:

1. The coarse forage types are apparently unique to each continent, with the North American group closely related and possibly indigenous to the respective country from which they were collected. The close relationship between African and Argentinean coarse types indicates introgression, probably from Africa to Argentina, e.g., PI 377709 and PI 5209018-2 have a similarity coefficient of 1 (Groups I, VI, VII).
2. The Argentinean fine to intermediate types (Group IV) are unique to that country, with expected similarity to the Sea Island, GA (Group II) and Hawaiian (Group III) material. Two of the AP types clustered in group II, indicating some introgression from Sea Island, GA to Alden Pines, near Ft. Myers, FL.
3. The Adalayd-Excalibur (Group V) of Australian origin obviously was the source material for the AP material. One accession from Hawaii clustered with this group, indicating a possible movement of material either from California or Florida into Hawaii. Adalayd and Excalibur are closely related with a coefficient of 0.97. The ecotypes can be split into two subgroups, one closely allied to Adalayd with coefficients near 1, and the second group related to Excalibur with coefficients near 0.85.

Ecotypes and Cultivars

In the United States, the oldest known fine-textured ecotypes have been found on Sea Island, GA. When the Sea Island golf course was built in 1925, the grass was already established in the salt marsh adjacent to the course. Coarse types can be found sporadically from North Carolina to Florida on the eastern United States coast, on Gulf of Mexico coastal sites and various Caribbean Islands, on the Hawaiian Islands, and along the Pacific Rim. The coarse types are apparently indigenous to each respective country.

No additional breeding work was conducted on Adalayd after its introduction into the United States. Adalayd was not well-adapted to the United States and its salinity tolerance was no better than the bermudagrasses. However, several cultivars (Table

Table 19.2. Genetic analysis of Paspalum spp.

Species	Analysis	Source
vaginatum	RAPDs[a]	Liu et al., 1994
vaginatum	SSRs	Liu et al., 1995
multiple	flow cytometry	Jarret et al., 1995
scrobiculatum	RAPDs	M'Ribu and Hilu, 1996
notatum	RFLPs/RAPDs	Ortiz et al., 1997
simplex	RFLPs	Pupilli et al., 1997
vaginatum	SSRs	Brown et al., 1998

[a] RAPD = Random Amplified Polymorphic DNA; SSR = Simple Sequence Repeat (microsatellites); RFLP = Restriction Fragment Length Ploymorphism.

19.1) have been selected and marketed by Environmental Turf Solutions and their association with Alden Pines golf course, Pine Island, FL. Very little actual research data has been accumulated on those cultivars (Table 19.1).

Southern Turf ® became a major supplier of paspalum for the Hawaiian Islands as well as the Pacific Rim. They eventually named their cultivar 'Salam,' which means peace in Arabic. This cultivar originated out of Sea Isle, GA vegetative material introduced into Hawaii in 1980s. Research data have been generated on this cultivar since 1993. Additional cultivars are becoming available (Duncan, 2001), but most have little research documentation.

The most research has been documented on Sea Isle 1 and Sea Isle 2000 cultivars out of the University of Georgia breeding program. These cultivars are the only certified seashore paspalums available in the world, with specific management protocols available on each cultivar (www.seaisle1.com and www.seaisle2000.com).

REFERENCES

Alderson, J. and W.C. Sharp. 1995. Grass Varieties in the United States. USDA Handbook 170. Lewis Publ., Boca Ration, FL.

Barnard, C. 1969. Herbage Plant Species. Austral. Herbage Plant Reg. Authority, CSIRO, Australian Division of Plant Industries, Canberra, Australia.

Barreto, I. L. 1957. The *Paspalum* species with two conjugate recemes in Rio Grande do Sul (Brazil). *Rev. Argentina Agron.* 24: 89–117.

Bashaw, E.C., A.W. Hovin, and E.C. Holt. 1970. Apomixis, its evolutionary significance and utilization in plant breeding. pp. 245–248. In *Proc. XI Int'l Grassland Congress*. Univ. Queensland Press, Australia.

Beehag, G.W. 1986. Paspalum turfgrass. *Bowling Greenkeeper* 20(11): 20–21.

Bor, N.L. 1960. *The Grasses of Burma, Ceylon, India, and Pakistan (excluding Bambuseae)*. Int'l Series Monog., Pergamon Press, Oxford, UK.

Bovo, O.A., L.A. Mroginski, and C. Quarin. 1988. *Paspalum* spp. pp. 495–503. In Y.P.S. Bajaj, Ed. *Biotechnology in Agriculture and Forestry*. Vol. 6, Crops II. Springer-Verlag, Berlin.

Brown, S.M., S.E. Mitchell, C.A. Jester, Z.W. Liu, S. Kresovich, and R.R. Duncan. 1998. DNA typing (profiling) of seashore paspalum (*Paspalum vaginatum* Swartz) ecotypes and cultivars. pp. 39–51. In M.B. Sticklen and M.P. Kenna, Eds. *Turfgrass Biotechnology: Cell and Molecular Genetic Approaches to Turfgrass Improvement*. Ann Arbor Press, Chelsea, MI.

Burson, B.L. 1981a. Genome relations among four diploid *Paspalum* species. *Bot. Gaz.* 142: 592–596.

Burson, B.L. 1981b. Cytogenetic relationships between *Paspalum jurgensii*, *P. intermedum*, *P. vaginatum*, and *P. setaceum* var. *ciliatifolium*. *Crop Science* 21: 515–519.

Burson, B.L. 1983. Phylogenetic investigations of *Paspalum dilatatum* and related species. pp. 170–173. In J.A. Smith and V.W. Hays, Eds. *Proc XIV Int'l Grassland Congress*. Lexington, KY. June 15–24, 1981. Westview Press, Boulder, CO.

Burson, B.L. 1985. Cytology of *Paspalum chacoense* and *P. durifolium* and their relationship to *P. dilatatum*. *Bot. Gaz.* 146: 124–129.

Burson, B.L. and H.W. Bennett. 1972. Cytogenetics of *Paspalum urvillei* x *P. juergensii* and *P. urvillei* x *P. vaginatum* hybrids. *Crop Sci.* 12: 105–108.

Burson, B.L., H.S. Lee, and H.W. Bennett. 1973. Genome relations between tetraploid *Paspalum dilatatum* and four diploid Paspalum species. *Crop Sci.* 13: 739–743.

Busey, P. 1977. Turfgrasses for the 1980s. *Proc. FL State Hort. Soc.* 90: 111–114.

Cardona, C.A. 1996. Development of a Tissue Culture Protocol and Low Temperature Tolerance Assessment of *Paspalum vaginatum* Sw. Ph.D. Dissertation. University of Georgia, Athens, GA.

Cardona, C.A., R.R. Duncan, and O. Lindstrom. 1997. Low temperature tolerance assessment in paspalum. *Crop Sci.* 37:1283–1291.

Cardona, C.A. and R.R. Duncan 1997. Callus induction and high efficiency plant regeneration via somatic embryogenesis in paspalum. *Crop Sci.* 37: 1297–1302.

Cardona, C.A. and R.R. Duncan. 1998. In vitro culture, somoclonal variation, and transformation strategies with paspalum turf ecotypes. pp. 229–236. In M.B. Sticklen and M.P. Kenna, Eds. *Turfgrass Biotechnology: Cell and Molecular Genetic Approaches to Turfgrass Improvement*. Ann Arbor Press, Chelsea, MI.

Carpenter, J.A. 1958. Production and use of seed in seashore paspalum. *J. Australian Inst. Agric. Sci.* 24: 252–256.

Carrow, R.N. and R.R. Duncan. 1998. *Salt-Affected Turfgrass Sites: Assessment and Management*. Ann Arbor Press, Chelsea, MI.

Chapman, G.P. 1992. *Grass Evolution and Domestication*. Cambridge Univ. Press, UK.

Chapman, G.P. and W.E. Peat. 1992. *An Introduction to the Grasses*. CAB Int'l., Wallingford, Oxon, UK.

Chase, A. 1929. The North American species of Paspalum. *Contrib. Nat'l Herb.* 28: 1–310.

Chase, A. 1951. Manual of the Grasses of the United States. USDA Miscellaneous pub. No. 200. Vol. 2, 2nd ed. pp. 928–929, 933.

Chippendall, L.K.A. 1955. A guide to identification of grasses in South Africa. In D. Meredith, Ed. *The Grasses & Pastures of South Africa Parov. (Cape Province)*, S. Africa, Central News Agency.

Clausen, A. M., S. I. Alonso, M. C. Nuciari, A. M. Echarte, and M. Pollio. 1989. FAO/IBPGR *Plant Genetic Resources Newsl.* 78/79:31.

Clayton, W.D. and S.A. Renvoize. 1986. Genera Gramium, grasses of the world. *Kew Bulletin, Add. Ser.* 13: 1–389.

Colman, R. L. and G.P.M. Wilson. 1960. The effects of floods on pasture plants. *Agric. Gazette, NSW* 71: 337–347.

Dirven, J.G.P. 1963. The nutritive value of the indigenous grasses of Surinam. *Netherlands J. Agric. Sci.* 11: 295–307.

Duncan, R.R. 1999a. Environmental compatibility of seashore paspalum (saltwater couch) for golf courses and other recreational uses. I. Breeding and genetics. *Int'l Turfgrass Soc. Res. J.* 8 (2): 1208–1215.

Duncan, R.R. 1999b. Environmental compatibility of seashore paspalum (saltwater couch) for golf courses and other recreational uses. II. Management protocols. *Int'l Turfgrass Soc. Res. J.* 8 (2): 1216–1230.

Duncan, R.R. 2000. Seashore paspalum: A turfgrass for tomorrow. *Diversity* 16(1–2): 45–46.

Duncan, R. R. 2001. All seashore paspalums are not created equal. *Golf Course Mgmt.* (June):54–60.

Duncan, R.R. and R.N. Carrow. 2000. *Seashore Paspalum—The Environmental Turfgrass*. Ann Arbor Press, Chelsea, MI.

Echarte, A.M. and A.M. Clausen. 1993. Morphological affinities between *Paspalum distichum* sensu lato and *P. vaginatum* (Poaceae). *BOL Soc. Argentina Botany* 29 (3-4): 143–152.

Echarte, A.M., A.M. Clausen, and C.A. Sala. 1992. Chromosome numbers and morphological variability in *Paspalum distichum* (Poaceae) in Buenos Aires province (Argentina). *Darwiniana* 31: 185–197.

Ellis, R. P. 1974. Comparative leaf anatomy of *Paspalum paspaloides* and *P. vaginatum*. *Bothalia* 11: 235–241. Also *Herb. Abstr.* 45: 31–42.

Espinoza, F. and C.L. Quarin. 1997. Cytoembryology of *Paspalum chaseanum* and sexual diploid biotypes of two apomictic *Paspalum* species. *Austral. J. Botany* 45: 871–877.

Fedorov A. 1974. *Chromosome Numbers of Flowering Plants*. Koenigstein, Germany, Otto Koeltz Science Pub.

Fosberg, F.R. 1977. *Paspalum distichum* again. *TAXON* 26: 201–202.

Gray, L.C. 1933. *History of Agriculture in the Southern United States to 1860*. Vol. I. Carnegie Institute, Washington, D. C.

Guedes, M. 1976. The case for *Paspalum distichum* and against futile name changes. *TAXON* 25(4): 512–513 (nomenclature).

Guedes, M. 1981. Against rejecting the name *Paspalum distichum* L.: Comment on proposal 528. *TAXON* 30:301.

Hafliger, E. and H. Scholtz. 1980. *Panicoid Grass Weeds*. Ciba-Geigy, Ltd. Basle, Switzerland, p. 108.

Hattersley, P.W. 1992. Significance of intra-C_4 photosynthetic pathway variation in grasses of arid and semi-arid regions. pp. 181–212. In G.P. Chapman, Ed. *Desertified Grasslands, Their Biology and Management*. Academic Press, London, UK.

Hill, J.C.R. 1978a. The use of range plants in the stabilization of phytotoxic mining wastes. pp. 707–711. In D.N. Hyder, Ed. *Proc. 1st Int'l Rangeland Congress*. Denver, CO. 14–18 Aug. 1978. Soc. Range Mgmt. Dept. Bot. Rhodesia Univ., Salisbury, Rhodesia.

Hill, J.C.R. 1978b. A root growth study used to examine the suitability of two grasses for stabilizing toxic mine wastes. *Proc. Grassland Soc. South. Africa* 13: 129–133. Dept. Conserv. Exten. Salisbury, Zimbabwe.

Hitchcock, A.S. 1971. *Manual of the Grasses of the United States*. 2nd ed. revised by A. Chase. Dover Publ. Inc., New York.

Hitchcock, A. S. and A. Chase. 1917. Grasses of the West Indies. In *Contribution from U.S. Nat'l Herbarium* 18 (7): 261–471. Smithsonian Institute, U.S. Nat'l Museum. Washington, DC. Govt. Print. Office.

Jarret, R.L., P. Ozias-Akins, S. Phatak, R. Nadimpalli, R. Duncan, and S. Hilliard. 1995. DNA contents in *Paspalum* spp. determined by flow cytometry. *Genetic Resources Crop Evaluation* 42: 237–242.

Jarret, R.L., Z.W. Liu, and R.W. Webster. 1998. Genetic diversity among *Paspalum* spp. as determined by RFLPs. *Euphytica* 104: 119–125.

Judd, B. I. 1979. *Handbook of Tropical Forage Grasses*. Garland STPM Press, New York, NY.

Lee, G. 2000. Comparative Salinity Tolerance and Salt Tolerance Mechanisms of Seashore Paspalum Ecotypes. Ph.D. dissertation, University of Georgia, Athens.

Liu, Z.W., R.L. Jarret, R.R. Duncan, and S. Kresovich. 1994. Genetic relationships and variation of ecotypes of seashore paspalum (*Paspalum vaginatum* Swartz) determined by random amplified polymorphic DNA (RAPD) markers. *Genome* 37: 1011–1017.

Liu, Z.W., R.L. Jarret, S. Kresovich, and R. R. Duncan. 1995. Characterization and analysis of simple sequence repeat (SSR) loci in seashore paspalum (*Paspalum vaginatum* Swartz). *Theoretical Applied Genetics* 91: 47–52.

Llaurado, M. 1984. El genero *Paspalum* L. A. Catalunya. Bulletin Inst. Catalana Historical Nature, *Session Botany* 51: 101–108.

Loxton, A.E. 1974. The taxonomy of *Paspalum paspaloides* and *P. vaginatum* as represented in South Africa. *Bothalia* 11: 243–245, Also *Herb. Abstr.* 45: 31–58.

Loxton, A. E. 1977. *Paspalum vaginatum* subsp. *nanum*. *J. South Africa Bot*. 43: 93–95.

Maddaloni, J. 1986. Forage production on saline and alkaline soils in the humid region of Argentina. *Reclam. Reveg. Res.* 5: 11–16.

Malcolm, C.V. 1962. *Paspalum vaginatum* for salty seepages. *J. Dept. Agric. West. Austr.* 3: 615–616.

Malcolm, C. V. 1983. Wheatbelt Salinity. A Review of the Land Problem in Southwest Australia. Tech. Bullen. #52. Dept. Agric. W. Austral., Perth.

Malcolm, C.V. 1986. Production from salt affected soils. *Reclam. Reveg. Res.* 5: 343–361.

McTaggart, A. 1940. Plant introduction. In Grassland Investigations in Australia. Herbage Publication Series Bulletin 29: 18–21. Imperial Bureau Pasture Forage Crops. Aberystwyth, UK.

Mejia, M.M. 1984. Scientific and common names of tropical forage species. *CIAT*, Cali, Columbia.

Morton, J. F. 1973. Salt-tolerant siltgrass (*Paspalum vaginatum* Sw.). *Proc. FL State Hort. Soc.* 86: 482–490.

M' Ribu, H.K. and K.W. Hilu. 1996. Application of random amplified polymorphic DNA to study genetic diversity in *Paspalum scrobiculatum* L. (Kodo millet, Poaceae). *Genetic Resour. Crop Eval.* 43: 203–210.

Norrmann, G.A., C.L. Quarin, and B.L. Burson. 1989. Cytogenetics and reproductive behavior of different chromosome races in six *Paspalum* species. *J. Heredity* 80: 24–28.

Okoli, B.E. 1982. IOPB chromosome number reports LXXIV. *TAXON* 31: 127.

Ortiz, J.P.A., S.C. Pessino, O. Leblanc, M.D. Howard, and C.L. Quarin. 1997. Genetic fingerprinting for determining the mode of reproduction in *Paspalum notatum*, a subtropical apomictic forage grass. *Theor. Appl. Genet.* 95: 850–856.

Pitman, M.W., B.L. Burson, and E.C. Bashaw. 1987. Phylogenetic relationships among *Paspalum* species with different base chromosome numbers. *Botanical Gazette* 148: 130–135.

Pupilli, F., M.E. Caceres, C.L. Quarin, and S. Arcioni. 1997. Segregation analysis of RFLP markers reveals a tetrasomic inheritance in apomictic *Paspalum simplex*. *Genome* 40: 822–828.

Quarin, C.L. 1992. The nature of apomixis and its origin in panicoid grasses. *Apomixis Newsletter* 5: 8–15.

Quarin, C. and B.L. Burson. 1983. Cytogenetic relations among *Paspalum notatum* var. *saurae*, *P. pumilum, P. indecorum,* and *P. vaginatum. Botanical Gazette* 144: 433–438.

Quarin, C.L. and B.L. Burson. 1991. Cytology of sexual and apomictic *Paspalum* species. *Cytologia* 56: 223–228.

Raynal, G. 1996. Presence en France de *Claviceps paspali* Stev. et Hall sur *Paspalum distichum* L. et de L'ergotisme correspondant sur du Betail. *Cryptogamie Mycol.* 17(1): 21–31.

Renvoize, S.A. and W.D. Clayton. 1980. Proposal to reject the name *Paspalum distichum* L. *System. Natural. Ed.* 10(2): 855, 1759 and *TAXON* 29: 339.

Report of the committee for Spermatophyta. 1983. Proposal 528. Rejection of *Paspalum distichum* L. (Gramineae). *TAXON* 32: 281.

Silveus, W.A. 1933. *Texas Grasses*. The Clegg Co., San Antonio, TX.

Skerman, P.J. and F. Riveros. 1990. *Tropical Grasses*. Food & Agric. Organization of the United Nations. pp. 119, 128, 130, 565–568.

Trumble, H.C. 1940. *Grass Investigations at the Waite Agricultural Research Institute,* University of Adelaide. pp. 57–60.

Valls, J.F.M. 1987. Recursos geneticos de especies de *Paspalum* no Brasil. p. 3–13. In J.F.M. Valls, Ed. *Encontro Internacional sobre Melhoramento de Paspalum.* Annals Institute de Zootecnia. Nova Odessa. Brasil.

Vegetti, A.C. 1987. Typological analysis of inflorescence in *Paspalum* Poaceae. *Kurtziana* 19: 155–160.

Watson, L. and M.J. Dallwitz. 1992. *The Grass Genera of the World.* CAB Int'l., Wallingford, UK.

Webster, R.D. 1987. *The Australian Paniceae (Poaceae).* J. Cramer. Berlin - Stuttgart, Germany. pp. 173–174, 181–182.

Chapter 20

St. Augustinegrass, *Stenotaphrum secundatum* (Walt.) Kuntze

Philip Busey

St. Augustinegrass, *Stenotaphrum secundatum* (Walt.) Kuntze, is widely used as a lawn and pasture grass in warm, subtropical, and tropical climate regions (Sauer, 1972; Judd, 1975). Other common names are "buffalo grass" in Australia and the Republic of South Africa, "Charleston" in some areas of the southeastern United States, and "San Augustin" in Latin America. St. Augustinegrass is well adapted to humid areas and where irrigation is provided. The world's first known record of planting St. Augustinegrass was on 11 November 1880, as a turf alongside an avenue at A. M. Reed's Mulberry Grove plantation, at Yukon, near Orange Park, Florida (Works Progress Administration, 1939), and is based only on a fragmentary reference to cultural practices, "George planting St. Augustine grass in avenue in afternoon." The species is expanding rapidly as a lawn grass, especially due to human migration to warm coastal regions. As an example, St. Augustinegrass turfgrass sod harvested in Florida was 3,100 hectares in 1974 (Florida Department of Agriculture and Consumer Services, 1976) compared with 13,400 hectares in 1991 (Hodges et al., 1994). By 2001, St. Augustinegrass was the primary turf grown in Florida (pop. 15 million) and St. Augustinegrass was grown on approximately 70% of the lawns (Busey, 2001, unpublished data).

BIOLOGY

Plant Characteristics

St. Augustinegrass has round-tipped leaf blades 5 to 14 mm wide and they are arranged in a strictly distichous manner. It is a perennial, and it spreads by branching stolons, forming a coarse and spongy canopy. Because of the absence of rhizomes or other protected stems, it recuperates poorly from defoliation and has poor wear tolerance. Cultivars with shorter internodes have higher wear tolerance (Busey, 1991, unpublished data). Crowns or basal shoot aggregates are absent. The stolon internodes are exposed, and slightly flattened dorsiventrally.

The leaf blades are generally glabrous, but genotypes showing possible introgression with *S. dimidiatum* (L). Brongn., pembagrass, are sparsely pubescent (Busey, 1990b). The midrib is conspicuous. The bases of the leaf blades are attenuated and subtended by constricted collars, which are conspicuously lighter than the blade or sheath, making the leaves pseudopetiolate. The ligule is a minutely ciliate membrane. The leaf sheaths are compressed laterally, nearly forming a keel. As a C_4 species, St. Augustinegrass has typical Krantz leaf anatomy (Krans et al., 1979), containing an inner parenchyma bundle sheath layer with centripetal chloroplasts (Figure 20.1). This anatomical characteristic of C_4 grasses facilitates the compartmentalization of photosynthetic processes in two different cellular regions, repressing photorespiration (Dengler et al., 1994).

The inflorescences of St. Augustinegrass are modified spike-like panicles, with the branches of the inflorescence contracted and often reduced to single spikelets. Branches are partially embedded in hollows on one face or the sides of a corky rachis. The rachis, which is terminated in a naked point, normally disarticulates at the branch nodes into squarish segments containing the spikelet(s). The inflorescence segments float in salt water for 7–10 days, which may not be sufficient for transoceanic dispersal (Sauer, 1972). The awnless spikelets are 3–6 mm long and have dissimilar glumes. The lower or first glume is scale-like, only about 1 mm long, and nerveless. The upper or second glume is similar to, and about the same length as the nerved, chartaceous lemmas. Spikelets contain a lower floret, which is most often staminate or is neuter, but is perfect and sets seed in some genotypes (Center and Busey, 1981, unpublished data). The upper perfect floret contains three anthers, which may vary among genotypes from orange-buff with flecks of purple to sulfur yellow. The two stigmata may be purple, or translucent-appearing white, or bicolor (purple shafts and translucent branches). Internodes may vary from purplish to green, in association with the color of anthers and stigmata. For example, plants with purplish internodes generally have purple stigmata and orange-buff anthers, while plants with green internodes generally have whitish translucent stigmata and sulfur-yellow anthers.

Environmental Adaptation and Management

St. Augustinegrass provides a tight leaf canopy, due to relatively prostrate leaf angle; therefore, it is highly resistant to weed infestation. Some cultivars, such as 'Floratam,' grow well in sandy coastal areas where zoysiagrasses, *Zoysia* spp., and bermudagrass, *Cynodon* spp., grow poorly due to parasitism by the sting nematode, *Belonolaimus longicaudatus* Rau (Busey et al., 1982b). St. Augustinegrass grows adequately across a wider range of soil conditions than other warm-season turfgrasses. It generally grows without problems in sand, loam, and humic soils, and across a pH range from 4.5 to 8.5. Under conditions of high pH and waterlogged soil, including production in plastic trays, interveinal chlorosis symptomatic of iron deficiency is sometimes observed, particularly in the Breviflorus Race (see Taxonomy and Geography).

Documented long-term nutritional management studies have not been done. Therefore, with increasing scrutiny of lawn maintenance practices as possible nonpoint sources of groundwater pollutants such as nitrate, the appropriate rates, timings, and nutrition sources for St. Augustinegrass turf fertilization are unresolved. St. Augustinegrass is often grown in warm coastal areas with shallow aquifers, and the appropriate nitrogen fertilization is therefore especially important in protecting groundwater.

Use of high rates of inorganic nitrogen has been associated with southern chinch bug, *Blissus insularis* Barber, outbreak in St. Augustinegrass (Busey and Snyder, 1993). Higher fertilization rates are associated with higher frequency of wilt in St. Augustinegrass turf, compared with lower fertilization rates (Busey, 1996). Higher fertilization rates are also associated with higher levels of thatch, a problem for St. Augustinegrass considering that it is entirely stoloniferous, and any accumulation of runners is aboveground.

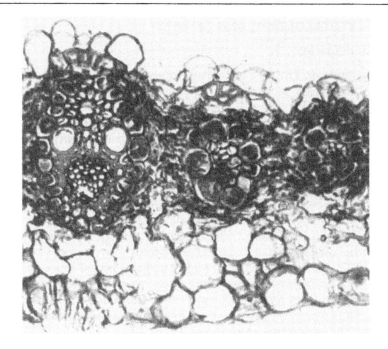

Figure 20.1. Leaf blade transverse section of 'Roselawn' St. Augustinegrass showing the dense Krantz bundle sheath cells surrounding each vascular bundle, an indication of the C4 photosynthetic pathway.

Several cultivars of St. Augustinegrass tolerate partial shade (Busey and Davis, 1991), a valuable trait for use in lawns, particularly in smaller residential landscapes and where trees are dominant. The shade tolerance of St. Augustinegrass is useful in mixed cropping systems of the tropics, where an herbaceous understory is grazed by animals, in the diminished illumination beneath tree crops. St. Augustinegrass has the least reduction in yield, and the largest yield, among eight grasses evaluated under the shade of coconuts, *Cocos nucifera* L., in the Solomon Islands (Smith and Whiteman, 1983); the coconuts transmitted 20% relative irradiance (full sunlight=100% irradiance). Among warm-season turfgrasses, St. Augustinegrass performs better under reduced illumination than bahiagrass, *Paspalum notatum* Flügge; bermudagrasses, *Cynodon* spp.; centipedegrass, *Eremochloa ophiurioides* (Munro) Hack.; and zoysiagrasses, *Zoysia* spp. (Beard, 1973).

St. Augustinegrass generally has 10 to 30% greater evapotranspiration than bermudagrass in minilysimeters under semiarid conditions (Casnoff et al., 1989; Kim and Beard, 1988; Kneebone and Pepper, 1982). However, electromagnetic measurement of soil moisture in unrestricted plot areas under humid conditions showed that the evapotranspiration of St. Augustinegrass is not significantly different from bermudagrass (Carrow, 1995). Drought resistance in St. Augustinegrass is due to drought survival through wilt avoidance due to deeper or more effective root systems and not by reduced evapotranspiration (see Physiology and Environmental Stresses). St. Augustinegrass is a model species for studying water relationships including evapotranspiration (Stewart and Mills, 1967) and leaf diffusive resistance (Johns et al., 1983).

DISTRIBUTION, CYTOTAXONOMY, AND GENETICS

Origin and Related Species

The genus *Stenotaphrum* Trin. is a primarily tropical member of the tribe Paniceae of the Panicoideae. Whereas *S. secundatum*, St. Augustinegrass, occurs on all continents except Antarctica, the six other species of *Stenotaphrum* are known naturally only from East Africa, the islands and coastlines of the Indian Ocean, and from southern China to the South Pacific (Busey, 1995b; Sauer, 1972). Most occupy restricted natural habitats, and three species are island endemics. Spikelet and inflorescence characteristics of *Stenotaphrum* are most similar to the monotypic genera *Thuarea* Pers. and *Uranthoecium* Stapf of Australia; *Thuarea* also occurs in coastal regions of tropical Asia (Webster, 1988). According to Webster (1988), the genus *Stenotaphrum* is probably not closely related to *Paspalidium* Stapf, as suggested by Sauer (1972).

Morphologically, pembagrass, *S. dimidiatum*, is the species most similar to St. Augustinegrass; the two species are separated primarily by number of spikelets per raceme (Sauer, 1972). Some polyploid St. Augustinegrass introductions show possible introgression with pembagrass. Inflorescence racemes of the presumptive introgressants, such as FX-10 and its relatives (Busey, 1993), produce three and occasionally four spikelets, which would be intermediate between the two species (Sauer, 1972). Pembagrass is used in lawns in Kenya (Bogdan, 1970), Ghana and Uganda (Sauer, 1972), and India (Sundararaj et al., 1971) and is also a useful pasture grass. The pembagrass USDA introduction PI-365031 is very coarse textured, and occasionally the leaf blades are plicate (Busey, 1977, unpublished observations).

Las-aga, *S. micranthum* (Desv.) C. E. Hubbard, is a widely distributed strand pioneer of the Indian Ocean and the South Pacific. It occurs on open sandy beaches, in the salt spray of coralline limestone, and other coastal habitats of small islands, but also extends inland to shaded woodlands and inhabited areas such as village streets and house yards (Sauer, 1972). In Guam it is considered an excellent pasture and lawn grass and is propagated by stolon cuttings (Safford, 1905). Other than *S. secundatum*, *S. dimidiatum*, and *S. micranthum*, the four remaining species of *Stenotaphrum* are described only from herbarium specimens, not from other firsthand accounts, and are not cultivated. *S. helferi* is distributed from Malaysia through Southeast Asia to southern China, including Hainan Island. It occurs along forest paths and serves as good pasture.

The origin of St. Augustinegrass is unknown. Its distribution has been described as "part of a larger migrational mystery involving ... other cosmopolitan seashore grasses that lack proven capability of longrange sea dispersal" (Sauer, 1972).

One hypothesis is that St. Augustinegrass originated in the Old World tropics, in the center of diversity for the genus, specifically the coastlines and islands of the Indian Ocean, and that Europeans later brought it to the New World during the post-Columbian era. The hypothesis of introduction by Europeans may not explain the diversity of St. Augustinegrass in the New World, unless there were multiple early accidental introductions by Europeans.

An alternative New World origin hypothesis for St. Augustinegrass is consistent with the early time of the first description of St. Augustinegrass, 1788, from a South Carolina collection (Sauer, 1972), and even earlier collections in the New World. For example, it was collected by Dale in the Bahamas, about 1730; by Browne in Jamaica, about 1750; and by Commerson in Brazil and Uruguay, in 1767 (Sauer, 1972). St. Augustinegrass has considerable diversity in cultivated and adventive populations in the West Indies and southern United States (Busey et al., 1982a). For example, based on herbarium specimens, by the 1800s St. Augustinegrass had a wide distribution in North America and showed racial divergence. The divergence of long-internode plants of the Longicaudatus Race in Florida, and short-spikelet plants of the Breviflorus Race in other southeastern states such as Louisiana (Busey et al., 1982a), suggests a long residence of St. Augustinegrass in the New World.

A third hypothesis is that St. Augustinegrass had an Old World origin and was brought to the New World before the time of European migration, by an early transoceanic dispersal predating the European voyages of discovery. This would be consistent with its early appearance in other distant places; for example, it was collected by Beauvois in 1787 in Ghana and Nigeria; by Menzies in 1798 from Kauai, Hawaii; and by Cunningham in 1822 in Australia.

Taxonomy and Geography

Any attempt to improve St. Augustinegrass genetically would be haphazard without an understanding of the existing genetic variation, which is not smooth, but punctuated into clusters of similar genotypes (Figure 20.2). If the clustering were ignored, quantitative expectations of genetic advance would be biased because assumptions underlying heritability would be violated. For example, while quantitative measures of genetic variation assume normal distribution, under clonal selection of clustered genotypes it is possible to make rapid initial genetic improvement as the number of taxonomic groups or ploidy levels is narrowed. But if there is not sufficient variation within genotype clusters, further advance may be difficult. In the extreme, attempts at genetic improvement may be confounded by the occurrence of duplicates of existing cultivars in the population under selection. Finally, by providing a natural classification, clustering helps predict the occurrence of useful alleles (Busey et al., 1982a) for documented traits such as chinch bug resistance (Busey, 1995a), disease resistance (Atilano and Busey, 1983), herbicide resistance (Busey, 1993), nematode resistance (Busey et al., 1993), drought resistance (Busey, 1996), and shade tolerance (Busey and Davis, 1991).

Morphotype clusters of St. Augustinegrass (Figure 20.2a–e) have been designated variously as "Groups," "Races" (Busey et al., 1982a; Busey, 1986), and "demes" (Sauer, 1972). As an overview to the classification system, ploidy levels, e.g., $2n=18$, are first subdivided into Races, and Races are subdivided into Groups, which contain multiple cultivars and breeding populations (Busey et al., 1982a). Most cultivars are diploids ($2n = 18$), and diploids are subdivided into the Breviflorus Race and the Longicaudatus Race.

The Breviflorus Race (Busey, 1986) is widely represented among weedy and adventive populations, and they have high (over 60%) seed set (Busey and Center, 1979, unpublished data). Within this race, the Gulf Coast Group (Figure 20.2d) is a moderately homogeneous assemblage of genotypes with green stolons and white stigmata, present since at least the mid-1800s in the southeastern United States, north of peninsular Florida. The Gulf Coast Group was first clearly represented in an 1868 collection [ALABAMA: Sandy shores of Mobile Bay, Point Clear, along the seashore from E. La. to N. Carol. August 1868, collector *Mohr s.n.* (AL)]. The Gulf Coast Group appeared more frequently by the 1890s in Louisiana. The Gulf Coast Group occurs in protected locations north of the Piedmont, such as old lawns in Memphis, Tennessee, and Corinth, Mississippi. These are sources of cold tolerant germ plasm (J. V. Krans, 1984, personal communication). The Gulf Coast group is endemic to the southeastern United States, and includes contemporary cultivars 'Texas Common' and 'Raleigh.'

The Dwarf Group (Figure 20.2c), another subcategory of the Breviflorus Race, includes genotypes with generally strong anthocyanin pigmentation in the stolons, purple stigmata, and dark green leaf blades (Busey et al., 1982a). Genotypes of the Dwarf Group generally have shorter leaves and inflorescences than the Gulf Coast Group.

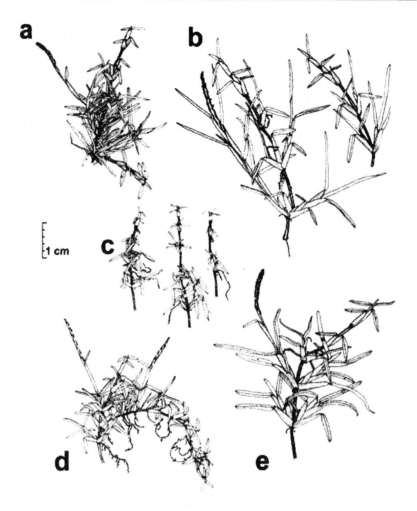

Figure 20.2. Races and Groups of St. Augustinegrass (Busey et al., 1982a; Busey, 1986). (a) Bitterblue Group, 'Bitterblue'; (b) Longicaudatus Race, 'Roselawn'; (c) Breviflorus Race Dwarf Group, FA-243; (d) Breviflorus Race Gulf Coast Group, FL-1933; (e) Floratam Group, 'Floratam.' Artist: Penny Longo-Silverman.

Artificial introgressants between the Gulf Coast Group and the Dwarf Group have produced suitable hybrids, some of which are represented by the shade tolerant cultivars developed by O.M. Scotts & Sons (Busey and Davis, 1991). One example is 'Seville' (Riordan et al., 1980), the first St. Augustinegrass released with a known pedigree; that is, both male and female parents are known.

Longicaudatus Race genotypes (2n=18) have elongate stolons (Busey, 1986) and long leaves (Figure 20.2b). This race is probably synonymous with the Natal-Plata deme (Sauer, 1972). Longicaudatus Race genotypes in older lawns and pastures have been assigned to 'Florida Common' (Busey et al., 1982a) and include the cultivar 'Roselawn' (Allen and Kidder, 1971). In Florida, Longicaudatus Race plants were collected by 1845 in Manatee County, Florida [FLORIDA: BM: Am Strande, Terraciera Bay, July 1845, *Rugel 370* (F,MO,US)], by 1848 in Key West [FLORIDA: Key West, Herb. Chap. "Prob. Torrey mis. 1848" (MO)], and by 1894 in central Florida, where it was regarded as "valuable in pastures" [FLORIDA: St. Augustine grass. Orlando, FL, 23 April 1894, *Northey*

2570 (US)]. This race grows in remote areas in Everglades National Park, e.g., from Highland Beach to East Cape Sable (Busey, et al., 1982), which was inhabited briefly by Anglo-Americans, around 1900 (Tebeau, 1968). Everglades Experiment Station, University of Florida, distributed the cultivar Roselawn in 1942 and 1943 (Allen and Kidder, 1971). It has an open habit of growth and does not form a dense sod (Busey, 1977, unpublished data). Although not making acceptable lawns, the Longicaudatus Race apparently has long-term survival ability in low maintenance habitats.

Polyploidy

Polyploid St. Augustinegrasses were first identified by Long and Bashaw (1961) who described sterile triploids (2n=27) with irregular meiosis. They were designated the Cape deme by Sauer (1972) who identified their first collection in 1791 at the Cape of Good Hope, and their use in lawns in the Republic of South Africa by 1900. In fact, use of polyploids in lawns occurred in Florida by 1892 [FLORIDA: Cultivated, Leesburg, 6 June 1892, *P. H. Rolfs 1008* (US)]. Since 1900, the polyploids have spread most often in association with intentional introductions and cultivation through vegetative propagation. 'Bitterblue,' a Cape deme genotype (Figure 20.2a), was the foundation for the commercial sod industry in Florida, starting in the 1920s (Busey and White, 1993). Although Bitterblue is a sterile clone, slight but detectable genetic variation exists (Busey, 1986). Another cytologically sterile variant, Floratam (2n = c. 32, Busey, 1979), was released for its combined resistance to the St. Augustine Decline Strain of Panicum Mosaic Virus (PMV-SAD) and the southern chinch bug (Horn et al., 1973). Floratam St. Augustinegrass (Figure 20.2e) became so popular that by 1980–81, it represented 77% of commercial sod in southeast Florida and 21% of lawn areas (Busey, 1986).

An unusual 2n=30 polyploid variation was discovered, among African introductions, with normal bivalent chromosome pairing (Figure 20.3) at diakinesis and normal set seed (Busey, 1990b). From this germ plasm, the cultivar FX-10 was developed with resistance against a virulent, Floratam-killing race of the southern chinch bug (Busey, 1993). The simplest cytological origin for the African polyploids would be allotetraploidy. A 2n=12 progenitor has not been discovered, and seems unlikely, considering that x=9 or 10 is the basic chromosome number of the Paniceae (Gould, 1968). Anomalous chromosome counts have been found, however, for *S. dimidiatum*: 2n=36 from Sri Lanka (Gould and Soderstrom, 1974), 2n=48 from Malagasy (Sauer, 1972), 2n = 54 for PI-365031 from the Republic of South Africa (Busey, 1990b), and 2n = c. 60 for FL-2195 from Mauritius (Busey et al., 1993). It is possible that polyploidy originated in *Stenotaphrum* occasionally and by different mechanisms. Polyploidy is important in the development of other warm-season turfgrasses in addition to St. Augustinegrass; examples are bahiagrass, *Paspalum notatum* Flügge and bermudagrasses, *Cynodon* spp. (Busey, 1989).

Encouragingly, taxonomic classifications based on cytology and chemistry are congruent, suggesting that they are natural. Polyploids have no detectable activity for uridine diphosphate (UDP) glucose pyrophosphorylase (Green et al., 1981). Polyploid genotypes studied included Bitterblue, 'Floralawn' (FA-108), Floratam, Floratine, FA-118, PI-290888, PI-300127, and PI-300130, based on direct counts of chromosomes (Busey, 1990b) and/or racial grouping (Busey, 1986). In contrast, 17 diploid St. Augustinegrasses have detectable UDP glucose pyrophosphorylase activity, and so do *S. dimidiatum* accessions PI-289729 and PI-365031 (Green et al., 1981). Among diploids in the latter study, all with low adenosine diphosphate (ADP) glucose pyrophosphorylase activity were of the Gulf Coast Group; most with high activity were of the Dwarf Group (Busey et al., 1982a).

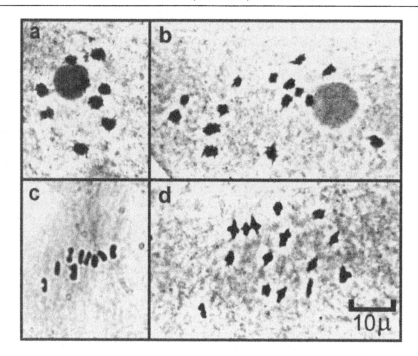

Figure 20.3. Pollen mother cells of St. Augustinegrasses showing entirely bivalent pairing in diploids (2n=18) and polyploids (2n=30). (a) FX-261 diakinesis (2n=18); (b) FL-1759 diakinesis (2n=30); (c) FA-243 diakinesis (2n=18); and (d) FX-10 metaphase (2n=30).

ADAPTIVE POLYMORPHISMS

Physiology and Environmental Stresses

Polymorphisms among St. Augustinegrasses have been detected for many physiological and morphological traits, including isozymes (Green et al., 1981), leaf extension rate and stomatal density (Atkins et al., 1991), leaf pubescence (Busey, 1990b), several morphological and pigmentation traits (Busey et al., 1982a; Busey, 1986), and lethal temperature and winter survival (Philley, 1994; Philley et al., 1998). No differences among genotypes have been observed for mowing energy requirement (Fluck and Busey, 1988).

Adaptive and morphological variations in St. Augustinegrass are associated with chromosome differences. The most conspicuous visible differences between ploidy levels are that diploids have narrower, thinner, more translucent, brighter green leaf blades, while polyploids have coarser, thicker leaf blades which are more opaque and less saturated in color (Busey, 1986, 1993). Compared with diploids, polyploid leaf blades look grayish blue-green in lawns. Diploids of the Breviflorus Race have lower growth habit and more rapid ground covering ability (Busey et al., 1982a). Their growth habit is more highly branched, which results in earlier sod maturity, earlier and easier harvest, but greater risk of thatch problems in the established landscape (Busey, 1979, unpublished data). In small experimental plots, such as those in the National Turfgrass Evaluation Program (NTEP), diploids receive higher turfgrass quality scores than polyploids, particularly during the first year of evaluation (Busey, 1985), which can be deceptive for estimating long-term performance. Polyploids such as Floratam, the main cultivar in Florida, perform unacceptably for turfgrass quality in small plots. Most population improvement has been done on diploids, while polyploid cultivars (e.g., Bitterblue and Floratam) are often selections or seedlings of unknown paternity (e.g., Horn et al., 1973).

Shade tolerance differences exist. Seville, DelMar, and Jade provide superior quality, compared with Floratam and Floralawn, under 21% relative irradiance (full sunlight = 100%). Shade was due to a mixed tree canopy (Busey and Davis, 1991). While photosynthetic rates among cultivars are similar at 45% or higher relative irradiance, at 29% relative irradiance, the photosynthetic rates of Floratam and Floralawn are reduced to less than half of maximum, which is also less (P<0.05) than Floratine and Seville, at the same shade level (Peacock and Dudeck, 1993). Some polyploids, such as Floratam, grow very poorly in the shade, and genotypic differences in shade adaptation are evident between 21% and 29% relative irradiance (Busey and Davis, 1991; Peacock and Dudeck, 1993). This could be largely an expression of leaf height, because polyploids are taller (Busey, 1991, unpublished data).

Compared with diploids, polyploids are more resistant to drought based on wilt avoidance due to deeper or more effective root systems, rather than reduced evapotranspiration. Evapotranspiration rates in an environmental chamber differ among cultivars, ranging from 6.7 mm day^{-1} to 8.1 mm day^{-1}; however, differences among cultivars are not detected in the field, based on the average of 3 years (Atkins et al., 1991). Likewise in weighing field lysimeters, St. Augustinegrasses do not differ in evapotranspiration (Miller and McCarty, 2001).

Among St. Augustinegrass genotypes differing in drought survival, extent of wilt is associated with canopy loss following irrigation suspension (Busey, 1986). Under conditions of unrestricted rooting in the field, where there is a water table at 1.45 m, 'FX-10' has significantly less wilt than Floratam and other cultivars. When the root systems are confined at 0.75 m, however, the number of days to wilt for FX-10 was 6.7, which is not significantly different than Floratam, 6.0 days, but is greater than 'Palmetto,' 4.8 days (Miller and McCarty, 2001).

These results are consistent with the hypothesis that FX-10 avoids wilt by deep rooting, provided there is room for deep rooting. Compared with Floratam, FX-10 was able to maintain a superior leaf water potential at the first end point (water exudation from the veins of the cut leaf edge), but not at the second end point (darkening of the leaf) (Miller and McCarty, 2001). FX-10 has a prominent, heavily suberized endodermis (Figure 20.4), which may be related to root permeability to water.

Floratam, the only polyploid extensively studied for freezing tolerance, has no detectable cold acclimation (Fry et al., 1991). Lethal temperatures for regrowth are −4.5°C and −6.0°C for Floratam and Raleigh, respectively; electrolyte leakage differences are similar, but smaller (Maier et al., 1994a, b). These differences are significant in the field. Winter-kill occurs to sensitive cultivars such as Floratam following temperatures of −9°C to −7°C (Busey, 1990a), yet Floratam also has been reported to survive as low as −15°C (Wilson et al., 1977). Raleigh St. Augustinegrass, with higher freezing tolerance than FX-332 or Floratam, is the only cultivar that acclimates to cold (Maier et al., 1994b). Differential thermal analysis (DTA) is highly correlated, r = 0.96, with field survival rating (Philley et al., 1995).

Cultivars differ in salinity response. For example, Seville is more tolerant of salinity than Floratam, Floratine, or Floralawn based on hydroponic culture (Dudeck et al., 1993; Peacock et al., 1993; Smith et al., 1993) but not based on whole plant microculture (Smith et al., 1993).

Biotic Stresses

Genotypic differences occur in resistance to the southern chinch bug (Reinert and Dudeck, 1974); resistance to the sting nematode, *Belonolaimus longicaudatus* (Busey et al., 1993); resistance to the St. Augustine Decline Strain of Panicum Mosaic Virus

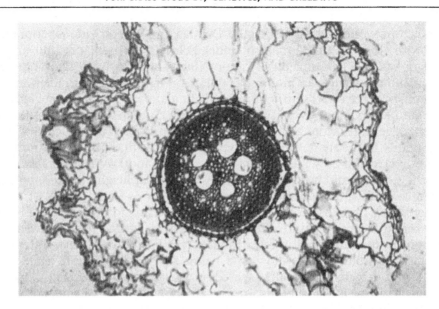

Figure 20.4. Root transverse section of St. Augustinegrass FX-10 showing the stele with five xylem elements, surrounded by a densely suberized ring of endodermis. The cortex cells are partially collapsed. Phloem cells are small and difficult to discern.

(PMV-SAD) (Horn et al., 1973); infectivity by *Sclerophthora macrospora* (Sacc.) Thirum., Shaw, & Naras. (Grisham et al, 1985), the cause of downy mildew disease; and susceptibility to *Pyricularia grisea* (Cke.) Sacc., the cause of gray leaf spot disease (Atilano and Busey, 1983).

No differences among genotypes have been observed for resistance to brown patch disease (Hurd and Grisham, 1983), caused by *Rhizoctonia solani* Kuhn; nor resistance to *Gaeumannomyces graminis* (Sacc.) Arx & D. Olivier var. *graminis*, the causal organism of take-all root rot of St. Augustinegrass (Elliott et al., 1993); nor resistance to the lance nematode, *Hoplolaimus galeatus* (Cobb) Thorne (Henn and Dunn, 1989; Giblin-Davis et al., 1995). Despite differences in suitability of St. Augustinegrass genotypes as hosts to the lance nematode, based on nematode reproduction, even at populations exceeding 10,000 nematodes g^{-1} soil, there is no measurable effect of lance nematodes on roots or shoots.

Much of the variation in resistance to biotic stresses is accountable by ploidy level. Compared with diploids, polyploids are more resistant to the southern chinch bug (Busey, 1990b; Busey and Zaenker, 1992; Reinert et al., 1986) and the sting nematode (Busey et al., 1993). Polyploids are less preferred by Lepidoptera than diploids (Busey et al., 1982a). Polyploids of the Bitterblue Group are highly susceptible to gray leaf spot disease (Atilano and Busey, 1983). Plant breeders should be encouraged by the large genotypic variations revealed in germ plasm screenings. Yet when variances in these studies are partitioned into ploidy levels, and genotypes nested within ploidy levels, often the vast majority of genetic variation is between ploidy levels (e.g., Busey and Zaenker, 1992; Busey et al., 1993). This suggests that some of the detectable genetic variation is not readily usable unless methods can be developed for gene exchange between ploidy levels.

Besides locating sources resistant to major pests, the dynamics of the host-pest relationship, and the most efficient method of screening need to be understood. This is illustrated by the southern chinch bug, an insect with variable populations. Floratam, released for its chinch bug resistance (Horn et al., 1973), remained free from economic damage by chinch bugs for 12 years, according to sod growers and commercial lawn applicators. The resistance of Floratam was confirmed by repeated laboratory screenings (reviewed by Quisenberry, 1990). In 1985, however, southern chinch bugs killed large areas of Floratam. The damaging chinch bugs were shown to be a population with virulence to Floratam (Busey and Center, 1987). Introduced African germ plasm provided the foundation for a new cultivar, FX-10, which remains resistant to different chinch bug populations (Busey, 1990b; Cherry and Nagata, 1997). Excreta residue deposited on aluminum foil is a rapid method for assessing chinch bug host suitability of St. Augustinegrass germ plasm (Busey and Zaenker, 1992). Both excreta residue and oviposition rate have high association with extent of field damage from natural infestations, $r^2 = 0.57$ and 0.67, respectively (Busey, 1995a).

Pathogens also vary in virulence, which may explain differences in disease incidence in different regions. St. Augustine decline isolates of Panicum Mosaic Virus (PMV-SAD) vary serologically, which may explain variable lethality to St. Augustinegrass lawns (Holcomb et al., 1989). Isolates may also represent mixtures of strains or serotypes, making resistance screening more unpredictable. Resistance screening based on a single isolate may be a poor representation of pathogen variation, and lead to inaccurate estimates of host susceptibility.

INTRODUCTION, SELECTION, AND BREEDING

Germ Plasm Resources

In 2001, 23 foreign introductions of *Stenotaphrum* were available for breeders in the National Plant Germplasm System (USDA, ARS, National Plant Germplasm Program, 2001). These clonal plants included two accessions (PI-289729 and PI-365301) of *S. dimidiatum* (incorrectly labeled *S. secundatum*), and the rest *S. secundatum*. The most recently introduced genotypes available for distribution were two collected by Dr. Milt Engelke from China, added in 1993; the next most recently added genotype was in 1979. In addition, four plants submitted by Mr. Tobey Wagner, and two plants from Dr. Jeffrey V. Krans, are awaiting release from quarantine. Much of the potential germ plasm of St. Augustinegrass occurs in pastures, especially in coastal Africa, from Kenya to the Cape of Good Hope (Chippindall and Crook, 1976), the West Indies (Busey et al., 1982a), and Oceania (Sauer, 1972).

Released cultivars and active breeding populations, outside the minuscule U.S. collection, represent most of the germ plasm available to breeders. Most of the *S. secundatum* genotypes used in breeding programs represent the Dwarf Group, with little attention to the African polyploids (Busey et al., 1982a). It is not known what other groups lay undiscovered. For the Breviflorus Race, which is so extensively used in breeding programs, genetic variation is readily available in adventive populations in the southeastern United States (Busey et al., 1982a).

St. Augustinegrass has been naturalized since at least the 1700s in North and South America, Africa, and the Pacific, and exhibits considerable phenotypic variation throughout its range. Because of this antiquity, there is probably no area where there is not some useful genetic variation. Even far outside the presumptive center of origin in the Indian Ocean area, germ plasm collections of St. Augustinegrass may be very useful, because they may represent relict types that no longer occur in the natural range. A caution in germ plasm collections of St. Augustinegrass is to be diligent to cull out duplicates that represent widely distributed clonal cultivars. Also, because St. Augustinegrass is propagated and marketed in an active, vegetative condition, breeders and germ plasm managers must also be aware of the perils of accidentally dispersing systemic and attached disease organisms, such *Sclerophthora macrospora, Gaeumannomyces graminis* var. *graminis*, as well as St. Augustine Decline Strain of Panicum Mosaic Virus (PMV-SAD), and the sting nematode.

Pembagrass, *S. dimidiatum*, PI-365031 has resistance to gray leaf spot disease caused by *Pyricularia grisea* (Cke.) Sacc. (Atilano and Busey, 1983) and the southern chinch bug (Busey, 1990b); *S. dimidiatum* FL-2195 has resistance to the sting nematode (Busey et al., 1993). Therefore *S. dimidiatum* is a good first candidate for wide crosses and other methods for gene transfer.

Breeding and Selection Techniques

St. Augustinegrass is easy to hybridize artificially (Philley et al., 1993). Inflorescences are photoperiod-controlled (Dudeck, 1974), and flowering occurs first in the center of the inflorescence, and progresses predictably in both directions. Anthesis in most genotypes occurs soon after sunrise, but anthesis of *S. dimidiatum* is at night. At the University of Florida, Fort Lauderdale, parchment pollinating bags were placed over inflorescences one day before anthesis, with the plants generally in containers in a greenhouse, although bagging of plants in field plots was also performed. The relatively large spikelets were easily emasculated with a pair of forceps, which was done in the morning as the anthers emerged. By also removing unused spikelets, and unused portions of the inflorescence, it was easier to keep track of crosses, and less likely to have stray pollen in a bag. Any spikelets with already dehisced anthers were removed from the inflorescences. Crosses were made using pollen transferred to the hand-emasculated florets. Pollinated spikelets were marked with an indelible marker. In addition, spikelet positions were numbered and recorded, so that a record of shriveled stigmata (an indication of effective crossing) could later be associated with individual seeds harvested. The reaction of the stigmata was recorded one day after pollination, and the bags removed.

Seed is set and easily produced within ploidy levels. Bivalent-pairing polyploids ($2n = 30$) from southern Africa produce 43% to 70% seed set (Busey, 1990b). Diploids ($2n = 18$) of the Breviflorus Race produce over 60% seed set. However, ploidy level differences impede the full use of germ plasm; intended crosses between different ploidy levels have not been successful (Busey, 1981, unpublished data). The most successful St. Augustinegrass in Florida, Floratam, normally produces no seed. However, in 1983 seed were obtained from Floratam growing in a greenhouse, open-pollinated by $2n = 30$ African parents. Among the progeny, several had laminar hairs similar to the putative male parents. One of the Floratam progeny, FX-5, had reduced oviposition by the Polyploid Damaging Population (PDP) southern chinch bug (Busey, 1990b), evidence for chinch bug resistance conferred by the African male parents.

Because antibiotic resistance to the southern chinch bug has not been discovered among diploids (Busey and Zaenker, 1992; Reinert et al., 1986), embryo rescue or protoplast fusion might be used to transfer this trait across the ploidy barrier. The caryopses of St. Augustinegrass mature more quickly than *Zea mays* L. At 9 days after pollination, the St. Augustinegrass embryo (Figure 20.5) has developed leaf primordia and vasculature, and is nearly half of its mature length (2.05 mm). By 10 days after pollination, radicle development has begun. In contrast, in *Z. mays* the leaf primordia form at 12 days, while in *Eragrostis curvula* this occurs at 5 days.

Somatic mutations are easily produced in St. Augustinegrass sprigs using gamma rays, and 3000 to 4500 rads is the appropriate dosage to generate high mutation rates and adequate sprig survival, depending on the cultivar (Busey, 1980). Complete plant regeneration has been accomplished for St. Augustinegrass from callus (Kuo and Smith, 1993).

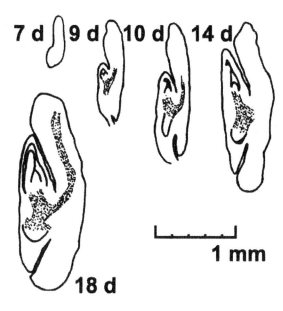

Figure 20.5. Seed development of St. Augustinegrass, *Stenotaphrum secundatum*, Scotts-1081, showing longitudinal sections at 7 to 18 days after pollination (Busey and Center, 1983, unpublished data).

The biggest challenge in breeding St. Augustinegrass is that it is a perennial, and evaluation is difficult. Field evaluation must be long-term, exposing genotypes to a range of chronic natural problems (e.g., nematodes and thatch) and acute environmental and biotic problems (injury from drought and chinch bugs). St. Augustinegrass does not exhibit some pest problems, such as sting nematode, for at least two years after establishment (Busey et al., 1991), and southern chinch bug infestation typically begins in susceptible cultivars about 1.5 years after plug planting. Attempts to accelerate the progress of evaluation by prescreening for plant characteristics in containers was not successful, as no correlation was found between container performance and field performance (Busey, 1981, unpublished data).

Inheritance

Diploid (2n = 18) St. Augustinegrass has normal paired-factor inheritance, based on Mendelian 3:1 ratios for stigma color observed in segregating progeny, consistent with an hypothesis that purple stigma is dominant to white (translucent) (Table 20.1). A white stigma irradiation-induced mutation was derived from a heterozygous purple-stigma genotype (Busey, 1980), which supports simple, diploid inheritance control.

Variegation is simply inherited. For example, the selfed progeny of normal green-leafed plant FA-243-39 were 7 variegated and 20 normal, consistent with an hypothesis that variegation is a single recessive, giving an expected 1:3 ratio. However, a second gene may also be involved, because the selfed progeny of normal green-leafed plant 365032-8F231 were 38 variegated and 42 normal, which is consistent with the variegated trait being recessive on two epistatic loci, giving an expected 7:9 ratio. Variegated St. Augustinegrass, which has invalidly been referred to in horticultural encyclopedias as *Stenotaphrum variegatum*, was documented by the famous agrostologist Dr. Agnes Chase from a hanging pot in a greenhouse in Garfield Park, Chicago [ILLINOIS: Chicago. 27 October 1915; *Chase, s.n.* (USNAT)]. Variegated St. Augustinegrass

Table 20.1. Inheritance of stigma color in selfed progenies of St. Augustinegrass parents (Busey and Center, 1982, unpublished data).

Parent	Stigma Color, Parent Phenotype	Frequency of Stigma Colors, by Progeny Phenotypes		
		Purple	"White" (Translucent)	Bicolor (Purple and White)
365032-8	Purple	175	0	1
300128-6	Purple	110	28	2
FL-1933-8	Purple	44	19	0
FA-243	Purple	88	32	0
Scotts 1081	Purple	111	42	1
Sum of segregating progeny		353	121	3

was used as a model species for studying chloroplast enzymes (Suzuki et al., 1986). The variegated mutation has appeared independently in different germ plasms, for example, in turf exposed to oxidizers such as laundry detergent and swimming pool water (Busey, 1980, unpublished observations). Other genetic traits are not well understood.

Reproduction

Deliberate propagation of St. Augustinegrass is usually vegetative, by stolon cuttings, plugs, and sod. The only commercially available cultivars are thus clones. Grown as a monoculture, St. Augustinegrass remains vulnerable to pest evolution (Busey and Center, 1987). Efforts to develop seeded cultivars might enhance genetic diversity, but have not been successful, despite repeated attempts. For example, in 1974, Curran L. Garrett received a plant patent for a heavily seed-producing St. Augustinegrass. Also, in the early 1990s Pennington Seed marketed seed from St. Augustinegrass, calling it 'Raleigh-S.' Unfortunately, genotypes that are prolific seed producers are often esthetically unacceptable in regions with an extended growing season (Busey, 1984, unpublished data). In addition, inbreeding depression occurs in St. Augustinegrass, and the seed produced from a clonal monoculture must, by nature, be inbred. Even ignoring the genetic problems of seed production in St. Augustinegrass, seed yield is low and it is very difficult to remove the caryopses from the corky rachis segments. An alternative to seeded cultivars would be clonal blends of cultivars differing in host resistance. An esthetically compatible blend might be protected from pest dispersal and outbreak, either directly because of the dilution of host density, or indirectly because the natural selection pest virulence would be delayed by the genetically heterogeneous host.

History of Breeding and Population Improvement

Organized breeding of St. Augustinegrass has occurred on few occasions. This is accountable in part because it is primarily a lawn grass, and not important for golf or sports turf, thus sources of research funds have been minimal. For example, between 1983 and 1997 the United States Golf Association (USGA) funded $3.86 million for turfgrass breeding of bermudagrass, *Cynodon* spp.; zoysiagrasses, *Zoysia* spp.; seashore paspalum, *Paspalum vaginatum* Swartz; buffalograss, *Buchloë dactyloides* (Nutt.) Engelm.; and creeping bentgrass, *Agrostis palustris* Huds. No funds were allocated, nor proposals solicited, for St. Augustinegrass improvement.

Commercial breeding development of St. Augustinegrass has also been limited because it is a clonal crop, which makes it harder to define and control the pathway to an effectively large market. St. Augustinegrass is produced on many independent sod farms. To recoup the cost of developing intellectual property in St. Augustinegrass, as well as marketing and quality control, requires effective licensing to many companies who are in competition with one another. Potential licensees may vary in size, experience, and production techniques (e.g., plug production versus sod) which makes it difficult to standardize licensing requirements and royalty basis. In contrast, for cultivars of seed-propagated turf species, such as a perennial ryegrass, *Lolium perenne* L., the developer can more readily control the stages of distribution by concentrating regulation on the more centralized seed production area, e.g., by subcontracting to growers who sell back to the developer, who then sells to consumers or brokers. A seed propagated species has two other advantages in intellectual property rights. Quality control can be more readily assured in a seed propagated species because there is a storage period for quality assessment. Quality control is more difficult in vegetatively propagated species such as St. Augustinegrass where the product is perishable and can vary in weed content and other characteristics during the time it takes to assess quality. Also, developers of some seed propagated turfgrasses, such as overseeded perennial ryegrass, expect a lucrative recurring market from users with annual budgets such as golf courses, whereas developers of vegetatively propagated turfgrasses do not expect frequent repurchases.

The Scotts Company has conducted the major commercial breeding development of St. Augustinegrass. In research at Scotts farm in Apopka, Florida, Dr. Terrence Riordan developed numerous clones, several of which were patented, and three were registered ('DelMar,' 'Jade,' and Seville). Mr. Tobey Wagner of Sod Solutions (South Carolina) patented Palmetto St. Augustinegrass, a clonal collection. A total of 18 plant patents for St. Augustinegrass have been awarded.

Efforts by public scientists have involved discovery of clonal types such as 'Floratine' and Raleigh, and discovery of seedlings of partially unknown pedigree, for example Floratam and 'Floralawn' St. Augustinegrasses. The author at the University of Florida-Fort Lauderdale did the only large-scale population improvement, from 1977 until 1996, when the program was assigned to Dr. Russell Nagata at the University of Florida-Belle Glade. The main basis for organized breeding of St. Augustinegrass at the University of Florida-Fort Lauderdale was a composite cross population.

From 1978 through 1982, an average of 28 parents per generation (Table 20.2) were hybridized randomly to produce offspring populations that were evaluated in the field in comparison with cultivar standards, Bitterblue, Floratam, Roselawn, and Seville. The turfgrass quality mean of cultivar standards was a constant reference to compare population changes due to composite crossing, selfing, recurrent selection involving the selection of elite parents, and vegetative repropagation of plants that had performed well in earlier trials.

Initial parents had been chosen to represent taxonomic groups classified from a worldwide population (Busey et al., 1982a). Parents of each succeeding generation were chosen based on phenotypic dissimilarity. In addition to four generations of composite crossing (C1, C2, C3, and C4), several selfed populations were also created (e.g., S1, C1S1, etc.). Recurrent selection populations were created (R1 and R2) from elite parents that were chosen based on individual plant performance or progeny performance, and vegetatively repropagated populations (V1, V2, and V3) were chosen for reevaluation based on their prior superior performance.

On several dates during the first 14 months after field planting, plots were evaluated for turf quality, a combination of adaptive and esthetic traits, with 10=complete coverage, deepest leaf color, and most dense, low, uniform habit; 7=acceptable coverage,

Table 20.2. Breeding progress from St. Augustinegrass population improvement based on turf quality of St. Augustinegrass breeding populations derived from random composite crossing (C1, C2, C3, and C4) of the original parents (P0 and P1), selfed progeny from various generations (S1, C1S1, C2S1, C1S2, C2S1, C3S1), recurrent populations selected based on parental performance (R1, R2, etc.), and vegetative selections of breeding lines (V1, V2, and V3). The C1S1C1 represents crosses of the C1S1 (selves,) which were selves of the C1. (Busey, 1986, unpublished data.)

Progeny Population Name	Nature of Origin	Parent Population	Size of Progeny Population (No.)	Size of Parents Population (No.)	Turf Quality, Cultivar Standards (10 = best)	Turf Quality, Progeny Population (10 = best)	Proportion of Progeny Plots Superior to Cultivars (%)
P0	Germ plasm	Wild	90	—	5.9	4.5	13
P1	Selected for diversity	P0	20	—	5.9	4.8	17
C1	Cross	P1 & P0	172	25	—	4.2	—
C2	Cross	C1	140	33	5.8	3.6	13
C3	Cross	C2	100	22	5.5	3.8	20
C3	Cross	C2	60	22	5.1	3.9	24
C1S1C1	Cross	C1S1	81	10	5.5	3.5	14
C4	Cross	C3	187	38	4.8	3.5	24
Total	Cross		680				20
S1	Self	P1	75	17	—	3.6	—
C1S1	Self	C1	807	7	—	2.8	—
C2S1	Self	C2	236	20	5.5	2.4	1
C1S2	Self	C1S1	117	11	5.5	2.1	2
C3S1	Self	C3	180	60	5.1	2.4	2
Total	Self		1415				2
R1	Cross	Various	66	22	5.3	4.3	28
R2	Cross	Various	163	29	4.2	3.8	37
Total			229				34
V1	Selected	Various	14		5.1	5.6	63
V2	Selected	Various	55		5.3	5.0	56
V3	Selected	Various	31		4.2	5.5	80
Total			100				67

color, and habit; and 1=plant dead. Because some populations were evaluated with only a single plot per genotype, and other populations were evaluated in randomized complete blocks, the comparison of populations to the mean of cultivars was on the basis of population individual plot values, rather than population genotype means. Cultivar standards were, however, replicated.

Composite crosses had 20% of plots with turf quality ratings exceeding the mean of cultivars, which was almost the same fraction as the initial parents, 17% (Table 20.2). Genotypic variances did not change under composite crossing in the absence of selection. For example, C3 (which was evaluated in three replicates) had a genotypic variance for turf quality of 1.30, compared with 1.43 for the P1 parents. Selfed populations were inferior, as would be expected for a normally cross-pollinated species; only 2% of plots exceeded the mean of four cultivars. In related work, gray leaf spot disease severity was higher for an open-pollinated and probably inbred offspring of a Gulf Coast accession than for the parent (Atilano and Busey, 1983), confirming the problems of inbreeding St. Augustinegrass.

Narrow-sense heritability for turf quality was estimated from midparent-offspring regression and was significant ($P < 0.05$) in two cases, C2 regressed on C1 (0.44) and C3 on C2 (0.66), but was not significant in two cases (C1 on P1 and C4 on C3). Recurrent selection based on crosses of elite parents was successful, as the R1 and R2 populations had a high proportion (34%) of plots superior to the mean of cultivars.

The broad-sense heritability for turf quality in 60 randomly selected, retested C3 clones was 0.45 (single-plot basis). The average broad-sense heritability of hybrids within single replicated experiments was 0.62. With such high heritabilities, little benefit would be obtained by replicating in first-stage clonal evaluations. In support of this conclusion, plots of vegetative selections that were reevaluated (V1, V2, and V3), and which had been chosen in most cases from no more than two replicates, were superior 67% of the time compared with the mean of cultivars. By not replicating in first-stage evaluations, a larger germ plasm can be screened and subjected to more intensive selection, even though heritability based on unreplicated selection is less than heritability based on genotype means.

Composite crossing was also successful in preserving genetic variation, because after recurrent selection and after vegetative selection, adequate genotypic variance was found compared with the original parents. Genotypic variances for the recurrent populations R1 and R2 were 1.26 and 1.52 units, respectively, and for the vegetative selections V1, V2, and V3 were 1.34, 1.22, and 0.85, respectively.

Other work on heritability under selection, based on the analysis of a diallel cross, resulted in estimated narrow sense heritability for lethal temperature of 0.58, and a range from 0.70 to 0.98 for winter survival. The two traits are correlated with one another (Philley et al., 1998). Specific combining ability was generally not significant.

CONCLUSIONS

Scientific attention to St. Augustinegrass has been sporadic. In the haste to get new cultivars to market, basic information such as pedigree and usable description have not been reported, if they are even known. Meanwhile, other cultivars have undergone unnecessarily lengthy test periods prior to release, e.g., 26 years in the case of Floralawn (Dudeck et al., 1986). The review process for scientific manuscripts and plant patent applications puts high emphasis on demonstrating cultivar differences, but a process for evaluating the applicability of the results to the field is not available. Repeatedly, field performance variation is poorly predicted based on laboratory evaluation, e.g., evapotranspiration, shade tolerance, and turfgrass quality. Even the process of evaluating St.

Figure 20.6. St. Augustinegrass in the landscape, FL-1997-6 developed by the writer, forming a dense turf under the shade of *Ficus* spp. trees and grapefruit, *Citrus paradisi* Macf. at the residence the late Paul Frank, Golf Course Superintendent, Wilderness Country Club, Naples, Florida.

Augustinegrass cultivars in tests including the National St. Augustinegrass Test by the National Turfgrass Evaluation Program (NTEP) has resulted in systematic biases against coarse-textured cultivars such as Floratam. Floratam is the best adapted and most widely used St. Augustinegrass in Florida, even though it receives poor turfgrass quality ratings in most field trials.

In other instances, e.g., differential thermal analysis for assessing freezing resistance and excreta residue for assessing host suitability to the southern chinch bug, laboratory criteria have high correlation with field traits, and are more efficient for screening than waiting for natural stresses to occur. Scientists have developed screening techniques for traits that are relatively easy to assess, while one of the most difficult adaptive problems in turf, i.e., shade, is infrequently studied. Most turf evaluation environments are in full direct sun. In the absence of accurate scientific information, marketers of proprietary St. Augustinegrass cultivars normally make the same claims of superiority, for drought resistance, shade resistance, and chinch bug resistance, for all new cultivars. At the least, landscape plantings of specific cultivars should be revisited a few years after establishment, to determine actual performance based on the original expectations.

Finally, there is the problem of limited funding of research for lawn grasses. Limited funds are available from state and governmental agencies to do targeted work on special problems, such as water conservation research funded by various water authorities. Such agencies have been ambivalent in recognizing the importance of turf in the environment, and often seek to replace turf with ground covers, rather than to distribute and promote useful irrigation technology. In other cases, proprietary interests have contracted for limited research on specific traits of interest in preparing patents and marketing of found cultivars, but they have not funded the breeding development of new genotypes. Meanwhile, the State Agricultural Experiment Stations, which are responsible for the development of publicly released cultivars, have in some cases failed to submit successful cultivars for evaluation in the National Turfgrass Evaluation Program, and in other cases have failed to maintain the original Breeder's Stock of released cultivars. A public commitment is needed to the study of St. Augustinegrass, as a versatile plant that provides the primary green landscapes and erosion control for tens of millions of people.

ACKNOWLEDGMENT

This research was supported by the Florida Agricultural Experiment Station, and approved for publication as Journal Series No. R-08303.

REFERENCES

Allen, R.J. Jr. and R.W. Kidder. 1971. Origin and history of Roselawn St. Augustinegrass. *Soil Crop Sci. Soc. Florida Proc.* 30:354–360.
Atilano, R.A. and P. Busey. 1983. Susceptibility of St. Augustinegrass germ plasm to *Pyricularia grisea*. *Plant Dis.* 67:782–783.
Atkins, C.E., R.L. Green, S.I. Sifers, and J.B Beard. 1991. Evapotranspiration rates and growth characteristics of ten St. Augustinegrass genotypes. *HortScience* 26:1488–1491.
Beard, J.B. 1973. *Turfgrass Science and Culture*. Prentice Hall, Inc. Englewood Cliffs, NJ.
Bogdan, A.V. 1970. Turfgrasses in Kenya. pp. 51–56. In *First Int. Turfgrass Res. Conf. Proc., Harrogate, England*. 15–18 July 1969. Sports Turf Research Institute, Bingley, England.
Busey, P. 1979. What is Floratam? *Florida State Hort. Soc. Proc.* 92:228–232.
Busey, P. 1980. Gamma-ray dosage and mutation breeding in St. Augustinegrass. *Crop Sci.* 20:181–184.
Busey, P. 1985. Preferences in St. Augustinegrass. *Florida Turf Digest* 2(7):3–6.
Busey, P. 1986. Morphological identification of St. Augustinegrass cultivars. *Crop Sci.* 26:28–32.
Busey, P. 1989. Progress and benefits to humanity from breeding warm-season grasses for turf. In D.A. Sleper, K.H. Asay, and J.F. Pedersen, Eds. *Contributions from Breeding Forage and Turf Grasses*. CSSA Spec. Publ. 15, Crop Science Society of America, Madison, WI.
Busey, P. 1990a. Empirical study of turfgrass adaptation. *Agron. Abstr.* p. 170.
Busey, P. 1990b. Polyploid *Stenotaphrum* germplasm: Resistance to the polyploid damaging population southern chinch bug. *Crop Sci.* 30:588–593.
Busey, P. 1993. Registration of 'FX-10' St. Augustinegrass. *Crop Science* 33:214–215.
Busey, P. 1995a. Field and laboratory resistance of St. Augustinegrass germplasm to the southern chinch bug. *HortScience* 30:1253–1255.
Busey, P. 1995b. Genetic diversity and vulnerability of St. Augustinegrass. *Crop Sci.* 35:322–327.
Busey, P. 1996. Wilt avoidance in St. Augustinegrass germplasm. HortScience 31:1135–1138.
Busey, P., T.K. Broschat, and B.J. Center. 1982a. Classification of St. Augustinegrass. *Crop Sci.* 22:469–473.
Busey, P. and B.J. Center. 1987. Southern chinch bug (Hemiptera: Heteroptera: Lygaeidae) overcomes resistance in St. Augustinegrass. *J. Econ. Entomol.* 80:608–611.
Busey, P. and E.H. Davis. 1991. Turfgrass in the shade environment. *Florida State Hort. Soc. Proc.* 104:353–358.
Busey, P., R.M. Giblin-Davis, and B.J. Center. 1993. Resistance in *Stenotaphrum* to the sting nematode. *Crop Sci.* 33:1066–1070.
Busey, P., R.M. Giblin-Davis, C.W. Riger, and E.I. Zaenker. 1991. Susceptibility of St. Augustinegrasses to *Belonolaimus longicaudatus*. *J. Nematol.* 23:604–610.
Busey, P., J.A. Reinert, and R.A. Atilano. 1982b. Genetic and environmental determinants of zoysiagrass adaptation in a subtropical region. *J. Amer. Soc. Hort. Sci.* 107:79–82.
Busey, P. and G.H. Snyder. 1993. Population outbreak of the southern chinch bug is regulated by fertilization. *Int. Turfgrass Soc. Res. J.* 7:353–357.
Busey, P. and R.W. White. 1993. South Florida: A center of origin for turfgrass production. *Int. Turfgrass Soc. Res. J.* 7:863–869.
Busey, P. and E.I. Zaenker. 1992. Resistance bioassay from southern chinch bug (Heteroptera: Lygaeidae) excreta. *J. Econ. Entomol.* 85:2032–2038.

Carrow, R.N. 1995. Drought resistance aspects of turfgrasses in the southeast: Evapotranspiration and crop coefficients. *Crop Sci.* 35:1685–1690.

Casnoff, D.M., R.L. Green, and J.B Beard. 1989. Leaf blade stomatal densities of ten warm-season perennial grasses and their evapotranspiration rates. pp. 129–131. In *Sixth Int. Turfgrass Res. Conf. Proc., Tokyo.* 31 July–5 August 1989. Japanese Soc. Turfgrass Sci., Tokyo.

Cherry, R.H. and R.T. Nagata. 1997. Ovipositional preference and survival of southern chinch bugs (*Blissus insularis* Barber) on different grasses. *Int. Turfgrass Soc. Res. J.* 8:981–986.

Chippindall, L.K.A. and A.O. Crook. 1976. *Grasses of Southern Africa.* M.O. Collins (Pvt.) Ltd., Salisbury, Zimbabwe.

Dengler, N.G., R.E. Dengler, P.M. Donnelly, and P.W. Hattersley. 1994. Quantitative leaf anatomy of C_3 and C_4 grasses (Poaceae): Bundle sheath and mesophyll surface area relationships. *Ann. Bot.* 73:241–255.

Dudeck, A.E. 1974. Flowering response of several selections of St. Augustinegrass. pp. 74–78. In E.C. Roberts, Ed. *Second Int. Turfgrass Res. Conf. Proc., Blacksburg, VA.* 19–21 June 1973. ASA and CSSA, Madison, WI.

Dudeck, A.E., C.H. Peacock, and J.C. Wildmon. 1993. Physiological and growth responses of St. Augustinegrass cultivars to salinity. *HortScience* 28:46–48.

Dudeck, A.E., J.A. Reinert, and P. Busey. 1986. Registration of 'Floralawn' St. Augustinegrass. *Crop Sci.* 26:1083.

Elliott, M.L., A.K. Hagan, and J.M. Mullen. 1993. Association of *Gaeumannomyces graminis* var. *graminis* with a St. Augustinegrass root rot decline. *Plant Dis.* 77:206–209.

Florida Department of Agriculture and Consumer Services. 1976. Florida Turfgrass Survey 1974. Florida Dep. Agr. Consumer Serv., Florida Crop Livestock Rep. Serv., Orlando.

Fluck, R.C. and P. Busey. 1988. Energy for mowing turfgrass. *Trans. ASAE* 31:1304–1308.

Fry, J.D., N.S. Lang, and R.G.P. Clifton. 1991. Freezing resistance and carbohydrate composition of 'Floratam' St. Augustinegrass. *HortScience* 26:1537–1539.

Giblin-Davis, R.M., P. Busey, and B.J. Center. 1995. Parasitism of *Hoplolaimus galeatus* on diploid and polyploid St. Augustinegrasses. *J. Nematol.* 27:472–477.

Gould, F.W. 1968. *Grass Systematics.* McGraw-Hill Book Co., New York.

Gould, F.W. and T.R. Soderstrom. 1974. Chromosome numbers of some Ceylon grasses. *Can. J. Bot.* 52:1075–1090.

Green, R.L., A.E. Dudeck, L.C. Hannah, and R.L. Smith. 1981. Isoenzyme polymorphism in St. Augustinegrass, *Stenotaphrum secundatum. Crop Sci.* 21:778–782.

Grisham, M.P., R.W. Toler, and B.D. Bruton. 1985. Effect of *Sclerophthora macrospora* on growth and development of St. Augustinegrass. *Plant Dis.* 69:289–291.

Henn, R.A. and R.A. Dunn. 1989. Reproduction of Hoplolaimus galeatus and growth of seven St. Augustinegrass (*Stenotaphrum secundatum*) cultivars. *Nematropica* 19:81–87.

Hodges, A.W., J.J. Haydu, P.J. van Blokland, and A.P. Bell. 1994. Contribution of the Turfgrass Industry to Florida's Economy, 1991/92: A Value Added Approach. Economics Report ER 94-1, Florida Agricultural Experiment Stations, Institute of Food and Agricultural Sciences, University of Florida, Gainesville.

Holcomb, G.E., T.Z. Liu, and K.S. Derrick. 1989. Comparisons of isolates of Panicum Mosaic Virus from St. Augustinegrass and centipedegrass. *Plant Dis.* 73:355–358.

Horn, G.C., A.E. Dudeck, and R.W. Toler. 1973. 'Floratam' St. Augustinegrass: A fast growing new variety for ornamental turf resistant to St. Augustine decline and chinch bugs. *Florida Agric. Exp. Stn. Circ.* S-224.

Hurd, B. and M.P. Grisham. 1983. *Rhizoctonia* spp. associated with brown patch of Saint Augustinegrass. *Phytopathology* 73:1661–1665.

Johns, D., J.B Beard, and C.H.M. van Bavel. 1983. Resistances to evapotranspiration from a St. Augustinegrass turf canopy. *Agron. J.* 75:419–422.

Judd, B.I. 1975. New World tropical forage grasses and their management. 4. Bermudagrass, giant stargrass, St. Augustinegrass, and jaraguagrass. *World Crops* 27(March/April):69–72.

Kim, K.S. and J.B Beard. 1988. Comparative turfgrass evapotranspiration rates and associated plant morphological characteristics. *Crop Sci.* 28:328–331.

Kneebone, W.R. and I.L. Pepper. 1982. Consumptive water use by sub-irrigated turfgrasses under desert conditions. *Agron. J.* 74:419–423.

Krans, J.V., J.B Beard, and J.F. Wilkinson. 1979. Classification of C_3 and C_4 turfgrass species based on CO_2 compensation concentration and leaf anatomy. *HortScience* 14:183–185.

Kuo, Y. and M.A.L. Smith. 1993. Plant regeneration from St. Augustinegrass immature embryo-derived callus. *Crop Sci.* 33:1394–1396.

Long, J.A. and E.C. Bashaw. 1961. Microsporogenesis and chromosome numbers in St. Augustinegrass. *Crop Sci.* 1:41–43.

Maier, F.P., N.S. Lang, and J.D. Fry. 1994a. Evaluation of an electrolyte leakage technique to predict St. Augustinegrass freezing tolerance. *HortScience* 29:316–318.

Maier, F.P., N.S. Lang, and J.D. Fry. 1994b. Freezing tolerance of three St. Augustinegrass cultivars as affected by stolon carbohydrate and water content. *J. Amer. Soc. Hort. Sci.* 119:473–476.

Miller, G.L. and L.B. McCarty. 2001. Water relations and rooting characteristics of three *Stenotaphrum secundatum* turf cultivars grown under water deficit conditions. *Int. Turfgrass Soc. Res. J.* 9:323–327.

Peacock, C.H., A.E. Dudeck, and J.C. Wildmon. 1993. Growth and mineral content of St. Augustinegrass cultivars in response to salinity. *J. Amer. Soc. Hort. Sci.* 118:464–469.

Peacock, C.H. and A.E. Dudeck. 1993. Response of St. Augustinegrass cultivars [*Stenotaphrum secundatum* (Walt.) Kuntze] to shade. *Int. Turfgrass Soc. Res. J.* 7:657–663.

Philley, H.W. 1994. Inheritance of Cold Tolerance in St. Augustinegrass. M.S. thesis. Mississippi State University.

Philley, H.W., J.V. Krans, J.M. Goatley Jr., V.L. Maddox, and C.E. Watson. 1993. The effect of nutrient solutions on seed set of excised St. Augustinegrass inflorescences. *Agron. Abstracts.* p. 163.

Philley, H.W., C.E. Watson Jr., J.V. Krans, J.M. Goatley Jr., and F.B. Matta. 1995. Differential thermal analysis of St. Augustinegrass. *HortScience* 30:1388–1389.

Philley, H.W., C.E. Watson Jr., J.V. Krans, J.M. Goatley Jr., V.L. Maddox, and M. Tomaso-Peterson. 1998. Inheritance of cold tolerance in St. Augustinegrass. *Crop Sci.* 38:451–454.

Quisenberry, S.S. 1990. Plant resistance to insects and mites in forage and turf grasses. *Florida Entomol.* 73:411–421.

Reinert, J.A., P. Busey, and F.G. Bilz. 1986. Old World St. Augustinegrasses resistant to the southern chinch bug (Heteroptera:Lygaeidae). *J. Econ. Entomol.* 79:1073–1075.

Reinert, J.A. and A.E. Dudeck. 1974. Southern chinch bug resistance in St. Augustine-grass. *J. Econ. Entomol.* 67:275–277.

Riordan, T.P., V.D. Meier, J.A. Long, and J.T. Gruis. 1980. Registration of Seville St. Augustinegrass. *Crop Sci.* 20:824–825.

Safford, W.E. 1905. The useful plants of the island of Guam. *Contrib. U.S. Nat. Herb.* 9:1–416.

Sauer, J.D. 1972. Revision of *Stenotaphrum* (Gramineae:Paniceae) with attention to its historical geography. *Brittonia* 24:202–222.

Smith, M.A.L., J.E. Meyer, S.L. Knight, and G.S. Chen. 1993. Gauging turfgrass salinity responses in whole-plant microculture and solution culture. *Crop Sci.* 33:566–572.

Smith, M.A. and P.C. Whiteman. 1983. Evaluation of tropical grasses in increasing shade under coconut canopies. *Exp. Agric.* 19:153–161.

Stewart, E.H. and W.C. Mills. 1967. Effect of depth to water table and plant density on evapotranspiration rate in southern Florida. *Trans. ASAE* 10:746–747.

Sundararaj, D.D., V. Ramakrishnan, and K.P. Vijayan. 1971. Plant introduction in Tamil Nadu—*Stenotaphrum dimidiatum* Brogn. Syn: *Stenotaphrum glabrum* Trin—a new colourful grass from the wild, for lawns and soil erosion control. *Madras Agric. J.* 58:337–341.

Suzuki, E., J. Ohnishi, M. Kashiwagi, and R. Kanai. 1986. Comparison of photosynthetic and photorespiratory enzyme activities between green leaves and colorless parts of variegated leaves of a C_4 plant, *Stenotaphrum secundatum* (Walt.) Kuntze. *Plant Cell Physiol.* 27:1117–1125.

Tebeau, C.W. 1968. *Man in the Everglades.* University of Miami Press, Coral Gables, FL.

USDA, ARS, National Genetic Resources Program. 2001. Germplasm Resources Information Network - (GRIN). Online Database] National Germplasm Resources Laboratory, Beltsville, MD. Available: www.ars-grin.gov/cgi-bin/npgs/html/stats/genussite.pl?Stenotaphrum (June 7, 2001).

Webster, R.D. 1988. Genera of the North American Paniceae (Poaceae: Panicoideae). *Systematic Bot.* 13:576–609.

Wilson, C.A., J.A. Reinert, and A.E. Dudeck. 1977. Winter survival of St. Augustine grasses in north Mississippi. *S. Nurserymen's Assoc. Res. Conf. Ann. Rep.* 22:195–198.

Works Progress Administration. 1939. Diary of A. M. Reed, 1848–1899, and a Portion of 1900 by others. Works Progress Administration, Sate Office, Historical Records Survey. (From the St. Augustine Historical Society Library.)

Chapter 21

Bahiagrass, *Paspalum notatum* Flügge

Philip Busey

Bahiagrass, *Paspalum notatum* Flügge, is widely used as a perennial drought-resistant pasture and utility turf in warm, tropical, and subtropical climate regions (Figure 21.1). It is adapted to humid areas where it survives on summer rain and can survive in a state of dormancy during a winter dry season. Bahiagrass originated in the New World, probably southern South America, and has been introduced since the late 1800s to the southeastern United States, Africa, Asia, Australia, and Europe. It was also present in Mesoamerica by the mid 1800s and the Type specimen is based on an 1802 collection from St. Thomas, West Indies. The diversity of common names for bahiagrass in Latin America is consistent with its long-term recognition and presence there (see Origin and Relatives). By 1974 bahiagrass was the principal turfgrass of Florida (Anonymous, 1976).

In the second half of the 1900s bahiagrass was expanded and developed for forage and soil improvement in Japan (Sakai, 1983), Kenya and Uganda (Bogdan, 1970), Zimbabwe (Mills and Boultwood, 1978) and the Republic of China, where continuous bahiagrass cover reduces annual soil loss from 31,200 g m^{-2} (with clean culture) to 300 g m^{-2} (Jean and Juang, 1979). Bahiagrass improves impoverished soils by increasing organic matter content and available cations (Beaty and Tan, 1972). Along the northern coast of New South Wales, Australia (Firth and Wilson, 1995) it has shown promise for erosion control under orchards. Soybean *Glycine max* (L.) Merr., field areas infested with the nematodes *Meloidogyne arenaria* (Neal) Chitwood and *Heterodera glycines* Ichinohe show 114% yield increase following rotation with bahiagrass (Rodríguez et al., 1991). Despite the biocontrol effect, living bahiagrass is harmful to newly planted peach trees (*Prunus persica* (L.) Batsch.) (Evert et al., 1992). Diazotrophic colonization and dinitrogen fixation occur in association with bahiagrass root systems in warm climates (Dobereiner et al., 1972), but have not been clearly verified under temperate and laboratory conditions (Vietor, 1982), nor is it clear whether the amount of nitrogen fixation that occurs is of practical value.

BIOLOGY

Plant Characteristics

Bahiagrass has erect or decumbent, pointed-tipped leaf blades to 35 cm long or longer and generally 3 to 8 mm wide. It spreads by stout (4–8 mm wide) aboveground

Figure 21.1. Large turf area of *Paspalum notatum* var. *latiflorum* bahiagrass in Fort-de-France, Martinique.

runners that are considered by botanists to be rhizomes (Silveus, 1942) because they are appressed to the soil and are covered by the persistent sheaths of dead leaves. Agronomists including the writer consider the runners to be stolons. In bahiagrass, belowground shoots do not exist. Distinguishable crowns or basal shoot aggregates are absent. The leaf blades are commonly only ciliate near the base, but in the cultivar Paraguay they are prominently pubescent. The midrib is conspicuous and 'Pensacola' has fissures between the veins (Burkart, 1969), which assist the leaf in rolling during wilt. The ligule is a short membrane with a densely puberulent region immediately behind it.

The flowering culm of bahiagrass typically reaches 40 to 90 cm height, including the inflorescence. Inflorescences of bahiagrass are racemose, with generally 2 racemes, uncommonly 3, and rarely 4, arising subconjugate. Racemes are 3 to 16 cm long. Spikelets are subsessile and solitary in two rows on each raceme, and are not paired as in many other species of *Paspalum*. The solitary state is a result of the loss or absence of the lower spikelet (Webster, 1988).

The 2.5 to 4.0 mm long spikelets are plano-convex, generally obtuse, but may range from ovate, ovate-elliptic, obovate, and orbicular. Spikelets are 1.7 to 3.0 mm wide and as with most members of *Paspalum*, the first glume is wanting. As with other members of the Paniceae, the proximal floret is reduced and sterile; its lemma is about the same length and chartaceous like the second glume. Together the sterile lemma and the second glume enclose the fertile floret. The lemma of the fertile floret is firm and inrolled at its margins onto the similarly firm palea, and is the main cause of postharvest seed dormancy in the cultivar Pensacola (West and Marousky, 1989).

In contrast to *Panicum* and most Paniceae, in *Paspalum* the spikelet is twisted 180 degrees, such that the rounded back of the fertile lemma faces the rachis. Disarticulation in bahiagrass occurs at the base of the spikelet about 13 days after anthesis and is due to the formation of an abscission layer (Burson et al., 1978). The floret contains three anthers, which are generally dark purple. There are two stigmas, also generally purple.

Drought Resistance

Bahiagrass has the highest level of drought survival among any sod-forming turfgrass for use in nonirrigated landscapes in warm humid areas. Compared with other warm-season turfgrasses, bahiagrass is drought resistant by means of avoidance due to better recovery from defoliation and due to deeper rooting, but not due to reduced evapotranspiration.

Because they are protected and extensive, bahiagrass stolons survive complete defoliation, due to desiccation and other stresses. Bahiagrass stolons have 11,358 kg ha^{-1} dry weight (Beaty and Tan, 1972). When bahiagrass undergoes severe wilt, it generally recovers by producing new leaves from the stolons. It also tolerates other kinds of defoliation. For example, bahiagrass tolerates use in Florida for turf parking lots (Busey, 1990). Sod harvested from areas defoliated by frost and shipped to warm areas often establishes well and refoliates within a few weeks.

The bahiagrass root system is more extensive than other turfgrasses. For example, Doss et al. (1960) reported root dry weight of 8,639 kg ha^{-1}, which was 82% more than common bermudagrass, *Cynodon dactylon* (L.) Pers., and Burton et al. (1954) reported 10,373 kg ha^{-1}, which was 46% more than common bermudagrass. Busey (1992) reported consistently higher root dry weight for 'Argentine' bahiagrass than Pensacola, e.g., 3,480 kg ha^{-1} for Argentine, compared with 2,670 kg ha^{-1} for Pensacola, 2.2 years after planting. By 4 years after planting, Argentine had 18,700 kg ha^{-1} root dry weight, compared with 13,470 kg ha^{-1} for Pensacola. During 3 years of irrigation curtailment, Argentine maintained higher turf quality ratings than Pensacola.

Although it is difficult to measure the effective root zone of plants, a shallow water table is sometimes within access of the bahiagrass root system, and must partly explain its drought survival. Burton et al. (1954) showed that 16% of the bahiagrass root mass is below 1.2 m, which exceeded the deep root mass of other species. In pine flatwoods soils of the coastal plains of the United States, there is often a perched water table within 1.5 m depth, where bahiagrass provides both pasture and turf without supplemental irrigation. In the urbanized wetlands of south Florida, specifically Broward and Miami-Dade Counties, the maintained water table averages 1.45 m deep (Busey, 1996) and there also the bahiagrass grows between rain events, often without wilting, when neighboring St. Augustinegrass, *Stenotaphrum secundatum* (Walt.) Kuntze, is wilted (Busey, unpublished data, 1988).

The evapotranspiration (ET) rate of bahiagrass has not been studied as frequently as that of other turfgrasses. In large pans, bahiagrass ET averaged 4.1 mm d^{-1} or a maximum of 6.1 mm d^{-1} in June in south Florida (Clayton et al., 1942), comparable to ET of 6.4 mm d^{-1} of the native sawgrass, *Cladium jamaicense* Crantz, in the same study. In minilysimeters, Argentine bahiagrass ET was not different from St. Augustinegrass; seashore paspalum, *Paspalum vaginatum* Swartz; centipedegrass, *Eremochloa ophiuroides* (Munro) Hack.; or 'Meyer' zoysiagrass, *Zoysia japonica* Steud. (Casnoff et al., 1989). But in the same study bahiagrass ET was greater than buffalograss, *Buchloë dactyloides* (Nutt.) Engelm., and bermudagrasses, *Cynodon* spp., and less than that of 'Emerald' zoysiagrass, *Zoysia* sp. In another study, ET rate of Argentine bahiagrass was 6.3 mm d^{-1} and in the same interval the ET rate of 'Texas Common' St. Augustinegrass was 6.3 mm d^{-1} (Kim and Beard, 1988). In weighing-lysimeters the ET rate of Pensacola bahiagrass was not different from three St. Augustinegrass cultivars, but days to wilt differed 9.0 days for bahiagrass, which was greater (P < 0.05) than the 4.8 to 6.7 days to wilt for St. Augustinegrass cultivars (Miller and McCarty, 2001). Leaf water potential of Pensacola bahiagrass was also greater than St. Augustinegrasses, indicating improved water status. The delayed wilting is consistent with the hypothesis of drought avoidance due to more effective rooting, even though root systems were confined at 0.75 m depth. If the root systems were not so confined one would expect an even greater potential advantage for bahiagrass.

Using the terminology of Levitt (1980), bahiagrass is an intolerant drought avoider, which survives drought either by spending (deep roots) or saving (defoliation). Although Pensacola bahiagrass, a diploid (2n=20), is considered to have greater heat tolerance than other cultivars, it has less epicuticular wax content compared with tetraploid (2n=40) genotypes of bahiagrass (Tischler and Burson, 1995). The use of drought avoidant turfgrasses is a worthwhile approach to sustainable turf culture in wet-dry climates in level, humid areas, such as the coastal plains of the southeastern United States, where soil moisture is naturally replenished (Busey, 1996). But in arid warm temperate regions, such as the southwestern United States, bahiagrass is replaced in nonirrigated conditions by members of the Eragrostoideae with thin canopy (e.g., buffalograss) and annual species. In arid subtropical regions such as the windward areas of the West Indies and the Florida Keys, bahiagrass is replaced by thin canopied grasses such as members of the Andropogoneae, such as Seymourgrass, *Bothriochloa pertusa* A. Camus, also called hurricanegrass.

Growth and Management

Bahiagrass has slow growth rate, only 2.1% daily growth rate (relative to initial fresh weight) compared with 9.2% for bermudagrass grown in sand in the same study (Busey and Myers, 1979). But the unmown height and rate of vertical extension of leaves and culms can be objectionable esthetically, and make it difficult to mow during warm and moist summers. For example, the primary energy consumption of electric and gasoline rotary mowers in bahiagrass is 15.2 and 35.8 Whm^{-2}, respectively, about three times the values for St. Augustinegrass (Fluck and Busey, 1988).

Bahiagrass has a relatively open habit of growth, which makes it vulnerable to weed infestation whenever the thatch layer is disturbed, and this also gives it poor density for turf quality. The extinction coefficient of light in Pensacola bahiagrass is 0.36, a way of saying that the leaf blades are erect (Agata, 1985). Looking directly down on the canopy, there is often more open area than green leaf cover. Despite the open habit of growth, and probably due to the extensive root system, bahiagrass is very competitive for nutrients. For example, in early lysimeter studies in Florida, 88% of the heavily applied nitrate fertilizer was taken up by the plant. In this study involving summer fertilization with soluble nitrate, the leachate had 12 ppm nitrogen (Leukel and Barnette, 1935), and bahiagrass was most effective in capturing nutrients when mown regularly. More recent work (Sveda et al. 1992) showed soil water nitrate concentrations average 0 to 0.2 ppm under fertilized bahiagrass.

Annual weeds usually outgrow bahiagrass under high soil fertilization (> 5 to 10 g N m^{-2} y^{-1}) and irrigation, particularly in the winter (Busey, unpublished observations, 1980), when it is naturally dormant. Weeds are also a serious problem in bahiagrass and grass herbicides are generally harmful. Triazine herbicides can be used to remove bahiagrass from centipedegrass (Johnson, 1979), and sulfometuron removes bahiagrass from both centipedegrass and bermudagrass (Hanna et al., 1989). Pensacola bahiagrass is more susceptible to triazine herbicides than the cultivar Wilmington (Smith, 1983).

In competition with annual ryegrass, *Lolium multiflorum* Lam., bahiagrass produces better growth and establishment at soil temperatures above 28°C (Ogata et al., 1980). Fertilization at the time of seed planting is associated with serious weed competition, while fertilization 5 weeks after seed planting, at about the three-leaf stage, fosters the highest establishment ratings and shoot development (Busey, 1992).

Bahiagrass has poor ability to survive in high (over 7.5) soil pH, and under those conditions it frequently shows iron deficiency in the spring (Busey, unpublished data, 1980). But at low (4.4) pH it benefits from liming (Blue, 1979). The largest determinate of bahiagrass turf quality, based on a survey of 79 highway soils, was soil depth; that is, depth to roadbed (Busey, unpublished data, 1977). Bahiagrass is very susceptible to salinity and is poorly tolerant of shade (Busey, unpublished observations, 1976). Therefore, it makes a poor choice for use in most densely populated urban and suburban areas, and a poor choice for areas near the ocean. It is well adapted to large, open landscapes with full sun. Bahiagrass has good submersion tolerance (Fry, 1991).

Long-term nutritional management studies have been conducted only for forage production and seed production, not for turf quality. Bahiagrass generally does not respond from fertilization rates over 100 kg N ha^{-1} yr^{-1}. Response to potassium and phosphorus has varied among investigations.

Bahiagrass inflorescences or seedheads are abundant for a brief period (about 6 weeks in summer), but some seedheads occur sporadically throughout the growing season during long days. Seed production of cultivar Argentine bahiagrass is enhanced most strongly in Florida by N fertilization during the critical period of inflorescence production; that is, daylength > 13 hours, or between the end of April and the end of May. But seed germination is depressed at high (200 kg N ha^{-1}) fertilization of the parent population, due to the suppression of floret development by excess vegetative growth (Adjei et al., 2000). Seed production of bahiagrass is meager; 225 to 350 kg ha^{-1} would be the high range (W. R. Cook, personal communication, 1989). Seed production of Argentine bahiagrass is hampered by the presence of ergotism caused by *Claviceps paspali* F. L. Stephens & J. G. Hall. Argentine seed production occurs several weeks later in the summer in Florida, compared with Pensacola bahiagrass, and further into the rainy season. Consequently, Argentine bahiagrass is more prone to ergotism and sometimes has only 25% viability due to the effects of the disease (Kelsey Payne, personal communication, 1983).

Bahiagrass is strongly photoperiod-dependent for inflorescence induction (Knight and Bennett, 1953), with flowering possible at a daylength of 14 hours, but not 12 hours. Flowering was initiated under short days whenever the night period was interrupted by red or far-red light (Marousky and Blondon, 1995). The writer has seen bahiagrass flowering in south Florida in January, when it was growing underneath street lamps (Busey, unpublished data, 1980). A survey of New World herbarium collections from the U.S. National Museum, from 1802 to 1977 (Figure 21.2) shows that 13 hours is the threshold for inflorescence occurrence among a broad germ plasm including different botanical varieties, and a range of latitudes (Busey, unpublished data; Figure 21.3). Actual induction probably occurs at slightly less than 13 hours daylength. Nonflowering bahiagrass may remain vegetative in equatorial latitudes; for example, in efforts to grow bahiagrass in East Africa starting by the 1940s, it was "usually established from stolons [cuttings], although seed can be obtained in the U.S.A." (Bogdan, 1970).

The distribution of botanical varieties within *Paspalum* has latitudinal discontinuity. The genotypes within the diploid (2n=20) *Paspalum notatum* var. *saurae* Parodi (Parodi, 1948), including Pensacola, have a historical temperate distribution (Figure

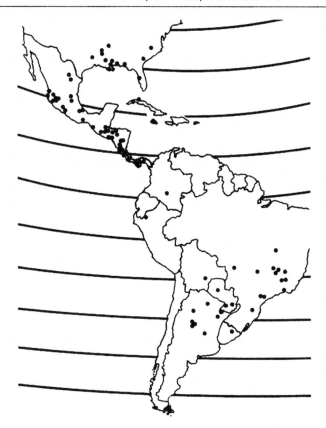

Figure 21.2. Geographic range of bahiagrass, *Paspalum notatum* Flügge, in the New World, based on herbarium collections from 1802 to 1977 (MO).

21.2) and better frost tolerance than genotypes within the tetraploid (2n=40) *P. notatum* var. *latiflorum* Doell, including Argentine and 'Common' bahiagrasses (Burton, 1946).

Seed Establishment

As a seed propagated crop, postharvest dormancy and the difficulty of establishment are problems. Mechanical and acid scarification have been the traditional methods to break dormancy (Burton, 1939) but temperature factors are also involved. Argentine has more seed dormancy than Pensacola. One year of storage at room temperature is usually sufficient to break dormancy. The seed of bahiagrass shows improved germination following short periods of digestion in cattle (Gardener et al., 1993) and ranchers indicate that cattle can spread bahiagrass seed to new areas.

Bahiagrass germination and seedling growth occurs best between 32 and 37.5°C (Marousky and West, 1988) and monthly seed plantings in south Florida show that suitable establishment occurs from seed plantings from March through August, but not from September through February (Busey, unpublished data, 1987). The use of a companion crop of red millet, *Panicum* spp., is extremely deleterious to bahiagrass establishment, probably by competing with the bahiagrass during the warm season (Busey, 1989). The bahiagrass seed must be incorporated in the soil—10 mm is the optimum depth—and pressed in (Burton, 1940). Postplanting fertilization of seed plantings at the three-leaf stage results in the highest first year establishment ratings (Busey, 1992).

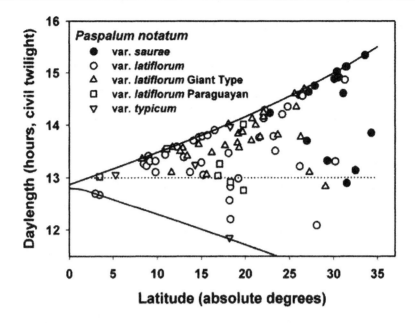

Figure 21.3. Distribution of daylength (at the time that inflorescences were observed and the plant was collected) as a function of absolute degrees latitude for bahiagrasses in the New World, based on bahiagrass, *Paspalum notatum* Flügge, herbarium collections from 1802 to 1977 (MO). Daylengths were derived by the CBM model of Forsythe et al. (1995) based on civil twilight. The solid lines represent the total range of daylight hours. The dotted line represents 13 hours. The botanical varieties var. *saurae*, var. *latiflorum*, and var. *typicum*, as well as the morphological groups "Giant Type" and "Paraguayan," were determined by the writer.

DISTRIBUTION, CYTOTAXONOMY, AND GENETICS

Origin and Relatives

Paspalum is in the tribe Paniceae of the supertribe Panicodae of the subfamily Panicoideae. Of the some 400 species of *Paspalum*, many are known only from the Old World, but the majority occur in tropical America. *P. notatum*, bahiagrass, has been placed in the Notata group (Silveus, 1942) and it has a genetic relationship to *P. pumilum* Nees, as their interspecific hybrids have considerable chromosome pairing (Quarin and Burson, 1983). Interspecific hybrids of bahiagrass and other sibling species, e.g., *P. vaginatum* and *P. intermedium* Munro ex Morong (Burson, 1981) and *Paspalum dilatatum* Poir. (Espinoza and Quarin, 2000) show no such affinity, because there is essentially no chromosome pairing between genomes.

Before intentional introduction and breeding improvement, the distribution of bahiagrass extended from northern Argentina to Mexico, with one collection from the West Indies (Figure 21.3). Early collections from Latin America and the West Indies were from the 1800s [MEXICO: Jalisco, common in marshy situations, solid stands, east of Guadalajara, 1855, *Leavenworth and Leavenworth, s.n.* (MO); CUBA: 1860-1864, *Wright 3438* (MO); and the Type specimen, WEST INDIES: St. Thomas, 1802, *Schrader and Ventanant s.n.* (US)]. The earliest collections from the United States were from the late 1800s [LOUISIANA: Along bayou, near St. Martinville, 28 June 1893, *A. B. Langlois* (US); NEW JERSEY: Ballast grounds at Camden, 24 Sep. 1880, Frank *Lamson-Scribner 2749* (US)].

Whether the northern hemisphere localities represent part of the pre-Colombian distribution, or early accidental introductions, is unknown. The gap in distribution in northern South America (Figure 21.2) suggests that there may have been a formidable barrier to migration, possibly partly due to photoperiod dependency, although bahiagrass reached the foothills of the Andes [PERU: Province Bongara, Department of Amazonas, sandstone area, growing in almost pure white sand between Rio Utcubamba and Shipasbamba, 4 km from Campomiento Ingenio, Alt. 1520 m, 1 Feb. 1964, *P. C. Hutchison and J. K. Wright 3990*].

Bahiagrass has many unique common names throughout Latin America, which is consistent with its natural distribution throughout. For example, it is called "pasto horqueta" and "gramilla" and sometimes "pasto de miel" in Argentina, though the latter is also used to refer to the genus. In Brazil, it is called "grama batatais," to distinguish it from "grama dulce," which is *P. urvillei* Steud., and rarely "forquinha," in obvious reference to the forked appearance of the two inflorescence racemes. In Costa Rica, it is called "gengibrillo;" in Mexico it is called "zacate bahía," and "cabeza de burro," and sometimes "seca sebo" or "zacate camalote" or "zacate de llanero." Names suggesting an intentional outside introduction are from Cuba, where it is called "pasto mexicano" and "Tejana;" from the United States, where it is called "bahiagrass;" and from Spain, where it is called "hierba de Bahía." Therefore bahiagrass was probably introduced to Cuba and the United States in historical times, possibly both accidentally and intentionally, and was given the names of the places from which it was thought to have originated. It is puzzling that the Type specimen was described from the West Indies, since the early 1900s writers were concerned with introductions from southern South America. Bews (1929) was aware of a distribution, "from Mexico and the W. Indies to the Argentine."

Some of the first intentional introductions to the United States were made around 1912 (Scott, 1920), but the first mention in the Bulletins of the Florida Agricultural Experiment Station was by Laird (1930). Bahiagrass was not mentioned as a turf in Florida in 1929 by Enlow and Stokes (1929) and received only slight research attention as a forage through the 1930s.

In 1938, Florida, Escambia County Agent E. H. Finlayson's attention was directed by dairyman Homer Diamond to what would be called Pensacola bahiagrass (*Paspalum notatum* var. *saurae*) on a sodded sand road bank about 6 miles north of Pensacola (Finlayson, 1941). He recognized the difference between Pensacola bahiagrass (Figure 21.4) and the other bahiagrasses that were being introduced, "Since there seems to be a difference in the Bahia grass growing in Pensacola and that of the West Indies and South America and since as far as I have been able to determine, it is growing in no other place, I am taking the liberty of referring to it in the balance of this article as Pensacola Bahia grass." The Pensacola of Finlayson seeded more profusely than the "broader bladed strain that is finding favor in the Southeast on sandy and drier soils," and it also crowded out weeds.

This excellent growing cultivar, which may have arrived as a stowaway on a fruit boat from Central or South America, was released in 1944 (Hanson, 1972) to become one of the major forage grasses of the southeastern United States (Burton, 1967). It is planted extensively from seed along highways in Florida, North Carolina, and other subtropical and mild climates. In 1944, P. I. 148996 (*P. notatum* var. *latiflorum*) was introduced from Argentina, and by 1950 it was released as 'Argentine' based on high forage production (Killinger et al., 1951). Argentine and Pensacola are quite different in chromosome number, method of reproduction, and performance traits. They are most easily distinguished by the width of the mature spikelet, which is 2 mm wide or wider for Argentine, and less than 2 mm wide for Pensacola. The same contrast applies

Figure 21.4. Typical appearance of *Paspalum notatum* var. *saurae* Parodi, commonly called Pensacola, the diploid (2n=20) cytotype of bahiagrass, *Paspalum notatum* Flügge. Pensacola is also a cultivar. It is widely used in highway grassing [FLORIDA: Walton County, along U.S. Hy. 90, 10 mi w. of De Funiak Springs, common; planted on sandy roadside, 01 June, 1964, *Richard W. Pohl, 9625.* (MO)].

to the botanical varieties, *P. notatum* var. *latiflorum* and *P. notatum* var. *saurae*, respectively, for Argentine and Pensacola.

Apomixis and Polyploidy

Burton (1946) discovered Pensacola bahiagrass is diploid (2n = 20) and reproduces sexually (Burton, 1955) based on variable progenies for traits such as plant morphology, seed and forage production, leafiness, disease resistance, and anther color. It is also highly self-incompatible, and shows severe inbreeding depression. Therefore, the Pensacola discovered by Finlayson in 1938 was most likely a naturalized population deriving from multiple seeds introduced; otherwise, if only one seed had been introduced, the resulting clone would have been sterile.

On the basis of crossing studies, Burton concluded (1948b) that the cultivar 'Common' bahiagrass, which he showed is tetraploid (2n=40), (thus probably *P. notatum* var. *latiflorum*) was most likely apomictic but that pollen was required for seed set. For example, when a white-stigma plant was surrounded by red stigma plants, all progenies from each population expressed the trait of their respective female parents. A male-sterile tetraploid bahiagrass was discovered in 1941 with light yellow anthers, and it produced seed only when exposed to foreign pollen, even when stimulated by pollen from the diploid Pensacola, confirming that the apomictic plants are pseudogamous, or exhibit false fertilization. The hundreds of progeny of the male-sterile plant were identical to the female parent, with rare exceptions that were recognized based on their red stigmas, similar to the pollen parent.

The exceptions to pseudogamy were two 2n=60 plants that must have arisen from unreduced Common female gametes X reduced Common male gametes, and three

2n=50 plants that must have arisen from unreduced Common female gametes X reduced Pensacola male gametes. Thus began a series of attempts to backcross between ploidy levels to attempt to bridge the barrier of obligate apomixis, which was identified based on embryology as apospory (Burton and Forbes, 1960). In apospory, a diploid embryo sac is formed from a nucellus or integument cell; in apomictic bahiagrass, the origin is the nucellus (Chen et al., 2000). Endosperm development in bahiagrass is independent of the embryo; effective fertilization of two unreduced polar nuclei occurs from a reduced male gamete (Quarin, 1999). The DNA content of Pensacola bahiagrass, 1.44 pg per diploid chromosome complement, is within the range of other warm-season turfgrasses (Arumuganathan et al., 1999).

Diploid (2n=20) Pensacola can be tetraploidized by the use of colchicine to produce sexual 2n=40 autotetraploid plants (Burton and Forbes, 1960), but one should be aware of the occurrence of diploid-tetraploid chimeral plants (Forbes and Burton, 1961a). Facultative apomicts also occur among naturally occurring tetraploids, and have been used to produce triploid (2n=30) hybrids (Hanna and Burton, 1986). A triploid was crossed with Pensacola to produce numerous 40-chromosome sexual plants, apparently arising from an unreduced female gamete and the reduced male gamete of Pensacola (Burton and Hanna, 1986). While the two methods of creating sexual tetraploids are successful, and the F_2s from sexual X apomictic crosses show that apomixis in tetraploid bahiagrass is mostly recessive, attempts to transfer obligate apomixis to 2n=20 Pensacola bahiagrass have not been successful (Burton and Hanna, 1992). The gene(s) for apomixis can exist at the diploid level in bahiagrass, but not be expressed (Quarin et al., 2001). In the case of obligate apomixis, whatever environmental factors might control facultative apomixis do not lessen obligate apomixis (Burton, 1982a). This fact suggests that breeding of sexual Pensacola cytotypes at the diploid level might be combined with tetraploidization, to hopefully bring out apomixis and the lower growth habit typical of the tetraploids. Tetraploids can be produced by soaking seeds in petri dishes on filter paper in an aqueous treatment of colchicine (Forbes and Burton, 1961b). The most effective treatments are 0.4 and 0.8% colchicine for 6 hours and 0.2, 0.4, and 0.8% cochicine for 48 hours, followed by rinsing.

Taxonomy of Cultivars and Natural Populations

The most definitive taxonomic works (Parodi, 1948; Burkhart, 1969) have described the typical botanical variety, *Paspalum notatum* var. *typicum*, as being small-seeded, while *P. notatum* var. *latiflorum* Doell is large-seeded, and *P. notatum* var. *saurae* is small-seeded. When the width of the spikelet is 2.0 mm or greater, the plant is usually *P. notatum* var. *latiflorum* and is a tetraploid (2n=40) apomict. When the width of the spikelet is less than 2.0 mm it is *P. notatum* var. *saurae* and is a diploid (2n=20) sexually reproducing genotype. Collections of *P. notatum* var. *typicum* are rare, and mostly from the West Indies, including the Type specimen, from St. Thomas.

The tetraploid bahiagrass cytotype *P. notatum* var. *latiflorum* (2n = 40) is the most common botanical variety (Figure 21.3) in tropical and subtropical America (Parodi, 1948). It has considerable variability (Moraes Fernandes et al., 1973) and includes several broad-leaved cultivars, such as Argentine, 'Paraguay-22', and Wilmington. The former is widely used in lawns and for solid sodding on highway embankments and the 0.8 m edge along the pavement. Argentine has broader leaf blades than Pensacola, yet it is superior as a turf because of its more prostrate, denser growth habit. This violates the common belief that finer-textured cultivars are necessarily superior turfgrasses. Native to eastern Argentina (Quarin et al., 1984), *Paspalum notatum* Flügge var. *saurae* Parodi (Parodi, 1948) includes the original Pensacola and the breeding lines and advanced forage cultivars such as Tifton Cycle 9, etc.

While Pensacola is a recognized cultivar, the name Pensacola is sometimes also used to describe any diploid, cross-breeding bahiagrass, that is, *Paspalum notatum* var. *saurae*. Although confusing, this is similar to the ambivalent use of the names of other landed races, such as "Common bermudagrass." The intended meaning can usually be discerned in the context.

Inheritance

Simple Mendelian genetic inheritance has been rarely reported in bahiagrass, with the exception of test characters such as stigma color. As another example, Hodgson (1949) used an albino trait that apparently was controlled either by a single recessive allele, or possibly by recessive alleles on two genes. The observed frequency of albinos, 36.92%, was between the expectations for a 3:1 ratio and a 9:7 ratio.

COLLECTION, SELECTION, AND BREEDING HISTORY

Germ Plasm Resources

In 2001, 181 foreign introductions of bahiagrass are available for breeders in the National Plant Germplasm System (USDA-ARS, 2001). The vast majority of historical introductions represent apomictic tetraploids (Busey, unpublished data, 1980). Because of the recent introduction to the Old World, virtually all the potential germ plasm of bahiagrass should be in the New World. Considerable genetic diversity exists in naturalized populations, and has already been the basis for the substantial forage improvement work at Tifton, Georgia, as well as turf improvement research in Florida. Because of the extensive distribution of bahiagrass cultivars as seed, it may be impossible to discern whether new introductions represent relict ancestral populations or duplicates of modern cultivars. Pozzobon and Valls (1997) found that only 11 bahiagrass accessions were diploids ($2n=20$), out of 127 collections from natural habitats in southern Brazil. Although collected in the wild, the diploids may have been derived from Pensacola bahiagrass introduced from the United States in the 1960s.

Breeding and Selection Techniques

Bahiagrass is easy to manipulate in breeding, but because it is perennial and grows as slowly spreading clones, selection is difficult. It may take one year or more for a seed planted stand to mature and display mature population characteristics (Busey, 1989). Consequently, much of the challenge of bahiagrass selection has been to find ways of predicting performance of clonal stands, without having to wait years. Spaced plant tests validly measure relative forage yield differences, but overestimate absolute yield potential of solid stands (Burton, 1985). By not replicating in first-stage evaluations, a larger germ plasm can be screened and subjected to more intensive selection, even though heritability based on unreplicated selection would be less than heritability based on genotype means.

Burton (1948a) showed that detached culms could be brought to the lab, placed in water with a short piece of rooted stolon, and allowed to cross-pollinate under controlled conditions and set seed. Bahiagrass is a prolific pollen producer, and up to 5% outcrossing may be expected within 125 m, and isolation distance of 175 m may be necessary for seed increase of pure cultivars (Hodgson, 1949). For purposes of controlled hybridization, flowering occurs first in the top of the inflorescence, or just below the top, and progresses predictably downward. Anthesis in most genotypes occurs soon after sunrise, but additional florets can open late in the day.

At the University of Florida, Fort Lauderdale, parchment pollinating bags were placed over inflorescences one day before anthesis, with the plants in containers in a greenhouse. The relatively large spikelets were easily emasculated with a pair of forceps, which was done in the morning as the anthers emerged. By also removing unused spikelets, and unused portions of the inflorescence, it was easier to keep track of crosses, and less likely to have stray pollen in a bag. Any spikelets with already dehisced anthers were removed from the inflorescences. Crosses were made using pollen transferred to the hand-emasculated florets. Pollinated spikelets were marked with an indelible marker. In addition, spikelet positions were numbered and recorded, so that a record of shriveled stigmata (an indication of effective crossing) could later be associated with individual seeds harvested. Stigma reaction was recorded one day after pollination, and the bags removed. Because the most obvious goal of breeding sexual *P. notatum* var. *saurae* bahiagrass is to develop a seeded crop with high genetic diversity, and because of the high degree of self-incompatibility, hand emasculation is usually of minimal value in breeding studies involving populations, and most efforts have involved controlled pollination without emasculation using whole inflorescences.

Modern tools to assist bahiagrass improvement have included plant regeneration from cultured inflorescences (Bovo and Mroginski, 1986) and seed (Marousky and West, 1990). Somaclonal variation has been detected in regenerated plants based on random amplified polymorphic DNA (RAPD) analysis (Akashi and Kawamura, 1998).

Breeding for Forage

Organized breeding of bahiagrass under the direction of Dr. Glenn Burton of the United States Department of Agriculture at the Coastal Plains Experiment Station, Tifton, Georgia has been extensive. The goal has been improved forage production, but the result is a practical and cytogenetic foundation for all other efforts to improve bahiagrass genetically, including for turf.

Initial efforts involving polycross testing resulted in parents with good general combining ability, which were planted vegetatively and allowed to intercross freely to produce the F1 hybrid cultivars 'Tifhi 1' and 'Tifhi 2' (Hanson, 1972). Despite improved forage yields and liveweight gains, the method of seed production was difficult, and the method of breeding did not lend itself to recurrent improvement. Recurrent Restricted Phenotypic Selection (RRPS), a modification of mass selection involving paternal as well as maternal selection based on phenotypic yields, and crossing in the laboratory, was successful (Burton, 1982b). Initial restrictions included local control of field gradients in plots, by selecting a certain number of parents from each small block. Other improvements were added: selecting seedlings to plant in the field, record keeping, and improving the uniformity of the field environment. These efforts resulted in taller, more erect, leafier plants. By cycle 16, plant weights were 382% greater than the initial Pensacola, but plant diameter decreased by 15% (Werner and Burton, 1991). Forage yields were increased consistently for 22 cycles. The use of an initially more narrow genetic base facilitated rapid response to RRPS and more quickly resulted in a high yielding population, compared with starting with a broad genetic base (Burton and Mullinix, 1998). On the basis of 18 cycles of genetic improvement, the number of alleles for high yield was believed to have increased in selected populations. While the results for yield have been outstanding, selection for tall plants may have reduced the amount of stolons, and this may explain the more severe winter injury of T14 (RRPS cycle 14) compared with Pensacola (Pedreira and Brown, 1996).

Mole crickets (*Scapteriscus* spp.) are sometimes a severe pest problem in bahiagrass used both as turf and forage, thus bahiagrass resistance would be very desirable. Resistance screening of clonal bahiagrasses yielded inconclusive results (Reinert and Busey, 1984). This pest damages turf mechanically while foraging, much as a rototiller would damage turf, and in some cases the damaged turf is not even the host. Thus in the absence of a close herbivore / plant association, it appears that it may be very difficult to develop mole cricket resistant cultivars.

Breeding for Turf

Attempts at turf improvement of bahiagrass have considered the detracting appearance (tallness and heavy seedhead production) and have attempted to shorten the height and/or reduce the seedhead count. But the correlations among traits have made it difficult to necessarily improve one trait without worsening plant performance in another trait. For example, the writer evaluated (Busey, unpublished data, 1980) diploid sexual bahiagrasses collected along roadsides from North Carolina to Texas. Replicated spaced plant evaluations were performed at the University of Florida Fort Lauderdale Research and Education Center and three other south Florida locations. Analyzing broad-sense heritability components, soil coverage rate of the sexual bahiagrass genotypes had positive ($P < 0.01$) genetic covariance with both foliage height and seedhead height; that is, shorter plants spread more slowly.

As proof, the third generation bulk (P3C1) of a prolific polycross population had culm height of only 54.4 cm compared with Pensacola at 75.9 cm, or a 28% reduction (Busey, 1989). The P3C1 had reduced coverage ratings as well as shorter foliage, narrower leaves, and 70% greater seedhead number, compared with Pensacola. The shorter bahiagrass population was less competitive and more seedy. In other work a Rapid Coverage Polycross (RCP) bahiagrass was developed, for highway rights-of-way and conservation areas (Anderson and Sharp, 1994). The cultivar achieves acceptable first-year establishment quality at lower seed planting rates than Argentine or Pensacola (Busey, 1989) and has improved establishment ratings and first year root and shoot growth compared with Pensacola (Busey, 1992). Despite initially high establishment of the RCP, during 3+ years of observations, the Argentine cultivar surpassed RCP and sustained better coverage ratings than RCP.

Seedhead height in bahiagrass is moderately heritable, $h^2 = 0.35$ based on midparent-offspring regression (Busey, 1985). By continuing polycross selection, in generation seven a population (P7C1) was developed with seedhead height averaging 40 cm. But based on the data on reduced cover ratings for the P3C1 short-culm population, the dwarf population was expected never to be sufficiently competitive under field conditions. Therefore, in abandoning this project, 354 plants of the dwarf bahiagrass population were interplanted and established under minimal maintenance in a dense stand of bermudagrass, and ignored for ten years. From 1991 to 2001, under no irrigation, no fertilization, and infrequent mowing, the bermudagrass and bahiagrass were invaded by other weedy grasses and a dense overstory of shrubs. In 2001 an accidental fire destroyed the overstory, and in the months that followed it was apparent that the bermudagrass had been eliminated by shading, while seven clones from the original dwarf bahiagrass population survived and spread under the conditions of weed infestation. These plants are now being propagated for development as a possible dwarf bahiagrass for turf.

Commercial breeding development of bahiagrass has been virtually nonexistent. No national bahiagrass test has been coordinated by the National Turfgrass Evaluation Program. Only cultivars Argentine and Pensacola are generally on the market for turf. Turfgrass sod producers have rarely picked up a dwarf bahiagrass genotype and tried to propagate it vegetatively, but the procedure is probably too slow to be profitable. For example, William E. Moran developed 'B-1' bahiagrass as a dwarf cultivar (U.S. Patent PP7,429, January 22, 1991), intending to propagate it through vegetative cuttings. Francis J. Marousky and Albert E. Dudeck developed 'MBA-1' bahiagrass based on slow growth rate, limited flowering, and short inflorescences (U.S. Patent PP8,788, June 14, 1994), stating "characteristics come true to form and are established and transmitted through secondary propagation as sprigs or sod." In neither case did the developers mention the chromosome number or the method of seed reproduction, either sexual or by apomixis. It would seem that based on the superior turf performance of the tetraploid bahiagrasses such as Argentine (Busey, 1992) and the presence of obligate apomixis (Burton and Forbes, 1960) that it would be relatively obvious to patent an apomictic bahiagrass as is frequently done for cultivars of Kentucky bluegrass, *Poa pratensis* L.

CONCLUSION

The vast majority of bahiagrass will probably continue to be used for rough turf and conservation, situations in which weed competition is a major problem. Land management agencies such as the Florida Department of Transportation and the South Florida Water Management District maintain tens of thousands of hectares of bahiagrass, and would like to reduce their mowing expenditures (Gary L. Henry, personal communication, 1980; Francois LaRoche, personal communication, 1997). Nevertheless, it is difficult to maintain a shade sensitive species such as bahiagrass in competition with unmown weeds, and a dwarf cultivar might have even less of a place, unless perhaps the dwarf habit is correlated with shade tolerance.

The biggest challenges in breeding bahiagrass are that it is a long-lived perennial, and evaluation is difficult. Field evaluation must be long-term, exposing genotypes to a range of chronic natural problems (e.g., competition with weeds). Attempts to accelerate the progress of evaluation by prescreening for plant characteristics in containers have not been successful, except for seedhead height and other obvious aspects of plant morphology.

A major concern in the development of any bahiagrass cultivar for turf is whether seed producers will accept it. Bahiagrass seed is produced traditionally in Florida as a by-product of the cattle industry (Frank Tomkow, personal communication, 1983). An improved turf bahiagrass, to be successful, would need either to be very competitive and useful as a forage, or it would need a potential market sufficiently large to justify specialized turf seed production areas. The production of seed for a new cultivar on any existing bahiagrass area entails a major difficulty in the eradication of existing bahiagrass, and isolation from pollen sources, which can contaminate and hybridize with a selected cultivar. The area would also have to be restricted from the movement of cattle, which can carry seeds. The fact that bahiagrass is extensively used for primarily low maintenance easements and conservation areas means that it may be difficult to justify the numerous expenses of development, yet the potential environmental benefits may be great. For this reason, an apomictic cultivar, reproducing true, has tremendous advantage.

REFERENCES

Adjei, M.B., C.S. Gardner, D. Mayo, T. Seawright, and E. Jennings. 2000. Fertilizer treatment effects on forage yield and quality of tropical pasture grasses. *Soil Crop Sci. Soc. Florida Proc.* 59:32–37.

Agata, W. 1985. Studies on dry matter production of bahiagrass (*Paspalum notatum*) sward. II. Characteristics of CO_2 balance and solar energy utilization during the regrowth period. pp. 1237–1238. In T. Okubo and M. Shiyomi, Eds. *Proc. Int. Grassland Cong., 15th, Kyoto, Japan. 24–31 Aug. 1985*. Kyoto, Japan.

Akashi, R. and O. Kawamura. 1998. Improvement of forage quality in bahiagrass (*Paspalum notatum* Fluegge) through genetic manipulation. 1. Detection of somaclonal variation using RAPD analysis in plants regenerated from suspension cultures. *Grassland Sci.* 44:203–207.

Anderson, J. and W.C. Sharp. 1994. Grass Varieties in the United States. USDA Agric. Hdbk. 170. U.S. Govt. Print. Office, Washington, DC.

Anonymous. 1976. Florida Turfgrass Survey 1974. Florida Department of Agriculture and Consumer Services, Tallahassee, FL.

Arumuganathan, K., S.P. Tallury, M.L. Fraser, A.H. Bruneau, and R. Qu. 1999. Nuclear DNA content of thirteen turfgrass species by flow cytometry. *Crop Sci.* 39:1518–1521.

Beaty, E.R. and K.H. Tan. 1972. Organic matter, N, and base accumulation under Pensacola bahiagrass. *J. Range Manage.* 25:38–40.

Bews, J.W. 1929. *The World's Grasses: Their Differentiation, Distribution, Economics and Ecology*. Russell and Russell, New York.

Blue, W.G. 1979. Forage production and N contents, and soil changes during 25 years of continuous white clover-Pensacola bahiagrass growth on a Florida Spodosol. *Agron. J.* 71:795–798.

Bogdan, A.V. 1970. Turfgrass in Kenya. pp. 51–56 in *Proc. First Int. Turfgrass Res. Conf., Harrogate, England. 15–18 July 1969*. Sports Turf Research Institute, Bingley, England.

Bovo, O.A. and L.A. Mroginski. 1986. Tissue culture in *Paspalum* (Gramineae): Plant regeneration from cultured inflorescences. *J. Plant Physiol.* 124:481–492.

Burkhart, A. 1969. Gramineas. In *Flora ilustrada de Entre Rios (Argentina)*, Vol. 6, Part II. Colección Cientifica del I.N.T.A., Buenos Aires, Argentina.

Burson, B.L. 1981. Genome relations among 4 diploid *Paspalum* spp. *Bot. Gaz.* 142:592–596.

Burson, B.L., J. Correa, and H.C. Potts. 1978. Anatomical study of seed shattering in bahiagrass and dallisgrass. *Crop Sci.* 18:122–125.

Burton, G.W. 1939. Scarification studies on southern grass seeds. *J. Amer. Soc. Agron.* 31:179–187.

Burton, G.W. 1940. The establishment of bahiagrass, *Paspalum notatum*. *J. Amer. Soc. Agron.* 32:545–549.

Burton, G.W. 1946. Bahia grass types. *J. Amer. Soc. Agron.* 38:273–281.

Burton, G.W. 1948a. A method for producing chance crosses and polycrosses of Pensacola bahia grass, *Paspalum notatum*. *J. Amer. Soc. Agron.* 40:470–472.

Burton, G.W. 1948b. The method of reproduction in common bahiagrass, *Paspalum notatum*. *J. Amer. Soc. Agron.* 40:443–452.

Burton, G.W. 1955. Breeding Pensacola bahiagrass, *Paspalum notatum*: I. Method of reproduction. *Agron. J.* 47:311–314.

Burton, G.W. 1967. A search for the origin of Pensacola bahia grass. *Econ. Bot.* 21:379–382.

Burton, G.W. 1982a. Effect of environment on apomixis in bahiagrass. *Crop Sci.* 22:109–111.

Burton, G.W. 1982b. Improved recurrent restricted phenotypic selection increases bahiagrass forage yields. *Crop Sci.* 22:1058–1061.

Burton, G.W. 1985. Spaced-plant-population-progress test. *Crop Sci.* 25:63–65.

Burton, G.W., E.H. DeVane, and R.L. Carter. 1954. Root penetration, distribution and activity in southern grasses measured by yields, drought symptoms and P^{32} uptake. *Agron. J.* 46:229–233.

Burton, G.W. and I. Forbes Jr. 1960. The genetics and manipulation of obligate apomixis in Common bahiagrass (*Paspalum notatum* Flügge). pp. 66–71. In *Proc. 8th Int. Grassland Cong.*, Alden Press, Oxford, England.

Burton, G.W. and W.W. Hanna. 1986. Bahiagrass tetraploids produced by making (apomictic tetraploid X diploid) X diploid hybrids. *Crop Sci.* 26:1254–1256.

Burton, G.W. and W.W. Hanna. 1992. Using apomictic tetraploids to make a self-incompatible diploid Pensacola bahiagrass clone set seed. *J. Hered.* 83:305–306.

Burton, G.W. and B.G. Mullinix. 1998. Yield distributions of spaced plants within Pensacola bahiagrass populations developed by recurrent restricted phenotypic selection. *Crop Sci.* 38:333–336.

Busey, P. 1985. Selection for seedhead characteristics in bahiagrass. *Agron. Abstr.* p. 114.

Busey, P. 1989. Genotype selection and seeding rate in bahiagrass establishment. *Transport. Res. Record* 1224:40–45. National Research Council (U.S.) Transportation Research Board.

Busey, P. 1990. Vehicular turf. *Proc. Florida State Hort.* Soc. 103:352–355.

Busey, P. 1992. Seedling growth, fertilization timing, and establishment of bahiagrass. *Crop Sci.* 32:1099–1103.

Busey, P. 1996. Wilt avoidance in St. Augustinegrass germplasm. *HortScience* 31:1135–1138.

Busey, P. and B.J. Myers. 1979. Growth rates of turfgrasses propagated vegetatively. *Agron. J.* 71:817–821.

Casnoff, D.M., R.L. Green, and J.B Beard. 1989. Leaf blade stomatal densities of ten warm-season perennial grasses and their evapotranspiration rates. pp. 129–131 in H. Takatoh, Ed., *Proc. Sixth Int. Turfgrass Res. Conf., Tokyo*, July 31–August 5, 1989. Japan. Soc. Turfgrass Culture, Tokyo.

Chen, L., L. Guan, A. Kojima, and T. Adachi. 2000. The mechanisms of appearance of aposporous initial cell and apomictic embryo sac formation in *Paspalum notatum*. *Cytologia (Tokyo)* 65:333–341.

Clayton, B.S., J.R. Neller, and R.V. Allison. 1942. Water Control in the Peat and Muck Soils of the Florida Everglades. Bull. 378, Univ. Florida Agric. Exp. Stn., Gainesville.

Dobereiner, J., J.M. Day, and P.J. Dart. 1972. Nitrogenase activity and oxygen sensitivity of the *Paspalum notatum-Azotobacter paspali* association. *J. Gen. Microbiol.* 71:103–116.

Doss, B.D., D.A. Ashley, and O.L. Bennett. 1960. Effect of soil moisture regime on root distribution of warm season forage species. *Agron. J.* 52:569–572.

Enlow, C.R. and W.E. Stokes. 1929. Lawns in Florida. Bull. 209, Univ. Florida Agric. Exp Stn., Gainesville.

Espinoza, F. and C.L. Quarin. 2000. 2n + n hybridization of apomictic *Paspalum dilatatum* with diploid *Paspalum* species. *Int. J. Plant Sci.* 161:221–225.

Evert, D.R., P.F. Bertrand, and B.G. Mullinix Jr. 1992. Nematode populations and peach tree survival, growth, and nutrition at an old orchard site. *J. Am. Soc. Hort. Sci.* 117:6–13.

Finlayson, E.H. 1941. Pensacola . . . a new fine-leafed bahia. *Southern Seedsman*, Dec. issue, 9, 28.

Firth, D.J. and G.P.M. Wilson. 1995. Preliminary evaluation of species for use as permanent ground cover in orchards on the north coast of New South Wales. *Tropical Grasslands* 29:18–27.

Fluck, R.C. and P. Busey. 1988. Energy for mowing turfgrass. *Trans. ASAE* 31:1304–1308.

Forbes, I. Jr. and G.W. Burton. 1961a. Cytology of diploids, natural and induced tetraploids, and intraspecies hybrids of bahiagrass, *Paspalum notatum* Flügge. *Crop Sci.* 1:402–406.

Forbes, I. and G.W. Burton. 1961b. Induction of tetraploidy and a rapid field method of detecting induced tetraploidy in Pensacola bahiagrass. *Crop Sci.* 1:383–384.

Forsythe, W.C., E.J. Rykiel Jr., R.S Stahl, H. Wu, and R.M. Schoolfield. 1995. A model comparison for daylength as a function of latitude and day of year. *Ecol. Modelling* 80:87–95.

Fry, J.D. 1991. Submersion tolerance of warm-season turfgrasses. *HortScience* 26:927.

Gardener, C.J., J.G. McIvor, and A. Jansen. 1993. Survival of seeds of tropical grassland species subjected to bovine digestion. *J. Appl. Ecol.* 30:75–85.

Hanna, W.W. and G.W. Burton. 1986. Cytogenetics and breeding behavior of an apomictic triploid in bahiagrass. *J. Hered.* 77:457–459.

Hanna, W.W., C.W. Swann, J. Schroeder, and P.R. Utley. 1989. Sulfometuron for eliminating bahiagrass (*Paspalum notatum*) from centipedegrass (*Eremochloa ophiuroides*) and bermudagrass (*Cynodon dacytlon*). *Weed Technol.* 3:509–512.

Hanson, A.A. 1972. Grass Varieties in the United States. USDA Agr. Hdbk. 170.

Hodgson, H.J. 1949. Flowering habits and pollen dispersal in Pensacola bahiagrass, *Paspalum notatum*, Flügge. *Agron. J.* 41:337–343.

Jean, S. and T. Juang. 1979. Effect of bahiagrass mulching and covering on soil physical properties and losses of water and soil of slopeland (First report). *J. Agric. Assn. China (Taipei)* 105:57–66.

Johnson, B.J. 1979. Bahiagrass (*Paspalum notatum*) and common lespedeza (Lespedeza striata) control with herbicides in centipedegrass (Eremochloa ophiuroides). *Weed Sci.* 27:346–348.

Killinger, G.B., G.E. Ritchey, C.B. Blickensderfer, and W. Jackson. 1951. Argentine Bahia Grass. Circ. S-31, Univ. Florida Agric. Exp. Stn., Gainesville.

Kim, K.S. and J.B Beard. 1988. Comparative turfgrass evapotranspiration rates and associated plant morphological characteristics. *Crop Sci.* 28:328–331.

Knight, W.E. and H.W. Bennett. 1953. Preliminary report of the effect of photoperiod and temperature on the flowering and growth of several southern grasses. *Agron. J.* 45:268–269.

Laird, A.S. 1930. A Study of the Root Systems of Some Important Sod-Forming Grasses. Bull. 211, Univ. Florida Agric. Exp. Stn., Gainesville.

Leukel, W.A. and R.M. Barnette. 1935. Cutting Experiments with Bahia Grass Grown in Lysimeters. Bull. 286, Univ. Florida Agric. Exp. Stn., Gainesville.

Levitt, J. 1980. *Responses of Plants to Environmental Stresses*. Vol 2: Water, Radiation, Salt, and Other Stresses. 2nd ed. Academic Press, New York.

Marousky, F.J. and F. Blondon. 1995. Red and far-red light influence carbon partitioning, growth and flowering of bahiagrass (*Paspalum notatum*). *J. Agric. Sci.* 125:355–359

Marousky, F.J. and S.H. West. 1988. Germination of bahiagrass in response to temperature and scarification. *J. Am. Soc. Hort. Sci.* 113:845–849.

Marousky, F.J. and S.H. West. 1990. Somatic embryogenesis and plant regeneration from cultured mature caryopses of bahiagrass (*Paspalum notatum* Flügge). *Plant Cell Tissue and Organ Culture* 20:125–130.

Miller, G.L. and L.B. McCarty. 2001. Water relations and rooting characteristics of three *Stenotaphrum* turf cultivars grown under water deficit conditions. *Int. Turfgrass Soc. Res. J.* 9:323–327.

Mills, P.F.L. and J.N. Boultwood. 1978. A comparison of *Paspalum notatum* accessions for yield and palatability. *Zimbabwe Agric. J.* 75:71–74.

Moraes Fernandes, M.I.B. de, I.L. Barreto, and F.M. Salzano. 1973. Cytogenetic, ecologic and morphologic studies in Brazilian forms of *Paspalum notatum*. *Can. J. Genet. Cytol.* 15:523–531.

Ogata, S., K. Kouno, and T. Ando. 1980. Studies on establishments of grasses: 1. Effects of soil temperature on establishments of oversown grasses into nontilled sowing. *J. Japanese Soc. Grassland Sci.* 26:59–66.

Parodi, L.R. 1948. Gramíneas Argentinas nuevas o críticas: 1. La variación en *Paspalum* notatum Fluegge. *Rev. Argent. Agron.* 15:53–61.

Pedreira, C.G.S. and R.H. Brown. 1996. Physiology, morphology, and growth of individual plants of selected and unselected bahiagrass populations. *Crop Sci.* 36:138–142.

Pozzobon, M.T. and J.F.M. Valls. 1997. Chromosome number in germplasm accessions of *Paspalum notatum* (Gramineae). *Brazilian J. of Genet.* 20:29–34.

Quarin, C.L. 1999. Effect of pollen source and pollen ploidy on endosperm formation and seed set in pseudogamous apomictic *Paspalum notatum*. *Sexual Plant Reprod.* 11:331–335.

Quarin, C.L. and B.L. Burson. 1983. Cytogenetic relations among *Paspalum notatum* var. *saurae*, *Paspalum pumilum*, *Paspalum indecorum* and *Paspalum vaginatum*. *Bot. Gaz.* 144:433–438.

Quarin, C.L., B.L. Burson, and G.W. Burton. 1984. Cytology of intra- and interspecific hybrids between two cytotypes of *Paspalum notatum* and *P. cromyorrhizon*. *Bot. Gaz.* 145:420–426.

Quarin, C.L., F. Espinoza, E.J. Martinez, S.C. Pessino, and O.A. Bovo. 2001. A rise of ploidy level induces the expression of apomixis in *Paspalum notatum*. *Sexual Plant Reprod.* 13:243–249.

Reinert, J.A. and P. Busey. 1984. Resistant varieties. pp. 35–40. In R.I. Sailer, J.A. Reinert, D. Boucias, P. Busey, R.L. Kepner, T.G. Forrest, W.G. Hudson, and T.J. Walker, Eds. Mole Crickets in Florida. Bull. 846, Univ. Florida Agric. Exp. Sta. Gainesville.

Rodríguez, K.R., D.B. Weaver, D.G. Robertson, E.L. Carden, and M.L. Pegues. 1991. Additional studies on the use of bahiagrass for the management of root-knot and cyst nematodes in soybean. *Nematropica* 21:203–210.

Sakai, K. 1983. New summer crop cultivars (I)—New cultivars registered in Ministry of Agriculture, Forestry and Fisheries in 1983 fiscal year. *Japan. J. Breed.* 33:499–506.

Scott, J.M. 1920. Bahia grass. *J. Amer. Soc. Agron.* 12:112–113.

Silveus, W.A. 1942. *Classification and Description of Species of Paspalum and Panicum in the United States.* W.A. Silveus, San Antonio, TX.

Smith, A.E. 1983. Differential bahiagrass (*Paspalum notatum*) cultivar response to atrazine. *Weed Sci.* 31:88–92.

Sveda, R., J.E. Rechcigl, and P. Nkedi-Kizza. 1992. Evaluation of various nitrogen sources and rates on nitrogen movement, Pensacola bahiagrass production, and water quality. *Comm. Soil Sci. Plant Analysis* 23:879–905.

Tischler, C.R. and B.L. Burson. 1995. Evaluating different bahiagrass cytotypes for heat tolerance and leaf epicuticular wax content. *Euphytica* 84:229–235.

USDA-ARS, National Genetic Resources Program. 2001. Germplasm Resources Information Network - (GRIN). [Online Database] National Germplasm Resources Laboratory, Beltsville, Maryland. Available at: www.ars-grin.gov/cgi-bin/npgs/html/stats/genussite .pl?Paspalum (09 July 2001)

Vietor, D.M. 1982. Variation of ^{14}C-labeled photosynthate recovery from roots and rooting media of warm season grasses. *Crop Sci.* 22:362–366.

Webster, R.D. 1988. Genera of the North American Paniceae (Poaceae: Panicoideae). *Systematic Bot.* 13:576–609.

Werner, B.K. and G.W. Burton. 1991. Recurrent restricted phenotypic selection for yield alters morphology and yield of Pensacola bahiagrass. *Crop Sci.* 31:48–50.

West, S.H. and F. Marousky. 1989. Mechanism of dormancy in Pensacola bahiagrass. *Crop Sci.* 29:787–791.

INDEX

A
'AberElf' perennial ryegrass, 94
'AberImp' perennial ryegrass, 94
'AberNile' perennial ryegrass, 95
Aberrants, 32, 33
'Aberstwyth S.23' perennial ryegrass, 81
Abiotic stresses, 7–8, 183
Acid phosphatase (APT) isozyme banding patterns, 275
Acid rain, 10
Acid soil tolerance
  Chewings fescue, 150
  colonial bentgrass, 188
  fine-leaved *Festuca*, 130
  hairgrasses, 225
  Kentucky bluegrass, 29
  tall fescue, 121
'Acme' velvet bentgrass, 203
Acremonium coenophialum endophyte, 120
'Adalayd' seashore paspalum, 296, 303
Adaptation, 6
  annual bluegrass, 43–45
  buffalograss, 257–258
  perennial ryegrass, 75–76
  rough bluegrass, 68–69
  St. Augustinegrass, 316–319
  seashore paspalum, 295–299
'Adelphi' Kentucky bluegrass, 33
Adenosine diphospate (ADP) glucose pyrophosphorylase, 315
Advanta Seeds, 132
'Adventure' tall fescue, 113
AFLP. *see* Amplified Fragment Length Polymorphism
Africa, 17
African bermudagrass *(Cyanodon transvaalensis)*, 238, 240–242, 245–249
Agricultural pest populations, 11
Agrobacterium tumefaciens-mediated gene transfer, 116
Agróóstide comúún, 191
Agrostide commune, 191
Agrostide ténue, 191
Agrostidinae, 202, 226
Agrostis canina. *see* Velvet bentgrass
Agrostis capillaris. *see* Colonial bentgrass
Agrostis castellana. *see* Highland bentgrass
Agrostis gigantea. *see* Redtop
Agrostis idahoensis. *see* Idaho bentgrass

Agrostis spp. *see* Bentgrasses
Agrostis stolonifera. *see* Creeping bentgrass
Agrostis tenuis, 190
Airfields
  buffalograss, 260
  red fescue, 141
Air pollution, 10
Air temperatures, 11
'A-20' Kentucky bluegrass, 28
Alaska Agricultural and Forestry Experiment Station, 229
Alden Pines golf course, 304
Alfalfa, 14, 243
Alkali grasses, 17
Alkali tolerance, 260
Allometric development, 12
Allopolyploids, 16, 30
Alpine regions, 17–18
'Alta' tall fescue, 107, 113
Aluminum
  fine-leaved *Festuca*, 131
  tall fescue, 121
  tufted hairgrass, 228
Alvamar CC, 278
'America' Kentucky bluegrass, 33, 34
American bison, 257
American Society of Agronomy, 107
Ampac Seeds, 151
Amphidiploids, 16
Amplified Fragment Length Polymorphism (AFLP), 63
  creeping bentgrass, 176
  perennial ryegrass, 87
  Texas bluegrass, 65
Anderson, Mel, 278
Andropogoneae, 288
Annual bluegrass *(Poa annua)*, 39–48
  characteristics, 40–41
  cultivated variety developmental limitations, 46–48
  cytotaxonomy, 41–42, 55–56
  distribution, 41
  environmental growth limitation, 42–43
  evolution, 43–45
  flowering habit of, 47–48
  perennial cultivars of, 45–48
  perennial *vs.* annual version of, 40–41
  reproduction, 43
  supina bluegrass *vs.*, 54–55
  University of Minnesota breeding program for, 48

weed *vs.* turf, 39
Annual meadow-grass, 39
Annual ryegrass. *see* Italian ryegrass
Anthesis date/time
  bahiagrass, 342
  St. Augustinegrass, 320
  tall fescue, 112
  velvet bentgrass, 204
Anthracnose *(Colletotrichum graminicola)*
  annual bluegrass, 42
  Festuca ovina, 156
  fine-leaved *Festuca*, 131
'Apache II' tall fescue, 112
'Apache' tall fescue, 113
Aphids, 76
Apomictic reproductive system, 13
  bahiagrass, 339–340
  Kentucky bluegrass, 28–34
  role and rule of, 31–32
'AP14' seashore paspalum, 303
APT (acid phosphatase) isozyme banding patterns, 275
Arabidopsis thalliana, 94
'Arctared' strong creeping red fescue, 146, 153
'Arena' perennial ryegrass, 95
'Argentine' bahiagrass, 333, 335, 336, 339, 340, 344
Argentine stem weevil *(Listronotus bonariensis)*, 76, 95
'Arid' tall fescue, 113, 122
Arlington Turf Gardens, 178, 201
Armyworm, 132. *See also* Fall armyworm
Arsenate, 10, 212
Arsenic, 228
Arthropods, 266
Artificial wear machines, 96
Asano, Yoshito, 219
Asia, 17
Asian supina bluegrass *(Poa supina ustulata)*, 55
'Astoria' colonial bentgrass, 191
Athletic fields. *see* Sports turfs
'Atlanta' bermudagrass, 244
Atmosphere, 10
'AU Centennial' centipedegrass, 291
'Aurora' hard fescue, 162
Australian National Herbarium, 275
Aveneae, 176, 190
Avoidance mechanisms, 10
Avoidance strategy, drought, 108

'Azay' sheep fescue, 161
Azores, 9

B
Back-cross method, 147–148
Bahiagrass (*Paspalum notatum* Flügge), 331–344
  apomixis and polyploidy, 339–340
  biology, 331–337
  breeding and selection techniques, 341–342
  characteristics, 331–337
  collection, 341
  distribution, 337–339
  distribution/cytotaxonomy/genetics, 337–341
  forage breeding, 342–343
  germ plasm sources, 341
  origin and relatives, 337–339
  and St. Augustinegrass, 311
  seed establishment, 336
  taxonomy of cultivars and natural populations, 340–341
  turf breeding, 343–344
Ballast, 295
Ball games, 9
Banding patterns
  APT isozyme, 275
  colonial bentgrass, 194
  esterase (*see* Esterase banding patterns)
  peroxidase isozyme, 277
'Banner' Chewings fescue, 151
'Banner II' fine-leaved *Festuca*, 132
'Barcampsia' hairgrass, 229
'Barcrown' slender creeping red fescue, 152
'Bardot' colonial bentgrass, 192, 194
Barenbrug, Inc., 132
'Barfalla' Chewings fescue, 150
'Barlexus' tall fescue, 115
'Barok' hair fescue, 158
'Barracuda' redtop, 217
Basal internodes, 8
Bateson, William, 86
'B-1' bahiagrass, 344
Bedding (slave boats), 295
'Bela' redtop, 217
Belonolaimum longicaudatus. see Sting nematode
Bentgrasses (*Agrostis* spp.), 7, 16, 176, 190, 202, 207–221
  and biotechnology, 219
  colonial bentgrass (*see* Colonial bentgrass)
  comparison of species, 209, 214

creeping bentgrass (*see* Creeping bentgrass)
cytotaxonomy, 216–217
distribution, 213–216
genetic improvement, 217–221
Highland bentgrass (*see* Highland bentgrass)
Idaho bentgrass (*see* Idaho bentgrass)
redtop (*see* Redtop)
unexploited species and hybrids, 219–221
velvet bentgrass (*see* Velvet bentgrass)
Bermudagrass (*Cynodon dactylon*), 8, 11, 235–251
  breeding, 242–247
  breeding progress, 250, 251
  chromosome numbers, 239–240
  chromosome pairing and karyotype, 240–241
  classification of, 237
  cultivar composition and breeding techniques, 245–247
  cytogenetic and reproductive characteristics, 239–242
  dactylon x *C. transvaalensis* Burtt-Davy variant, 16
  distribution, 236–239
  early germ plasm dispersal, 242–243
  early turf development, 243–245
  flowering characteristics, 247
  germ plasm sources, 242–243
  hybridization potential, 241
  morphology, 248
  organized breeding of turf, 245
  and redtop, 217
  reproduction, 241–242
  and rough bluegrass, 70
  selection criteria, 248–250
  taxonomy, 236–239
Bermudagrass decline (*Gaeumannomyces graminis*), 249
Berner, P.H., 57
Bessey, C.E., 258
Big Flats Plant Material Center, 218
'Bighorn,' 160
B$_{III}$ hybrids, 33
'Biljart' hard fescue, 162
Billbug (*Spenophorus* spp.)
  and endophytes, 120
  fine-leaved *Festuca*, 132
  hairgrasses, 225
  perennial ryegrass, 76
  Zoysiagrasses, 279

Biomass, 96
Bioremediation of disturbed lands. *see* Reclamation areas
Biotechnology, 219
Biotic stresses, 6–7, 317–319
Bipolaris cynodontis, 249
Bipolaris sorokiniana (leaf spot disease)
  Kentucky bluegrass, 30
  rough bluegrass, 69
Bison, American, 257
Bitterblue Group (St. Augustinegrass), 318
'Bitterblue' St. Augustinegrass, 315, 316
Black beetle (*Heteronychus arator*), 95
Black stem rust (*Puccinia graminis*)
  perennial ryegrass, 93
  red fescue, 147
Blissus insularis. see Southern chinch bug
Blissus leucopterus, 157
Blissus leucopterus hirtus. see Hairy chinch bug
Blissus occiduus, 266
Blissus spp. *see* Chinch bug
Bluebunch fescue, 158
Blue fescue (*Festuca glauca*), 132
Blue grama grass (*Bouteloua gracilis*), 257–259, 261
Bluegrass billbugs (*Sphenophorus parvulus* Gyllenhal), 132
Bluegrasses (*Poa* spp.), 16
  annual (*see* Annual bluegrass)
  rough (*see* Rough bluegrass)
  supina (*see* Supina bluegrass)
  Texas, See Texas bluegrass
Bluestems, 258
'Bonanza' tall fescue, 113, 121
'Bonsai' tall fescue, 113, 123
'Boral' colonial bentgrass, 191
'Boreal' strong creeping red fescue, 153
'Borfesta' red fescue, 146
Boulevards, 17
Bowling, 9
Bowling greens
  annual bluegrass, 40–41
  red fescue, 141
  seashore paspalum, 296
'Bradley' bermudagrass, 244
Breeding
  annual bluegrass, 45–48
  bahiagrass, 341–344
  bermudagrass, 242–247
  buffalograss, 266–268
  centipedegrass, 290–292, 291
  colonial bentgrass, 191–197
  creeping bentgrass, 178–180

false sheep fescue, 163–164
Festuca ovina, 155–157, 156–157
fine-leaved *Festuca*, 132–137
hairgrasses, 229–230
hard fescue, 161–162
Highland bentgrass, 218
Idaho bentgrass, 218–219
Kentucky bluegrass, 31–36
perennial ryegrass, 79–81, 84–89
red fescue, 142–148
red fescue complex, 141–148
redtop, 217–218
seashore paspalum, 301–304
St. Augustinegrass, 319–325, 320–321, 322–325
supina bluegrass, 57
tall fescue, 114–115, 122–123
Texas bluegrass, 63–65
velvet bentgrass, 203–205
Zoysiagrasses, 278–283
Breviflorus Race (St. Augustinegrass), 313–314, 316, 319
'Brightstar' perennial ryegrass, 90
Brown bentgrass, 201
Brown patch *(Rhizoctonia solani)*, 63
 bermudagrass, 249
 colonial bentgrass, 192
 comparison of resistance to, 215
 creeping bentgrass, 175, 181
 description of, 117
 Festuca ovina, 156
 hard fescue, 162, 163
 Highland bentgrass, 212
 perennial ryegrass, 92
 rough bluegrass, 69
 St. Augustinegrass, 318
 tall fescue, 109, 113, 115, 117–119, 122–123
 Texas bluegrass, 65
 Zoysiagrasses, 279, 282
Browntop, 187, 191, 210
Buffalo, 257
Buffalograss *(Buchloe dactyloides* (Nutt.) Engelm), 17, 257–269
 adaptation and distribution, 257–258
 breeding and selection, 266–268
 cultivars of, 268–269
 distribution, 257–258
 drought resistance, 259
 and erosion control, 258–259
 establishment methods, 263–265
 improvements, 266–269

management, 265
 morphology, 261–262
 potential pest problems with, 265–266
 sex expression, 262–263
 taxonomy, 259
 turf-type, 260–261
 uses, 260–266
Buffalograss mealybug, 266
Buffalo grass (St. Augustinegrass), 309
Buffalograss webworm, 266
Bulletins of the Florida Agricultural Experiment Station, 338
Burning, 28
Burs, 261
Burton, Glenn, 245, 342
Buru-bubu, 271

C
Cabeza de burro, 338
Cadmium, 10, 131
Canada, 18
Canada bluegrass *(Poa compressa* L.), 33
Canada Department of Agriculture Research Station, 153
Cañuela de oveja, 159
'Career' sheep fescue, 160
Carnegie Institute of Washington, 229
Carpetgrass *(Axonopus affinis* Chase), 17
'Cascade' Chewings fescue, 150
Catalogue of New World Grasses, 226, 227
Cattail disease. *see* Choke disease
'Cavalier' Zoysiagrasses, 279, 283
Cebeco International, 132, 195
Cell suspension protoplasts, 116
Cemeteries
 bermudagrass, 235
 buffalograss, 260
Centipedegrass *(Eremochloa ophiuroides)*, 287–292
 breeding techniques, 291
 distribution, 288–289
 genetic markers, 292
 germ plasm sources, 290–291
 selection criteria, 291–292
 taxonomy, 289–290
Centipedegrass mosaic virus, 292
Central Park (Manhattan), 11
Cercospora seminalis. see False smut
'Cezanne' slender creeping red fescue, 152
Charleston (St. Augustinegrass), 309

Charlotte Country Club, 245
Chase, Agnes, 321
'Checker Chewings fescue, 150
Chepica, 298
Chewings, George, 149
Chewings fescue *(Festuca rubra* ssp. *commutata)*, 130, 131, 132, 147, 148–151
 cytology, 141–142
 taxonomy, 136–141
Chiba University, 219
China, 9
Chinch bug *(Blissus* spp.). *See also* Hairy chinch bug; Southern chinch bug
 buffalograss, 261
 fine-leaved *Festuca*, 132
 perennial ryegrass, 76
 St. Augustinegrass, 318–319
Chloridoideae, 274
Chlorophyll catabolism, 94
Choke disease *(Epichloe typhina)*, 132
Chorology, 299
Chromatography, thin layer, 139
Chromosome numbers
 Agrostis, 216, 220
 annual bluegrass, 42
 bermudagrass, 239–240
 buffalograss, 259
 centipedegrass, 290
 colonial bentgrass, 214
 creeping bentgrass, 214
 Festuca ovina, 154
 hairgrasses, 226, 227
 Highland bentgrass, 214
 Idaho bentgrass, 214
 Kentucky bluegrass, 30
 red fescue complex, 136, 138, 141–142
 redtop, 214
 rough bluegrass, 68
 St. Augustinegrass, 315
 seashore paspalum, 300
 Texas bluegrass, 61
 Zoysiagrasses, 275, 276
Chromosome pairing, 240–241
'Cimarron' tall fescue, 113
'Clatsop' red fescue, 141
Clay soils
 supina bluegrass, 54
 Texas bluegrass, 61
Cleistogamy, 43
Clemson University, 122
Climate, 10
 changes in, 6, 11
 and colonial bentgrass, 188–189
 perennial ryegrass, 75–76, 93
 supina bluegrass, 54
 tall fescue, 108–109

Climatic stress tolerances, 15
Closely-grazed sods, 8
Clumpiness, 282
CO_2, 11
Coastal environmental stabilization, 296
Coastal Plains Experiment Station, 245
Cobalt
  redtop, 210
  tufted hairgrass, 228
Coconuts, 311
Code of Botanical Nomenclature, 207–208
Cold seed storage, 56
Cold tolerance, 7, 14, 16
  centipedegrass, 287–288, 292
  creeping bentgrass, 175
  Festuca ovina, 156
  fine-leaved *Festuca*, 130
  supina bluegrass, 53, 54
  velvet bentgrass, 201
  Zoysiagrasses, 282
Collection
  bahiagrass, 341
  centipedegrass, 290–292
  colonial bentgrass, 191–197
  fine-leaved *Festuca*, 133
  hairgrasses, 229–230
  Kentucky bluegrass, 30–31
  perennial ryegrass, 81–84
  seashore paspalum, 301
  supina bluegrass, 57–58
  Texas bluegrass, 63
  velvet bentgrass, 203
  Zoysiagrasses, 278–283
Colletotrichum graminicola. see Anthracnose
Colonial bentgrass *(Agrostis capillaris)*, 11, 187–197, 209
  botanical and physiological descriptions, 187–190
  characteristics, 214
  climatic and edaphic limitations and flexibility, 188–189
  collection/selection/breeding history, 191–197
  cultivar developmental progress, 192, 194–197
  cytology and cytogenetics, 191
  distribution and cytotaxonomy, 190–191
  germ plasm sources, 191–194
  morphology, 189–190
  origin and natural distribution, 190
  and red fescue, 141
  taxonomy, 190–191
  worldwide use and management, 187–188

Color, 13
  bermudagrass, 248
  centipedegrass, 287, 290
  colonial bentgrass, 189, 191
  comparison of, 215
  Festuca ovina, 130
  hard fescue, 162
  perennial ryegrass, 77, 90, 91, 94, 96
  rough bluegrass, 71
  St. Augustinegrass, 322
  supina bluegrass, 57
  Zoysiagrasses, 282
Colorado State University, 64
'Colt' rough bluegrass, 71
Columbus, Christopher, 243
Commercial sod farming, 9
'Common' bahiagrass, 339
Common bentgrass, 191
'Common' centipedegrass, 291
Common Kentucky bluegrass, 28
Competitive ability, 16
Competitive experiences, VII
Complex polyploids, 15
Conflicts-of-interest, 14
Consistency, 8
Cool-season grasses, 11, 16, 17, 25–231
  Agrostis species, minor, 207–221
  annual bluegrass, 39–48
  colonial bentgrass, 187–197
  creeping bentgrass, 175–183
  fine-leaved *Festuca* species, 129–164
  hairgrasses, 225–230
  Kentucky bluegrass, 27–36
  perennial ryegrass, 75–99
  rough bluegrass, 67–72
  supina bluegrass, 53–58
  tall fescue, 107–123
  Texas bluegrass, 61–65
  velvet bentgrass, 201–205
Copper, 10
  Highland bentgrass, 212
  redtop, 210
  tufted hairgrass, 228
Copper spot *(Gloeocercospora sorghi)*, 205
Coquiole, 159
'Coronado Gold' tall fescue, 118, 120
'Coronado' tall fescue, 114–115
Cosmetically appealing, VII
Couchgrass, 235
Couch paspalum, 298
'Countess' Chewings fescue, 150
'Count' slender creeping red fescue, 152
'Covar' false sheep fescue, 163
'Covar' *Festuca ovina*, 154

Crambus spp. (sod webworm), 76
Creeping bentgrass *(Agrostis palustris)*, 10, 45
Creeping bentgrass *(Agrostis stolonifera)*, 175–183, 209
  breeding and cultivar development, 178–180
  characteristics, 214
  cytotaxonomy, 176–177
  disease resistance, 181–183
  distribution, 176
  flowering requirements, 180
  and gene escape, 179–180
  germ plasm sources, 178–179
  inheritance mode, 177
  management, 214
  marker-assisted selection, 179
  photoperiod sensitivity and genetic variation for maturity, 180–181
  plant transformation potentials, 179
  selection criteria, 181–183
Creeping bentgrass breeding programs, 11
Creeping fescue *(Festuca rubra* ssp. *tricophylla)*, 136
'Crenshaw' creeping bentgrass, 178, 181
Crickets, mole, 343
Cricket fields
  Chewings fescue, 149
  perennial ryegrass, 75, 76
Cropping systems, healthy, 11
Crop Science, 82
Cross-compatibility, 6, 275
Cross-fertilization, 32
Cross-pollination
  annual bluegrass, 44
  rough bluegrass, 68
  seashore paspalum, 301–302
'Crowne' Zoysiagrasses, 279
Crown height, 7
Crown rot, 249
Crown rust *(Puccinia coronata* Corda)
  perennial ryegrass, 88, 93, 95
  tall fescue, 107
C-series creeping bentgrasses, 178
Cultivars (defined), 136
Cultivars Eligible for Certification (OECD), 141
Cutting height. *see* Mowing height
Cyanodon arcuatus, 238–241
Cyanodon barberi, 239–241
Cyanodon dactylon var. *polevansii*, 239–241
Cyanodon incompletus, 238, 240, 241

INDEX

Cyanodon transvaalensis. see African bermudagrass
Cyanodon X magennisii, 238, 241
Cyctocephala, 64
Cynodon, classification of, 237
Cynodon dactylon. see Bermudagrass
'Cypress' rough bluegrass, 71
Cytogenetics
  bermudagrass, 239–242
  colonial bentgrass, 191
  Festuca ovina, 155–157
  perennial ryegrass, 78–79
  red fescue complex, 141–148
Cytology
  Festuca ovina, 155–156
  red fescue complex, 141–142
Cytotaxonomy
  Agrostis, 216–217
  annual bluegrass, 41–42, 55–56
  bahiagrass, 337–341
  colonial bentgrass, 190–191
  creeping bentgrass, 176–177
  Festuca ovina, 155–157
  hairgrasses, 226–229
  Kentucky bluegrass, 29–30
  perennial ryegrass, 77–81
  red fescue complex, 141–148
  rough bluegrass, 68
  St. Augustinegrass, 313–315
  seashore paspalum, 300–301
  supina bluegrass, 55–57
  tall fescue, 109–110
  Texas bluegrass, 61–63
  velvet bentgrass, 202–203
  Zoysiagrasses, 275, 276

D

Dactylis glomerata L. see Orchardgrass
Danish Common, 70
'Darbysire' fine-leaved *Festuca*, 129
'Darius' perennial ryegrass, 94
'Darkhorse' rough bluegrass, 71
'Dasas' rough bluegrass, 70
'Dawson (E)' slender creeping red fescue, 152
'Dawson' slender creeping red fescue, 151, 152
Daylengths, 47, 65
'DeAnza' Zoysiagrasses, 278
Decomposing tissue, 6
Defoliation, 8
DeFrance, J.A., 191, 204
'DelMar' St. Augustinegrass, 317, 323
Density, 13, 118–119
Density-dependent environments, 44–45
Density-independent environments, 44
Deoxyribose nucleic acid (DNA), 7
Deschampsia caespitosa. see Tufted hairgrass
Deschampsia elongata (slender hairgrass), 229
Deschampsia flexuosa (wavy hairgrass), 226, 227
Deschampsia setacea, 226, 227
Deschampsia spp. see Hairgrasses
Dessication, 99
Deutsche Saatveredlung Lippstadt-Bremen GmbH, 132, 163
Diamond, Homer, 338
'Diamond' Zoysiagrasses, 279, 282
Diazotrophic colonization, 331
Differential grasshopper *(Melanoplus differentialis)*, 279
Differential thermal analysis (DTA), 317
Dinitrogen fixation, 331
Direct gene transfer, 116
'Discovery' hard fescue, 163
Disease resistance
  annual bluegrass, 42
  creeping bentgrass, 181–183
  Festuca ovina, 156
  fine-leaved *Festuca*, 131, 132
  Kentucky bluegrass, 29
  perennial ryegrass, 88, 92, 97–98
  Texas bluegrass, 64
Dispersion, 295–298
Distribution (of grasses)
  Agrostis, 213–216
  annual bluegrass, 41
  bahiagrass, 337–341
  bermudagrass, 236–239
  buffalograss, 257–258
  centipedegrass, 288–289
  Chewings fescue, 149
  colonial bentgrass, 190
  creeping bentgrass, 176
  false sheep fescue, 163
  fine-leaved *Festuca* species, 129–130
  hair fescue, 157–158
  hairgrasses, 226–229
  hard fescue, 161
  Highland bentgrass, 215
  Idaho bentgrass, 215–216
  Idaho fescue, 158–159
  Kentucky bluegrass, 29
  perennial ryegrass, 77
  red fescue complex, 139–140
  redtop, 213, 215
  rough bluegrass, 68–69
  St. Augustinegrass, 312–313
  shade fescue, 148
  sheep fescue, 159
  slender creeping red fescue, 151
  strong creeping red fescue, 152
  supina bluegrass, 55
  tall fescue, 107–108, 122
  Texas bluegrass, 61
  velvet bentgrass, 202
  Zoysiagrasses, 271–273
DNA amplification fingerprinting, 291
DNA (deoxyribose nucleic acid), 7
Dollar spot *(Sclerotinia homeocarpa)*, 11
  annual bluegrass, 42
  bermudagrass, 249
  colonial bentgrass, 192
  creeping bentgrass, 175, 181
  Festuca ovina, 156, 157
  fine-leaved *Festuca*, 131, 132
  hard fescue, 162, 163
  Highland bentgrass, 212
  Idaho bentgrass, 213
  perennial ryegrass, 92
  red fescue, 147
  rough bluegrass, 69, 71
  sheep fescue, 156
  slender creeping red fescue, 152
Domestication
  of livestock, 8
  and natural variation, 11–12
  of turfgrasses, 11–12
Doobgrass, 235
Doubled haploids, 116
Downy mildew *(Sclerophthora macrospora)*
  colonial bentgrass, 192
  fine-leaved *Festuca*, 131
  St. Augustinegrass, 318, 319
Drechslera dictyoides
  hard fescue, 163
  sheep fescue, 156
Drechslera poae (leaf spot disease)
  comparison of, 215
  Festuca ovina, 156
  fine-leaved *Festuca*, 131
  Kentucky bluegrass, 30
  perennial ryegrass, 93
  rough bluegrass, 69
  slender creeping red fescue, 152
  strong creeping red fescue, 153–154
  tall fescue, 109, 122
Drought strategies (in grasses), 98, 108

Drought tolerance, 7, 14, 16
  bahiagrass, 331, 333–334
  bermudagrass, 236
  buffalograss, 258, 259
  centipedegrass, 292
  Chewings fescue, 149
  colonial bentgrass, 191
  and endophytes, 120
  hair fescue, 158
  Kentucky bluegrass, 28, 29
  perennial ryegrass, 88, 93, 98–99
  St. Augustinegrass, 311, 317
  seashore paspalum, 295
  sheep fescue, 160
  supina bluegrass, 57
  tall fescue, 107, 123
  Texas bluegrass, 63, 65
  velvet bentgrass, 201
  Zoysiagrasses, 282
Dryland bent, 210
Dryland bentgrass, 187, 191
Dryland conditions, 17
DTA (differential thermal analysis), 317
'Duchess' Chewings fescue, 150
'Duchess' colonial bentgrass, 192, 194, 197
Dudeck, Albert E., 344
Duplicate-gene asynchrony model, 35
'Durar' hard fescue, 161–162
'Duraturf' red fescue, 141
Dwarf Group (St. Augustinegrass), 313, 314, 315, 319
Dwarf-type cultivars, 113, 119

E

Early semidwarf tall fescue, 119
Early standard tall fescue, 118
Eastern Oregon Branch Experiment Station, 161
East Lake Country Club, 244
'Eclipse' Kentucky bluegrass, 33
'Ecostar' hard fescue, 162
Edaphic adaptation
  colonial bentgrass, 188–189
  perennial ryegrass, 75–76
  supina bluegrass, 54
Edaphic stress tolerances, 15
'Egmont' colonial bentgrass, 192
Egypt, 9
'18th Green' cb, 181
'Eldorado' tall fescue, 113
Electrophoresis gels, 133, 135, 155
Electroporation, 116, 179
'Elfin' slender creeping red fescue, 152
'Elizabeth' vb, 203

'Elka' perennial ryegrass, 82
Ellis, Henry, 243
'El Toro' Zoysiagrasses, 278
'Emerald' creeping bentgrass, 181
'Emerald' Zoysiagrasses, 277, 283
'Endeavor' tall fescue, 112
Endophytes
  Acremonium coenophialum, 120
  Chewings fescue, 150–151
  Festuca ovina, 157
  fine-leaved *Festuca*, 131–132
  red fescue, 147–148
  redtop, 210
  tall fescue, 119–122
'Endurance' perennial ryegrass, 95
Endurance strategy, drought, 98
Engelke, Milt, 278, 319
'Ensylva' strong creeping red fescue, 153
Environment, human effects on, 10–11
Environmental adaptation
  fine-leaved *Festuca*, 130–132
  St. Augustinegrass, 310–311
Environmental limitations, 6, 14
  annual bluegrass, 42–43
  Festuca ovina, 155
  red fescue complex, 140–141
  St. Augustinegrass, 316–318
Environmental Turf Solutions, 304
Epichloe spp.
  fine-leaved *Festuca*, 131, 132
  red fescue, 147
Epidermal ridging, 7
Eradication, invasive species, 17
Eremochloa, 288
Eremochloa ophiuroides. see Centipedegrass
Ergotism, 335
Eriophyid mite, zoysiagrass, 279
Eriophyte slykhuisi, 266
Erisyphe graminis
  Chewings fescue, 151
  Festuca ovina, 156
Erosion control
  bahiagrass, 331
  buffalograss, 258
  and buffalograss, 258–259
  colonial bentgrass, 187
  fine-leaved *Festuca*, 130
  perennial ryegrass, 82–84
  seashore paspalum, 296
Erysiphe graminis. see Powdery mildew
Escape strategies, drought, 98
Establishment capacity, 17
Establishment methods, 263–265

Estacion Experimental Agropecuaria, 301
Esterase banding patterns
  fine-leaved *Festuca*, 135
  red fescue complex, 139
  Zoysiagrasses, 275
ESTs (Expressed Sequence Tags), 87
Eternity grass, 298
EuroAsia, 29
Europe, 9, 17
  annual bluegrass in, 39
  Kentucky bluegrass origins in, 28
  perennial ryegrass in, 93–95
  shade fescue, 148
European Central Crop Databases, 83
European colonialization, 10
  colonial bentgrass, 187
  tall fescue, 107
European Community Common Catalogue, 95
European Core Collection, 82
European Union, 140–141
Evaluation criteria, 14
Evaluation trials, 34
Evapotransporation rate, 259
  bahiagrass, 333
  St. Augustinegrass, 311, 317
Everglades Experiment Station, 315
'Excalibur' seashore paspalum, 303
'Exeter' colonial bentgrass, 191, 194
Expressed Sequence Tags (ESTs), 87

F

Facultative apomict, 30
'Fairway' crested wheatgrass [*Agropyron cristatum* (L.) Gaertn.], 17
Fairways (golf courses)
  bermudagrass, 235
  colonial bentgrass, 187
  creeping bentgrass, 178
  fine-leaved *Festuca*, 129
  Highland bentgrass, 211
  perennial ryegrass, 75, 76
  supina bluegrass, 54
  Zoysiagrasses, 283
'Falcon II' tall fescue, 118, 121, 122
'Falcon' tall fescue, 112, 113, 118
Fall armyworm (*Spodoptera frugiperda* J. E. Smith), 63
  bermudagrass, 250
  perennial ryegrass, 76
  Texas bluegrass, 64

Zoysiagrasses, 279
'False crowns,' 211, 212
False sheep fescue *(Festuca valesiaca),* 132, 137, 163–164
False smut *(Cercospora seminalis),* 265
'FA-118' St. Augustinegrass, 315
'Fawn' tall fescue, 118
'Fenway' strong creeping red fescue, 153
Fergus, E. N., 107
Fertility, soil. *see* Soil fertility
Fertilization, VII
  bahiagrass, 334
  St. Augustinegrass, 310
Fescue-ryegrass hybrids, 16
Fescues *(Festuca* spp.), 7, 16
  Chewings fescue *(see* Chewings fescue)
  and colonial bentgrass, 188
  false sheep fescue *(see* False sheep fescue)
  hair fescue, 157–158
  hard fescue, 161–163
  Idaho fescue *(see* Idaho fescue)
  meadow fescue *(see* Meadow fescue)
  and perennial ryegrass, 79
  red fescue complex *(see* Red fescue complex)
  sheep fescue *(see Festuca ovina* complex)
  tall fescue *(see* Tall fescue)
Festuca arundinacea. see Tall fescue
Festuca glauca (blue fescue), 132
Festuca idahoensis. see Idaho fescue
Festuca-ovelha, 159
Festuca ovina complex, 129–130, 154–164
  adaptations, 131
  breeding, 142–145, 156–157
  classifications, 134, 135
  cytology, 155–156
  false sheep fescue, 163–164
  hair fescue, 157–158
  hard fescue, 161–163
  hybidization, 145
  Idaho fescue, 158–159
  morphology, 137, 148
  and red fescue, 143
  salt tolerance of, 141
  sheep fescue, 159–161
  taxonomy, 154–155
  use/management/ environmental limitations, 155
Festuca pratensis. see Meadow fescue
Festuca pseudovina, 163

Festuca richardsonii, 139, 140
Festuca roemeri, 136, 139
Festuca rubra. see Red fescue complex
Festuca rubra ssp. *commutata. see* Chewings fescue
Festuca rubra ssp. *litoralis. see* Slender creeping red fescue
Festuca rubra ssp. *rubra. see* Spreading fescue
Festuca rubra ssp. *rubra* Gaudin. *see* Strong creeping red fescue
Festuca rubra ssp. *tricophylla. see* Creeping fescue
Festuca spp. *see* Fescues
Festuca tenuifolia, 157
Festuca trachyphylla, 132, 137–139, 156, 157
Festuca valesiaca. see False sheep fescue
Festuceae, 109
Festucoideae, 109
Fétuque des moutons, 159
Fétuque ovine, 159
Ficus, 326
'Fidalayel' seashore paspalum, 303
Fine bentgrass, 191
'Finelawn 5GL' tall fescue, 113
'Finelawn' tall fescue, 113
Fine-leaved *Festuca* species, 129–164
  breeding, 132–133
  classification of, 134
  distribution, 129–130
  environmental adaptation, 130–132
  Festuca ovina complex, 154–164
  morphology, 135
  red fescue complex, 136–154
  taxonomy, 133–135
Finer textured grasses, VII
'Finesse' tall fescue, 112
Finlayson, E. H., 338
'Fireball' redtop, 217, 218
Flax, 77
Flooding tolerance, 175
'Floralawn' St. Augustinegrass, 315, 317, 323
Flora Suecica (Linneaus), 207
'Floratam' St. Augustinegrass, 310, 315–320, 323, 326
FloraTex™ bermudagrass, 250
'Floratine' St. Augustinegrass, 315, 317, 323
'Florida' bermudagrass, 244
'Florida Common' St. Augustinegrass, 314
Flow cytometry, 34, 240
Flower induction
  colonial bentgrass, 195

Festuca ovina, 156
Highland bentgrass, 218
red fescue, 146
Flowering characteristics
  bermudagrass, 247
  centipedegrass, 290
  creeping bentgrass, 180
  supina bluegrass, 55
  Texas bluegrass, 65
Flowering control
  creeping bentgrass, 180
  tall fescue, 111–112
  Zoysiagrasses, 277–278
Flowering date/time, 8
  Agrostis, 216
  supina bluegrass, 55
'Fl-1997-6' St. Augustinegrass, 326
'Flyer II' strong creeping red fescue, 153
'Flyer' strong creeping red fescue, 153
Football, 9
Forage and fodder crops, VII
  bahiagrass, 338, 342–343
  bermudagrass, 243
  buffalograss, 257, 260, 267
  hairgrasses, 225
  Kentucky bluegrass, 27
  perennial ryegrass, 75, 84, 88
  seashore paspalum, 296
  shade fescue, 148
  tall fescue, 118
  Texas bluegrass, 64
Forage and Range Section (USDA), 107
Forage Breeding Center of the Italian Research Council, 195
Forquinha, 338
Fort Hays Branch Experiment Station, 259, 260, 263
'Fortress' strong creeping red fescue, 153
Foxes, 9
Frank, Paul, 326
Frankenwald Botanical Research Station, 244
Freezing tolerance, 7
  annual bluegrass, 42
  perennial ryegrass, 99
  St. Augustinegrass, 317
  supina bluegrass, 55
'Freja' redtop, 217
'FR-1' seashore paspalum, 303
Functionality, 14
Functional stress tolerance, 13–14
Funk, C. Reed, 31, 112, 113

Fusarium blight *(Fusarium roseum)*
  creeping bentgrass, 175
  fine-leaved *Festuca*, 131
  rough bluegrass, 69
  sheep fescue, 156
Fusarium nivale. see Pink snow mold
Fusarium roseum. see *Fusarium* blight
Fusarium tricinctum, 175
'Futurf' seashore paspalum, 296
'FX-10' St. Augustinegrass, 315, 317, 319

G

Gaeumannomyces graminis
  bermudagrass, 249
  rough bluegrass, 69
  St. Augustinegrass, 319–320
Gaeumannomyces incrustans (take-all patch), 131
Galvanized (Zn-coated) electricity pylons, 10
Games, ball, 9
Garden City Branch of the Kansas Experiment Station, 264
Garrett, Curran L., 322
Gebr. Van Engelen, 160
Gemeines Straußßgras, 191
Gene escape, 179–180
Gene mapping, 87–88, 116
'Genesis' tall fescue, 121
Genetics
  bahiagrass, 337–341
  centipedegrass, 288–290
  fine-leaved *Festuca*, 132–137
  St. Augustinegrass, 313–316
  supina bluegrass, 55–56
Genetic analysis, 302–303
Genetic diversity, 82–84
Genetic erosion, 83–84
Genetic markers, 292
Genetic redundancy, 16
Genetic Resources Unit, 83
Genetic transformation, 116
  creeping bentgrass, 179
  red fescue, 146
Genetic vulnerability, 121
Gengibrillo, 338
Genome size, 7
Genomic *in-situ* hybridization (GISH), 89
Genotypic recurrent selection, 13
Gentian violet, 264
Geographic adaptation, 68–69
Germination, 264, 302
Germ plasm exchange, 57

Germplasm Resource Information Network (GRIN), 133, 191, 217
Germ plasm sources
  bahiagrass, 341
  bermudagrass, 242–243
  centipedegrass, 290–291
  colonial bentgrass, 191–193
  creeping bentgrass, 178–179
  Festuca ovina, 157
  fine-leaved *Festuca*, 133
  Kentucky bluegrass, 35
  perennial ryegrass, 81–82
  red fescue, 146–147
  rough bluegrass, 70–72, 71
  seashore paspalum, 301
  St. Augustinegrass, 319–320
  tall fescue, 113–114
  Zoysiagrasses, 278
Gharib, 298
Gibeault, Vic, 278
GISH (genomic *in-situ* hybridization), 89
Global climate changes, 11
Gloeocercospora sorghi (copper spot), 205
Glucuronidase (GUS), 89
Goats, 8, 9
Golden Grass, 272
Golf, VII, 9, 10, 16
Golf courses. See also Fairways; Putting greens; Roughs; Tees
  buffalograss, 260, 268
  Chewings fescue, 149
  fine-leaved *Festuca*, 129
  rough bluegrass weed, 70
  seashore paspalum, 296
  supina bluegrass, 53
'GolfStar' Idaho bentgrass, 212, 215, 218–219
'Gosta' redtop, 217
Governmental testing agencies, 14
Grama, 298
Grama batatais, 338
Grama colorada, 298
Grama de aqua, 298
Grama dulce, 338
Gramilla, 298
Gramilla blanca, 298, 338
Gramineae, 40, 55, 77, 176, 274, 288, 300
Gramon, 298
Grasshoppers, 266, 279
Grasslands, North American, 5
'Grasslands Cook' Chewings fescue, 150
'Grasslands Coronet' perennial ryegrass, 95
'Grasslands Tasman' Chewings fescue, 150

'Grasslands Trophy' perennial ryegrass, 95
Gray leaf spot *(Pyricularia grisea)*
  perennial ryegrass, 92
  St. Augustinegrass, 318, 320
  tall fescue, 114, 121, 123
Gray snow mold *(Typhula incarnata)*
  creeping bentgrass, 175, 182
  rough bluegrass, 69
Grazing, 8, 17
  buffalograss, 257
  Festuca ovina, 155
  sheep fescue, 160
Great Plains, 257, 259, 260
Greece, 9
Green couchgrass, 235
Greenhouse crossing technique, 31
Greenhouse screening, 146
Greens
  bowling, 9
  putting (see Putting greens)
Greenspace, VII
GRIN germ plasm repository, 63
Groffe doeba, 298
Groundwater, 10, 310
Growth habit, 8
Grubs, soil-inhabiting, 123
Gulf Coast Group (St. Augustinegrass), 313, 314, 315
GUS (glucuronidase), 89

H

Haekrott, H. A., 245
Hair fescue, 157–158
Hairgrasses *(Deschampsia* P. Beauv.), 225–230
  collection/selection/breeding history, 229–230
  distribution and cytotaxonomy, 226–229
Hairy chinch bug *(Blissus leucopterus hirtus)*
  fine-leaved *Festuca*, 131
  red fescue, 147
Hall, D. Lester, 244
Hanna, W. W., 245
Hard fescue, 161–163
Hay harvesting, 8
'Hays' buffalograss, 260
'Hays' treatment, 264
Healthy cropping systems, 11
Heat tolerance, 14
  annual bluegrass, 42
  bahiagrass, 334
  bermudagrass, 236
  buffalograss, 259
  Chewings fescue, 151
  and endophytes, 120

fine-leaved *Festuca,* 130
Kentucky bluegrass, 29
Texas bluegrass, 63
velvet bentgrass, 201
Heat treatment, 85
Heavy metals tolerance, 10
Heavy metal tolerance, 10–11, 17. *See also specific metals*
fine-leaved *Festuca,* 131
hairgrasses, 225
Idaho bentgrass, 212
red fescue, 146
redtop, 210
slender creeping red fescue, 151
tufted hairgrass, 228–229
Helicotylenchus, 249
Helminthosporium, 35
Helminthosporium inconseicuum (leaf blotch), 265
Henderson Research Station, 244
Herbage yields, 63
Herbicides
annual bluegrass, 46
bahiagrass, 334
buffalograss, 265
Chewings fescue, 150
Herbivore dependence, 6
Herpetogramma phaeopteralis. see Tropical sod webworm
Heterogeneity, 12
Heteronychus arator (black beetle), 95
Hierba de Bahía, 338
Hierba fina, 191
High altitudes, 17–18
Idaho fescue, 158–159
rough bluegrass, 67
supina bluegrass, 53, 57
Highland bentgrass *(Agrostis castellana),* 209–212, 215, 218
botanical description, 211, 214
breeding, 218
characteristics, 209
distribution, 215
management limitations, 211–212, 214
Highland Bentgrass Commission, 211
'Highland' bentgrass (cultivar), 210
'Highland' colonial bentgrass, 187, 189, 191, 194
'Highland velvet' bentgrass, 203
High latitudes, 67
'Highlight' Chewings fescue, 150
Hitchcock, A. S., 207–208
'HJA 166' colonial bentgrass, 197
Hockey, 9
Holcus mollis L., 8

'Holfior' colonial bentgrass, 191, 192
Home lawns, VII, 8
Hoplolaimus, 249
Hoplolaimus galeatus (lance nematode), 318
Horses, 5
pasture for, 272
race, 9
'Houndog' tall fescue, 113
'Houndog V' tall fescue, 123
Humans
and ball games, 9
and domestication of livestock, 8
environment affected by, 10–11
evolution of, 6
Humidity, 92
Hunting billbug, 279
Hurricanegrass, 334
Hybridization. *See also* Interspecific hybridization; Intraspecific hybridization
Agrostis, 220
bermudagrass, 241, 245
Festuca ovina, 156
Kentucky bluegrass, 16, 32–36
perennial ryegrass, 16, 79, 84–86, 88–89
red fescue, 142–146
St. Augustinegrass, 320–321, 323–325
seashore paspalum, 301
tall fescue, 16
Texas bluegrass, 61
Zoysiagrasses, 279

I

Ice-encasement, 99
Idaho Agricultural Experiment Station, 161
Idaho bentgrass *(Agrostis idahoensis),* 17
botanical description, 212–214
breeding, 218–219
characteristics, 209
distribution, 215–216
management, 214
Idaho fescue *(Festuca idahoensis),* 136, 137, 157, 158–159
IGER. *see* Institute of Grassland and Environmental Research
ILGI (International *Lollium* Genome Initiative), 88
'Illahee' red fescue, 141
Industrial complexes, VII
Industrial pollution, 10
Inflorescence
bahiagrass, 335
Idaho fescue, 159

perennial ryegrass, 76, 77, 78
St. Augustinegrass, 310, 320
Texas bluegrass, 65
'Ino' rough bluegrass, 70
Insects, 7
bermudagrass, 250
and endophytes, 120
fine-leaved *Festuca,* 131–132, 132
tall fescue, 108
Texas bluegrass, 64
Institute of Grassland and Environmental Research (IGER), 83, 84
Instituto Forestal de Investigaciones y Experiencias, 218
Intellectual property rights, 323
Intercrossing, 13
International *Lollium* Genome Initiative (ILGI), 88
International Seeds, Inc., 150
Internodes, 12
Interspecific hybridization
Agrostis, 217
Kentucky bluegrass, 29, 33–34
red fescue, 144
tall fescue, 117
Texas bluegrass, 64
Zoysiagrasses, 275, 277
Intraspecific hybridization
Kentucky bluegrass, 33
red fescue, 144
Introgression, 88–89
Invasive species, eradication of, 17
In vitro propagation, 302
Ionizing radiation, 246
Iron, 210
Irrigation
buffalograss, 260
colonial bentgrass, 189
tall fescue, 113
Isoelectric focusing extracts, 139
Isozyme variation, 132
Italian ryegrass *(Lolium multiflorum),* 8, 13
and bahiagrass, 335
and fescue hybrid, 16
Italy, 8

J

Jacklin Seed Company, 156, 163
'Jade' St. Augustinegrass, 317, 323
'Jaguar' tall fescue, 113, 122
'Jaguar 3' tall fescue, 121, 122, 123, 215
'Jamestown' Chewings fescue, 150

'Jamestown II' Chewings fescue, 150–151
'Jamestown II' fine-leaved *Festuca*, 131–132
Japan, 9
'Jasper II' red fescue, 147
'Jasper' strong creeping red fescue, 153
Jensen, Ray, 291
JOINMAPÔ, 87
Joint grass, 298

K

Kansas Agricultural Experiment Station, 245
Kansas State University, 245
'Karmos' redtop, 217
Karyotype, 240–241
Keen, R. A., 245
'Kenhy' tall fescue, 117
Kentucky Agricultural Experiment Station, 107
Kentucky bluegrass *(Poa pratensis)*, 10, 11, 27–36
  apomictic reproductive system, 31–33
  breeding, 31–36
  classifications, 35
  collection of, 30–31
  cytotaxonomy, 29–30
  distribution, 29
  future breeding, 34–35
  germ plasm sources, 35
  and hard fescue, 162
  hybridization, 32–34
  induced variation, 35–36
  interspecific hybridization, 33–34
  intraspecific hybridization, 33
  and perennial ryegrass, 96
  selection, 30–31
  and Texas bluegrass hybrid, 16
  and tl, 109
'Kentucky-31' tall fescue, 107–108, 112–114, 118, 121–123
Kernwood Country Club, 203
'Kernwood' velvet bentgrass, 203
Kew Gardens (UK), 40
Kew Herbarium, 289
'Kingstown' velvet bentgrass, 204
Kitale Research Station, 244
'Kita' redtop, 217
Kneebone, W. R., 245
Knotgrass, 298
'Koket' Chewings fescue, 150
Krans, Jeffrey V., 319
Kweekgrass, 235

L

Laetisaria fuciformis. see Red thread
Lägerrispe, 55
Lance nematode *(Hoplolaimus galeatus)*, 318
Landscaping
  centipedegrass, 287
  perennial ryegrass, 75
  seashore paspalum, 296
Large chromosomes, 42
'Las-aga' St. Augustinegrass, 312
'Laser II' rough bluegrass, 71
'Laser' rough bluegrass, 71
Latitude of origin, 7
Lawns
  annual bluegrass, 39, 40
  bermudagrass, 235
  buffalograss, 260
  centipedegrass, 287
  Festuca ovina, 155
  hair fescue, 157–158
  hard fescue, 161
  Kentucky bluegrass, 29
  perennial ryegrass, 75, 76
  red fescue, 141
  rough bluegrass, 69
  rough bluegrass weed, 70
  St. Augustinegrass, 312, 322
  sheep fescue, 159
  strong creeping red fescue, 152
Lazy man's grass, 287
Lead tolerance, 10
  Festuca ovina, 155
  fine-leaved *Festuca*, 131
  tufted hairgrass, 228
Leaf blades, 12
Leaf blotch *(Helminthosporium inconseicuum)*, 265
Leaf diffusive resistance, 311
Leafhoppers, 266
Leaf osmotic potential, 7
Leaf rot, 249
Leaf rust *(Puccinia* spp.)
  fine-leaved *Festuca*, 131
  Texas bluegrass, 63
Leaf spot disease
  comparison of, 215
  Festuca ovina, 156
  fine-leaved *Festuca*, 131
  hard fescue, 162
  Kentucky bluegrass, 30, 35
  perennial ryegrass, 93
  red fescue, 147
  rough bluegrass, 69
  slender creeping red fescue, 152
  strong creeping red fescue, 153–154
  tall fescue, 109, 114–115, 122, 123

Leaf water conductance, 7
'Legacy' buffalograss, 269
Leisure, VII
Leptosphaeria korrae (necrotic ring spot), 131
'Liget' shade fescue, 148
Lighting, artificial, 31
Liming, 335
Limonomyces roseipellis (pink patch), 122
'Limousine' Kentucky bluegrass, 215
Linkage calculation program, 87
Linneaus, 207
'Linn' perennial ryegrass, 81, 90, 92, 93
'Listra' redtop, 217
Listronotus bonariensis. see Argentine stem weevil
Litter, 6
Livestock
  bahiagrass, 336
  and endophytes, 120
  redtop, 210
'Liwally' false sheep fescue, 163
L.L. Patten, 291
'Logro' slender creeping red fescue, 152
Lolium multiflorum. see Italian ryegrass
Lolium perenne. see Perennial ryegrass
Lolium spp. (ryegrasses), 7
Longevity, 16
  of individual genotypes, 12
  of individual plants, 8
'Longfellow' red fescue, 147
Longicaudatus Race (St. Augustinegrass), 314–315
'Loretta' perennial ryegrass, 82
Lowe, A.E., 264
Low-maintenance turf, 17

M

'Magennis' bermudagrass, 244
Magnaporthe poae. see Summer patch
Maize, 94
Male sterility, 84
Management
  buffalograss, 265
  colonial bentgrass, 187–188
  creeping bentgrass, 175
  Festuca ovina, 155
  Highland bentgrass, 211–212, 214
  perennial ryegrass, 76
  red fescue complex, 140–141
  redtop, 210, 214
  St. Augustinegrass, 310–311

'Manhattan' perennial ryegrass, 11, 81
Mannitol, 99
'Manoir' red fescue, 148
'Manoir' slender creeping red fescue, 152
Mapping, gene, 87–88, 116
Marker-assisted selection (MAS), 14, 15
  creeping bentgrass, 179
  tall fescue, 116
Marousky, Francis J., 344
MAS. *see* Marker-assisted selection
'Matador' tall fescue, 112, 119, 123
Max Planck Institute, 56
Mayans, 9
'MBA-1' bahiagrass, 344
Meadow fescue *(Festuca pratensis)*
  and perennial ryegrass, 88–89, 94, 95
  and ryegrass hybrid, 16
  tall fescue *vs.*, 109–110
Mediterranean Basin, 8
Melanoplus differentialis (differential grasshopper), 279
Meloidogyne, 249
Mendel, Gregor, 86
'Menuet' Chewings fescue, 150
Mercer grass, 298
Merion Cricket Club, 30, 203
'Merion' Kentucky bluegrass, 11, 28, 30
'Merion' velvet bentgrass, 203
'Merlin' fine-leaved *Festuca*, 131
'Merlin' slender creeping red fescue, 151
'Mesa' tall fescue, 113
Methodical selection, 5, 11–15
Meyer, Frank N., 278, 290
'Meyer' Zoysiagrasses, 283
Michigan State University, 54, 132
Microdochium nivale. *see* Pink snow mold
Microprojectile bombardment, 179
'MIC 18' tall fescue, 123
'Midfield' bermudagrass, 245
'Midiron' bermudagrass, 245, 246
'Midlawn' bermudagrass, 245, 250
'Midnight' Kentucky bluegrass, 33, 34
'Midway' bermudagrass, 245
'Midwest' Zoysiagrasses, 283
'Millennium' tall fescue, 115, 119, 121–123

Miller, Robert, 243
Mine spoils, 10
'Minotaur' *Festuca ovina*, 156
Miocene epoch, 5
Mite, zoysiagrass erophyid, 279
'Mocassin' slender creeping red fescue, 152
Mole crickets, 343
Molecular genetics, 65
Molecular markers, 34–35
Monographella nivalis. *see* Pink snow mold
'Monterey' perennial rye, 215
Moran, William E., 344
Morphology
  bermudagrass, 248
  buffalograss, 261–262
  Chewings fescue, 149–150
  colonial bentgrass, 189–190
  false sheep fescue, 163
  Festuca ovina, 154–155
  fine-leaved *Festuca*, 137
  hard fescue, 161
  Idaho fescue, 159
  perennial ryegrass, 77
  rough bluegrass, 68
  seashore paspalum, 298–299
  shade fescue, 148
  sheep fescue, 160
  slender creeping red fescue, 151–152
  strong creeping red fescue, 153
  supina bluegrass, 54–55
  tall fescue, 110–111
  Zoysiagrasses, 273–274
Mound-building prairie ants, 266
'Mountain Ridge' velvet bentgrass, 203
Mowing height
  annual bluegrass, 39–41
  and biomass, 97
  buffalograss, 259
  Chewings fescue, 150, 151
  colonial bentgrass, 188, 214
  creeping bentgrass, 214
  hair fescue, 158
  Highland bentgrass, 214
  Idaho bentgrass, 214
  Kentucky bluegrass, 29
  perennial ryegrass, 76, 93
  red fescue, 141
  redtop, 214
  sheep fescue, 160
  supina bluegrass, 53, 54, 56
  tall fescue, 109
Multiple stress tolerances, 12
'Murietta' tall fescue, 113
Murray, Jack, 154, 278
Musser, H. B., 178
'Mustang' tall fescue, 113, 123
'MX-86' sheep fescue, 156

N

Nagata, Russell, 323
Narrow-leaf meadowgrass *(Poa pratensis angustifolia)*, 29
National Fine Fescue test, 132
National Genetic Resources Program (NGRP), 82, 83
National Plant Germplasm System (NPGS), 187, 192, 193, 319, 341
National St. Augustinegrass Test, 326
National Tall Fescue Test, 120
National trials, 14
National Turfgrass Evaluation Program (NTEP)
  bermudagrass, 245, 251
  colonial bentgrass, 196
  Idaho bentgrass, 218
  Kentucky bluegrass, 34
  perennial ryegrass, 90, 92, 93
  redtop, 217
  St. Augustinegrass, 316, 326
  tall fescue, 108, 122, 123
Native species, 17, 257
Natural selection, 5–8
Natural variation, 11–12
NCI endophytes. *see* Nonchoke-inducing endophytes
Necrotic ring spot *(Leptosphaeria korrae)*, 131
Nematodes
  bahiagrass, 331
  bermudagrass, 249–250
  St. Augustinegrass, 318
Neotyphodium coenophialum
  Chewings fescue, 150–151
  fine-leaved *Festuca*, 131–132
  hard fescue, 163
Neotyphodium endophyte, 157
Neotyphodium lolii, 76
Neotyphodium spp.
  fine-leaved *Festuca*, 131
  red fescue, 147
New Jersey Agriculture Experiment Station, 71, 82, 112, 150, 153, 162, 203
'Newport velvet' bentgrass, 203
New York Botanical Garden, 40
New Zealand, perennial ryegrass in, 95
New Zealand bentgrass, 191
'New Zealand Chewings' fescue, 150
NGRP. *see* National Genetic Resources Program
Niche-adapted grasses, 17
'Nichol Ave. No. 1' velvet bentgrass, 203
'Nichol Ave. No. 2' velvet bentgrass, 203

Nickel, 10
  redtop, 210
  tufted hairgrass, 228
Ninepins, 9
Nitrogen
  centipedegrass, 287
  perennial ryegrass, 76
'NK100' perennial ryegrass, 81
'NK200' perennial ryegrass, 99
Nonchoke-inducing (NCI) endophytes
  Festuca ovina, 156, 157
  fine-leaved *Festuca*, 132
  red fescue, 147
Non-NPGS Security Backup Collection (NSSB), 192, 193
'Norcoast' hairgrass, 229
'Nordic' hard fescue, 162
North, H.F.A., 204
North America, 5, 9
'Nortran' hairgrass, 229
NPGS. *see* National Plant Germplasm System
NSSB. *see* Non-NPGS Security Backup Collection
NTEP. *see* National Turfgrass Evaluation Program
'NuMex Sahara' bermudagrass, 247

O

OECD, 141, 155
Off-types, 32
Oklahoma, 17
Oklahoma State University, 244, 245, 247
'Oklawn' centipedegrass, 291
'OKS 95-1' bermudagrass, 247
Old Deer Park (UK), 40
'Olds' red fescue, 141
'Olympic' tall fescue, 113, 118
O.M. Scotts & Sons, 314
'One-Eye' buffalograss, 267
Ophiobolus patch (*Gaeumannomyces graminis*), 69
Ophiobolus patch (*Ophiobolus graminis*), 192
Ophiosphaerella herpotricha, 249
Ophiosphaerella korrae, 249
'Orbica' colonial bentgrass, 189
Orchardgrass (*Dactylis glomerata* L.), 79, 108
Oregon Agriculture Experiment Station, 107, 150, 161, 191
Organ size, 12
Ornamental plantings, 155
Ovina, 159
'Oxford' hard fescue, 162

P

'Palisades' Zoysiagrasses, 279
'Palmer III' perennial ryegrass, 90
'Palmetto' St. Augustinegrass, 317, 323
'Panella' colonial bentgrass, 195
Paniceae, 300
Panicodae, 300
Panicoideae, 300
Panicum Mosaic Virus. *see* St. Augustinegrass Decline Strain of Panicum Mosaic Virus
'Paraguay-22' bahiagrass, 340
Parapediasia spp. (sod webworm), 131
Parapediasia teterella (sod webworm), 54
Parks
  bermudagrass, 235
  buffalograss, 260
  fine-leaved *Festuca*, 130
  hard fescue, 161
  perennial ryegrass, 76
  strong creeping red fescue, 152
Park Grass Experiments at Rothamstead, 7
Parking lots, grass, 333
Paspalum, 298
Paspalum, 300
Paspalum notatum Flügge. *see* Bahiagrass
Paspalum vaginatum. *see* Seashore paspalum
Pasto de miel, 338
Pasto dulce, 298
Pasto horqueta, 338
Pasto mexicano, 338
Pasture grass
  bahiagrass, 331
  Highland bentgrass, 211
  redtop, 210
  St. Augustinegrass, 312, 314
  Zoysiagrasses, 272
Pata de gallina, 298
'Pathfinder' strong creeping red fescue, 153
Pathogens, 7
Pea, 94
Peach trees, 331
PEG. *see* Polyethylene glycol
'Pelo' perennial ryegrass, 82
'Pembagrass' St. Augustinegrass, 320
'Penncross' creeping bentgrass, 178, 181
'Pennfine' perennial ryegrass, 81, 90, 92, 93
Pennington Seed, 322
'Pennlawn' red fescue, 141

Pennsylvania Agriculture Experiment Station, 90
Pennsylvania State University, 42, 46, 132, 178, 194
'Pensacola' bahiagrass, 334–336, 338–341, 344
Perennial ryegrass (*Lolium perenne*), 6–8, 10, 13, 75–99
  adaptation, 75–76
  breeding, 79–81
  breeding objectives and progress, 90–99
  breeding techniques, 84–89
  climatic and edaphic adaptation, 75–76
  collection of, 81–84
  comparison of, 75, 76
  cytogenetics, 78–79
  cytotaxonomy, 77–79
  distribution, 77
  and fescue hybrid, 16
  and genetic diversity/erosion, 82–84
  germ plasm sources, 81–82
  management, 76
  morphology, 77
  number of cultivars of, 82, 83
  and red fescue, 143
  reproduction, 79–81
  and tall fescue, 108
  taxonomy, 77–78
Peroxidase isozyme banding, 277
Persistence, 27
'Peru Creek' hairgrass, 229
Pests, 11, 13, 265–266
Peterson Seed Company, 48
PGQO. *see* Plant Germplasm Quarantine Office
'P-2517' hair fescue, 161
Phenotypic plasticity, 12, 16, 43
Phenotypic selection, 13
  perennial ryegrass, 84
  red fescue, 148
Phleospora idahoensis (stem eyespot), 156
Photoperiod sensitivity
  annual bluegrass, 47
  bahiagrass, 335, 338
  creeping bentgrass, 180–181
  St. Augustinegrass, 320
  tall fescue, 112
Photosynthesis rates
  perennial ryegrass, 96
  supina bluegrass, 53
Phyllophaga, 64
Phyllophaga crinita. *see* White grubs
'P.I. 148996' bahiagrass, 338
Pickseed, Inc., 150, 162
Pickseed West, Inc., 132, 151
Picloram, 265

INDEX

'P-6435' Idaho fescue, 159
Pink patch *(Limonomyces roseipellis)*, 122
Pink snow mold *(Fusarium nivale)*
  colonial bentgrass, 192
  creeping bentgrass, 182
Pink snow mold *(Microdochium nivale)*
  creeping bentgrass, 175, 182
  Festuca ovina, 156
  fine-leaved *Festuca*, 131
  perennial ryegrass, 92
  sheep fescue, 160
  tall fescue, 109, 122
Pink snow mold *(Monographella nivalis)*
  annual bluegrass, 42
  fine-leaved *Festuca*, 131
Piper, C. V., 244
'Piper' velvet bentgrass, 203
'PI-239729' St. Augustinegrass, 319
'PI-289729' St. Augustinegrass, 315
'PI-290888' St. Augustinegrass, 315
'PI-300127' St. Augustinegrass, 315
'PI-365031' St. Augustinegrass, 315
'PI-365301' St. Augustinegrass, 319
'P-104' Kentucky bluegrass, 33
'Plantation' tall fescue, 115, 119, 121–123
Plant Breeding Institute of the University of Perugia, 195
Plant Germplasm Quarantine Office (PGQO), 192, 193
Plant Introduction Service, 112
Plant Materials Center of the Soil Conservation Service, 159, 160
Plant Variety Protection, 141
Ploidy
  in apomictic reproductive systems, 32
  seashore paspalum, 300–301
PMV-SAD. *see* St. Augustinegrass Decline Strain of Panicum Mosaic Virus
Poa, 7
Poa annua. see Annual bluegrass
Poa arachnifera. see Texas bluegrass
Poaceae, 16, 67, 109, 190, 202, 226
Poa infirma H.B.K., 41–42
Poa pratensis. see Kentucky bluegrass

Poa pratensis angustifolia (narrow-leaf meadowgrass), 29
Poa supina. see Supina bluegrass
Poa trivialis. see Rough bluegrass
Poeae, 40, 55, 67, 77, 176, 202, 226
Point and nonpoint source pollutants, 10
'Polis' rough bluegrass, 70
Pollen shed
  red fescue, 142
  Zoysiagrasses, 277–278
Pollen-stigma response, 80, 81
Pollination, 31–32
Pollution, 10
  air, 10
  industrial, 10
Polyethylene glycol (PEG), 116, 179
Polymorphisms, 316–319
Polyploidy, 16
  bahiagrass, 339–340
  Kentucky bluegrass, 29, 30
  St. Augustinegrass, 315–316
Poodae, 40
Pooideae, 40, 55, 67, 77, 176, 190, 202, 226
Poor man's grass, 287
Population improvement, 322–325
Powdery mildew *(Erysiphe graminis)*
  fine-leaved *Festuca*, 131
  hard fescue, 162
  rough bluegrass, 69, 71
  strong creeping red fescue, 153
  Texas bluegrass, 63
Pozarnsky, Tom, 263
'Prairie' buffalograss, 268, 269
Prairie regions, 17–18
'Prelude III' perennial ryegrass, 90
Primary meristem size, 12
Prince Edward Island bentgrass, 191
Pseudogamous apospory, 30
'P-274' sheep fescue, 160
Puccinellia distans (L.) Parl., 17
Puccinia coronata Corda. *see* Crown rust
Puccinia crandallii
  hard fescue, 163
  slender creeping red fescue, 152
  strong creeping red fescue, 153
Puccinia graminis. see Black stem rust
Puccinia kansensis, 265
Puccinia spp. *see* Leaf rust; Rust fungi; Stem rust
Pullman (WA), 35

Pure Seed Testing, Inc., 122, 132, 153, 162, 163, 195, 196
Putting greens, 9, 10
  annual bluegrass, 39–40, 41, 44
  bermudagrass, 235, 236
  creeping bentgrass, 175, 178
  red fescue, 141
  redtop, 217
  rough bluegrass, 70
  supina bluegrass, 56
  velvet bentgrass, 201
Pyricularia grisea. see Gray leaf spot
Pythium blight
  creeping bentgrass, 175
  hard fescue, 163
  perennial ryegrass, 92
  rough bluegrass, 69
  tall fescue, 123

Q

QTL. *see* Quantitative trait loci
Quality control, 323
Quality of life, VII
Quantitative trait loci (QTL), 14–15
'Quatro' sheep fescue, 160
Queen's University (Northern Ireland), 132
Quickgrass, 235

R

Rabbits, 9
Racehorses, 9
Racing, 9
'Radiant' perennial ryegrass, 90
Radiation
  bermudagrass, 246
  centipedegrass, 291
  Kentucky bluegrass, 35
Railway banks
  Festuca ovina, 155
  hard fescue, 161
  sheep fescue, 159
Rainfall, 75
'Rainier' red fescue, 141
'Raleigh' St. Augustinegrass, 317, 323
'Raleigh-S' St. Augustinegrass, 322
Randomly Amplified Polymorphic DNA (RAPD)
  bahiagrass, 342
  creeping bentgrass, 179
  Kentucky bluegrass, 34
  perennial ryegrass, 87
  seashore paspalum, 303
  supina bluegrass, 56
  Texas bluegrass, 63, 65
Range (grazing), 17

RAPD. *see* Randomly Amplified Polymorphic DNA
Rapid microevolution, 44
'Raritan' velvet bentgrass, 203
Rating scale, 13
RDNA (ribosomal DNA), 78
Read, James, 33, 63
'Rebel II' tall fescue, 113
'Rebel Jr.' tall fescue, 118
'Rebel' tall fescue, 112–114, 122
Reclamation areas, 17
  Festuca ovina, 155
  hairgrasses, 225
  hard fescue, 161
  redtop, 210
  seashore paspalum, 295
  sheep fescue, 159
  tufted hairgrass, 228
Recreation, VII, 10
Recurrent Restricted Phenotypic Selection (RRPS), 342
Recurrent selection, 13
Red fescue complex (*Festuca rubra*), 8, 10, 11, 129–154
  breeding, 142–148
  Chewings fescue, 148–151
  creeping vs. noncreeping, 138
  cytology, 141–142
  distribution, 139–140
  endophytes in, 131–132
  hybridization, 142–146
  and perennial ryegrass, 75
  selection criteria, 146–147
  shade fescue, 148
  slender creeping red fescue, 151–152
  strong creeping red fescue, 152–154
  subspecies of, 139–140
  taxonomy of, 136–140
  use/management/environmental limitations, 140–141
Red millet, 336
Red thread (*Laetisaria fuciformis*)
  colonial bentgrass, 192
  Festuca ovina, 156
  fine-leaved *Festuca*, 131
  hard fescue, 162
  perennial ryegrass, 92, 93
  red fescue, 147
  sheep fescue, 156
Redtop (*Agrostis gigantea*), 17
  botanical description, 208, 214
  breeding, 217–218
  characteristics, 209
  distribution, 213, 215
  management limitations, 210, 214
Reed, A. M., 309
Reintroduction, native plant, 17

'Reliant' hard fescue, 162
'Rembrandt' tall fescue, 115, 119, 121–123
Renfoe, Riley, 291
Reproduction
  annual bluegrass, 43
  bahiagrass, 339–340
  bermudagrass, 241–242
  buffalograss, 262–263
  Kentucky bluegrass, 31–33
  perennial ryegrass, 79–81
  seashore paspalum, 300–301
  St. Augustinegrass, 322
'Rescue 911' hard fescue, 163
Resistance, 6–7
Restriction Fragment Length Polymorphism (RFLP) analysis
  fine-leaved *Festuca*, 135
  perennial ryegrass, 79, 86
  seashore paspalum, 302
  tall fescue, 110
  Zoysiagrasses, 276, 277
'Reton' redtop, 210, 217
Revegetation, 17
'Reveille' Kentucky bluegrass, 33
'Reveille' Texas bluegrass, 61, 64
Reversed-phase high-performance liquid chromatography (RP-HPLC) analysis, 135
RFLP. *see* Restriction Fragment Length Polymorphism analysis
Rhizoctonia solani. *see* Brown patch
Rhode Island Agricultural Experiment Station, 191, 196
Rhode Island bentgrass, 191
Rhodesgrass mealybug, 266
Ribosomal DNA (rDNA), 78
Rights-of-way, 17, 260
Riordan, Terrence, 323
Riparian plantings, 225
Roadsides, 17
  bermudagrass, 235
  centipedegrass, 287
  Chewings fescue, 149
  colonial bentgrass, 188
  Festuca ovina, 155
  fine-leaved *Festuca*, 130
  hard fescue, 161
  Kentucky bluegrass, 27
  perennial ryegrass, 76
  red fescue, 141
  sheep fescue, 159
Roberts Seed, 151
Robinson, B.P., 291
Roman Circus, 9
Romans, 9
Root fluorescence, 135

Root plasticity, 15
Root systems
  bahiagrass, 333
  colonial bentgrass, 189
  St. Augustinegrass, 317
  supina bluegrass, 53
  tall fescue, 108
'Roselawn' St. Augustinegrass, 311, 314, 315
Rothamstead, 7
Rotstraußßgras, 191
Rottboelliinae, 288
Roughs (golf courses)
  bermudagrass, 235
  Festuca ovina, 155
  fine-leaved *Festuca*, 129
  perennial ryegrass, 76
  sheep fescue, 159
  strong creeping red fescue, 152
Rough bluegrass (*Poa trivialis*), 67–72
  commercial use, 69–72
  cytology, 68
  geographic adaptation, 68–69
  growth characteristics, 69
  and Kentucky bluegrass, 32–33
  morphology, 68
  naturalization, 70
  seed sources and cultivar development, 70–72
  taxonomy, 67–69
  as weed, 70
Rough meadowgrass, 67
Roughstalk bluegrass, 67
Roughstalked meadowgrass, 67
Royal Botanic Gardens, 236, 275
'Royal' Zoysiagrasses, 279
RP-HPLC (reversed-phase high-performance liquid chromatography) analysis, 135
RRPS (Recurrent Restricted Phenotypic Selection), 342
'Ruby' strong creeping red fescue, 153
Rust fungi (*Puccinia* spp.), 7, 69
Rutgers University, 11, 31, 35, 92, 114, 122, 123, 132, 133, 146, 150, 153, 162, 189, 195, 229
Ryegrasses
  Italian ryegrass (*see* Italian ryegrass)
  perennial ryegrass (*see* Perennial ryegrass)
  and tall fescue, 16

S

'Sabre' rough bluegrass, 71
Sacasebo, 298

SAES (State Agricultural
  Experiment Stations), 245
St. Augustinegrass Decline Strain
  of Panicum Mosaic Virus
  (PMV-SAD), 315, 317–320
St. Augustinegrass *(Stenotaphrum
  secundatum)*, 309–327
  adaptive polymorphisms,
    316–319
  biology, 309–312
  biotic stresses, 317–319
  breeding and selection
    techniques, 320–321
  characteristics, 309–310
  cytotaxonomy, 313–315
  distribution, 312–313
  environmental adaptation and
    management, 310–311
  genetics, 313–316
  germ plasm sources, 319–320
  history of breeding/population
    improvement, 322–325
  and inheritance, 321–322
  introduction/selection/
    breeding, 319–325
  origin and related species,
    312–313
  physiology and environmental
    stresses, 316–318
  polyploidy, 315–316
  reproduction, 322
  taxonomy and geography,
    313–315
'St. Lucie' bermudagrass, 243
Salaillo, 298
'Salam' seashore paspalum, 304
Saltpeter, 264
Salt tolerance, 7–8, 17, 18
  creeping bentgrass, 175, 183
  hair fescue, 158
  Kentucky bluegrass, 29
  red fescue, 141, 146, 147
  redtop, 210
  St. Augustinegrass, 317
  seashore paspalum, 295, 298
  slender creeping red fescue,
    151
  Zoysiagrasses, 278, 283
Saltwater couch, 298
San Augustin (St.
  Augustinegrass), 309
Sand dune stabilization, 296
Sand knotgrass, 298
Sandy soils, 61
'Saturn' perennial ryegrass, 90
'Sawa' shade fescue, 148
'Scaldis' hard fescue, 162
Scapteriscus borellii. see Southern
  mole cricket
Scapteriscus vicinus. see Tawny
  mole cricket

Schafschwingel, 159
School grounds, 130
Sclerophthora macrospora. see
  Downy mildew
Sclerotinia homeocarpa. see Dollar
  spot
Scotland, 9
Scotts Company, 323
SDS-PAGE. *see* Sodium
  dodecylsulphate-
  polyacryalamide gel
  electrophoresis
SEA, 153
'Seabreeze' slender creeping red
  fescue, 152
'Sea Isle 1' seashore paspalum,
  303, 304
'Sea Isle 2000' seashore
  paspalum, 303, 304
Seashore paspalum *(Paspalum
  vaginatum)*, 8, 11, 295–304
  adaptation and morphology,
    295–299
  breeding approach, 301–302
  chorology, 299
  classification, 300
  collection, 301
  cultivars, 297
  cytotaxonomy, 300–301
  ecotypes and cultivars,
    303–304
  genetic analysis, 302–303
  morphological descripton and
    differentiation, 298–299
  nomenclature, 298
  ploidy, 300–301
  propagation, 299
  selection, 301–302
  worldwide dispersion and use,
    295–298
'Seaside' creeping bentgrass, 178,
  183
Seca sebo, 338
Seed, stripping, 28
Seed development, 321
Seed dormancy
  bahiagrass, 336
  supina bluegrass, 56
Seed formation
  colonial bentgrass, 190
  Kentucky bluegrass, 34
Seeding vigor, 17
Seed production, 12
  annual bluegrass, 46–47
  bahiagrass, 335
  centipedegrass, 287
  perennial ryegrass, 79–80, 80
  rough bluegrass, 68
  seashore paspalum, 302
  supina bluegrass, 56–57
  tall fescue, 112, 113, 118

Zoysiagrasses, 279
Seed propagation (St.
  Augustinegrass), 323
Seed Research of Oregon, Inc.,
  132, 151, 154, 161–163, 195
Seed shattering
  bermudagrass, 247
  rough bluegrass, 68
Seed storage
  buffalograss, 264
  supina bluegrass, 56
  tall fescue, 121–122
'Sefton' colonial bentgrass, 192
Selection, 5
  bahiagrass, 341–342
  bermudagrass, 248–251
  buffalograss, 266–268
  centipedegrass, 291–292
  colonial bentgrass, 191–197,
    192
  creeping bentgrass, 181–183
  hairgrasses, 229–230
  Kentucky bluegrass, 30–31
  perennial ryegrass, 81–99
  red fescue, 146–147
  St. Augustinegrass, 320–321
  seashore paspalum, 301–302
  supina bluegrass, 57–58
  tall fescue, 117–121
  Texas bluegrass, 63–65
  velvet bentgrass, 203–204
  Zoysiagrasses, 282–283
Self-fertilization
  in apomictic reproductive
    systems, 32
  supina bluegrass, 56
Self-incompatibility, 6
  Agrostis, 216
  bermudagrass, 242
  centipedegrass, 290
  perennial ryegrass, 77, 79
  tall fescue, 111
  tufted hairgrass, 228
Self-pollination
  annual bluegrass, 43
  velvet bentgrass, 203
Self-seed setting, 80
Self-thinning line, 97
Self-thinning rule, 96
Semidwarf tall fescue, 119
Semiquantitative rating scale, 13
Senescence, 94
Setariinae, 300
'Seville' St. Augustinegrass, 314,
  317, 323
Sex expression, 267
Seymourgrass, 334
Shade bluegrass, 67
Shade fescue, 148
'Shademaster' strong creeping
  red fescue, 153

Shade tolerance
  Chewings fescue, 150
  fine-leaved *Festuca*, 130
  hairgrasses, 225
  Kentucky bluegrass, 29
  moderate, 17
  red fescue, 141
  rough bluegrass, 68, 69
  St. Augustinegrass, 311, 317, 326
  supina bluegrass, 53, 54, 57
  tall fescue, 109
  velvet bentgrass, 201
  Zoysiagrasses, 282
'Shadow' Chewings fescue, 151
'Sharps Improved' buffalograss, 267, 268
Sheaths, 12
Sheep, 8, 9
  Festuca ovina, 155
  perennial ryegrass, 81
Sheep fescue. *see Festuca ovina* complex
'Shenandoah' tall fescue, 122
Shoot density
  annual bluegrass, 39
  and biomass, 97, 98
Shoot-feeding insects, 76
Side-oats grama, 258
'Silhouette' Chewings fescue, 151
Silicon carbide fiber method, 179
Siltgrass, 298
'Silverado' tall fescue, 113
Simple Sequence Repeats (SSRs)
  creeping bentgrass, 179
  perennial ryegrass, 87, 88
  seashore paspalum, 302–303
'Simplot' Zoysiagrasses, 279
'609' buffalograss, 268–269
Skogley, C.R., 191, 204
Slave boats, 295–296
Slender creeping red fescue (*Festuca rubra* ssp. *litoralis*), 130–132, 137, 141, 142, 151–152
Slender hairgrass (*Deschampsia elongata*), 229
Smelters, 10
Snowmold resistance, 11
  fine-leaved *Festuca*, 131
  perennial ryegrass, 93, 99
Snowmold (*Typhula* spp.), 11
Snows, 17
Soaking and chilling method (germination), 264
Soccer, 9
Sod farms, 323
Sod-forming ability, 132

Sod harvesting
  bahiagrass, 333
  St. Augustinegrass, 309
Sod industry, 315
Sodium dodecylsulphate-polyacrylamide gel electrophoresis (SDS-PAGE), 133, 155
Sod production
  centipedegrass, 287, 291
  Kentucky bluegrass, 27
  rough bluegrass weed, 70
Sod propagation, 299
Sod Solutions, 323
Sod webworm
  fine-leaved *Festuca*, 131
  perennial ryegrass, 76
  supina bluegrass, 54
Soil, 10
  bahiagrass, 331
  contamination of, 10
  Festuca ovina, 155
  perennial ryegrass, 75
  pH of, 7
  tall fescue, 108
  Texas bluegrass, 61
Soil Conservation Service, 161, 258
Soil fertility, 7
  centipedegrass, 287, 292
  Chewings fescue, 150
  fine-leaved *Festuca*, 130
  Idaho fescue, 158
  Kentucky bluegrass, 28
Soil-inhabiting grubs, 123
Soil pH
  bahiagrass, 335
  buffalograss, 260
  colonial bentgrass, 188
  fine-leaved *Festuca*, 130
  tall fescue, 121
Soil stabilization
  bermudagrass, 243
  perennial ryegrass, 76
Somaclonal variation, 115–116
'Sonesta' bermudagrass, 247
Sorghum, 94
SO_2 tolerance, 10
Southern chinch bug (*Blissus insularis*), 310, 315, 317–321
Southern corn leaf blight disease, 244–245
Southern Illinois University, 279
Southern mole cricket (*Scapteriscus borellii*), 246, 250
Southern Turf®, 304
South German bentgrass, 203
South German Mixed Bentgrass, 178
'Southshore' creeping bentgrass, 181

Soybean, 43, 94, 331
Spaced-plant progeny nurseries, 33
Spalding, Thomas, 243
'Spartan' hard fescue, 162
Spenophorus spp. *see* Billbug
Sphenophorus parvulus (bluegrass billbug), 132
Spodoptera frugiperda J. E. Smith. *see* Fall armyworm
Spodoptera mauritia Boisduval. *see* Armyworm
Sports turfs, VII
  annual bluegrass, 41
  bermudagrass, 235
  buffalograss, 260
  hard fescue, 161
  perennial ryegrass, 75, 96
  red fescue, 141
  seashore paspalum, 296
  supina bluegrass, 53
  Zoysiagrasses, 283
Sports Turf Research Institute (STRI), 83, 93–94
Sprague, H. B., 203–204
Spreading fescue (*Festuca rubra* ssp. *rubra*), 136–138
Sprigging rates, 299
Spring dead spot, 249
Spring greenup, 92
'SR 5000' Chewings fescue, 151
'SR 5100' Chewings fescue, 151
'SR 7100' colonial bentgrass, 192, 197
'SR1020' creeping bentgrass, 178, 181
'SR 5200E' red fescue, 147
'SR 5200E' strong creeping red fescue, 154
'SR 6000' hairgrass, 229
'SR 3000' hard fescue, 162
'SR 3100' hard fescue, 163
SSRs. *see* Simple Sequence Repeats
Standards, 14
Standard tall fescue, 118–119
Starch gel electrophoretic method, 139
'Stardust' rough bluegrass, 71
Starvation, 99
State Agricultural Experiment Stations (SAES), 245
Stay-green gene, 16, 94–96
Stem eyespot (*Phleospora idahoensis*), 156
Stem rot, 249
Stem rust (*Puccinia* spp.)
  fine-leaved *Festuca*, 131
  tall fescue, 115, 122
  Texas bluegrass, 63
Stems, 12

Stenotaphrum secundatum. see St. Augustinegrass
'Sterling' slender creeping red fescue, 151
Sting nematode *(Belonolaimum longicaudatus)*
  bermudagrass, 250
  St. Augustinegrass, 310, 317, 318, 320, 321
Stomata, 10
Stomatal resistance, 7
'Streaker' redtop, 210, 217
Stresses, biotic *vs.* abiotic, 6–8
STRI. *see* Sports Turf Research Institute
Stripe smut *(Ustilago striiformis)*
  Festuca ovina, 156
  fine-leaved *Festuca*, 131
  rough bluegrass, 69
"Stripping" seed, 28
Strong creeping red fescue *(Festuca rubra* ssp. *rubra* Gaudin), 131, 152–154
Submersion tolerance, 335
Sulfometuron, 334
Summer dormancy, 65
Summer patch *(Magnaporthe poae)*
  fine-leaved *Festuca*, 131
  red fescue, 147
Super dwarf bermudagrass, 250
'Supernova' supina bluegrass, 57
Supina bluegrass *(Poa supina)*, 53–58
  breeding history, 57
  collection, 57
  comparative traits, 54–55
  cytotaxonomy, 55–57
  distribution, 55
  genetics, 55–56
  morphology, 54–55
  seed production, 56–57
  selection, 57–58
'Supra' supina bluegrass, 57
Surface water, 10
Svoboda, J.L., 264
Synaptospermy, 262
Synthetic cultivars, 11–13
  colonial bentgrass, 196
  of perennial ryegrass, 84
  Texas bluegrass, 64
Systematic Botany Laboratory, 40

T

TAES. *see* Texas A & M Agricultural Experiment Station
Take-all patch *(Gaeumannomyces incrustans)*, 131

Take-all root rot *(Gaeumannomyces graminis)*, 318
Tall fescue *(Festuca arundinacea)*, 7, 107–123
  breeding progress of, 122–123
  breeding techniques, 114–115
  climatic and physiological limitations, 108–109
  cultivar compositions, 121–122
  cultivar evaluation, 122–123
  current challenges and future prospects, 123
  cytotaxonomy, 109–110
  distribution, 107–108
  early cultivars, 112–113
  flowering control, 111–112
  genetic vulnerability, 121
  geographic distribution of breeding efforts, 122
  germ plasm sources, 113–114
  history and worldwide use, 107–108
  molecular approach, 115–117
  morphological description, 110–111
  and ryegrass hybrid, 16
  selection criteria, 117–121
  taxonomy and cytology, 109–110
  turf-type cultivar development, 112–122
  wild germ plasm, 114
Tall grass prairie, 258
Tawny mole cricket *(Scapteriscus vicinus)*
  bermudagrass, 250
  Zoysiagrasses, 279
Tees (golf courses)
  annual bluegrass, 41
  bermudagrass, 235
  perennial ryegrass, 75
Tejana, 338
Temperatures, 11, 16
  air, 11
  annual bluegrass, 42
  perennial ryegrass, 76, 88, 92, 95
  rough bluegrass, 69
  tall fescue, 108
'Tendenz' colonial bentgrass, 192
Tennis courts
  Chewings fescue, 149
  perennial ryegrass, 75
'Tenn Turf' centipedegrass, 291
Testing procedures, 14
Texas, 17
Texas A & M Agricultural Experiment Station (TAES), 63, 65, 268, 275, 278, 279

Texas A&M University, Dallas, 33, 268
Texas bluegrass *(Poa arachnifera)*, 17, 61–65
  breeding, 63–65
  collection of, 63
  cultivars, 64–65
  cytotaxonomy, 61–65
  distribution, 61
  and Kentucky bluegrass hybrid, 16, 33
  molecular genetics and tissue culture, 65
  selection, 63–65
'Texoka' buffalograss, 267
Texture, 13, 39
Thin layer chromatography, 139
'TifBlair' centipedegrass, 291, 292
'Tifdwarf' bermudagrass, 246, 250
'TifEagle' bermudagrass, 246
'Tiffany' Chewings fescue, 151
'Tiffine' bermudagrass, 245
'Tifgreen' bermudagrass, 244, 245, 246, 250
'Tifgreen II' bermudagrass, 246
'Tifhi 1' bahiagrass, 342
'Tifhi 2' bahiagrass, 342
'Tiflawn' bermudagrass, 245
'TifSport' bermudagrass, 246, 250
'Tift 94' bermudagrass, 246
'Tifton Cycle 9' bahiagrass, 341
'Tifway' bermudagrass, 244, 246, 250
'Tifway II' bermudagrass, 246
'Tiger' colonial bentgrass, 192, 197
Tiller densities, 8
Tissue culture, 65
'Tomahawk' tall fescue, 121
'Torpedo' tall fescue, 121
Toxicities, 10, 120
'Tracenta' colonial bentgrass, 189, 192
'Trailblazer' tall fescue, 113
Traits, 12
Transgenic technology, 15
Transgressive segregation, 31
Trials, 14
Triazine herbicides, 334
'Tribute' tall fescue, 113
Triclopyr, 265
'Trident' Idaho fescue, 159
Tridiscus sporoboli, 266
Trionymus sp., 266
Triploid hybrids, 16

Tropical sod webworm
  (Herpetogramma phaeopteralis)
  bermudagrass, 250
  Zoysiagrasses, 279
"Tsu Chu," 9
'Tufcote' bermudagrass, 246
Tufted hairgrass *(Deschampsia caespitosa)*, 225–230
Turf, 17
Turf breeding
  annual bluegrass, 39
  bahiagrass, 343–344
  bermudagrass, 243–245
Turfgrasses
  abiotic stresses, 7–8
  and ball games, 9–10
  biotic stresses, 6–7
  form *vs.* function, 13–14
  hard fescue, 161
  and human effects on environment, 10–11
  interspecific hybrids of, 15–16
  and livestock domestication, 8
  methodical selection, 11–15
  natural selection, 5–8
  natural variation and domestication, 11–12
  new germ plasm sources, 15–18
  new technologies, 14–15
  novel species, 16–18
  origins of, 5–18
  recurrent selection, 13
  unconscious selection, 8–11
Turfgrass seed industry, 12
"Turfgrass Seed" (STRI booklet), 94, 95
Turf quality, 13–14
  bermudagrass, 251
  comparison of, 215
Turf-Seed, Inc., 151, 160, 162
Two-locus multiallelic gametophytic incompatibility system, 80
Typhula blight, 175, 182
Typhula incarnata. see Gray snow mold
Typhula ishikariensis, 175, 182
Typhula itoana, 182

U

'U-3' bermudagrass, 244
UDP (uridine diphosphate) glucose pyrophosphorylase, 315
Unconscious selection, 5, 8–11
Uniformity, 8
United States, 14
United States Golf Association (USGA), 178, 203, 244, 268, 322

U.S. Department of Agriculture (USDA), 107, 153, 278
U.S. National Herbarium, 275
U.S. National Museum, 335
U.S. Patent Office Book of Agriculture, 261
USDA. *see* U.S. Department of Agriculture
USDA-ARS, 245
USDA Kentucky bluegrass collection, 35
USDA-NPGS-GRIN system, 204, 278
University of Alaska Agricultural Experiment Station, 153
University of Arizona, 64, 245
University of California, 278
University of California, Davis, 268
University of Florida, 279, 320, 323, 342, 343
University of Georgia, Griffin, 122, 299
University of Georgia breeding program, 304
University of Illinois, Urbana, 122
University of Minnesota breeding program, 47, 48, 54, 57
University of Nebraska, 264
University of Nebraska Agricultural Experiment Station, 268
University of Nebraska-Lincoln, 263
University of Rhode Island, 132, 150, 195, 196, 204
University of Rhode Island Research, 194
University of Wisconsin, 11, 195
University of Witwatersrand, 244
Uptake systems, 11
Urban areas, VII
Urban parks and recreation, 10
Uridine diphosphate (UDP) glucose pyrophosphorylase, 315
Ustilago striiformis. see Stripe smut

V

Valentine, Joe, 30, 203
'Valentine No. 2' velvet bentgrass, 203
'Valiant' *Festuca ovina*, 157
Van der Have, D.J., 192
Variability, 6, 14
Variegated St. Augustinegrass, 321–322
Various-leaved fescue, 148

Vegetative propagation, 13
  buffalograss, 263
  velvet bentgrass, 204
Velvet bentgrass *(Agrostis canina)*, 9, 201–205
  breeding history, 203–205
  collection, 203
  cytotaxonomy, 202–203
  distribution, 202
  selection, 203–204
'Venture' redtop, 217
'Verdi' perennial ryegrass, 94
Vernalization
  annual bluegrass, 47–48
  hairgrasses, 230
  supina bluegrass, 57
  tall fescue, 112
  Texas bluegrass, 65
'Victoria' Zoysiagrasses, 278
'Victory' Chewings fescue, 150
Vigor, 16, 27
Visual appearance, 14
Volcanoes, 9
Vulnerability, 13
Vulpia, 142–143

W

Wagner, Tobey, 319, 323
Waipu, 191
'Waldina' hard fescue, 162
'Waldorf' Chewings fescue, 150
Warm climates, 17
Warm-season grasses, 17
  bahiagrass, 331–344
  bermudagrass, 235–251
  buffalograss, 257–269
  centipedegrass, 287–292
  St. Augustinegrass, 309–327
  seashore paspalum, 295–304
  Zoysiagrasses, 271–283
'Warwick' fine-leaved *Festuca*, 129
Washington Agricultural Experiment Station, 161
Water
  ground-, 10, 310
  movement of, VII, 10
Water couch, 298
Waterlogging, 99
Water-soluble carbohydrate content, 88
Wavy hairgrass *(Deschampsia flexuosa)*, 226, 227
Wear tolerance, VII
  Chewings fescue, 149
  fine-leaved *Festuca*, 132
  Kentucky bluegrass, 29
  perennial ryegrass, 76, 96–97
  sheep fescue, 160
  slender creeping red fescue, 151

supina bluegrass, 53, 57
Weed competition
    bahiagrass, 334
    buffalograss, 259
    St. Augustinegrass, 310
    Zoysiagrasses, 282
Weed pests
    annual bluegrass as, 39
    hair fescue as, 157
    Highland bentgrass as, 215
    perennial ryegrass as, 77
    rough bluegrass as, 70
Welsh Plant Breeding Station, 81
Western Regional Plant Introduction Station, 35, 193
Wet fields
    rough bluegrass, 68
    supina bluegrass, 53, 54
Wetland grasses, 225
Wheat, 43, 77
White, Don, 48, 57
White and yellow seedling disease, 121
White grubs *(Phyllophaga crinita)*, 64, 266
Wilderness Country Club, 326
Wildflower mixtures
Festuca ovina, 155
    sheep fescue, 159
Wild germ plasm, 114
"Wild" grasses, 8, 16–17
'Wilmington' bahiagrass, 334

Wilt avoidance, 311
Windblown mine waste, 10
'Wind Dancer' perennial ryegrass, 90
Wind-pollination, 79
'Wintergreen' strong creeping red fescue, 153
Winter hardiness, 93, 99
Winter overseeding
    Chewings fescue, 149
    rough bluegrass, 70
'Winterplay' rough bluegrass, 71
Woodland areas
    hair fescue, 158
    hard fescue, 161
    shade fescue, 148
'Wrangler' tall fescue, 113
'Wykagyl' vb, 203

Y
Yellow-stem, 290, 292
Yellow tuft. *see* Downy mildew
"Yield lag" phenomenon, 15
'Yorkshire' velvet bentgrass, 203
Youngner, Vic, 278
'Yukon' bermudagrass, 247

Z
Zacate bahía, 338
Zacate camalote, 338
Zacate de llanero, 338
'Zeon Royal' Zoysiagrasses, 283

Zinc, 10
    fine-leaved *Festuca*, 131
    Highland bentgrass, 212
    tufted hairgrass, 228, 229
'Zorro' Zoysiagrasses, 279, 283
Zoysia, 16, 274
Zoysiagrass eriophyid mite, 279
Zoysiagrasses (*Zoysia* spp.), 271–283
    collection/selection/breeding history, 278–283
    commercially available varieties, 280–281
    cytology, 275, 276
    distribution, 271–273
    flowering control, 277–278
    general development criteria, 283
    interspecific hybridization, 275, 277
    key to, 274–275
    morphology, 273–274
    selection criteria, 282–283
    taxonomy and nomenclature, 274–275
Zoysia japonica, 274
Zoysia japonica L., 149
Zoysia matrella, 275
Zoysia pacifica, 274
Zoysieae, 274
'Zygma' redtop, 217

CPSIA information can be obtained at www.ICGtesting.com
Printed in the USA
BVOW09*1049131016

464678BV00024B/49/P